SCIENTISTS AND SWINDLERS

Johns Hopkins Studies in the History of Technology

MERRITT ROE SMITH, SERIES EDITOR

THE JOHNS HOPKINS UNIVERSITY PRESS   *Baltimore*

SCIENTISTS &

SWINDLERS

CONSULTING ON COAL AND OIL

IN AMERICA, 1820–1890

Paul Lucier

© 2008 The Johns Hopkins University Press
All rights reserved. Published 2008
Printed in the United States of America on acid-free paper

9 8 7 6 5 4 3 2 1

The Johns Hopkins University Press
2715 North Charles Street
Baltimore, Maryland 21218-4363
www.press.jhu.edu

Library of Congress Cataloging-in-Publication Data
Lucier, Paul, 1962–
    Scientists and swindlers : consulting on coal and oil in America,
1820–1890 / Paul Lucier.
        p. cm. — (Johns Hopkins studies in the history of technology)
    Includes bibliographical references and index.
    ISBN-13: 978-0-8018-9003-1 (hardcover : alk. paper)
    ISBN-10: 0-8018-9003-9 (hardcover : alk. paper)
    1. Science and industry—Moral and ethical aspects—United States—
History—19th century.   2. Petroleum industry and trade—United States
—History—19th century.   3. Petroleum industry and trade—Canada—
History—19th century.   4. Coal trade—United States—History—19th
century.   5. Coal trade—Canada— History—19th century.   6. Science
and law—United States—History—19th century.   I. Title.
    Q127.U6L83 2008
    509.73'09034—dc22   2008006551

A catalog record for this book is available from the British Library.

Special discounts are available for bulk purchases of this book.
For more information, please contact Special Sales at 410-516-6936
or specialsales@press.jhu.edu.

The Johns Hopkins University Press uses environmentally friendly book
materials, including recycled text paper that is composed of at least 30
percent post-consumer waste, whenever possible. All of our book papers
are acid-free, and our jackets and covers are printed on paper with
recycled content.

For Oliver and Rosalind

Contents

List of Illustrations   ix

Acknowledgments   xi

Introduction · Money for Science   1

PART 1   **COAL**

Chapter 1 · Geological Enterprise   11

Chapter 2 · The Strange Case of the Albert Mineral   41

Chapter 3 · The American Sciences of Coal   69

Chapter 4 · Mining Science   108

PART 2   **KEROSENE**

Chapter 5 · The Technological Science of Kerosene   143

Chapter 6 · The Kerosene Cases   162

PART 3   **PETROLEUM**

Chapter 7 · The Rock Oil Report   189

Chapter 8 · The Elusive Nature of Oil and Its Markets   208

Chapter 9 · The Search for Oil and Oil-Finding Experts   244

Chapter 10 · California Crude   273

Epilogue · Americanization of Science   313

Notes   325

Essay on Sources   397

Index   411

Illustrations

Abraham Gesner's geological map of Nova Scotia (1846)   16

Abraham Gesner's geological survey of New Brunswick (1838–1842)   24

Richard Taylor's map of the coal fields of British North America (1848)   36

Charles Jackson's stratigraphy of "the regular coal series" in the region of the Albert mine   50

Left, Charles Jackson's view of the crushed Albert "coal" on the ninth level of the Albert mine. *Right,* Richard Taylor and James Robb's view of the Albert "asphaltum" as an injected vein within the crest of an anticline on the eighth level of the Albert mine.   52

Richard Taylor and James Robb's traverse section from the Albert mine, New Brunswick, to the Joggins, Nova Scotia   53

Amos Eaton's geological nomenclature for North American rocks (1828)   82–83

Idealized section of Paleozoic formations (circa 1860)   88

Charles Lyell's interpretation of the growth-on-the-spot theory of the origin of coal (1852)   90

Top, Charles Lyell's ideal geological section from the Atlantic coast to the great Appalachian coal field of the Midwest. *Bottom,* Coal regions of Pennsylvania exhibiting Henry Darwin Rogers's law of gradation.   92

Top, Erect *Sigillaria* containing the fossil reptile, *Dendrerpeton acadianum.* *Bottom,* Section of the Joggins, Nova Scotia.   96

J. Peter Lesley's "zigzag" topography in Pennsylvania (1856)   101

James Dwight Dana's depiction of the Carboniferous Age (1863)   105

Coal areas of various countries (1848)   106

James Dwight Dana's Geological map of the United States (1860)   107

Bristol Mining Company stock certificate   135

Coal oil plant (1860)   148

Samuel M. Kier's label for a bottle of "Petroleum"   205

The oil region of Pennsylvania (1866)   212

"Kicking Down a Well"   214

Left, Different depths of oil wells along the Burning Spring Creek in West Virginia. *Right,* Vertical oil fissure.   219

Oil well setup (1866)   223

Left, Scientific schematic of a large oil cavern (1864). *Right,* Popular schematic of a large oil cavern (1866).   226

Petroleum still and condenser   230

Petroleum refinery (1865)   231

Hillside drilling along Pioneer Run (1865)   235

Pithole (1865)   236

Ideal cross section of oil wells (circa 1865)   237

Prospecting for oil   253

Advertisement for oil-finding engineers (1865)   255

Alexander Winchell's ideal geological section of the Neff Petroleum Company's property in Ohio (1866)   258

Various geological positions of rock oil (circa 1865)   266

E. B. Andrews's illustrations of the anticline or "break" along the Ohio River (1866)   267

Map of the Ojai rancho in Santa Barbara County and surrounding ranchos in southern California (1865)   282–283

Benjamin Silliman Jr.'s ideal cross section (1865)   284

Acknowledgments

The first step in thanking everyone is finding the right metaphor. In a book about commercial science it would seem appropriate that I should say something about paying debts and obligations, but I am going to choose a metaphor that reveals something about me and the way I work. It is soccer, a sport that I have enjoyed playing all my life and one that I am now coaching for boys and girls. Finishing (in soccer and in writing) is all about the team behind you. And this book has needed a very large one.

I would never have ventured onto the field of history of science and technology in the first place were it not for my teachers at Princeton University. Charles Gillispie, Gerald Geison, and Michael Mahoney taught me to write and to think like a historian. Charles Gillispie, especially, has never let me lose hope that my story is worth telling, and in my own words. My graduate education was also enhanced by a cohort of skilled scholars who I am honored to call my friends: Ken Arnold, Ann Blair, John Carson, Kevin Downing, Marybeth Hamilton, Tama Hasson, Stuart McCook, Bo Sanitioso, Molly Sutphen, Emily Thompson, and Jeffery Westbrook. The History Department at Princeton University got this book started in another way by awarding me a postdoctoral fellowship. The committee was also instrumental in getting the rules changed to allow a free-range scholar to receive such an award.

Switching fields is a common tactic in soccer, and this book, too, has moved forward by going from side to side. The Wellcome Institute for the History of Medicine has afforded me many kindnesses, from research grants to office space, during my visits to London; particular thanks go to William Bynum and Sally Bragg. The incomparable Roy Porter pointed out how much more there is to the history of geology, and Janet Browne, who edited an article of mine for the *British Journal for the History of Science,* showed me what graceful writing looks like. I would also like to thank John Komlos and the National Endowment for the Humanities for the chance to study economics and economic history at a summer seminar on the Industrial Revolution held at the University of Munich. Although it is called the miserable science, I had a great time discussing it over liters of beer with John Murray, Philip Pajakowski, Christine Rider, Robert Schultz, Gennady Shkliarevsky, and Micheal Thompson.

Much of the research for this book was underwritten by a grant from the National Science Foundation (SBR-9711172). Edward Hackett and Michael Sokal made that grant work, and the anonymous reviewers suggested ways to make the research workable. Some of them cautioned me against an overly optimistic view of the relations between knowledge and money, a view I displayed in an early *Isis* article. Margaret Rossiter, Nathan Reingold, Marc Rothenberg, and Michele Aldrich provided much needed correction to that rosy view as well as useful insights into the characters and interests of nineteenth-century men of science. This book has also benefited from the advice and encouragement of P. Thomas Carroll, Robert Silliman, David Spanagel, Hugh Torrens, and Julie Newell, whose research often overlaps mine. I have also been lucky to have many generous and thoughtful patrons who have been instrumental in finding support for a consulting historian, in particular, Shirley Gorenstein, Deborah Johnson, Jennifer Phillips, Ronald Rainger, and Marie Schwartz.

The ideas in this book have been sharpened and enriched by the comments and queries of numerous audience members at History of Science and History of Technology meetings, but I would like to single out David Edge and John Carson, who always asks the best questions. I would also like to thank the participants and organizers of several colloquia to which I was invited to speak. At the Davis Center in Princeton, William Jordan, Charles Gillispie, and Ed Tenner provided a welcome homecoming; at the University of Pennsylvania, Emily Thompson, Robert Kohler, and Susan Lindee pushed me to rethink my perspective on the historiography of American science. At the Chemical Heritage Foundation in Philadelphia, Christopher Hamlin and later Theodore Porter offered invaluable expertise on professionalization. And at the Max Planck Institute in Berlin, I had the chance to discuss my thoughts on law and science with Ken Alder, Tal Golan, Daniel Kevles, Michael Hagner, Ann Johnson, and Simon Schaffer. I will always be thankful to (and amused by) those swift and lighthearted players of the "duck game."

Like any sports team, this book has had the backing of a number of big institutions. The interlibrary loan offices at Rensselaer Polytechnic Institute and the University of Rhode Island have provided a constant supply of hard-to-get titles. The librarians at URI also provided timely assistance in getting some of the images, as did Joy Shipman, who drew the map of New Brunswick and sailed FJs with me. The archivists and staff at the New York State Library and Archives, American Philosophical Society, Drake Well Museum, Geological Society of London, Hagley Museum and Library, Huntington Library, Rensselaer Polytechnic Institute, Smithsonian Institution, University of New Brunswick at Fredericton, University of Glasgow, University of Strathclyde, Wellcome Institute, and Yale University have made my research that much more pleasurable and productive. John Dojka at RPI and Ronald Brashear at the Huntington and then at the Smithsonian deserve special appreciation.

The book manuscript was completed during a postdoctoral fellowship at the

Dibner Institute for the History of Science and Technology. I am indebted to Jed Buchwald for inviting me and to Richard Sorrenson for help with the proposal. David Cahan, Robert Friedel, Slava Gerovitch, Arne Hessenbruch, Mary Jo Nye, Robert Post, Leonard Rosenband, Robert Seidel, George Smith, Andre Wakefield, and Benjamin Weiss made the Dibner an intellectually enjoyable place. The Dibner also sponsored a couple of summer seminars at Woods Hole, Massachusetts, to which I was invited, and I would like to thank the participants and organizers, especially Jane Maienschein.

The manuscript, although complete, was too fat to get past an anonymous reader, and my editor, Robert J. Brugger, put up a very stiff defense. The finished result, thanks to them and to Brian R. MacDonald and Courtney Bond, is a stronger and more fit narrative that I hope fulfills the promise projected by Merritt Roe Smith, who first recommended this book for publication.

But it was at that crucial moment, when the game did seem to be lost, that two friends and admirable scholars stepped up to take on the editing challenge. I will be forever grateful to John Servos and James Secord. They have coached my work from the very beginning, and when I called on them, they read the entire manuscript and brought this book within striking range.

It has thus been a very long run, and down the sidelines the whole time has been my family. My brother- and sister-in-law, Paul Rusnock and Elizabeth Melanson, made the Canadian part of my research so much easier and enjoyable. Andrew and Alice Rusnock have my heartfelt appreciation for helping to make a two-career household work. And to my mother, Lynn Roberson, and to my sisters, Georgia Adams and Gabrielle Lucier, I offer a lifetime of loving thanks. Although you have not seen most of the action on the scholarly field, you have cheered me along regardless of my stumbles and wayward runs.

But there is one person who has been with me every step of the way. Andrea Rusnock is my best friend, my fellow traveler, my wife, and my muse. Together we share the joy of this and many other adventures. But now that I have reached my goal, I realize, quite happily, that I will not take it. I will pass it on to my son Oliver and to my daughter Rosalind. To them I dedicate this book.

SCIENTISTS AND SWINDLERS

Introduction

Money for Science

I T IS A TRUTH UNIVERSALLY ACKNOWLEDGED that a scientist in possession of experience and expertise must be in search of funding. Today support-
ing someone to do science is routine. Scientists abound in universities, private
foundations, government agencies, and corporate research and development
laboratories. Science is a job, and scientists are professionals. For most of the
nineteenth century, neither was true. Science had few established sources of
support, and the descriptive noun *scientist,* a term coined in Britain in the 1830s,
was rarely used in America until late in the century. Nineteenth-century men of
science were becoming professionals, and key to that process was money—
money to live on and money to do science. How American men of science made
a living *by doing* science is a crucial historical question.[1]

One place to start looking for an answer is the geological and natural history
surveys of the first third of the nineteenth century. Organized at both the state
and federal levels, surveys were a form of government patronage providing
salaried positions to a large number of geologists, chemists, botanists, zoolo-
gists, and mineralogists. During the 1830s and 1840s, surveys were the training
grounds, so to speak, for the first generation of American men of science.
(Women at this time were consciously excluded from surveys and largely from
science in general.)[2] Surveys stimulated the growth of scientific disciplines as
well as the development of a national scientific community.[3] They also provided
a measure of legitimation by putting science at the service of government and
the public. Surveys were thus seeds around which a form of professional American science began to crystallize.[4]

The process, however, was halting, largely because surveys were short-lived.
Once completed, surveyors faced the daunting task of finding further employment as *men of science.* And there were not many options. In this regard, the situation was different from that in Britain, where there was an array of opportunities for getting paid to do science, including writing, editing, reviewing,
instrument making, and public lecturing.[5] That kind of scientific work was not
available in America because there was not such a well-developed market for the

display of knowledge. Nor was there a large class of wealthy patrons. As Alexander Dallas Bache, director of the U.S. Coast Survey, put it, "we have no rich aristocracy or extended middle ranks to be enlisted in the cause [of science]."[6] Bache's comments on class point to another distinction. In Britain and continental Europe, men of science were often men of means—in a word, gentlemen.[7] In America, men of science came from modest backgrounds; they were the sons of ministers, lawyers, physicians, teachers, and small farmers.[8] If Americans were going to do science, they were going to have to work at it.

Making money doing science required initiative, luck, and no small degree of self-promotion, besides the requisite experience and expertise. In other words, nineteenth-century American men of science had to be entrepreneurs. Accordingly, many became active lobbyists for the creation of, or appointments to, new state and federal surveys. Some sought teaching positions, although these did not provide much time, money, or encouragement to do research. Still others opted for the path chosen by William W. Mather and James Hall. In 1838 Mather and Hall were geologists working on the New York State Geological and Natural History Survey. Because they knew the survey would come to an end, they decided to set up a business doing geology, and they issued an advertisement of their professional services.

Minerals, Mines, Ores, &c. Examined

Messrs. W. W. Mather, and James Hall, Geologists, have established an office for the analysis and assay of minerals and ores; for the examination of mines, mining districts, mineral beds, quarries and quarry stones; for communicating information upon the best methods of smelting and working ores and minerals to bring them to a marketable state; and for imparting all the various knowledge which is a necessary preliminary to the successful prosecution of mining enterprises. So many mining operations are undertaken through mistaken views of their probable productiveness, and even of the nature of the mineral ore, that it is deemed necessary for the public interest that an office similar to that mentioned should be established. This professional knowledge is as important to the community, to prevent the undertaking of mining and metallurgic operations where they would be unproductive, as to guide and direct enterprise to the most economical and profitable methods of working mines and preparing their marketable products.

Messrs. Mather and Hall have had experience of several years in the different branches of their profession, and now solicit the patronage of the public. The office will impart information, not only upon the subjects above mentioned, but upon the applications of all mineral substances to the various useful purposes of life.

Letters, post-paid, addressed to Messrs. Mather & Hall, Mining Engineers, corner of South-Market and Hudson sts., Albany, soliciting information, and enclosing a fee of five dollars, will be promptly attended to. Should it be necessary to examine the locality of the mineral or ore, or make an assay or chemical analysis, or make drawings, and give descriptions of machinery, furnaces, &c., &c. an additional fee will be charged, varying in amount according to circumstances.

January 1, 1838[9]

Setting up an office for the business of science went by another name, *consulting*,[10] and it would prove to be the most profitable and popular avenue of scientific entrepreneurship in nineteenth-century America.[11]

The development of consulting marks a historically significant, yet hitherto unexamined, broadening of the patronage for science.[12] Patronage, as scholars have shown, is critical to the pursuit of science.[13] It involves complex exchanges of credit (both social and scientific) and of capital (cultural and, of course, financial). In moving from government employment to "the patronage of the public," Mather and Hall were among the first men of science to try to capitalize on their survey experience and scientific expertise. Many others would follow, including geologists like Charles Thomas Jackson, Abraham Gesner, Henry Darwin Rogers, J. Peter Lesley, Josiah Dwight Whitney, and John Strong Newberry; chemists such as Benjamin Silliman (father and son), Augustus A. Hayes, and Thomas Antisell; and a handful of renowned naturalists, for example, John Torrey and Joseph Leidy.

Geologists, though, were the most prominent and prosperous consultants, and the centrality of their science highlights some distinct features of the American context. For, unlike France and Germany, where state-trained engineers provided mining expertise, or Britain, where land surveyors, canal builders, and coal prospectors proffered such advice (gentlemanly geologists, by virtue of class and custom, had little to do with British mining),[14] in the United States the best-trained and most numerous experts available were geologists. In fact, the proportion of geologists to men of science was greater in America than in any other country.[15] By contrast, the physical sciences were not well developed in America before the late nineteenth century. And noticeably absent from the ranks of consultants were astronomers, physicists, and mathematicians. Scholars, however, have focused closely on the small coterie in the physical sciences and, as a result, have failed to recognize the profound impact that consulting had on American science and society.[16]

As it developed in the three decades before the Civil War, consulting came to form a central part of the practice and identity of American men of science. That Mather and Hall felt the need to advertise spoke to the novelty of this type of professional science in the 1830s. Within a decade, however, no such publicity was needed, and by the 1850s consulting was commonplace. Numerous men of science offered various services and information, especially regarding the land and its minerals, to any and all members of the interested public. Broadly defined, consulting is what all active men of science did, regardless of whether they offered their experience and expertise to enterprising individuals, private companies, or state governments. Whatever the employment, men of science had the chance to contribute actively and directly to the common weal and wealth. Such civic engagement reflected well on consultants' entrepreneurship and reinforced the widely held assumption that science was useful and therefore should be used. In theory, then, consulting was another means by which science promoted material improvements and general enlightenment.[17]

In practice, consultants solicited the patronage of a particular segment of the public—capitalists. In midcentury America, the ideal capitalist was a gentleman, well informed (if not also well educated), well connected (if not also well-bred), and well-off (if not downright rich). Model capitalists had money to invest and plans to build, but the projects on which capitalists might consult men of science were not so large as railroads or canals, the kinds of internal improvements requiring large amounts of government funding, but rather on the scale of a mine or manufactory, investments needing other capitalists or, after 1850, funds raised by joint-stock companies.[18] Capitalists and their companies engaged consultants for more than mere promotion or puff. Consultants provided the science needed to purchase lands, dig mines, design and build equipment, improve manufacturing processes, and market new products. On occasion, consultants were called upon to testify in court cases involving patent rights. By midcentury, patents were being taken out for processes and products that involved demonstrable degrees of scientific research. Expert witnessing was courtroom consulting by another name and, hence, another profitable branch of scientific entrepreneurship.[19]

Whatever their specific responsibilities, consultants proved to be major proponents of invention and innovation. They were agents of midcentury technological change and economic growth. Admittedly, change and growth are roomy concepts that can be applied to just about any time or place. The decades during which consulting developed have been described by various labels—Industrial Revolution, Transportation Revolution, Market Revolution, Age of Enterprise—each purporting to capture a fundamental transformation in the American economy and society. In none of these familiar interpretations do men of science figure.[20] Nor for that matter do such great transformations seem to have much of an impact on American science.[21] Nineteenth-century science and industry seem to be two trains headed in the same direction but running on separate tracks. By most scholarly accounts, convergence occurred at some distant point late in the century with the emergence of professional engineering and science-based industry, the hallmarks of the so-called Second Industrial Revolution. The popularity of consulting, however, points to contacts that occurred earlier than and differently from previously thought.

The development of consulting defines a crucial phase in the relations among science, technology, and industry, relations that have troubled scholars for quite some time.[22] There remains much debate, for instance, about the contributions of science to the British Industrial Revolution of the late eighteenth century, although new studies have highlighted the busy activities of natural philosophers in projects for wealthy landowners intent on improvement.[23] American consultants had similar interests in improvements and progress, but unlike their eighteenth-century English predecessors, they undertook more numerous, widespread, and impersonal engagements. By design and definition, consulting was a short-term, advisory activity and, in this way, was also different from the kind of industry-based science that appeared in the laboratories of chemical and electrical firms toward the end of the nineteenth century.[24] Consultants worked alone or in partnerships, not in corporate confines with their concomitant physical plants and brand-name identities. Consultants were not permanent employees or parts of product development; they were independent, professional men of science.

"By professional I meant geological," explained Lesley, "as geology is my profession."[25] If geology were the science with the most to offer American mid-century industry, it only stood to reason that geology had the most to gain. Private engagements provided geologists with access to unexplored lands, unexamined minerals, and unexplained phenomena. In consulting lay the routine of normal fieldwork as well as the new materials for research. Private engagements thus had much in common with government surveys. But consulting had a social location different from a survey, or a university, or a government bureau. It was a professional practice pointing toward the union of science and capitalism.

To assert that consulting emerged in concert with the growth of nineteenth-century industry, or with a distinct form of professional science, or with the development of American geology, does not, however, fully address the question of how the practice wove together these strong historical cords. To do so, this study tracks the development of American coal and petroleum, material resources that were simultaneously subjects of immense scientific and industrial interest and hence the objects of commercial engagements for enterprising men of science. Consulting on coal and petroleum epitomized practical science, and to nineteenth-century Americans, practical meant both "scientific" and "economical." It did not mean untheoretical, as some scholars have asserted.[26] In studying coal and petroleum, American geologists, chemists, and naturalists were anything but indifferent to basic research *or* to the needs of industry.[27]

This peculiar combination of science, mining, and consulting led to the American specialties of coal and petroleum geology. In the early and mid-nineteenth century, coal was the most heavily researched subject (in both private and public surveys) because it was unquestionably the most valuable mineral.[28] And not just because it was needed to fuel steam engines, to light buildings, or to heat homes. Coal was key to explaining the earth's structure and to delineating the earth's history—specifically, to how rocks were laid down in oceans or thrown up into mountains and to how rocks were identified and ordered from oldest to youngest. Coal was also key to understanding the transformations that occurred when plants and animals died. Finally, the complex geometry of coal fields forced changes in the methods of doing geological fieldwork; the contour map and topographic surveying became the essential elements of a new specialty that Americans named structural geology. In similar fashion, Americans created petroleum geology, a scientific specialty combining chemical research into petroleum's organic constituents with field studies of the structural conditions required for its accumulation.[29] Petroleum geology relied on the new technologies of well boring and well records to push the contours of structural geology deep underground. In the cases of coal and petroleum, consulting flourished because the interests of science and industry were complementary.

To illustrate the extent of this complementarity and to help in navigating the shifting scientific landscape, this study follows the extraordinary career of the Albert mineral. Discovered in Albert County, New Brunswick, in 1849, this volatile and valuable substance became famous as the source of a twenty-five-year scientific, legal, and commercial controversy over whether it was coal-like or petroleum-like. Men of science consulted for companies wanting to mine and process the mineral, and they served as expert witnesses in celebrated court cases between capitalists wanting to own and patent it. Substantively and symbolically, the Albert mineral grounds the theoretical arguments about scientific entrepreneurship and the relations among science, technology, and industry in midcentury America. In a sense, the mineral is treated as a character, not unlike the men of science themselves, in the development of American consulting.

The principal use for the Albert mineral was lighting. While the history of electrical lighting is well known, that of coal gas and coal oil is not. Coal-based lighting required its own scientific, technological, legal, and commercial system—or, more precisely, systems, for coal gas and coal oil, although closely related, evolved differently.[30] This study pays particular attention to coal oil, or kerosene as it was often called. Kerosene was the first manufactured mineral oil. As it became increasingly popular and profitable during the late 1850s, kerosene became the subject of a series of important public and legal debates about the meaning of invention and innovation. As products of "technological science"—a new concept of the midcentury—mineral oils raised challenging questions about the role that science *could* and *should* play in industry.[31]

Part of that debate concerned the dark side to the coin of consulting. Men of science, along with men of business and politics, worried about what would happen if, and when, the interests of science and industry were not complementary. The danger with fees-for-expertise was that it smacked of interest. Money might uproot the good of science and the goodwill of the public. The excesses of Gilded Age capitalism, particularly in the oil industry, brought into bold relief the ethical and moral troubles intrinsic to this form of professional science. Consulting was at the heart of one of the most troubling scandals in nineteenth-century American science. The setting was the National Academy of Sciences, and the subject was California petroleum. The controversy pitted Josiah Dwight Whitney, a former consulting geologist who in 1860 became the director of the California Geological Survey, against Benjamin Silliman Jr., the consulting chemist for several California oil companies. On the surface, theirs was a disagreement over the origin and occurrence of petroleum, a scientific dispute that went back to questions about the classification of the Albert mineral. At a deeper level, the controversy was about where American science should be practiced (on government surveys or corporate assignments), who should do it (public surveyors or private consultants), and who should pay for it (the people or the companies). All these unresolved issues were brought to the fore when Whitney accused Silliman of swindling.

Scientist or swindler? It is a perfect endpoint for a study that traverses North America from early nineteenth-century New Brunswick to Gilded Age California. In following the careers of the most influential consultants, from their work on government surveys to their engagements for mining and manufacturing companies, the story moves within, as well as across, the contours of scholarship in the history of science and technology, law and ethics, business and economic history, and environmental and science studies. Coal and oil serve as stepping-stones between the diverse places, personalities, scientific theories and methods, and technologies and industries. Perforce as much as per design, it is a complex tale about the imperfect fit between the human attempts to order the world and the intransigent materials they confront along the way.[32] The extended efforts to classify coal and oil, to explain their origin and occurrence, and to explore and exploit their deposits reflected larger problems of social definition and intellectual categorization: what are the differences, if any, between theory and practice, technology and science, discovery and invention, and ethical and unethical behavior? Even the coolest-headed consultants realized that commercial engagements could lead to moral entanglements. Mixing money and science was (and still is) a volatile concoction. The search for truth and the pursuit of the almighty dollar were not (and are not) always and everywhere compatible. Still, scientists have to work. And if they work for money, is the science they do less trustworthy and true?

PART 1
COAL

CHAPTER 1

Geological Enterprise

Coal is power, it is the foundation of manufacturing industry, the greatest source of national wealth; and administers largely to the comforts of man. . . . It hurls the train along the rail-road, the boat across the mighty deep; it lights the city traveller along his midnight way, and warms the shivering peasant after his daily toil.
—Abraham Gesner, *Remarks on the Geology and Mineralogy of Nova Scotia* (1836)

Nova Scotia might not seem the obvious place to begin a history of American coal, but it is a natural one. Since at least the eighteenth century, observers had commented on the splendid exposures of coal along its coasts, especially at the Joggins on Chignecto Bay, an arm of the Bay of Fundy.[1] Beginning in the 1830s, Nova Scotia was the largest producer of coal in British North America and a major exporter to the United States. To seaboard cities like Boston and New York, Nova Scotia's coal was closer, easier to transport, and hence better priced than Pennsylvania's. To American and British geologists, Nova Scotia's coasts were among the best sites in the world to study the age and origin of coal.

The geology of Nova Scotia, however, was much like its coal, something to be possessed and shipped around the North Atlantic as well as to be explored and explained. When the American geologist-chemist Charles Jackson began to study the province in the late 1820s, he started a contest over who would create and control the scientific knowledge of Nova Scotia. In that struggle, Jackson's principal adversary would be Abraham Gesner, a country surgeon turned geologist; but others, like the British gentlemanly geologist Charles Lyell, would soon join in. Which one was to be the expert on Nova Scotia?

The answer depended on what, exactly, a geologist had to offer, and to whom. For Gesner, geology was the necessary prerequisite to finding the all-powerful coal and sending it on its civilizing mission. Such scientific optimism was welcomed in Nova Scotia as well as in neighboring New Brunswick, where in 1838 Gesner was appointed the provincial geologist, the first such position in any British colony.[2] The context and consequences of that unprecedented initiative warrant historical examination; for Gesner found a great coal field. "It exceeds in its dimensions any found in Great Britain," he declared, "and is one of the largest ever discovered upon the globe."[3] Gesner's discovery caused quite a stir in scientific and commercial circles; but it also raised poignant problems about how to reconcile geological information with mining investments and about the proper role a surveyor should and could play in developing the coal he finds.

As it turned out, Gesner's role was not what some in New Brunswick anticipated, and so, by the early 1840s, he, like many former surveyors, was in search of ways to make a living doing science. By accompanying Gesner on his various employments, historians can gain some new insights into scientific entrepreneurship as well as the culture of coal in early nineteenth-century America.

Plagiarism, Patronage, and Expertise

Abraham Gesner was born in Cornwallis, Nova Scotia, "the garden of Acadia," on 2 May 1797.[4] One of twelve children of Henry Gesner and Sarah Pineo, he received little formal education, nothing more than "the ordinary instruction of the grammar schools of the day," although he was "a great reader."[5] As a young man, Gesner experimented with a number of moneymaking schemes, including a partnership in an ill-fated venture to trade Nova Scotia horses for West Indies rum. When that business ended in shipwreck, Gesner settled down to farming, and in 1822 he married Harriet Webster, the daughter of Isaac Webster, a prominent physician from Kentville. Gesner, however, proved as adept at farming as he was at horse trading. Attempts at scientific agriculture put him in debt, and soon he was under house arrest, much to the chagrin of the wealthy Websters, who were forced to bail out their son-in-law. In return, Gesner agreed to give up farming, and in 1825, with his father-in-law's backing, he traveled to London to study surgery at Guy's Hospital and St. Bartholomew's Hospital. There Gesner received his only formal instruction in science, most likely chemistry and *materia medica*. Within a year, though, he had returned home upon learning of the death of his second son.[6] He chose to settle in Parrsboro, on the north shore of the Minas Basin, across the bay from his in-laws, where he served as surgeon to the local military garrison. Gesner, however, showed more concern for the area's natural history than for the soldiers. In the tradition of a country physician, he collected minerals, rocks, and fossils on his rambles.[7] As the local expert, he was probably not surprised when two Bostonians dropped by.

Charles Thomas Jackson (1805–1880) and Francis Alger (1807–1863) were "tourists" in search of rare and beautiful curiosities of natural history. Back in the summer of 1826, they had made a reconnaissance of Nova Scotia and the following year had coasted along the peninsula from Brier Island to the Minas Basin and then around Cape Chignecto and into Cumberland Bay. An account of their travels appeared in 1828 as "A Description of the Mineralogy and Geology of a Part of Nova Scotia." By the summer of 1829, when they visited Gesner, Jackson and Alger were planning an expanded edition.[8]

Their object was a mineral geography, "to describe only those [minerals] which are peculiar to the place, or which possess singular beauty, or present remarkable phenomena."[9] Jackson and Alger found many valuable kinds, including amethyst, jasper, stilbite, and zeolite, and they noted where to pick up samples along the shore. Bypassing locations with "no objects of natural history worthy of description,"[10] they instead focused on areas that promised prized specimens, such as the trap rocks of the North Mountains, a range running along the Bay of Fundy and skirting the southern shore of the Minas Basin.[11] In Jackson and Alger's view, the rocks were not as interesting as the minerals they contained.

The distinction between rocks and minerals went back at least to the mid-eighteenth century. Minerals were defined as natural kinds, individual species best exemplified by crystals, although not all minerals formed crystals. Rocks were mixtures of minerals. Granite, for instance, was composed primarily of quartz, feldspar, and mica crystals. Moreover, rocks occurred as large, independent masses with a thickness and geographical extension. According to Abraham Gottlob Werner (1750–1817), the famed professor of mineralogy at the Königliche Sächsische Bergakademie at Freiberg, all rocks on the Earth's surface were of aqueous origin. They had precipitated out of a primal ocean once covering the entire globe. Consequently, rocks were layered one on top of the other in an orderly and sequential stack. Trap, a rock that Jackson and Alger thought was very contentious, was at the top of Werner's stack.[12]

In contrast to the Wernerian or "Neptunist" theory, the Scottish philosopher James Hutton (1726–1797) had proposed an igneous origin for rocks. Rather than precipitate down, rocks had risen up from the seabed where they had been consolidated by heat. By this "Vulcanist" theory, trap was the product of heat and, according to some of Hutton's followers, the lava of ancient volcanoes.[13] Jackson and Alger favored the Vulcanist theory. They believed the Nova Scotia trap had been ejected suddenly and violently from the "unfathomable depths of the Bay of Fundy." In their opinion, the bay was a huge volcanic crater.[14]

In treating Nova Scotia's other rocks, Jackson and Alger relied on Werner. Besides his Neptunist theory, Werner had proposed a classification system for rocks toward the end of the eighteenth century. In brief, Werner's system was based on the idea of a *formation*—a group of rocks that shared specific characters.[15] In the early nineteenth century, formations were usually characterized by their composition or lithology (the rock type), which was how Jackson and Alger identified them. Thus, for example, the formation Granite (capital G) consisted entirely of the rock granite (little g). Jackson and Alger identified two other similar formations in Nova Scotia, Quartz Rock and Clay Slate, each composed of only one rock, namely, quartz rock and clay slate. Identifying formations composed of only one rock type was relatively straightforward compared to identifying stratified formations composed of several different rock layers or strata. Such was the case with Jackson and Alger's fourth formation. The Red Sandstone formation of Nova Scotia contained layers of shale, red sandstone, and conglomerate. Red sandstone was the predominant one, meaning the thickest and the one usually found on the surface. Accordingly, it was chosen for the name.[16]

After identifying and naming some thirty rock formations, Werner had arranged them in chronological order, from oldest to youngest. The Primary or Primitive class comprised the oldest formations, the ones composed of a single rock type. Werner thought they were the first to precipitate out of the universal ocean and hence were to be found on mountaintops. As the universal ocean subsided, the next class of formations, the Transition, was deposited on the flanks of the Primary mountains. The Transition class was so named because it represented an intermediate position between the Primary and the Secondary classes. Secondary formations resulted from the deposition of sediments (bits and pieces worn away from Primary and Transition rocks). They contained layers of sedimentary rocks such as sandstone, shale, and conglomerate. The Secondary also contained limestone (a precipitate), although it required keen judgment to distinguish Primary, Transition, and Secondary limestones. Werner had characterized Secondary formations by their flat-lying position (for which he had originally named the class Floetz) and by their fossils, which were becoming increasingly important by the early nineteenth century. In practice, fossils distinguished the three classes of rocks: Primary formations contained none, Transition a few, and Secondary the most.

Using Werner's system, Jackson and Alger outlined the geological structure of Nova Scotia (see table 1.1). The South Mountains were composed of Granite, the only Primitive formation on the peninsula. Overlying the Granite and tilted at a sharp angle of 50–60 degrees lay the Clay Slate formation, which belonged to the Transition because it contained a few fossils and some iron ore deposits. (Metals were considered another characteristic of Transition formations.) Around Halifax, Jackson and Alger mapped the Quartz Rock formation, which they also attributed to the Transition. Above this and in the Secondary class, they placed the Red Sandstone formation, which dipped at a slight 10–30 degrees and contained many fossils along with seams of coal and gypsum. Jackson and Alger thought the gypsum was of "practical worth," but they regarded

Table 1.1. Charles Jackson and Francis
Alger's Formations of Nova Scotia
(1832) (Oldest to Youngest)

Primitive Class
    Granite
Transition Class
    Quartz Rock
    Clay Slate
Secondary Class
    Red Sandstone
    Trap

the coal as very soft and friable, "a fault which greatly injures it for the market."[17] Lastly, at the very top, they placed the Trap of the North Mountains.

Overall, Jackson and Alger described a simple and remarkably regular structure, one that could not "fail to excite the admiration of every geologist."[18] Indeed, their survey, one of the first to be undertaken in North America, was well received and widely read.[19] The New York State geologist Ebenezer Emmons (1799–1863), for one, felt a "strong desire" to visit "the same interesting region," and in the summer of 1836, he led a group of students from Williams College on an expedition. They, too, visited Gesner, who was also very excited having just published his own *Remarks on the Geology and Mineralogy of Nova Scotia* (1836).[20]

Gesner had no doubt been impressed by Jackson and Alger's book, but they had not mentioned him in it, probably because they regarded Gesner as a mere amateur, someone with useful information but without scientific or cultural standing beyond Parrsboro. Gesner might well have been spurred to write his own book because of the slight, but he had also wanted to publish his observations for "the love of science" and "the gratification of curiosity."[21] In the event, Gesner's book sparked a heated dispute with Jackson, the first of many, over priority, patronage, and the proper behavior of men of science.

Gesner's title was nearly identical to Jackson and Alger's, but the extent to which he copied their work was (and is) a matter of debate.[22] Gesner admitted that he had "received some information" from the Americans, just as they had received some information from him, but Gesner also relied on many other sources.[23] More important than the details was Gesner's design. In putting geology first in the title, Gesner signaled that mineralogy was decidedly inferior. He restricted minerals mostly to his section on the Trap and kept their descriptions short.[24] In contrast to Jackson and Alger, Gesner thought gems were of little practical value. They would never "afford an article very important to com-

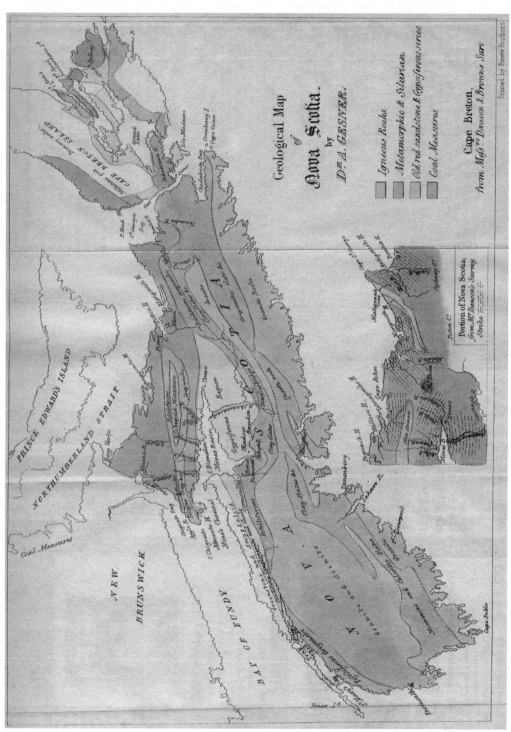

Abraham Gesner's geological map of Nova Scotia (1846). Between his 1836 survey and the publication of this map, Gesner changed the names of his four major districts: Primary to Igneous; Clay Slate to Metamorphic & Silurian; Red Sandstone to Old Red & Gypsiferous; and Coal. The Trap district was deleted *Source:* Abraham Gesner, "A Geological Map of Nova Scotia, with accompanying memoir," *Proceedings of the Geological Society of London,* 4 (1846):186–190, facing page 186.

merce, or make up the source of an abundant revenue" for Nova Scotia.[25] Economic resources, not ornamental curiosities, were Gesner's main concern.

Another reason for Gesner's emphasis on geology was the way he conducted his survey. Jackson and Alger had sailed along the shore gathering minerals and studying their enclosing rocks as "viewed from the sea."[26] They wrote as scientific tour guides for "traveller[s] proceeding from the United States to Halifax."[27] Gesner surveyed on the ground, along the roads, by foot, on horse, or in a gig. He explored many areas ignored by or inaccessible to Jackson and Alger.[28] This was no easy task because most of the province, Gesner observed, was "covered in dense forests, and trackless mountains, where the moose and carriboo still enjoy quiet repose from the yell of the Indian, or the sound of the woodsman's axe."[29] Native Americans, moose, and "lazy" bears popped up a lot in his book. The colorful details reminded readers of how vastly different the environment of "new countries" was from that of "old" Britain, where geology, although likewise pursued in the field, was less taxing and dangerous.[30] In Nova Scotia, maps were nearly unknown; even measuring "the distances from one place to another" was frequently impossible.[31] Despite such hardships, Gesner succeeded in exploring most of the peninsula.

Gesner divided Nova Scotia into four major "districts"—Primary, Clay Slate, Red Sandstone, and Trap. These corresponded to geographical regions, parts of the province where such rocks were exposed. Although Gesner's districts recalled Jackson and Alger's formations, the similarity disguised a confusing classification system. "District" sometimes meant "class" and sometimes "formation." Gesner was inconsistent, and nowhere in his book did he step outside the thicket of details to give a clear overview of Nova Scotia's rocks (see table 1.2).

Basically, Gesner wrote a geological travelogue—his trip from township to township and the rocks he saw on the way. Around Halifax, for instance, he found Granite, which he classed as Primary. He did not find Clay Slate, as Jackson and Alger had. In fact, Gesner found no Transition formations and rejected the term *transition*, so called "by older Geologists."[32] (To be fair, Jackson and Alger had also hesitated about using the term.)[33] When dismissing the Transition, Gesner deferred to Charles Lyell (1797–1875), who, in his *Principles of Geology* (1830–1833), had placed *all* formations above and younger than the Primary (i.e., ones with fossils or some signs of stratification) within the Secondary.[34] Yet despite this expressed rejection, Gesner repeatedly used the term "transition" to describe the four rocks (Clay Slate, Argillite, Greywacke, and Quartz Rock) he found in his Clay Slate district. In effect, Gesner undermined Lyell's theoretical point that there was no transitional period between the Primary and Secondary.

Inconsistency and imprecision were also apparent in Gesner's discussion of the Red Sandstone district, which contained *two* red sandstones, Old and New,

Table 1.2. Abraham Gesner's Geological Divisions of Nova Scotia (1836) (Oldest to Youngest)

Primary District
    Granite
    Gneiss
    Mica Slate
Clay Slate District
    Clay Slate
    Argillite
    Greywacke
    Quartz Rock
Red Sandstone District
    Old Red Sandstone
    Coal
    New Red Sandstone
Trap District

but often Gesner did not specify which one. In certain locations, such as Horton Bluff at the mouth of the Windsor River, he identified New Red Sandstone by its characteristic salt springs and gypsum beds.[35] Elsewhere he left the reader to wonder whether the all-important coal was above or below his unspecified red sandstone.

Coal was the key to Gesner's scientific perspective on Nova Scotia. He devoted nearly a quarter of his book to a discussion of its geology, its fossil plants, and theories of its origin, but practically all of this information came from English sources, primarily William Conybeare and William Phillips's *Outlines of the Geology of England and Wales* (1822). Gesner's contributions amounted to detailed descriptions of the coal fields at Pictou and the Joggins. Pictou was the larger. Ten seams, from one to three yards thick, were being mined and the coal was being exported. At the Joggins, Gesner counted eight seams, from half a foot to three feet thick, perfectly exposed from the cliff top to the beach, but none were being mined. "The strata are extremely regular, parallel and equal," Gesner observed, "affording a section of a coal basin, not surpassed by any in the world."[36] This might have been taken for hyperbole were it not for the numerous fossils, especially the upright trees, some as tall as forty feet and piercing several rock layers. Gesner identified them as giant varieties of *Lepidodendron*, whose roots branched into the rock beneath. "Great care should be taken in removing pieces," he warned, because of the "danger of being killed from the unexpected launch of a huge fossil."[37] Gesner avoided the falling trees, but he missed the theoretical importance of the fossil forest. He did not dwell on the connections between the trees, the roots, and the coal. He spent most of his time on how to mine the Joggins. Geology was useful only when it acquainted the government "with the sources of public wealth and economy."[38] Profit, not theory, was the dominant theme of Gesner's book.

Beyond its economic power, coal represented moral progress. It was evidence of "the resources Providence has placed within [Nova Scotians'] reach."[39] Like many at the time, Gesner saw the hand of God in nature. He was a deeply religious man who found no conflict between the account of creation in Genesis and the geological record. "There is no necessity for making the world appear older than its date given by Moses," he insisted.[40] Gesner viewed the erratic boulders scattered across the landscape as evidence for Noah's Deluge, as had Jackson and Alger. All three referred to the high authority of the Reverend William Buckland's *Reliquiæ Diluvianæ* (1823).[41] All three relied on sudden and great catastrophes to explain certain geological features. Gesner explicitly rejected Lyell's theory of "one class of causes" and asked rhetorically, "have all these changes taken place by the influence of causes now operating upon the surface of the earth?"[42] Gesner, as well as Jackson and Alger, thought the answer was definitely no.

Gesner's combination of piety, practical geology, and folksy prose proved very popular. According to another native of Nova Scotia, the geologist John William Dawson (1820–1899), Gesner's book was "extensively circulated in the province" and was "of great service in directing popular attention" to geology and mineralogy.[43] Gesner had intended his *Remarks* "for the perusal of the general reader," which is why he had organized the book by townships, for easy reference, and added short pieces on local history, like his own shipwreck near Brier Island in December 1821.[44] He sent a copy to the lieutenant governor, Sir Colin Campbell (1776–1847), and presumably one to the House of Assembly. In return he received a vote of thanks and £100 from the House "as a mark of the estimation in which they held his services."[45]

More than anything else, this gift of political and financial support angered Jackson and Alger. They had not thought to send their essay to Campbell or the legislature. Instead of a popular book, they had written "purely for the advancement of Science without any expectation of pecuniary reward." "[The] Memoir," they later explained to Campbell, "was contributed freely to the Scientific World no copies having been offered for sale on their account." One hundred copies were distributed to learned societies around the world. "Since we have served the Province as pioneers in its Geology," they continued, "we should be wanting in justice to ourselves did we not humbly claim that our services should be acknowledged in a similar manner to those of the Gentleman [Gesner] whose essay we have noticed."[46] Campbell elected not to respond, which only infuriated them. In a follow-up letter to the House of Assembly, Jackson and Alger exploded, and this time they hit hard upon Gesner's ethics. "[The Authors] must add that a large portion of his work has been borrowed from them without a candid acknowledgement—that their work has served as the model and basis of his, that discoveries & observations made by them, either appear as his own, or are referred to others."[47] Accusing a native son of surreptitious science surely did not endear Jackson and Alger to the Nova Scotians. The "Bluenoses," as descendants of the Loyalists were known, were not particularly predisposed to Yankees. Not surprisingly, Jackson and Alger's demands went unattended.[48]

Frustrated in Nova Scotia, Jackson turned to the scientific community at home. In a biting letter to Benjamin Silliman Sr., editor and founder of the *American Journal of Science,* he denounced Gesner for "literary larceny." "Pray you have seen 'Gesner's Geol. of Nova Scotia'? The creature has swallowed our whole Memoir & then brought it up again in an undigested state & styled it a new book!! . . . He actually knows nothing about either Mineralogy or Geology."[49] Silliman advised his friend to make the matter public. Jackson decided against this. Undoubtedly it would have been difficult to prove plagiarism, and it certainly would have done him no good in Nova Scotia, where he most coveted recognition.[50] Jackson contented himself with the knowledge that Gesner's *Remarks* would not be reviewed in Silliman's *Journal,* and it would remain missing from many other American reviews.[51]

From a historical perspective, the Jackson-Gesner dispute reveals the difficulties surrounding priority and patronage faced by many American geologists. Gesner had undertaken his survey without official government support but had skillfully courted the lieutenant governor and the House of Assembly. Jackson and Alger could lay claim to pioneering science, but that did not produce political or pecuniary thanks. Years later, Jackson tried to explain what he thought might be the consequences of this mismatch between recognition and rewards. "It is not to be expected that men of science will devote their lives to increasing the sum of human knowledge, if their labors are not appreciated and honored [and] if ingratitude and base envy meet them at every step."[52] Priority could be conferred within the confines of an elite group; patronage could not. For all the rhetoric about the diffusion of scientific benefits, in this instance Nova Scotians disregarded Jackson and Alger. Gesner, by contrast, spoke directly to an enlightened public and a commercially minded government. And it paid off.

Besides the money, Gesner gained another asset, legitimation. As part of professionalization, scholars have treated legitimization as an internal, peer-reviewed process of setting standards, recognizing priority, and distributing credit and kudos.[53] Gesner's legitimation was of a different sort. By valuing his survey, Nova Scotia bestowed a good measure of authority on him and esteemed him as reliable, competent, and useful. In a word, Gesner became an expert.

The Great Coal Field of New Brunswick

In contrast to Nova Scotia, the economic outlook for early nineteenth-century New Brunswick was overcast. The province had no coal mines, railroads, or manufactories; its sole export was timber. Under the Navigation Acts, New Brunswick enjoyed a privileged trading position, but so dependent had it grown upon timber exports to Britain that it had to import foodstuffs, manufactured goods, and practically everything else. New Brunswick was basically a community of lumberjacks and timber merchants. Fredericton was the political capital, but Saint John was the largest commercial city and cultural center. Its Loyalist, mercantile elite dominated the House of Assembly.[54]

In the 1830s Britain's Parliament began debating whether to continue protective policies or to open British markets to free trade. For New Brunswick, this would mean removing discriminatory tariffs on Baltic timber. Parliament's debate over free trade coincided with another equally important one over responsible government. As a crown colony, New Brunswick's executive government—lieutenant governor, judges, and other prominent officeholders—were appointed (or approved) and paid for by London. In 1837 the House of Assembly agreed to assume the costs of executive government; in return London relinquished the revenues and responsibilities for crown lands—about 26,000 square miles (or 16.5 million acres) of wooded preserve in the north. For the moment, tariffs remained on Baltic timber, but it would only be a matter of time before New Brunswick lost that protection. To a small group of reform-minded politicians, bankers, and capitalists, the impending storm required a new economic tack. A survey of mineral and agricultural resources might be just the thing.

Gesner's patron in New Brunswick was the lieutenant governor Sir John Harvey (1778–1852). Appointed by the Liberal government in London to supervise the handover of crown lands and revenues in June 1837, Harvey identified with colonial reformers and encouraged commercial interests. He was a leader who understood coal, or at least Gesner assumed so when he addressed several open letters to him that summer. Published in the *New Brunswick Courier* (Saint John), the letters depicted the province's potentially rich, yet unsurveyed resources.[55]

Harvey, however, did not have the power to create a geological survey. The lieutenant governor neither initiated nor controlled the expenditure of public funds; the House of Assembly determined how and where money was spent. Harvey did have a key ally in Charles Simonds (1783–1859), Speaker of the

House and president of the Bank of New Brunswick, one of the province's largest and based in Saint John.[56] Simonds shared Harvey's views on the need for a broader commercial base. The House of Assembly did not; the legislators disliked provincial plans, such as hospitals, schools, and surveys. They preferred to direct public money to local projects, particularly road building within their own districts. But in 1838, the House of Assembly got its hands on a big pot of money. Revenues more than doubled to over £200,000 with the financial windfall from the crown lands.[57] New Brunswick could now indulge in a survey. But how to organize one?

Surveys, as functions of government, were just beginning to take shape in the 1830s in both Britain and the United States, the two models for New Brunswick. In Britain, the gentlemanly specialist Henry De la Beche (1796–1855), through personal connections with prominent politicians and gentlemen of science, had been able to turn an ad hoc job of coloring geological features onto Ordnance survey maps of Devonshire into a secure, salaried, and permanent position as director of the Geological Survey of Great Britain from 1835 until 1855.[58] American government support for geology had likewise begun on an experimental basis, albeit somewhat earlier. In the 1820s, North Carolina and South Carolina supported summertime reconnaissances by resident professors of natural history. The first statewide survey, involving annual appropriations and progress reports, was initiated by Massachusetts in 1830. By 1836 eight states, including New York and Pennsylvania, had surveys. Unlike Britain, however, American surveys were designed as short-term government expenses.[59]

The New Brunswick survey confected a bit of both models—a dash of British gentlemanly preference with a dollop of American legislative accountability—to produce an unprecedented colonial organization of a different flavor.[60] In April 1838 Harvey appointed Gesner the provincial geologist. The House of Assembly grudgingly appropriated £200 for the first year, but only token support thereafter.[61] In return, legislators wanted annual progress reports. They also specified the survey's goals—"the discovery and application of such [mineral] substances as have been found most important to the interest and support of commerce, agriculture, and manufacture."[62] The principal ones were coal and iron. Further, they instructed Gesner *where* to explore. They directed him to the coast, extending southward and westward of Saint John, from the river to the American border. Such geographical constraints contrasted sharply with American surveys where funding flowed from the promise that geologists would cover the *entire* state and examine *all* the rocks.[63] New Brunswick, like Britain, began in a targeted location.

For his first season of fieldwork, Gesner followed the pattern he had developed in Nova Scotia. Starting in Saint John, he went from township to township looking for useful minerals. In effect, he retraced the path of a consulting trip he made in 1837, when a coal mining company had hired him to explore the lower St. Croix River, the islands off the coast, and other parts of Charlotte County.[64] This overlap between private and public surveying was not uncommon. Wealthy landowners and prospective capitalists, if not in government themselves, had enough political clout to ensure their properties received attention. Gesner visited the estates and farms of several prominent individuals, although he chose *not* to investigate the salt springs "on lands owned by his Honor the Chief Justice."[65] In this instance, complementary interests might have been misconstrued as a conflict of interest.

Gesner's ability to fulfill the House of Assembly's instructions resulted in funding for another year and emboldened him to broaden the survey's scope beyond mining and agriculture. He thought the "the discovery of rocks" with a view to ordering them and explaining their origin might also be useful. He presented a plan with equal weight on "scientific" and "economical" geology. New Brunswick now fell more in line with contemporary American surveys.[66]

Yet the distance between Gesner's broad geographical view and the House of Assembly's myopia remained, and it undermined the survey.[67] During the first three years, Gesner was told to concentrate on a handful of counties—Charlotte, Saint John, Kings, Queens, and Westmorland. Located along the Bay of Fundy (with the exception of Queens), these counties had the largest populations, the most cleared and cultivated land, and the best roads. They were the places most likely to support a survey.[68] They were also relatively close to Saint John, where Gesner had moved in 1838, and which served as the survey's headquarters. The northern timber counties and central ones like York, where Fredericton was located, made few, if any, demands for geology.

During his second season (1839), Gesner began in Saint John County, ventured up the coast to the head of Shepbody Bay, and then turned inland via the Petitcodiac River. In Westmorland County he found a big coal field, "a tract of country, one hundred and fifty miles long, and, upon average, forty miles in breadth."[69] The size of this surprising coal field "scarcely require[d] remark."[70] In fact, Gesner did not get much chance to study the structure, extent, or quality of the coal, even after he returned to Westmorland County the following year. Instead, the legislature required him to spend his third season reexamining Saint John County. Only after completing his resurvey of the coastal counties did Gesner get to go north and west. In Queens County, he discovered an iron ore deposit and more coal.

Coal had been known to exist in a few places along Grand Lake since at least the mid-1830s. A seam about two feet thick had been opened by the Salmon River Coal Company, a relatively large mining operation and one Gesner had close contacts with.[71] The coal was difficult to trace, however; the seam lay covered by sandstones or loose soil, and the strata were very nearly horizontal.

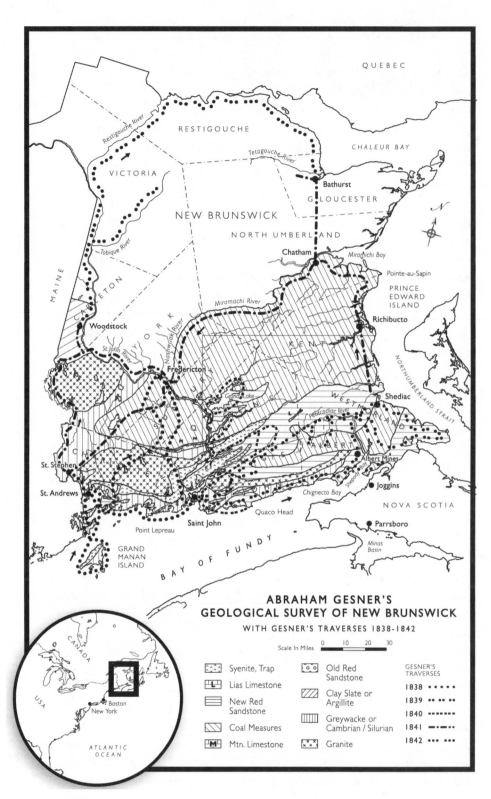

Abraham Gesner's geological survey of New Brunswick (1838–1842).

Moreover, it was impossible to estimate the field's size because the entire northern part of the province had yet to be explored.[72] Nonetheless, Gesner believed the coal of Grand Lake might extend northward to Bathurst, a distance of 150 miles, eastward to Shediac, a distance of 70 miles, and westward beyond Fredericton. It was a big coal field. "This tract," Gesner speculated, "may bear the reputation of being the largest coal-field ever discovered on the globe."[73] He dubbed it the "Great New-Brunswick Coal-Field."

Gesner planned a thorough exploration, but before he could get started, the political situation changed dramatically. In 1841 Sir William Colebrooke (1787–1870) replaced Harvey as lieutenant governor. Harvey's dismissal came in response to complaints about his failure to repel American encroachments along the upper St. John River. In that area, the exact boundary between the United States and British North America had been in dispute ever since the 1783 Treaty of Paris had ended the American Revolution. The treaty defined the boundary as the "high lands" separating rivers flowing north into the St. Lawrence from those flowing south into the Atlantic. The British had vital military interests in controlling the highlands; they provided the only overland route for moving troops between Halifax and Québec. Some in New Brunswick were also looking to build a railroad through the region. American interests were primarily economic; Maine wanted to expand logging operations into the Aroostook River valley. The climax of the "Aroostook War" came in late 1840 when Maine militia forces encountered British infantry. Not a shot was fired. Neither London nor Washington wanted war, but Harvey's position had been compromised by indecision. (John Fairfield, the Democratic governor of Maine, also lost his bid for reelection because of alleged military hesitation.) In 1842 a settlement was reached through the Webster-Ashburton Treaty. Britain retained the highlands, but Maine got the majority of the Aroostook pine.[74]

Maine acquired other potentially valuable resources. The boundary was surveyed by none other than Jackson. In 1836 Maine had organized a geological survey under Jackson's direction, but unlike Gesner, Jackson worked in the disputed territory and described in detail the mountains, rivers, lakes, and resources. According to Gesner, Maine's government "spared no pains or expense in obtaining accurate knowledge." Jackson's survey thus "afforded the United States a great advantage."[75]

Given the new political climate and Jackson's exploratory encroachments, Colebrooke instructed Gesner to move into the timber regions along the boundary. "In consequence of the late dispute," Gesner explained, "it was necessary to obtain as much information as possible in regard to the nature, resources, and value of the district claimed by each of those powers."[76] Gesner traveled up the St. John River to Woodstock, where Jackson had recently uncovered iron ore.[77] Gesner then followed the Mirimachi and Nashwaak rivers.

What he found in the forests astonished him—the great New Brunswick coal field. "From all the observations I have been able to make, it is now known to embrace an area of seven thousand five hundred square miles!!! If the Westmoreland Coal Field also belongs to this carboniferous deposit, the whole area may be computed at eight thousand seven hundred square miles."[78] Gesner could barely contain his enthusiasm. He listed coal's many uses, all of which were very familiar, but apparently not of great interest to the House of Assembly. In 1842 legislators cut the survey's funding.

Colebrooke, nonetheless, commissioned Gesner to continue his work, without pay, and sent him, again, to "the territory recently in dispute between Great Britain and the United States."[79] Prominent merchants in Saint John loaned money to him, and the directors of the Saint John Mechanics' Institute planned to purchase his collections of minerals, fossils, and other specimens for £1,000. (Gesner was, at the time, the institute's vice-president.)[80] So for another year Gesner explored the northern timber counties. He followed the upper St. John River, the Tobique River, and, in the most northern reaches, the Restigouche River and the Bay of Chaleur. "The knowledge heretofore possessed of some vast and important tracts of land within the Province," he reported, "was confined to a few lumbermen." The land was "unoccupied even by the native Indians."[81]

Gesner asked for support for "the labour of another season, which would complete the geological survey of the whole Province."[82] In his argument, he summarized the numerous advantages already accruing from his work. First, there were the marl and limestone deposits, mineral fertilizers that had not previously been used in New Brunswick. Gesner admitted that New Brunswick farmers were predisposed against such fertilizers, perhaps because they were not widely used in Britain, but he noted that American farmers reaped good returns from judicious trials. More generally, Gesner thought his description of "millions of acres of excellent lands" in the province's interior would be a great benefit to farmers.[83]

As for minerals, Gesner spoke with pride of his discovery of granite and other building materials. "Formerly, all the granite was imported from the United States or Nova-Scotia."[84] But by 1841, he declared, "upwards of six thousand tons" had been quarried and shipped to Fredericton, Kingston, and Saint John, which was busily rebuilding after a devastating fire in 1839. Gesner listed "extensive" deposits of iron, copper, lead, and manganese, in addition to "deposits of salt [and] gypsum," which, he predicted, "may hereafter become the sources of wealth."[85]

Finally, there was coal. Prior to his survey, coal had to be imported from Nova Scotia. Gesner's discoveries would change all that, *if* New Brunswick could attract investors. "There is scarcely any quarter of the globe where British capital and science are not employed in mining pursuits," he explained. "If it be enquired why foreign countries have received so many advantages from this source, while the British North American Provinces have been neglected, it will be seen, that in the former, the objects of wealth have been made known, while in the Colonies the discovery of similar objects has been neglected."[86] In short, further study of the great New Brunswick coal field would lead to British investment.[87]

In the meantime, local capitalists, inspired by Gesner's glowing reports, had taken mining leases on properties within his great coal field and started to dig. But they did not get far before the ventures failed. And it is not entirely clear why. Gesner suspected they were not begun in the right fashion and advised that, "before much capital is expended in opening these mines, the most careful, patient, and scientific examination should be made."[88] Mining companies must first consult geologists. Gesner was not touting for custom; in fact, he endorsed the "accomplished" English geologist W. J. Henwood.[89] Rather he was arguing for an extension of his survey and, simultaneously, delimiting the boundaries of his position. As provincial geologist, he was "to assist" New Brunswick capitalists by pointing out the "new and most important objects of enterprize and wealth," although he warned "it is not the object of a Geological Survey to extend its operations to the absolute working of any mine."[90]

Despite his efforts to forestall criticism, complaints about coal mining reached the House of Assembly. Whether it was the failed ventures or the loss of Harvey's patronage, Colebrooke did not reappoint Gesner. The New Brunswick survey came to its official end in 1843.

Gesner anticipated it would. The tone of his last government report was remarkably subdued. The usually upbeat Gesner made no mention of his great coal field and said very little about coal in general. His resigned economy of words reflected a gloomy outlook. New Brunswick was in recession. British timber preferences had been sharply reduced, and revenues from timber leases had declined. The unrestrained spending of the previous four years had put the province in debt by 1841. Gesner, however, did not think the fault for any business failures lay with his survey or his science, but with bad commercial practices. Mining leases had been secured "for the sole objects of speculation, and not for their actual workings," he retorted. He could not be held responsible for recklessness; "such things [were] common to all new discoveries."[91] Moreover, Saint John capitalists, in contrast to those in Halifax where commercial caution prevailed, had a reputation for "rash and improvident speculation."[92] Perhaps this was why Gesner favored British ones. In the end, the survey was not discontinued because of sour mining deals. As Gesner flatly declared, "the embarrassed state of the finances of the Province" did him in.[93]

Nonetheless, "[t]here is," Gesner observed, "in the examination of the natural productions of the country, an object to be gained beyond the benefits arising from the discovery of useful minerals."[94] The scientific results of the New Brunswick survey were substantial—the best geology in British North America—and much clearer than his Nova Scotia account. Gesner identified Primary formations in an extensive and moderately elevated chain of mountains running along the Bay of Fundy coast. In the province's southern part, he found rocks belonging to the Transition, a term he used in his first report, but soon replaced with Greywacke, and later Cambrian and/or Silurian. Secondary formations, principally the Coal and New Red Sandstone, occupied most of the large, central part of New Brunswick. Tertiary rocks were found in the northern counties of Restigouche and Gloucester.

In describing these formations, Gesner relied on lithology. He mentioned fossils and even listed a few from the Coal, but as for a systematic study of animal and plant remains, he confessed, they were "too numerous, and frequently too complex in their characters."[95] In identifying the formations, he turned to British authorities. Gesner listed eight, sometimes nine, "systems"; the number fluctuated as he debated whether to adopt Adam Sedgwick's "Cambrian" or Roderick Impey Murchison's "Silurian" for the oldest fossiliferous strata of the Secondary.[96] Gesner did not try to correlate New Brunswick rocks with American formations, and he rejected the practice, recently adopted by New York geologists, of using local names for formations. The idea of coining names, such as "Nerepis, Mispeck or Quaco rocks," simply invited ridicule and confusion. Formations, Gesner declared, should be recognizable "in every quarter of the globe."[97] He thus deferred to British geologists. "I have been induced to use the names employed by those gentlemen," he reasoned, "not only because their important divisions have been sanctioned in Europe, but from the applicability of the descriptions attached to them to the rocks of New-Brunswick."[98]

Overall, Gesner's five annual reports (1839–1843) were not designed to deal with problems of correlation. Basically, he had made detailed lists of what he saw, not theoretical explanations of how it got there. His geology was more inventory than inventive.[99] In fairness to Gesner, this was what was expected of him. Surveying for the sake of science was not his job. On occasion, and when appropriate, he did theorize about geological causes. In his account of river terraces and boulders scattered over the land, for example, he addressed "the glaciel [sic] theory" of Louis Agassiz (1807–1873).[100] He admitted that in other countries, such as Agassiz's Switzerland, mountains might be covered with "eternal snow," but he did not think the New Brunswick highlands were high enough for glaciers. To explain the terraces, Gesner relied on sudden drops of the riverbed, although he proposed no explanation for the change in elevation. As for the scattered boulders and the parallel scratches and grooves (striæ) on some rocks, Gesner invoked a sudden and violent rush of water over the land.

In contrast to his Nova Scotia geology, he did not explicitly refer to Noah's Flood. Gone too were other references to the Bible. Gesner, of course, did acknowledge "the power and goodness of [the] Creator."[101]

His reports were written in a familiar travelogue style. They contained practically no diagrams, sections, or maps. Although he continually spoke of completing a geological map, he never published one. He did include numerous woodcuts depicting nature, which coupled with his account of his explorations could be quite moving.

> One mile above the Indian village there is a dangerous rapid, called the Narrows. The [Tobique] river at this place passes through a gorge a mile long, and upon average only one hundred and fifty feet wide, and between perpendicular cliffs from fifty to one hundred feet high. . . . In descending, the Indians were anxious to "shoot the rapid," as it is called, and we were carried through the opening with almost inconceivable swiftness.[102]

Shooting the rapids was a relatively rare and exciting event in Gesner's reports. Native Americans, along with their villages and terms for local landmarks, appeared frequently, because they guided him through the thick forests. Moose, caribou, and lazy bears, on the other hand, were missing.

His general tone was more serious than his Nova Scotia book, a reflection of change in his intended audience—the New Brunswick government rather than the reading public. Gesner was also keenly aware that his reports were to be distributed in Britain, and occasionally he addressed "persons abroad" who might emigrate to New Brunswick and take up and clear the tracts of ungranted land in the interior. At other times, he targeted his discussions to British geologists.

In 1840 his pioneering efforts were rewarded with a fellowship in the Geological Society of London. This honor, however, required some political finesse. At the start of the survey, Gesner had written to the society requesting information on how to become a member. He was told that because he resided in New Brunswick he could not become one.[103] This was to be regretted, Gesner replied, especially because he was the first official geologist of a crown colony. Somewhat peevishly he added that he would not be forwarding New Brunswick minerals and fossils. "It will be but right to bestow them upon such an Institution as will allow their donor the most favorable consideration."[104] Gesner soon received confirmation from Lyell, then vice-president, that he would be elected a fellow.[105]

Gesner's fellowship illustrates how a geologist, far removed from the metropolis, could leverage the specimens of a distant land and the authority of a government appointment into scientific rewards. It also reflects Lyell's fascination with America. The two might have met during Gesner's brief stay in London. Lyell certainly read Gesner's account of the fossil trees at the Joggins and was "particularly desirous" to see them. So he arranged to meet Gesner in July 1842.[106]

Lyell had arrived in America in August 1841 and had spent the year geologizing in the United States. When he contacted Gesner, Gesner was busy shooting the rapids, but directly afterward went to meet Lyell at Parrsboro. They spent two days studying the 100-foot cliffs of the Joggins. Lyell was impressed. Along two miles of coast, they measured a section of sandstone, shale, and coal strata amounting to more than 2,000 feet of deposition. They counted nineteen coal seams, some as much as four feet thick, and seventeen vertical trees.

Immediately below this coal section, Lyell found beds of gypsum, limestone, and red sandstone, which he called the Gypsiferous formation. Lyell located this gypsum group *below* the Coal formation, an interpretation that conflicted with Gesner's (and Jackson and Alger's).[107] Gesner had placed the gypsum group as part of the New Red Sandstone formation, which lay *above* the Coal. In separate articles to the Geological Society, Lyell and Gesner debated the proper stratigraphical location of the gypsum.[108] Much depended on the interpretation; if Gesner were correct, large tracts of coal might still lay hidden. If Lyell were correct, the search for coal in parts of Nova Scotia *and* New Brunswick (especially in Westmorland County) would be fruitless.

Gesner, the provincial surveyor, did not stand alone against Lyell, the cosmopolitan geologist. William Logan (1798–1875), a Montreal-born geologist who had apprenticed in the Welsh coal fields in the 1830s, toured Nova Scotia in 1841, and he too placed the gypsum group with the New Red formation. Logan's stratigraphy, like Gesner's, was based largely on lithology and relative position. But Logan also collected fossils, which he brought back to London to show to Murchison and the French geologist Édouard de Verneuil. After studying them, Murchison and de Verneuil confirmed the location of the gypsum group in the New Red formation.[109]

But this did not end the debate. During the 1840s, Dawson took up the correlation question. In October 1847 he presented a detailed survey of the gypsum group, which he argued should be placed in the lower part of the Coal formation, that is, *below* the Coal, as Lyell had thought.[110] In 1849 Gesner agreed. In a brief note to the Geological Society, Gesner took "pleasure in confirming the views held" by Lyell.[111] The Gypsiferous formation of Nova Scotia lay *below* the productive Coal Measures. The debate between Lyell, Dawson, and Gesner thus highlights how far, scientifically speaking, Gesner had come from the early 1830s (when Jackson and Alger avoided mentioning him in print) to the early 1840s (when Lyell sought him out). Gesner had established a reputation as a man of science based on his pioneering work in New Brunswick.

The robustness of that reputation was tested in later discussions of the most contentious aspect of Gesner's survey—the great New Brunswick coal field.[112] L. W. Bailey (1839–1925), a member of the Geological Survey of Canada in the 1890s, explained that Gesner's coal field was very thin, and although it extended over a large area, as Gesner had outlined, its shallow depth resulted in very few mining possibilities.[113] G. F. Matthew (1837–1923), another survey member in the 1890s, wrote a very sympathetic sketch of Gesner in which he pointed out "the error" of expressing a too "favorable" opinion on mineral wealth.[114] In contrast, Gesner's contemporary, James C. Robb (1815–1861), the first professor of chemistry and natural history at King's College (later renamed the University of New Brunswick) in Fredericton, criticized him for unwarranted optimism. "The Great New Brunswick Coal Field," Robb chided, "certainly is very considerable, although it is not 'one of the largest area[s] discovered on the Globe.'" Rather, "it is sufficiently obvious that the importance of the [coal beds] has been over-stated, while the probability of finding others of greater thickness and improved quality, has been much exaggerated."[115] Robb's opinion appeared in an agricultural report on New Brunswick prepared by the Scottish chemist James F. W. Johnston (1796–1855), who had a subtler understanding of the inherent difficulties of survey geology.

> [L]ike all men whose fate it is to pioneer the way to new views, new studies, and new habits of thought, [Gesner] evidently writes as if he felt his work to be very much up-hill—as if he were laboring for men who did not generally understand or appreciate his task, and he was therefore induced occasionally to minister a little too strongly to the vulgar views of immediate profit from scientific inquiry, and thus create expectations which his own labours did not realize.[116]

Commercial predictions were risky, though at times necessary. Optimism might be confused with misrepresentation, a behavior unsuited to a man of science.

From a historical perspective, the significance of Gesner's New Brunswick survey is mixed. He undoubtedly provided the first detailed descriptions of the rocks of a hitherto unexplored region, but he studied less than half of the province, an unfinished job he sorely wished to complete.[117] As far as coal went, the field was great geographically, not commercially.

The final irony may be that New Brunswick got what it paid for. Parochial politics produced patchy science. Unlike American surveys, where democratic governments required more representative coverage, including, in the case of Maine, lands not yet belonging to the state, the New Brunswick House of Assembly could not see the value of a province-wide view. Colebrooke called the way legislators distributed money a highly organized system of corruption.[118] Gesner thought it "a system of extravagance."[119] That is not to say American state governments were not prone to corruption or wasteful expenditures when it came to internal improvements. But in New Brunswick's system of narrow-based patronage where each faction and every community were concerned with promoting their own immediate interests, geographic range and scientific autonomy were limited.

Scientific Entrepreneurship

The circumstances of Gesner's dismissal were peculiar to New Brunswick, but the survey's termination would not have surprised any American geologists. Whether a survey was completed or not, unemployment eventually came to all surveyors. They then faced the arduous task of earning a living by pursuing their science. Gesner, like many American men of science, was not independently wealthy and thus did not have the leisure to geologize. Nor did he have an academic position to fall back on. He had tried to secure an appointment as lecturer in chemistry and natural history at King's College, New Brunswick, in 1837, but that position had gone to Robb.[120] In general, teaching positions in the first half of the nineteenth century were scarce, especially in British North America. Gesner thus needed another means to a scientific end.

Scientific entrepreneurship refers to the ways in which men of science like Gesner tried to make a livelihood using their knowledge, experience, and social contacts.[121] It is a broad category encompassing a variety of employments and career paths chosen for scientific and personal advancement, including other government surveys, academic positions, public lecturing, popular writing, and consulting. The usefulness of scientific entrepreneurship as an analytical concept lies in its highlighting the temporary nature of these varied positions and the required effort of self-promotion to obtain them. Scientific entrepreneurs had to convince others of their expertise and the advantages of their science. Further, the concept forces a serious evaluation of the degree and kind of accommodations men of science made and were willing to make, as they attempted to weigh the issue of how much "science" versus how much "enterprise." What balance was to be struck between the pursuit of research and the provision of practical results? These questions are similar to those generated by government surveys, and, not surprisingly, surveys fitted geologists for enterprising careers afterward. But not all scientific entrepreneurs had the benefit of government appointment and hence the legitimation derived from such authority. Creating and maintaining a reputation for honesty and competence was a new and challenging ethical problem for many midcentury American men of science.

Gesner was forty-six years old when he lost his New Brunswick position. He returned to his father's home in Cornwallis, Nova Scotia, where he made another desultory effort at farming and a reluctant stab at doctoring. Gesner, however, considered himself a man of science. And between 1843 and 1849 he undertook a number of scientific projects. Public lecturing, for example, suited his interests and outgoing personality very well. He demonstrated the latest scientific findings and fads, such as galvanism, and drew large audiences and widespread publicity. One of his most famous performances, delivered at the Saint John Mechanics' Institute in March 1843, involved the animation of the head of a freshly killed ox. In dramatic fashion he showed that muscles could be stimulated by an electrical jolt; he made the dead beast's eyelids open and close. In this instance, and in many others, Gesner effectively mixed spectacle with respectability and thus entertained as well as enlightened his audiences.[122]

Besides lecturing, Gesner continued surveying, both public and private. In the summer of 1846, he made a reconnaissance of Prince Edward Island and wrote a brief fifty-page report, for which the government granted him £200.[123] That same year, he consulted for a group of Halifax capitalists about iron ore near Londonderry. He considered it "the richest and most valuable deposit" in Nova Scotia.[124] In general, though, surveying presented limited opportunities. There were only six colonies in British North America, and the number of mining companies was distressingly small, despite Gesner's untiring efforts.

In contrast, the situation in the United States was booming, as illustrated by Jackson's career. As mentioned previously, Jackson had surveyed Maine while Gesner worked in New Brunswick. In 1839, after three years of fieldwork, a financial downturn brought that survey to an end. Jackson, however, managed to land two other government positions. In Rhode Island, with the help of the Society for the Encouragement of Domestic Industry, he received a one-year appointment (1839–1840) to make a geological and agricultural survey, and in New Hampshire he organized a four-year geological and mineralogical survey (1839–1842).[125] In 1847 the federal government appointed him to the post of geologist-in-charge of surveying the public lands around Lake Superior. Jackson's serial surveying was not unusual. Many American geologists worked on multiple surveys (state and federal) during their careers. Prior to the Civil War, twenty-eight of thirty-four states began geological surveys, a large number that reflected both the ever-increasing territorial size of the United States and its growing industry, especially mining. Jackson did have the advantage of experience and of location; Boston was home to several capitalists looking to consult a geologist about their mines and properties. During the 1840s and 1850s, consulting became such a lucrative, professional practice in the United States that it emerged as the dominant expression of scientific entrepreneurship.

Barring a move to the United States, Gesner had to find another way to make a living as a geologist. He chose popular writing. Authorship was a familiar kind of scientific entrepreneurship in Britain. Lyell, for one, had shown how to make money selling geology books. Gesner, of course, was no Lyell in terms of scientific or social standing. And neither was Nova Scotia anything like the British market for books and popular culture. As a result, the two books Gesner published in the 1840s were less about geology than about political economy, and they were directed more at a British audience than one at home.[126] As a self-described reformer, Gesner presented his books on New Brunswick and Nova Scotia as templates for change. Science, generally, and geology, specifically, would be guides to progress and improvement.

In his sweeping guidebook entitled *New Brunswick; with Notes for Emigrants* (1847), Gesner presented the most comprehensive account of the province's natural phenomena as well as a history of its native peoples, European settlements, Loyalist families, and recent British immigrants. It contained "all the information acquired during the performance of public service," Gesner explained.[127] He described the land, town by town, and addressed all possible employments —hunting, trapping, fishing, farming, lumbering, and, of course, mining. Published in London, the book painted a bright picture of British North America, a part of the world often mistakenly characterized by its frigid climate, dense fogs, and endless wilderness—what Gesner called "the great moose yard."[128]

Within the fabric of facts and figures, Gesner wove a plan to capture the "redundant population and dormant wealth" of Britain. He wanted to increase investment in New Brunswick and to encourage emigration of "respectable" persons, those with capital of £25, £50, or £100. He opposed the Irish. Those impoverished immigrants had been arriving in large numbers, although they tended to leave New Brunswick because of paltry opportunities. What few jobs existed were in lumbering, a regrettable situation because it signaled continuing reliance on one export. "The enterprise of the country," as Gesner called it, was bound up to such an extent with the timber trade that all other industry scarcely existed.[129] His favorite example of a neglected industry was, predictably, mining. He praised the "great mineral wealth" of New Brunswick, but his own survey had failed to encourage that industry. The great New Brunswick coal field supported just two mines in Queens County.[130] Gesner conceded that mining "will probably remain inactive until the timber resources have failed."[131]

Only foreign investment could develop the mining industry, a theme Gesner developed more fully in his next book, *The Industrial Resources of Nova Scotia* (1849). From a cursory glance, it looked a lot like the one on New Brunswick, a business prospectus for the province based on a thorough inventory of its "natural wealth"—plant, animal, and mineral. (It was dedicated to Harvey, his former New Brunswick patron, who in August 1846 became the lieutenant governor of Nova Scotia.) On closer examination, it presented Gesner's mature views on geology, coal, and progress through a serious and sustained critique of the kind of capitalism promoted by the Nova Scotia and British governments. What Gesner wanted to explain (and therefore correct) was why the province was "not advanced in a degree proportionate to her resources."[132]

Like many British and American Whigs, Gesner saw advancement in spreading the fruits of science across the landscape in the form of mines, canals, railroads, and telegraphs.[133] Internal improvements made nations prosperous and progressive. Governments were therefore responsible for promoting them.[134] The reason Nova Scotia languished, while Britain, the United States, and even Upper and Lower Canada moved forward, was directly related to the *lack* of government encouragement. Gesner thought that Britain bore the brunt of responsibility. Not only had London failed to grant large sums of money for improvement—as it had for Canada—but individual capitalists had betrayed the British Empire by investing in American mines, canals, railroads, and telegraphs. This preference for the United States was doubly galling because many of those companies and a few state governments, most prominently Pennsylvania, had gone bankrupt.[135]

Nova Scotia deserved better, and Gesner set out to inform those at home and abroad of investment opportunities and needs. He projected a "New Britain on this side of the Atlantic." Oddly enough, his model was not Britain, but the United States, a country undertaking "colossean public improvements."[136] He admired the "superiority" that Americans displayed through their vigorous spirit of enterprise.[137] Nova Scotians, by contrast, showed "little enterprize"; they followed "extreme caution" and lacked "general and persevering industry."[138] Americans exceeded Nova Scotians in wealth and improvement because they had been weaned from the patronage of the mother country, a paradoxical explanation in light of Gesner's demands for *increased* British investment. Nevertheless, American advancement was "not from the peculiarity of their government" (which Gesner thought was essentially British) but because "almost every natural production of their country has been collected."[139] This was the heart of Gesner's argument. Nova Scotia had sponsored "no public inquiry in reference to the natural productions and resources of the country."[140] The province needed a geological survey.

> The objects to which money can be profitably applied, must first be made known, before any hope can be entertained of their being rendered useful in themselves, and doubly useful by the capital they draw in from abroad. With such views the Americans have made examinations and surveys of every part of the Union from the St. Croix to Oregon. . . . Every state has its geologists, botanists, chemists, agriculturalists, and civil engineers, who are paid from the public treasury.[141]

If science were the basis of improvement, then it was no wonder Nova Scotia stagnated.

Absent a new survey, *Industrial Resources* contained the best overview of Nova Scotia geology. In concise fashion Gesner outlined the major stratigraphic

Richard Taylor's map of the coal fields of British North America. Taylor relied on Abraham Gesner's surveys of Nova Scotia and New Brunswick, but he did not use Gesner's report on Prince Edward Island, which showed no coal there. *Source:* Richard Cowling Taylor, *Statistics of Coal* (Philadelphia: J. W. Moore, 1848), facing page 208.

systems—Granite, Cambrian, Silurian, Devonian, Carboniferous, New Red Sandstone, Intrusive, and Drift. In identifying various outcrops and tracing their extent, he followed Lyell's 1842 reconnaissance. But when it came to mineral resources, Gesner asserted his own expertise. He repeated his claim to the discovery of the great New Brunswick coal field and even added a discussion of the "great Nova Scotia coal field." Taken together, Gesner assured his readers that British North America contained 10,000 square miles of coal, which could supply fuel, sustain furnace and forge, and produce steam for "multifarious operations of manufacture."[142]

But even if Nova Scotia got the necessary British capital for internal improvement, it faced one more hurdle—foreign monopoly. Nova Scotia's minerals did not belong to the people. The Duke of York had acquired a crown lease in 1827 to all Nova Scotian mines and minerals. He then granted sole rights under this lease to a group of London capitalists, who formed the General Mining Association (GMA).[143] The reservation of all mines and minerals to the GMA, Gesner concluded, had "retarded discoveries by the inhabitants, and checked that kind of inquiry which has been so beneficial to neighboring colonies."[144]

In Nova Scotia, coal was the principal mineral, but it was the one most "retarded" from reaching its full potential. The GMA had opened only two collieries: one in Pictou (Albion Mines) and the other in Cape Breton (Sydney Mines). There were certainly other coal fields, as Gesner had shown, but the GMA refused to mine them. Throughout the 1830s and 1840s, Gesner steadfastly opposed the GMA. "The Mining Association possesses an entire monopoly, which has prevented every kind of mining enterprise in the province . . . a colony that will never thrive until her resources are liberated from the fetters of unyielding monopolists."[145] In 1838 and again in 1844, he petitioned the House of Assembly to open a mine at the Joggins. He even applied for a mining license himself using a legal technicality to force a hearing. According to the terms of its lease, the GMA was required to open a mine in places where coal had been found within a year of its discovery.

Gesner's petition resulted in the formation of a government committee of inquiry set up in March 1845 to investigate the GMA's monopoly. Gesner and Dawson were the star witnesses. They testified to the GMA's deleterious effects on coal mining. The committee found that foreign countries (i.e., United States, where the GMA's coal was being exported) received lower prices than did citizens in western Nova Scotia. The House of Assembly decided therefore not to reduce the royalty paid by the GMA on Nova Scotia coal and to force the GMA to begin mining at the Joggins, which it did, in 1846.[146]

The opening of the Joggins mine had a decided effect on Gesner. By the time *Industrial Resources* appeared in 1849, Gesner had tempered his hostility toward the GMA. He now thought the GMA had brought much needed capital into the province. Still, he hoped it would do more, especially in developing iron ore resources. There was still not a single furnace smelting iron ore in any of the British American provinces. That changed when the Londonderry Iron Works was established, one of the very few examples of a profitable mining enterprise before Canadian Confederation in 1867.[147]

The publication of *Industrial Resources* marked the end of Gesner's career as a geologist. In the late 1840s, he began to turn his attention toward chemistry and the commercial interests of his new patron, the audacious and famous Thomas Cochrane (1775–1860), tenth Earl of Dundonald and admiral of the British North American and West Indian fleet stationed in Halifax. Gesner had become acquainted with Dundonald while writing the book and had relied on him for information about Cape Breton. Moving to Halifax in 1849, Gesner took up coal chemistry, and he and Dundonald began work on an ambitious project—gas lighting for Halifax. By 1850, then, Gesner had moved from government geologist, through lecturing and writing, to businessman. Gas lighting would test his scientific knowledge, political experience, and personal connections (as well as those of Jackson) in the coming years.

> The occupations and habits of the inhabitants of any civilized country will ever be influenced by the geological structure of the districts they inhabit.
> —Abraham Gesner, *The Industrial Resources of Nova Scotia* (1849)

GESNER'S ASSERTION was both grandiose and banal. In reducing society to bedrock, he seemed to make an extraordinary case for his science, if not his hubris. But he was not alone. Geology was the acknowledged queen of the sciences in the early nineteenth century. Romantic, adventurous, and practical, it encompassed time, land, and the raw materials of progress. Gesner's oft-repeated slogan that coal "lay the foundation" for industry would have struck contemporaries as perfectly obvious. It is, however, historically important to understand why and how this connection between science and improvement became commonplace.

Part of the answer turns on the individuals Gesner most readily identified with progress—capitalists. When Gesner spoke of capitalists, he meant a class of wealthy, upright, and enterprising gentlemen—mostly British, but also American or British American—willing to invest monetary and moral capital in developing coal mines, canals, railroads, telegraphs, manufactories, furnaces, and forges. These big industries or "technologies" had several common features. They were expensive, they were complex, and they were risky. It was never certain that these improvements would return the investment, to individuals or to governments. It was therefore crucial that investors had the best advice available, precisely the kind of science Gesner could provide through government reports or private consulting engagements. In either case, science could mitigate the risks.

Industry was regarded as an improvement because it returned a profit on investment and benefited the public. In this ideal framework, industry was *not* speculation, an ephemeral financial bubble; and capitalists were *not* speculators, impetuous individuals like those who did not mine the great coal field of New Brunswick. Nor was industry intended to be a monopoly, like the General Mining Association. To Gesner's mind, improvement meant many industries and many companies—multiple capitalists and their consultants. Speculators and monopolists marked the dangerous banks of the economic mainstream. The former were too rash and too many; the latter were too stubborn and too few. There was a superior middle way to prosperity—scientifically informed investment. The link between science and progress was not abstract; it was personal and direct, between the geologist and the capitalist.

There were definite blind spots in this vision of scientific capitalism. Gesner, for one, did not have much to say about the laborers who mined coal, fired furnaces, laid tracks, or dug canals. He disliked Irish immigrants, precisely those who did the hard and dirty work.[148] He was much more concerned with share prices, production, and equipment costs. To that extent, he shared the prejudices of the capitalists he associated with. Science entered the industrial equation at the top and informed big decisions about scope and scale, not everyday toil.

It would be unfair, however, to class Gesner with the mercantile elite that ran places like Saint John. He opposed timber merchants as much as he did the General Mining Association; both restricted growth. Gesner advocated change of a steady and stable variety, although he knew very well that industry, especially coal mining, could promote such economic unrest as workers' strikes. He avoided making the connection between industrial and political protests. His was a thoroughly Whiggish vision of an orderly, yet prosperous and expanding society. Democratic urges were largely alien to his reforms for New Brunswick and Nova Scotia. To Gesner's mind, democracy should remain in Upper and Lower Canada or in the United States.

It is therefore difficult to position Gesner's science on one side or the other of a dichotomy, invoked by some historians, between a commercial elite and a manufacturing-labor alliance in British America. Gesner's geology was undoubtedly more suitable to gentlemanly landowners and enterprising manufacturers than it was to miners and mechanics. But it was equally clear that industry allied with science could promote stability *and* instability; and both could serve conservative and radical ends.[149]

Behind Gesner's economic vision lay a political one. He wanted Nova Scotia and New Brunswick to prosper. Instead of exporting coal, Nova Scotia and New Brunswick should use it to smelt their own iron ore. "[A]mong the subjects of our most gracious Queen, none are more loyal," Gesner declared, "[but] it would be a silly patriotism for them to sacrifice the natural advantages they possess to the benefit of England."[150] He stopped short of connecting economic with political independence. Gesner firmly believed that a domestic industry would strengthen the colonies *and* Britain. He also had no doubts about the advantages the United States and Canada held by virtue of their western expanse. New Brunswick and Nova Scotia did not have vast lands to exploit. That was precisely why they must remain part of the British Empire.

A final point with respect to Gesner's geology draws attention to the land. When Gesner looked at it, he saw more than mountains, rivers, trees, soils, or Native Americans. He peered beneath—"open the earth beyond the depth of the soil." He wanted others to find the riches that lay under foot. Prosperity meant mining, not just in the practical sense of opening coal pits, but in the metaphorical sense of digging deep to find a foundation on which to base it. If geologists and capitalists failed to do that, the consequences were dire. To someone like Gesner, the connection between civilization and science was direct, and it was built on the bedrock of geology.

CHAPTER 2

The Strange Case of the Albert Mineral

No mineral found in New Brunswick has awakened
more interest than this. None is so peculiar in its nature
and associations, none has been the subject of greater
controversy, both scientific and legal.
—L. W. Bailey, *The Mineral Resources of the Province
of New Brunswick* (1893)

I N EARLY 1850, IN ALBERT COUNTY, New Brunswick, John and Peter Duffy
opened a mine. The mine lay within the bounds of Abraham Gesner's great coal
field of New Brunswick, but what the brothers dug out did not seem like ordinary coal. To find out what it was and its value, they shipped some to Halifax,
and from there samples went to the Boston Gas-Light Company, which sent
them to its consulting chemist Charles Jackson. In his private laboratory, Jackson studied the mineral, and on 17 April 1850, at a meeting of the Boston Society of Natural History, he announced "a new kind of fuel recently discovered in
New Brunswick."[1]

The Albert mineral (as it will be called) was unusual. Jet black, glossy, and
"free from smut," a sort of black dirt, it broke with a "broad conchoidal fracture
like obsidian," was a little softer than rock salt, and had a specific gravity of 1.107.
When heated, the mineral softened and melted and then gave off "an abundance
of bituminous liquid analogous to Petroleum." Jackson ran two distillation trials (see table 2.1)—heating the mineral until no further gases (volatile matter
or bitumen) could be drawn off from the residue (fixed carbon or coke).

The high percentage of volatile matter was remarkable. The only known
mineral containing more than 50 percent was asphaltum. Jackson thus decided
that, although "[i]t was regarded as Cannel Coal of a particular kind," the Albert mineral was actually "a very beautiful variety of Asphaltum." "This sub-

Table 2.1. Charles Jackson's Experiments on a New Fuel from New Brunswick (April 1850)

	1st Trial	2nd Trial
Volatile matter or bitumen	58.5%	58.8%
Coke or fixed carbon	41.5%	41.2%

stance," he predicted, "is particularly valuable for the production of gas for illumination."[2]

Jackson's assessment would prove to be the source of much controversy, both scientific and legal. The controversies arose when Gesner tried to use the mineral for what Jackson had suggested. In the winter of 1850, Gesner organized a gas company in Halifax and a mining company in New Brunswick. Gesner's entrepreneurial forays precipitated two celebrated legal cases. The cases, one in Nova Scotia and the other in New Brunswick, turned on the issue of ownership. To the litigants, the key to rightful possession lay in the ability to say what, exactly, the Albert mineral was. Gesner, the plaintiff in both cases, wanted to classify the mineral as asphaltum, whereas the defendants wanted to call it coal. Legal rights thus seemed to rest on mineral identity.

To strengthen their legal positions, both sides called upon men of science to testify in court. Because the employment (and deployment) of scientific expert witnesses was a relatively new thing in America, the importance of the Albert cases can be judged by the large number and high prominence of the experts willing to serve.[3] More than a dozen well-known British and American chemists, geologists, and naturalists testified. Within the courtroom, these experts soon discovered that their reputations as arbiters of science, along with their social standing and personal integrity, were as much on trial as the mineral rights. For historians, the Albert trials present excellent case studies of the making (and unmaking) of scientific authority. The high-stakes cases also reveal the degree to which the law depended on men of science and, conversely, how scientific expertise was reinforced (or undermined) by the courts.

Beyond the courtroom, the Albert mineral was equally contentious. It raised fundamental issues about how to classify minerals and how to determine the age and identity of rocks. Complex and often heated arguments erupted over the relevance and value of various geological, paleontological, chemical, and physical criteria. At a deeper level, the differing criteria reflected the contested disciplinary domains among geologists, chemists, and mineralogists. Laying claim to the Albert mineral meant extending and defending the limits of disciplinary boundaries. Thus what began as a legal disagreement over the owner-ship of a peculiar kind of fuel escalated into a far-reaching scientific controversy.

The Coal Gas System

There would have been no Albert controversy had not the gas industry been de-pendent upon science. The meaning and extent of that dependence can be as-sessed by examining the entrepreneurship of two men of science: Gesner and the Yale chemist Benjamin Silliman Jr. (1816–1885).

Gesner's interest in gas lighting stemmed from a series of public lectures he gave at Charlottetown, Prince Edward Island, in the summer of 1847. In his demonstrations on "caloric," he showed how to distill gas from various miner-als, including Trinidad pitch, which he had encountered two decades earlier during his West Indian horse-trading farce.[4] Newspaper accounts of Gesner's lectures caught the attention of the entrepreneurial Admiral Thomas Cochrane, tenth Earl of Dundonald. Dundonald was in the process of acquiring the min-ing rights to Trinidad's famous "Pitch Lake"[5] and decided to consult Gesner. By June 1849, Gesner had invented a technique for producing gas using pitch or as-phaltum and had applied for a Nova Scotia patent on it. He called it Kerosene gas. The name derived from the Greek words *keros,* meaning wax, and *olene,* for oil. Wax and oil were produced when asphaltum was heated.[6] The prospects for Kerosene gas seemed good. The government of Nova Scotia, for example, was willing to try it in lighthouses.[7] So, in early 1850, Gesner went to the United States to secure another patent.[8] The timing could not have been better.

Gas was the largest and fastest-growing sector of the American lighting in-dustry. Since the first company had been established in Baltimore in 1817, its popularity had increased steadily, and by 1850 gas had "nearly superseded every other kind of artificial light," including all plant and animal oils.[9] More than fifty cities had public gas companies, while many smaller communities and factories relied on private firms.[10] Among urban, middle- and upper-income households, gas lighting was preferred. It was cleaner and safer than the cheaper and poorer-quality lamp oils, such as camphene and "burning fluid," and gas was more affordable than better-quality illuminants like sperm whale oil.[11] *Sci-entific American* predicted that gas would soon be available to the working class once rates followed lower manufacturing costs.[12] Yet costs had not appreciably declined during the 1840s. Setting up a plant was expensive, especially when the quality of the raw materials varied so much.[13]

The majority of gas consumed in the United States, and in Britain and Eu-rope, was obtained from common bituminous coal. The coal was heated in an iron retort to a high temperature, between 1,000° and 1,200° F, at which point permanent gas (which would not condense into a liquid upon cooling) was driven off. During the process, noxious gases were also released; chief among these were ammonia, "sulphuretted hydrogen," cyanogen, and carbonic acid. All were potentially dangerous and required removal before the gas could be burned in homes or offices. In addition, creosote, a solid impurity, usually became intermixed during manufacturing. When burned, creosote smoked and smelled foul. Companies drained money into expensive techniques to remove these impurities; however, they often discovered that purifying processes tended to diminish the gas's illuminating power. To maintain a brilliant and long-lasting flame, companies acquired assorted patented apparatus to enrich the gas with "carburetted hydrogen."[14]

Different coals presented another way around manufacturing obstacles. Companies in Boston, New York City, and Philadelphia used Nova Scotia coal because it contained fewer impurities, especially sulfur, than did domestic bituminous coal.[15] But the Manhattan Gas-Light Company, one of the largest in the country, was interested in other coals. The company's president, Charles Roome, consulted Silliman about "Boghead & Several other Scotch Cannels, also some from Wigan-Lancashire [England] & several American Cannels from Pennsylvania, Virginia, & Missouri." Roome even suggested the "cannel" from Albert County, New Brunswick. At Yale's analytical laboratory, Silliman ran several experiments to determine which variety contained the most bitumen and fewest impurities. Manhattan Gas then chose an English cannel.[16]

Whichever coal a company selected, it became the basis of a technological system—distillation retorts, purification chemicals, enrichment methods, distribution pipes, lamps, measuring equipment, and a knowledge of all these operations.[17] There were many places for improvement in this large system, and all were subject to possible patent protection. By far the most controversial patents were those covering the *use* of a particular raw material. Specific kinds of coal, for example, could be included in the same patent for the technique and apparatus for distilling it or purifying its gas. Companies like Manhattan Gas opted for English cannels because they were part of a tested, patented system. In this way, technological transfer came with coal imports.

Gesner arrived in New York City ready to sell his patented Kerosene gas, and he met with some success.[18] *Scientific American* reported on its introduction in New York City, and the Philadelphia Gas Company tried it.[19] Upon his visit to Philadelphia, Gesner was elected a corresponding member of the Academy of Natural Sciences and invited to lecture on Kerosene gas. "[My] discovery," he modestly proposed, "brings [asphaltum] into operation for illuminating purposes, to which it is admirably adapted." Gesner thought gas companies would realize savings in purification costs. "Sulphur, not being one of [asphaltum's] constituents, injurious and noxious gases are not produced in its decomposition, and the absence of nitrogen prevents the formation of ammonia." Besides, asphaltum yielded "double the quantity of gas," and by substituting it for coal, consumers paid "less than one half its present cost." Asphaltum thus combined the advantages of Nova Scotia coal (few impurities) with the benefits of English cannels (lots of bitumen). Best of all, Gesner's patented retort allowed asphaltum to be used within already existing gas systems.[20]

Back home in Nova Scotia, Dundonald began laying plans for the Kerosene Gas Company of Halifax. He and Gesner petitioned the House of Assembly for a charter of incorporation and recruited more than 200 subscribers among the city's most prominent citizens, including Sir John Harvey, Gesner's former New Brunswick patron and now lieutenant governor of Nova Scotia. Kerosene Gas was Gesner's big commercial step; using his patents, his knowledge, and his know-how, he was organizing his own company.

There were precedents for such scientific entrepreneurship. Three years earlier, in 1847, Silliman had helped found the New Haven Gas-Light Company.[21] At the time, he was "Professor of Chemistry and the Kindred Sciences as Applied to the Arts" at Yale College. He was also a member of the New Haven Common Council, and through his lobbying efforts, New Haven, the smallest city in the United States to plan a gas system, gained a charter from the Connecticut legislature. Silliman was elected to the company's board of directors and given 20 shares of stock (out of 4,000). He also became the company's consulting chemist, responsible for overseeing the works, recommending improvements, and supervising construction. By 1850 New Haven Gas was paying a 6 percent dividend, and by 1856 it had 1220 customers.[22]

That Silliman was the prime mover in the organization and operation of New Haven Gas would have come as no surprise to his friends (or enemies), for in many ways entrepreneurship extended naturally from his science. Much like Gesner, Silliman believed in the useful application of knowledge, and his research directly profited from his company connections. Silliman carried out many studies of gas lighting and published them in the *American Journal of Science,* of which he was, conveniently, coeditor.[23] Silliman, however, did not patent his improvements.

Gesner treated patents as natural extensions of his chemical experiments. The press likewise did not regard his Kerosene patent as anything unusual. Perhaps this reflected the close ties between chemists and gas companies. On the other hand, asserting priority through a patent was uncommon among mid-nineteenth-century men of science. The practice of patenting, and whether that meant patenting scientific *knowledge,* would be issues that would take root and grow with the Albert controversy.

Mining Licenses and Land Rights

On returning to Nova Scotia, Gesner faced two setbacks to his Kerosene gas plans.[24] In March 1851 Dundonald returned to England when his assignment as commander of the North American and West Indian station ended. Gesner lost a generous and encouraging patron, and the business partnership suffered. More serious was the asphaltum supply. According to Gesner, asphaltum occurred in "abundance" along "the whole coast" of South America, Mexico, and Texas, not to mention Cuba, with a vein of "no less than 144 feet in perpendicular thickness," or Trinidad's "Pitch Lake," to which Dundonald had secured mining rights.[25] The problem was transportation. Asphaltum tended to fuse in a ship's hold thereby requiring "mining" before it could be unloaded. Captains flatly refused to carry asphaltum for fear it would burst the ship's bottom.[26] The solution, Gesner thought, lay in New Brunswick.

The Albert mineral was found on the property of John Steves. In January 1850 the Duffys had acquired a mining license from the British crown to extract coal from Steves's land. In December 1850 they assigned their license to William Cairns, a New Brunswick businessman, who continued to raise coal.[27] That same month, Gesner purchased a lease to Steves's property and opened his own mine. Gesner claimed to be mining asphaltum; therefore, he avoided infringing Cairns's mining license. Cairns thought otherwise, and soon a dispute arose over who could mine on the property.

Early in January 1851 Cairns and a group of about twenty men confronted Gesner's miners, forced them out of the mine, and took 200 cauldrons of Gesner's asphaltum. Cairns subsequently sold the cauldrons to the Halifax Gas-Light Company. On 4 February 1851 Gesner filed two suits to prevent further confiscations. The first, entered in Saint John, accused Cairns of trespassing on Gesner's land and taking possession of his asphaltum. The second suit, entered in Halifax against the gas company, sought to recover £500 for the cauldrons Cairns sold to them.[28] These two cases, *Abraham Gesner v. William Cairns* and *Abraham Gesner v. Halifax Gas-Light Company,* mark the beginning of the Albert controversy.

From a legal standpoint, both cases hinged on British land grant laws. In New Brunswick, Nova Scotia, and other British colonies, land was granted with the following reservation: "except and reserved, nevertheless, to the Crown, all coals and also all gold and silver, and other mines and minerals."[29] The crown reservation meant that one person could own the surface of the land (Steves), while the crown maintained ownership of the subsurface. The crown could grant a mining license to anyone (e.g., Cairns), and the landowner could lease the land to anyone (e.g., Gesner). Gesner built his cases on the claim that Cairns's mining license explicitly covered coal but not asphaltum.[30] At issue then was a question of law and a question of fact. Which minerals did the reservation cover? Was the Albert mineral coal or not?

In the opening phase of legal arguments, Gesner took the high ground. The separation of surface and subsurface rights was confusing, counterproductive, and perhaps unconstitutional. He demanded nothing less than a complete overhaul of British land grant laws, and the advent of responsible government, which by 1850 had come to New Brunswick, provided the leverage for him to pry open this issue. The crown reservation was a vestige of when the colony was subject to London, and Gesner believed his legal action would set a precedent for New Brunswick by securing an individual's right to the soil *and* subsoil against the usurpation of crown-licensed "foreigner speculators."[31] He predicted dire consequences if he lost. New Brunswick would suffer the fate of Nova Scotia, where the General Mining Association had purchased a mining license in 1827 to *all* crown reservations in the *entire* province for sixty years. Gesner called the monopoly "mineral slavery." "If the Crown has rights, the people also have rights, and a right to the quiet and undisturbed possession of property, without the interference of the Crown, distinguishes liberty from slavery. . . . The minerals of New Brunswick are free."[32] Either New Brunswick would follow its "agitated" neighbor or pursue a path of freedom. For Gesner, the model was the United States, where the people owned the soil, subsoil, and everything found or captured on their land. New Brunswick suffered under an "old feudal system."[33]

Experts on Demand

When Gesner's two cases went to preliminary hearings in the summer of 1851, the defense's attorneys petitioned for a postponement. They wanted extra time to collect evidence and experts, and Gesner's attorneys did, too. Both sides thought the trials would hinge on the question of fact: was the Albert mineral coal or not? The defendants, Cairns and Halifax Gas, contacted Boston Gas, Manhattan Gas, and New Haven Gas. Those companies bankrolled the cases and supplied the lawyers and scientific experts. In Gesner's words, the "American auxiliaries" furnished "the sinews of war."[34]

The principal expert for the defense was none other than Charles Jackson. In April 1851 he was consulted about the mineral. On first glance, this seems unremarkable; Jackson was the consulting chemist for Boston Gas. But only a year before he had described the mineral as asphaltum, and that classification had prompted Gesner to write to him in March 1851. "I have taken the liberty to republish [in a Halifax newspaper] your analysis of the New Brunswick asphaltum," Gesner explained. "I have deemed it but due to you to make my acknowledgements to your superior discernment and ability." If it were meant as flattery, Jackson took it as guile and published Gesner's private letter as evidence of an attempt to influence him.[35] Gesner thought that Jackson's actions had "violate[d] every rule of decency and honor."[36] Gesner, it seems, had no idea of the anger Jackson harbored over the alleged Nova Scotia plagiarism. Jackson would later testify in court that he and Gesner "had no quarrelling, no controversy in [the] papers." "I never published a word about him," Jackson cunningly admitted, "but I may have said a great deal."[37]

From the moment Jackson accepted the commission from Cairns and Halifax Gas, his scientific opinion began to change. On 16 April 1851 he made a second presentation before the Boston Society of Natural History. He noted that although the mineral's appearance "much resembles Asphaltum," its "action under the influence of heat" did not. Without elaborating, Jackson concluded with a striking suggestion: "From the combination in the specimen of properties resembling those of Asphaltum and Coal, I propose for this new material the name of 'Asphaltic Coal.'"[38] Here then was a new mineral, part asphaltum, part coal.

Gesner charged Jackson's change of mind to a reversal of fortune. "To the man of science the material is asphaltum," he observed, "to the hungry chemist it is changed into *Asphaltic Coal.*" Jackson's "self-contradiction" revealed his "intima[cy] with the speculators."[39] To others, Jackson's new classification did not reflect a conflict of interest but a reasonable reconsideration of the facts. James Dwight Dana (1813–1895), America's foremost mineralogist, thought asphaltic coal was a good name for a mineral so rich in volatile matter. Jackson later explained that it was Dana who coined the term. As for his initial identification, Jackson later asserted under oath that he had been "wrong." "I don't pretend to be infallible," he huffed.[40] But, the reason he had been wrong about the Albert mineral was Gesner's treachery. The first specimen (from April 1850), Jackson decided, had been sent by Gesner "I believe to mislead me." Jackson thought that the sample must have been taken from "the scum of an old petroleum spring" and not from the Albert mine.[41] Gesner remained silent on the salted sample allegation, a somewhat suspicious response given his other outbursts, but it does seem unlikely that he could have interrupted the shipment from the Duffys to Halifax Gas to Boston Gas to Jackson. Nowhere in that transmission did Gesner have an ally. But raising the suspicion of salting, while effectively disparaging Gesner's character, allowed Jackson to distance himself from his first announcement and, simultaneously, to explain why he had to visit the mine himself, namely, to ensure the authenticity of a sample.

The Albert mine was located in the hills a few miles from the Petitcodiac River in Hillsborough Parish. (In 1845, Albert County had been carved out of the southwestern portion of Westmorland County.) Gesner had surveyed the area during his second season of fieldwork in 1839 and had reported, optimistically, on coal prospects.[42] But there was no mining until the Duffys arrived. The experts who examined the mine all agreed on one thing: the strata were very crumpled and contorted, which made it difficult to measure the dip or general direction of the beds. In the summer of 1851, the mine was roughly 150 feet deep with at least nine levels about 300 to 400 feet long.[43] At several locations, the mineral appeared nearly vertical, an orientation requiring careful consideration about which rocks, if any, lay above (commonly called the "roof") and which below (the "floor"). Other levels exposed the mineral in a position nearly horizontal and running parallel to the surrounding strata. No two levels seemed to present the same structure, and no two opposing experts seemed to agree on theories, interpretations, measurements, or even observations.

Jackson visited the mine twice: 3–13 May and 23 May–12 June 1851. His first visit was cut short by deep snow, and he could only investigate underground. During his second visit, he examined the general area, including parts of Nova Scotia.[44] When Jackson returned to Boston, he had changed his mind, again. The mineral was now coal, and the reason was geology. "[T]he Albert coal is a true coal bed," he explained, "included in bituminous shales of the true coal formation."[45] Down the mine, Jackson had found the mineral to be "in reality *parallel* to the including strata." It was layered horizontally in a conformable bed typical of coal seams and did not cut across other beds like asphaltum veins.[46]

Jackson went on to describe the structure and stratigraphy of the area. He found a "curiously crushed coal basin."[47] Although he had no explanation for the crushing, he did think the highly bituminous nature of the Albert mineral was "owing to its *never having been subjected to heat,* which would have removed a portion of the bitumen."[48] Coal usually occurred in basins, and Jackson was "perfectly sure" that the Albert mineral belonged "in the regular coal formation, and not in the old red sandstone, nor below it."[49] As evidence, he presented fossils, all the "usual coal plants," identical with those from the Joggins on the opposite shore of the Bay of Fundy.[50] He had also unearthed fish, "the first found in the coal formation in America."[51] This important scientific discovery "cut off the last ray of hope in the opposing party," Jackson declared. "It is well known that *not one of the fossils found in the shales of the Albert coal mine, was ever seen in any rock below the regular coal formation,* and that no law of nature is more certain than the order in which fossil animals and plants are disposed in the earth's crust. Fishes, though, proverbially dumb, are good witnesses in this case."[52] Jackson ended his consulting report with some supplementary chemical experiments, one for fusibility and another for solubility. The mineral, he reported, did not melt at temperatures below 700° F, and it did not entirely dissolve in naphtha, benzole, or turpentine, in contrast to asphaltum, which did melt and was "instantly and entirely soluble."[53]

The Boston scientific community was impressed with Jackson's study of the

Dip 10° *to* 15° *S. W.*

L.—Fine gray Grit—Fossil Trees.　　O.—Coal.
M.—Coarse brown Grit.　　　　　　P.—Under Conglomerate.
N.—Fine blue Sandstone.　　　　　　Q.—Peticodiac River.

Charles Jackson's stratigraphy of "the regular coal series" in the region of the Albert mine. Jackson thought the New Brunswick series was equivalent to the coal series at the Joggins, Nova Scotia. *Source: Reports on the Geological Relations, Chemical Analyses, and Microscopic Examination of the Coal of the Albert Coal Mining Co., Situated in Hillsboro, New Brunswick* (New York: George F. Nesbitt & Co., 1851), 10.

Albert mineral.[54] The geologist James G. Percival (1795–1856) and the chemist Augustus A. Hayes (1806–1882) decided to work for Cairns and Halifax Gas. Percival had conducted the geological survey of Connecticut between 1835 and 1841, and since then he had been employed, "professionally," to report on various mineral properties, including some in Nova Scotia.[55] In early August 1851, he visited the Albert mine. Confining himself strictly to "geological relations," he described in excruciating detail each and every rock layer. He decided the mineral was "a bed of deposit" conformable to the surrounding strata.[56] Its vertical orientation he attributed to a "disturbing force act[ing] from the southwest," which had contorted all the beds in the immediate area.[57] This force meant the mineral had been "originally in a fluid or semi-fluid state," which then cooled and hardened, giving it a glossy appearance and conchoidal fracture. Percival concluded the Albert mineral was a kind of cannel coal or jet.[58] Hayes reached a similar conclusion, but he did not need to travel to New Brunswick to do so. As a consulting chemist to Boston Gas, he worked in his laboratory where he repeated Jackson's experiments on fusibility and solubility and decided that the Albert mineral was coal.[59]

The commissioned studies of Jackson, Percival, and Hayes appeared in the fall of 1851 in a forty-eight-page pamphlet aptly entitled *Reports on the Geological Relations, Chemical Analyses, and Microscopic Examination of the Coal of the Albert Coal Mining Co.* The pamphlet also contained short reports from the Scottish chemists Andrew Ure and Frederick Penny and from the American chemists Benjamin Silliman, Sr. and Jr., and James Booth of Philadelphia. John Bacon Jr., a physician and Boston Society of Natural History member, added a report on his investigations of microscopic traces of vegetable structure in the Albert mineral. The pamphlet was an exemplar of the new genre of scientific consulting reports and presented a well-organized defense of Albert coal.[60]

Gesner had not been idle while Cairns and Halifax Gas were busily soliciting scientific expertise. And he was not without his own American auxiliaries, particularly the Philadelphia Gas Company.[61] He also called on his newly acquired colleagues at the Academy of Natural Sciences, foremost among whom was Richard Cowling Taylor (1789–1851), arguably the world's authority on coal.

Taylor was an Englishman, who described himself variously as a surveyor, mineral agent, and geological and mining engineer. A self-proclaimed "original disciple of William Smith," the celebrated English land surveyor and canal engineer who had introduced the practice of characterizing formations by their fossil content, Taylor had also worked briefly for the Ordnance Survey under Henry De la Beche.[62] Since as early as 1810, Taylor had been accepting professional engagements as a mining geologist and had worked for numerous companies and capitalists. In 1830 he emigrated to the United States, where his professional career prospered, financially and scientifically. After three years consulting for the Dauphin and Susquehanna Coal Company, in whose employ he surveyed the eastern coal regions of Pennsylvania,[63] Taylor undertook several foreign engagements, including trips to Central America and the West Indies. A prolific writer, he published numerous treatises on "the science of practical geology," the most famous of which was his encyclopedic *Statistics of Coal* (1848).[64] In that book, Taylor praised Gesner for his discoveries in New Brunswick. "[Dr. Gesner] has explored the whole of this vast region," Taylor declared; "the result of this geological survey is, that the coal formation is found to occupy, in New Brunswick, no less than eight thousand square miles."[65]

In late May 1851 Gesner consulted Taylor along with the New Brunswick geologist James Robb. Over the next few weeks, Taylor and Robb investigated the "geological circumstances under which [the mineral] occurs" and wrote a joint report.[66] The evidence they adduced was strikingly at odds with that of Jackson and Percival. Taylor and Robb described the vertical orientation of the mineral "in the form of a wedge" cutting across the surrounding strata. "We found that this Hillsborough vein was *unconformable*."[67] Coal, in contrast, "never traverses the contiguous strata, but continues parallel to it." The Albert mineral had "neither a true floor nor roof." Taylor and Robb preferred to use the term "walls."[68]

Walls. Walls.

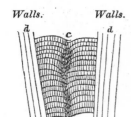

9th Level.

d. Shale including the coal.
c. Crushed coal.

7 feet 10 inches width of bed ; underlie of strata on left 5°, or about 7 feet in 25 feet.

Fig. 65.—*Relation of the " Albert Coal" to the containing beds, as seen near the shaft of the mine.*

Left, Charles Jackson's view of the crushed Albert "coal" on the ninth level of the Albert mine. Source: *Reports on the Geological Relations, Chemical Analyses, and Microscopic Examination of the Coal of the Albert Coal Mining Co., Situated in Hillsboro, New Brunswick* (New York: George F. Nesbitt & Co., 1851), 7. *Right,* Richard Taylor and James Robb's view of the Albert "asphaltum" as an injected vein within the crest of an anticline on the eighth level of the Albert mine. Note that the image is from John William Dawson, who, in 1855, tentatively classified the Albert mineral as "coal." Source: John William Dawson, *Acadian Geology* (Edinburgh: Oliver and Boyd, 1855), 201.

Another feature of the "Hillsborough vein [which] must not be overlooked" was its tendency to throw off small branches, a characteristic of asphaltum, not coal.[69]

As to the geological structure of the surrounding area, Taylor and Robb described a very different picture from that of Jackson and Percival. There was no basin, "but rather the reverse," an anticlinal axis with strata tilted off in opposite directions.[70] The beds had been thrown upward into an arch by "a powerful force . . . exerted from below."[71] This force had opened a fissure or fracture along the axis of the arch into which the liquid asphaltum had been injected from below and then hardened. The Albert mineral was "an intrusive substance."[72] Regarding the age of the enclosing rocks, Taylor and Robb found no fossil plants "characteristic of true coal seams." The rocks belonged to the Old Red Sandstone formation; they were older than the Coal formation. In conclusion, Taylor and Robb declared the Albert mineral an "asphalte, or a variety of asphalte, and not coal or a variety of coal."[73]

Taylor and Robb had reached that conclusion "independent of any chemical questions or opinions."[74] And so Gesner turned to Charles Wetherill (1825–1871), a young chemistry professor at the Franklin Institute, who had received a PhD from the University of Giessen in 1848. He was also a consulting chemist for the Philadelphia Gas Company. Wetherill submitted the Albert mineral to an "organic analysis," which resolved it into its ultimate elements—carbon, hydrogen, oxygen, and nitrogen—rather than into such large and vague proximate constituents as "volatile matter or bitumen" and "fixed carbon or coke." Wetherill determined the composition of the Albert mineral to be $C_{68}H_{42}ON$. He then examined the question of solubility, basing his experiments explicitly

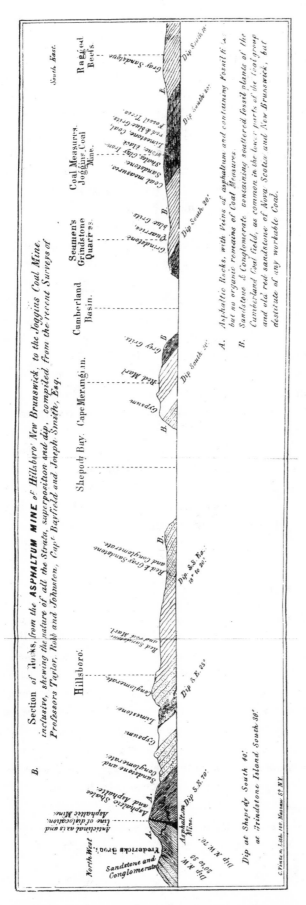

Richard Taylor and James Robb's traverse section from the Albert mine, New Brunswick, to the Joggins, Nova Scotia. The "Asphaltum-Asphaltic Mine" was located at the anticlinal axis and line of dislocation (*on the left*). *Source*: [Abraham Gesner], *Review of* "*Reports on the Geological Relations, Chemical Analyses, and Microscopic Examination of the Coal of the Albert Coal Mining Company, Situated in Hillsboro, Albert County, New Brunswick*," *as Written and Compiled by Charles T. Jackson, M.D., of Boston* (New York: C. Vinten, 1852), frontispiece.

on those of Jackson and Hayes. Comparing the mineral to Egyptian asphalt and cannel coal, Wetherill recorded its *partial* solubility in different liquids, while noting that Egyptian asphalt dissolved entirely and cannel coal remained insoluble.[75] "The substance from Hillsborough," Wetherill reasoned, "is not coal, nor any variety of coal, but a true and a new variety of asphalt [which] I propose for it the name melan-asphalt."[76] Wetherill published his findings in the *Transactions* of the American Philosophical Society, and the article serves as a nice illustration of how scientific publication was often intimately connected with specific commercial (or legal) questions.

Gesner, on the other hand, published his scientific views anonymously. In a "review" of the pamphlet by Jackson, Percival, and Hayes, Gesner attacked Jackson's science and his character. Gesner wondered, sarcastically, what Jackson meant by a "crushed coal basin," a "new term in science." Gesner was also curious about Jackson's fish; no one familiar with fossil fish had actually seen them. Jackson had merely stated that "any ingenious person" could identify them. "Now why did he not lay the fossil fishes . . . before Professor [Louis] Agassiz, near at hand [at Harvard], and whose decision would have been conclusive upon the subject?"[77] Gesner thought Jackson had confused fossils of the Old Red Sandstone with those of the Coal formation. In closing, he remarked: "Whatever Dr. Jackson's knowledge may be of chemistry, he certainly can have but few pretensions to geology."[78]

By the spring of 1852, the battle lines were drawn: Gesner and his scientific allies against Cairns, Halifax Gas, and their consultants. Wetherill aptly summarized the situation: "Owing to a law suit in which great interests were at stake, the mineral in question has been examined by a large number of experts, who are singularly divided as to their conclusions."[79] The upcoming court trials would bring the two parties face-to-face and the Albert controversy to a boil.

Experts on Trial

In his suit against Halifax Gas, Gesner sought to recover damages for the property Cairns had sold to them. There was, however, much more than a few cauldrons involved. The company admitted to using the Albert mineral in its manufacturing process, and so if the mineral were asphaltum, Halifax Gas was guilty of infringing Gesner's patent and would have to pay fines and royalties to Gesner. Gesner, in fact, had offered to Halifax Gas an exclusive license to his patented process in the spring of 1850, but the company ignored him. Gesner thus regarded the company's claim to Albert *coal* as a transparent attempt to pirate his invention and accused the "gas monopoly" of "barefaced acts of dishonesty" and "assassin-like means" to defame his scientific reputation.[80] Halifax Gas returned the insult by dismissing Gesner's invention as of "no service" to them and denouncing Gesner as "an empiric, a very juggler, a mountebank in science and a dolt in literature."[81]

Besides the vicious war of words waged in the press, the trial of *Gesner v. Halifax Gas* had acquired the aura of political theater by April 1852. Gesner's lead counsel was James W. Johnston (1792–1873), head of Nova Scotia's Conservative Party, while Halifax Gas's lead counsel was James B. Uniacke (1799–1858), head of the Liberal Party, currently in power.[82] A year earlier, Uniacke, then president of Halifax Gas, had blocked Gesner's attempts to lay pipe to distribute Kerosene gas. For Gesner, Johnston, and other Conservatives, the trial presented the perfect opportunity to deal a blow to the gas monopoly and to the ruling Liberals.

The trial, however, did not live up to its billing. On the first day, Uniacke moved to dismiss the suit on the grounds that Gesner had not shown original ownership of the mineral, regardless of its classification. Justice Henry B. Bliss (1795–1874), a Tory by family tradition (he was the son of the former chief justice Jonathan Bliss), agreed, but reserved the point and directed the trial to proceed.[83] Justice Bliss, however, had derailed the case. The trial sputtered on for ten days, and in the words of a gas company spokesman, the scientific evidence was "all on the side of Dr. Gesner." Robb played the decisive role, while the gas company's expert witnesses did not cut very impressive figures.[84] On the final day, Justice Bliss instructed the jury to reach its decision on the question of fact: was the Albert mineral coal or not? The jury decided it was asphaltum. Naturally, Gesner was elated and complimented the "intelligent jury."[85] Justice Bliss, as indicated at the outset, put aside the verdict, and the case of *Gesner v. Halifax Gas* was scheduled to be heard before the full Supreme Court of Nova Scotia the following spring. Thus, even though Gesner won the trial, it proved to be more symbolic than substantial.

The case of *Gesner v. Cairns* began three months later, on 20 July 1852, before Judge Lemuel Allan Wilmot (1809–1878) and a special jury in Albert Circuit Court. It was lengthier, legally and commercially more significant, and the expert witnesses played much larger roles. The presence of distinguished men of science provided the opportunity for each side to make forceful and extended scientific arguments. The trial also allowed the participants to confront each other face-to-face, although the confrontation followed the rules of the court. Judge Wilmot and the attorneys, in effect, mediated the scientific debate.

For historians, the Saint John courtroom can serve as a sort of laboratory for examining a scientific controversy. Controversies are often favored sites for studying science in the making.[86] This courtroom controversy provides scholars with another opportunity, perhaps an even better one because of its controlled circumstances, for studying the manner in which argument and evidence are marshaled, deployed, and engaged. The self-consciously agonistic language reflects the adversarial environment, especially the cross-examinations,

---

designed to undermine scientific arguments and evidence and the authority of experts. The courtroom thus put on display the moral economy of midcentury American science. Who were these experts? Were they trustworthy? The witnesses themselves were acutely aware of their exposure. "The eyes of the whole scientific world are upon us," exclaimed one. "I felt honored to have been selected to make [an] investigation."[87]

The case was specifically about forcible entry and removal of asphaltum from the property Gesner had leased. The majority of the trial, however, addressed the question of fact: was the mineral coal or not? While the scientific experts carried the bulk of the argument and advanced most of the evidence, each side also introduced nonscientific experts. Gesner relied on several Scots miners with experience in the Pictou and Cape Breton collieries to describe conditions in the Albert mine and to contrast them with known coal mines. "This mine smells different from coal," one recalled, "a strong nasty smell."[88] Cairns called on blacksmiths to explain how the mineral served the same purposes as ordinary coals. The daily experiences of laborers and artisans thus grounded the esoteric and theoretical points about classification.

Gesner then called four men of science to testify: Robb, Wetherill, Thomas Antisell (1817–1893), and Joseph Leidy (1823–1891). Taylor had died in October 1851, but Gesner had had the foresight to take a sworn deposition during his visit in June 1851.[89] The first question usually asked of each one was to describe his profession or occupation. Their own introductions provide insights into the rhetoric and strategies of self-presentation (table 2.2).

All of Gesner's experts were from the New York–Philadelphia area, with the exception of Robb. Most—Robb, Wetherill, Antisell, and Leidy—were trained in medicine, a reflection of the historically close connection between medicine and natural history, but only Leidy identified himself as a physician. Nonetheless, they continually referred to each other as "doctor," even if, like Taylor and Gesner, they had no medical degree. Robb and Wetherill were the only ones to mention institutional affiliations, although all of them, except Taylor, indicated that they had taught science. Robb was the most distinguished in terms of age and standing within the New Brunswick community. Wetherill and Leidy were relatively young, and although Wetherill had a PhD from a German institution, it was of no significance to either his attorneys or the defense. What was left unmentioned, at least in the opening statements, were scientific credentials. All were members of the American Association for the Advancement of Science, the American Philosophical Society, and the Academy of Natural Sciences.

Gesner did not testify in his own defense. According to Taylor, he wanted to avoid any appearance of conflict of interest. Cairns's attorneys, however, submitted Gesner's New Brunswick survey, wherein he had described the Albert region as part of the great coal field. Under cross-examination, Robb corrected

Table 2.2. Expert Witnesses for the Plaintiff Abraham Gesner

RICHARD TAYLOR: "My profession is that of geological and mining engineer. [I am] a resident of Philadelphia."

JAMES ROBB: "I am Professor of Geology and Chemistry at King's College, Fredericton. I have been so 12 or 14 years. I am from the north of Scotland."

CHARLES WETHERILL: "I am a native of Philadelphia. I am 27—will be 28. I am Analytical Chemist and Lecturer on Chemistry at [the] Franklin Institute. [I] have lectured there three years: educated partly in United States, partly in France, partly in Germany."

THOMAS ANTISELL: "I reside in New York. I am an analytical chemist, teacher of chemistry, and lecturer on geology. I am engaged as a chemist since 1843. I was educated a medical man. I belong to [the] Medical College in Woodstock, Vermont, as Professor of chemistry. I have been concerned in geology since 1846–7."

JOSEPH LEIDY: "I am a Physician. I reside in Philadelphia; at present I am engaged in teaching physiology and anatomy, and microscopic anatomy."

Source: Report of a Case, tried at Albert Circuit, 1852, before his Honor Judge Wilmot, and a Special Jury. Abraham Gesner vs. William Cairns. Copied from the Judge's Notes (Saint John: William L. Avery, 1853), Taylor (appendix A); Robb (30, 37); Wetherill (43); Antisell (49); Leidy (56, 59).

Gesner's enthusiastic estimates, but, once again, Gesner's grand geography came back to haunt him.[90]

During the second week of the trial, seven men of science took the stand for Cairns: Silliman, Hayes, Jackson, James E. Teschemacher (1790–1853), John Torrey (1796–1873), William H. Ellet (1806–1859), and Isiah Deck (1821–1854). Percival did not attend the trial, for unknown reasons, but he too gave a sworn deposition during his consulting engagement the previous summer (table 2.3).[91]

Their introductions were obviously much longer than those of Gesner's witnesses. (Jackson, the star, apparently could not afford excess modesty.) Most of them had been in the audience for the first week, although Silliman had been busy examining the mine and doing experiments.[92] The majority—Percival, Jackson, Torrey, Ellet, and Deck—once again had medical degrees. Hayes was introduced as "doctor," although his degree was an honorary one from Yale. None identified his occupation as physician, however. All were from either Boston or New York City, and they were generally older than Gesner's witnesses. Teschemacher, the oldest, was sixty-two; Deck (thirty-one) and Silliman (thirty-six) were the youngest. More importantly, they represented the highest rank of American science, a position that Jackson and Silliman did not fail to impress upon the jury. They were also among the most active consultants of the time, a professional practice that Jackson, Silliman, and Percival made clear from the start. Others revealed their commercial connections after further questioning;

Table 2.3. Expert Witnesses for the Defendant William Cairns

JAMES E. TESCHEMACHER: "I am an Englishman. I have been twenty years in the United States. I have acquired a knowledge of all minerals in the United States; but during the last eight or ten years, I have turned my attention to the structure of coal."

BENJAMIN SILLIMAN JR.: "I am a chemist: have devoted my attention especially to chemical mineralogy and geology. In winter I am in the medical department of the University of Louisville; in summer at Yale College, in technical chemistry, and am one of the editors of the *American Journal of Science*. Since 1837 I have been entirely devoted to them. I have been in the habit of examining coals and coal fields, and have studied their geology. A fair proportion of my engagements have been of that nature."

JAMES PERCIVAL: "I am a Doctor of medicine, of New Haven, Connecticut. I was employed in the geological survey of the State of Connecticut, and prepared a report thereon which was published in 1841. I have since been employed in exploring different mineral localities, in the same and adjoining States."

AUGUSTUS A. HAYES: "My occupation is that of a practical chemist. Employed in special geology and mineralogy. I have been 28 years in all, engaged in study and practice. I have been sub-Professor in a literary institution in New Hampshire. I hold [the] office of assayer in [the] State of Massachusetts. I mean by special geology the study of internal structure of mines, and this includes all minerals."

CHARLES JACKSON: "I am 47 years of age. I am a geologist and chemist. My residence is Boston, U.S. My earlier studies [were at Harvard] in 1824. I continued to study till 1829, when I graduated as M.D. For twenty years, I made chemistry, geology, and surveying mines, my particular business. I am a member of the Boston Society of Natural History, and am now Vice-President of it. I am [a] fellow of [the] American Academy of Arts. One of the Presidents of [the] American Association of Geologists and Naturalists—Gentlemen engaged in geological surveys, and interested in botany, geology, & c., for mutual improvement for their observations. I am [a] member of [the] Geological Society of France, and [a] corresponding member of others. I made a pedestrian tour in France, Austria, Switzerland, Italy, Bavaria, and other parts, to study geology; which covers [the] whole of middle and southern Europe. I have made geological surveys of several states. Maine, New Hampshire, Rhode Island, public lands of Massachusetts, and of mineral lands of U.S. in Michigan. I examined geology of Nova Scotia, parts of Canada, and portions of this Province. I published [the] first geological memoir of geology of Nova Scotia. I am one of [the] Assayers of [the] State of Massachusetts, and Assay Master for Boston; [and am] engaged constantly in chemical and mineralogical researches."

JOHN TORREY: "I reside in New York. I am Professor of Chemistry in College of Physicians and Surgeons, New York. I have held office since 1827. I am 55 years of age."

WILLIAM ELLET: "I reside in New York. Resided there till 1835; went to [South] Carolina, remained there fourteen years, returned to New York in 1849, and there since. I am a chemist. I have been professor of chemistry. I was engaged for many years in chemical studies. Professor of Columbia College [in New York City] two years, and in College [of] South Carolina [in Columbia, S.C.] fourteen years. I am between forty five and forty six years of age."

(continued)

Table 2.3. Continued

ISIAH DECK: "I am an Englishman. I am [a] practical chemist, and studied geology and mineralogy. I studied at Cambridge, England, under [Adam] Sedgwick; and chemistry, under [James] Cumming. I studied chemistry at St. Thomas' hospital, under Dr. Leeson, a celebrated chemist. I afterwards practised four or five years in Leamington—about fifteen years in all. Eighteen months since I came to America. I reside in New York. . . . I am now 31 years of age."

Source: Report of a Case, tried at Albert Circuit, 1852, before his Honor Judge Wilmot, and a Special Jury. Abraham Gesner vs. William Cairns. Copied from the Judge's Notes (Saint John: William L. Avery, 1853), Teschmaker (76–77); Silliman (82); Percival (152); Hayes (94); Jackson (107–108); Torrey (118); Ellet (121, 123); Deck (125, 128).

Teschemacher was employed in a sugar refinery and Deck in a New York City chemical firm. As for gas lighting, Silliman mentioned his position with New Haven Gas, while Torrey and Ellet said they were consulting chemists for the Manhattan Gas-Light Company.

After the introductions, each expert explained his conclusions about classification, but there was not much that was new. Robb, for example, reiterated his joint report with Taylor and emphasized, again, the vertical orientation of the vein within an anticlinal axis. Jackson likewise rehearsed his report, although during questioning he explained how some coal beds, like those in eastern Pennsylvania, occurred vertically but remained conformable to the enclosing strata, that is, all the beds were tilted on end.

Perhaps the most significant geological evidence introduced at the trial was the increased depth of the mine. The tight contortions once visible near the surface seemed to have blended into the general dip of the surrounding strata. In the course of digging, some more fossil plants had been exhumed, but not many fish. Robb nevertheless maintained that he could not find fossils characteristic of the Coal formation.[93] Jackson, however, stood by his identification of the fossil fish as characteristic of the Coal formation.[94] Jackson did revise his structural interpretation. "I do not say it is a coal basin, but a coal deposit." This subtle distinction was based on his glaring inability to trace the basin's margins. Jackson admitted that he was "puzzled to know" what became of the edges of his theoretical "wash bowl."[95] Still, he was sure Taylor's and Robb's anticline was "imaginary."[96] Similarly, Hayes remarked, "[i]t would be a miracle for that mine to be in an anticlinal axis."[97]

Wetherill and Antisell then presented the chemical evidence. Wetherill was the more cogent and contentious; he submitted a sample he claimed to have melted. This sample came under a barrage of objections from Cairns's attorneys, but eventually it was admitted in evidence. Nearly all of Cairns's chemical experts concentrated on Wetherill's sample, and it must have been apparent to those in the courtroom that the defendant's case rested heavily on chemistry. The most articulate advocate was Silliman. He faulted Wetherill for failing to measure the exact temperature at which the Albert mineral began to melt. He also questioned whether it melted at all and took time to explain to the jury the distinction between melting and decomposition. Jackson, on the other hand, rather arrogantly asserted that "[i]f 15 or 16 witnesses were to swear that they melted it so as it could be poured, I would not believe them, because I have tried it over and over again carefully."[98] Teschemacher underscored the point: "If Liebig were to say that he had melted the Albert coal, I would not believe him."[99] So much for scientific authority, it would seem.

Leidy then took the stand with the most compelling evidence for the plaintiff. He showed under a microscope the absence of any vegetable structure in the Albert mineral. This contradicted the published report of Bacon, who was supposed to appear at the trial but inexplicably did not.[100] Cairns's attorneys turned, immediately, to Teschemacher, who claimed to be *the* authority on coal, to rebut Leidy. Under cross-examination, however, Teschemacher admitted that he could not use a microscope (it "injured" his eyes). Nevertheless, Teschemacher asserted that he could see vegetable impressions with the unaided eye.[101] Hayes next took up the defense, but under cross-examination he too acknowledged that he had never used a microscope (he used a lens). Still, Hayes contended that Leidy's negative finding was *not* "a fact gained." It was therefore "no proof" of the mineral being asphaltum.[102] Finally, Torrey confronted Leidy's evidence. "No man has a higher reputation than Dr. Leidy, of his age, as a naturalist," he testified. The problem was "Dr. Leidy's results are negative: he may find it if he looks long enough in other cases. It was a long time before [vegetable structure] was discovered in other coals." Under cross-examination, Torrey also admitted that "my microscope examinations are so slight, they are not worth anything." And most damaging, Torrey said he could not see Teschemacher's plant impressions nor was he "aware that [Teschemacher] had been studying coal."[103]

Attorneys on both sides were quick, clever, and well coached in science. According to Gesner, Jackson had prepared the questions for Cairns's legal team. Given the specificity of their questions, Gesner probably briefed his.[104] Conducted in an aggressive manner, cross-examinations chiseled away at the cracks of doubt, and the constant attack often eroded the experts' professional and gentlemanly demeanor. What was left of Leidy and Hayes, for example, was hardened stubbornness. Percival became downright rude during his deposition and declined to answer more than thirty questions (table 2.4). Gesner wryly remarked that such intransigence made it difficult to see what expertise Percival could offer.[105] In contrast, Silliman and Torrey managed to remain friendly, informative, unassuming, and seemingly objective about the whole matter.

Table 2.4. Cross-Examination of James Percival

Q. What is your opinion of the origin and geology of coal?

A. I decline to answer that question.

Q. Are not fossil vegetables very abundant in the vicinity of coal mines and touching the coal?

A. I decline answering the question.

Q. Do not metallic veins run and ramify in all directions, and does coal ever occur in this manner?

A. I decline answering, as I cannot do so from personal observation.

Q. Of what is asphaltum composed?

A. That being a chemical question, I decline to answer it.

Q. What is the difference between coal and asphaltum?

A. That being a chemical question, I decline to answer it.

Source: [Abraham Gesner], *Gas Monopoly* (Halifax, 1851), 28.

In general, the attorneys' blunt questioning disclosed a tightly knit scientific community acutely aware of the standards of behavior for its members. Take the question of expert witnessing, for example. Each man of science had been hired specifically to support one side or the other. Deck, for instance, explained that he had been engaged two or three months prior to the trial by Cairns. He had in fact been a key witness for the gas company at the Halifax trial. When Gesner's attorneys asked him about his financial obligations, Deck stated frankly: "I expect to be paid."[106] Exactly how much was never divulged. Only Jackson stated his professional fee—$10 a day for his three-week survey of the mine. A fee of $200 would have been reasonable compensation for Cairns's other prominent experts such as Silliman and Torrey; however, three weeks would have been an excessive amount of time, especially for the busy Silliman, who, ever the scientific entrepreneur, had scheduled another consulting engagement for the week before the trial *and* arranged to give a series of lectures at the Saint John Mechanics' Institute after the trial![107] How much Gesner paid his witnesses was not recorded, but Wetherill testified that his services had been arranged and paid for by Gesner's aptly named agent "Mr. Legal."[108] Taylor, in a strictly "professional capacity," had been "employed and paid by Dr. Gesner or his agents at New York."[109] Thus, despite some reluctance to disclose actual amounts, the witnesses treated their commissions as commonplace; they expressed no scruples about accepting fees for expertise.

Intertwined with the subject of money was the issue of interest. Whether testifying in court or consulting for a company, it was crucial for the man of science to appear impartial. During the trial, the expert witnesses exhibited a defiant rectitude. When asked about a possible conflict, Percival responded indignantly that he "had no interest whatever in the result of my examination of [the mine]."[110] Ellet echoed the sentiment: "I have no interest, direct or remote, in this cause."[111] Torrey replied in a more congenial tone that he had a "great interest in this matter," but it was "purely scientific."[112] Similarly, Jackson responded that he had made all his experiments "as a scientific man." And Taylor made sure to defend his behavior in such a way as to protect Gesner's reputation: "Dr. Gesner has invariably preserved that propriety towards me, which left me at liberty to form my own conclusions from the facts I had to investigate."[113] In short, neither money nor interest swayed these men of science from their proper relation to one another or toward their science.

When the attorneys elicited the opinion of each witness about the competence of his fellow experts, the men of science again comported themselves in accordance with an unarticulated, yet well-understood, standard of behavior. They were very careful not to denigrate each other; instead, they commented effusively on their colleagues' "high authority" or "very high authority." Ellet, for example, considered Robb "a gentleman of great scientific attainments."[114] Jackson mentioned that he often recommended Taylor as a consultant when he could not go himself.[115] Taylor thought Jackson had a "very high standing," although, he added adroitly, he would rely upon what Jackson would say, "unless contradicted by higher professional authority."[116] The only one who broke ranks was Wetherill.[117] Under cross-examination, he said that "Dr. Jackson has not the reputation in Philadelphia of an accurate experimenter." Wetherill concluded all-too-frankly that "Dr. Jackson, don't stand high with us."[118]

The response to Wetherill was not to attack his character directly but to go after his evidence. Silliman, Hayes, and Deck, together, tried to replicate the fusibility experiment. "We worked two days and two evenings, until we broke up all the vessels we had," Silliman testified. Silliman, however, remained generous and composed: "I should decline drawing any inference. Dr. Wetherill is a person in whom I have every confidence. I decline stating any conclusions."[119] Hayes was not so circumspect or kind: "I have no knowledge of Dr. Wetherill by his works, and we judge of chemists by their works. If I had tried experiments as he did, and could not produce the same result, I should be bound to say I could not put confidence in his experiments."[120] Hayes concluded that if three chemists could not melt a sample, then "the article produced by Dr. Wetherell [*sic*] could not be produced from the Hillsborough coal."[121]

Given the conflicting nature of the evidence and the reluctance of the experts to criticize one another, Gesner's attorneys tried to get the men of science to state which field—geology or chemistry—should take precedence. Because so much of Gesner's case was built upon geological evidence, it is not surprising that his attorneys argued that a geologist with practical experience, ideally Taylor, was the one best suited to classify. Accordingly, Taylor declared that classification "should be decided by a geologist rather than by a chemist."[122] Gesner's attorneys pushed the priority of geology to the limit when they got Torrey to say that he would not put his opinion "against that of persons who have turned their attention to [geology] more particularly."[123] Hayes, however, fired back at Gesner's attorneys; geology, he declared, had "nothing to do with determining whether coal or asphaltum."[124] Percival, the geologist, in one of his very few answers, admitted that classification may be a chemical question, thereby undermining his own expertise.[125]

Cairns's attorneys tried *not* to emphasize one discipline over another. They pounced on Taylor for his overreliance on geology. Would not an opinion based on geological *and* chemical evidence be more valuable? "If a man possessed a thorough knowledge of chemistry and geology," Taylor conceded, "his opinion would have greater weight than that of one possessing a knowledge of geology only, and the same *vice versa*."[126] Jackson and Silliman presented themselves as wide-ranging experts. In contrast, Wetherill came off as a chemical specialist and Leidy as merely a "professed microscopist."[127] During their testimony, both Wetherill and Leidy, the youngest witnesses, seemed to suggest that the general expert was a thing of the scientific past. But in the courtroom, as in consulting, a knowledge of science, broadly defined, seemed more valuable than narrowly focused research. As Silliman put it, "I form my opinion on all [geological, mineralogical, and chemical] as concurrent evidences."[128]

Overall, the strategy of both sides, regardless of evidence or expertise, was to draw a clear distinction between asphaltum and coal. Definitive categories of classification needed to be established and defended. Asphaltum, Hayes asserted, "is separated by a wide interval, under mineralogical arrangement, from coals."[129] This strategy was for the most part adhered to by all the witnesses, with the notable exception of Robb, who admitted under cross-examination that it was "a well-known difficulty to know where bituminous coal ceases and [asphaltum] begins."[130] He agreed that the Albert mineral possessed "a great many properties in common with coal."[131] Given the adversarial nature of the proceedings, Robb's reasonable doubts became concessions to Cairns: if not definitively asphaltum, then possibly coal. It had to be one *or* the other; there could be no compromise. The science and the men of science split into two. The classification of the Albert mineral recapitulated the division of legal and commercial interests.

The Verdict

On 2 August 1852, after fourteen days of testimony, the jury heard closing arguments. Gesner's attorneys reached for the highest ground, a perch staked out by Gesner at the very beginning. This was a conflict between the government and the people. The government, in this case, was literally represented by the attorney general John H. Gray (1814–1889), Cairns's lead counsel. If Cairns won, farmers would become "mere serfs." The jury's big question was "whether you are to be freemen or slaves."[132] In response, Attorney General Gray stressed the legal point: did Gesner's lease cover mining? The answer was no. To the question of fact, Gray thought a decision on coal or asphaltum was "not absolutely necessary in this case."[133]

In charging the jury, Judge Wilmot directed the jurors' attention, first to the question of ownership (did a land grant provide mining rights?) and then to the question of fact. As to the character of the mineral, Wilmot summarized the scientific points: the plaintiff's geological evidence followed by the defendant's chemical evidence. The mineralogical facts were not in much dispute (table 2.5).

The jury retired for two hours and returned a verdict for the defendant Cairns. It decided the Albert mineral was coal. Taylor had predicted as much: "strange as it may appear to geologists . . . to depend on the decision of a jury!"[134] Gesner was embittered. After spending in excess of $30,000, "a jury of

Table 2.5. Judge Lemuel Allen Wilmot's Abstract of Points in Charging the Jury (*Gesner v. Cairns*, Albert Circuit Court, 1852)

I. Geological.
    1. *General.* Position of mines and surrounding strata.
    2. *Special.* 1. Internal structure of mines. 2. Structure of mineral.

II. Mineralogical.
    Shewing the difference or resemblance between asphaltum and coal in
    1. Density. 2. Fracture. 3. Cleavage. 4. Odour. 5. Electricity. 6. Lustre.
    7. Charcoal dust.

III. Chemical.
    Fusibility and Solubility.
    1. Positive.
    2. Comparative.

Source: Report of a Case, tried at Albert Circuit, 1852, before his Honor Judge Wilmot, and a Special Jury. Abraham Gesner vs. William Cairns. Copied from the Judge's Notes (Saint John: William L. Avery, 1853), 139.

farmers" had misjudged the most important scientific, legal, and commercial question in the history of New Brunswick.[135]

Not surprisingly, Gesner appealed the decision, and the case of *Gesner v. Cairns* went to the Supreme Court of New Brunswick the following year. The other case, *Gesner v. Halifax Gas,* also reached the Supreme Court of Nova Scotia in the spring of 1853. In hearing the cases, both supreme courts chose not to decide on the question of fact—that is, whether the mineral was coal or not. As the chief justice of the New Brunswick court explained, "On this purely scientific question we do not feel ourselves called upon to give any opinion, nor do we think it was one in which it was necessary for the jury to give an opinion."[136] Instead, the justices concentrated on the question of law, which meant the wording of the crown reservation. But in effect, science, once again and in spite of the justices' aversion, came to the fore. Gesner's attorneys argued that the wording of the reservation, "all coals, and also all gold and silver, and other mines and minerals," should be construed to distinguish two clauses. The first referred to "all coals" and the second to "all gold and silver." "Mines and minerals" referred to the second clause, and therefore it meant all mines and minerals *like* gold and silver. This construction, they asserted, was consistent with mineralogy. Gold and silver belonged to the class of metals, and metals were minerals. Coals belonged to the class of inflammables, and until "the last few years" inflammables were not considered minerals.[137] Hence, "all coals" had to be specified separately from "mines and minerals." Because asphaltum was neither coal, nor gold, silver, or any other metal, it was not reserved to the crown.

The attorneys for Cairns and Halifax Gas argued that this construction did violence to the ordinary and popular understanding of the term mineral. "The language must be taken according to the common sense interpretation."[138] A mineral was an article dug out of the ground. The Albert mineral was dug out of the Albert mine. Moreover, "[n]o authority had been shewn for limiting the word 'minerals' to metallic minerals."[139] All scientific men "agreed that it [the Albert mineral] was a combustible mineral."[140] But, even so, "[s]cience can put no definite construct upon them [the terms mines and minerals], for even in the present day scientific men differ widely from this very question."[141]

In reaching their verdicts, the New Brunswick and Nova Scotia justices concluded that the term mineral referred to coal *and* asphaltum. "All we are called on to decide," wrote the New Brunswick chief justice, "is whether *carbonaceous minerals* are excepted. We think they are; and that therefore the plaintiff is not entitled to recover the value of the mineral taken away."[142] The Nova Scotia chief justice agreed: "I am of the opinion that the subject matter in dispute, is a mineral, included in the exception ... consequently, the plaintiff can deduce no ownership."[143] The Albert mineral was a mineral. As far as the New Brunswick and Nova Scotia courts were concerned, it did not matter what kind.

Few men professing a love of the objects of nature have ever been found willing to sacrifice science and character to mercenary gain; to merge their philosophy into mining speculations, and bury their reputation in the rubbish of a mine of either coal or asphaltum.

—Abraham Gesner, *Gas Monopoly* (1851)

THERE ARE at least three ways to interpret the Albert mineral cases. The first would be to regard science and the law as independent of each other. Accordingly, the decisions handed down by the courts were entirely predictable. How else would professionals trained and practiced in the law reach a verdict other than by legal reasoning? Attorneys treated questions of land grants and mining rights; men of science dealt with questions of classification. The predominance of the question of law can be illustrated by subsequent events. In 1852 the New Brunswick legislature incorporated the Albert Mining Company with the stipulation that mining licensees (i.e., Cairns and Halifax Gas) had the right of entry, subject to payment of fair compensation for any damages to property. This action sparked fierce debate in communities in Queens, Westmorland, and Albert counties (the counties encompassing Gesner's great coal field) over the legality of licenses issued to mining promoters by the government in Fredericton. In 1855 the Conservatives triumphed in the legislature by passing an act giving owners of the land exclusive rights to open mines and to dig for minerals, precisely what Gesner had argued for.[144] By this independent-professions interpretation, then, Gesner wasted his money on expert witnesses and should have put it into politics.

Another interpretation might portray a conflict between science and the law. Each profession had its own standards for evidence, modes of explanation, and criteria for truth. From the law's perspective, decisions could not be contingent upon science because science could not be depended upon. There was no certainty in it. "Chemists and geologists appear much divided in opinion, respecting the proper character of this mineral," noted one Nova Scotia justice, "and the scientific world may hereafter place it in [another mineral] class."[145] Any scientific claim to classification was temporary. By their own admission, men of science were continually searching for truth. The justices needed to reach a decision; science confounded the matter.

From science's perspective, courtroom controversies were too contrived by lawyers, judges, and juries to be of any good to science or men of science. Science was supposed to be impartial, whereas law was designed to be adversarial. The former sought consensus through facts, while the latter wanted to win cases by any means whatsoever. The aggressive act of cross-examination, designed to undermine authority, would seem a most compelling illustration. It was not evidence that attorneys wanted but self-contradiction and doubt. The question of the Albert mineral's classification was so artificially concocted that it was only natural for men of science to disagree. Experts and expertise were expedient and expendable.

In this reading of the Albert cases, the men of science paid a far higher price than whatever fees they received. Gesner's attack on Jackson is evidence that courtroom controversies could be detrimental to scientific reputations. In New York City newspapers, Gesner accused Jackson of sacrificing science to mercenary gain. According to Gesner, Jackson had prepared his scientific opinions "for the occasion" of the trial and published his consulting report "under the cloak of science." Jackson's opinions were, in short, "greatly influenced by the price paid for them."[146] Gesner's opinions might have been equally suspect. His embrace of asphaltum, as the classification for the Albert mineral, might appear to be influenced by his own commercial interests.

No other man of science spoke out, in public, about any untoward influence of money on science. This in itself would have been regarded as highly improper behavior. Besides, it would have been very difficult to prove (as it is today) that an expert's change of mind was a quid pro quo for his fees. Any hint of mercenary science was damaging enough, hence the adamant denials by *all* the expert witnesses at the New Brunswick trial. Men of science shared an interest in avoiding the appearance of a conflict of interest. They did not want to be seen compromising themselves for their commercial patrons. This is not to suggest a conspiracy of silence, only to highlight the seriousness of the situation. Fees for expertise smacked of interest, which was why men of science had to protest their disinterestedness so much.

In this conflict-of-interest reading of the Albert mineral cases, the employment of expert witnesses would also seem to have been a colossal waste of time and money. So why were men of science willing to serve as consultants and expert witnesses, and why were lawyers eager to consult them and have them testify? Both parties to the suits believed they needed men of science. Gesner's attack on Jackson might be treated as evidence of how much influence Jackson's testimony had on the trial. To put it another way, imagine if Jackson had not testified. Would Cairns's defense have been as strong? The answer would be no. At the first trial in Halifax, Jackson had not appeared. All the scientific evidence had been on Gesner's side, and he won. That lesson was not lost on Cairns. He recruited Jackson specifically, and he won the New Brunswick trial.

In the practice of law, expert witnesses were essential. That the supreme court justices in New Brunswick and Nova Scotia chose not to decide the matter of classification was not a sign of the interference or irrelevance of science, but of deference to it. In spite of hostile cross-examination, the one fact gained during the trial was the trustworthiness of the men of science themselves. The justices did not question their impartiality; on the contrary, they reaffirmed their authority. From the justices' perspective, neither side had won the scientific debate over classification. It was a toss-up, an honest disagreement among men of science. And, if those experts could not reach consensus on classification, the justices would not presume to do so, unlike the juries in the two cases. Instead, the justices reached for consensual science, namely, the *fact* that the Albert mineral was a mineral. Science and law were thus mutually dependent.

An even more optimistic corollary to this interpretation might be to regard the courtroom controversies as beneficial to science. Experts not only got the chance to display their knowledge, experience, and disinterestedness but also had the opportunity to play a central role in deciding legal cases and in commercial innovations. They also got the chance to do science. In the Albert mineral trials, calling expert witnesses to testify meant calling for new surveys of the Albert mine and its vicinity, new chemical analyses, and new studies of fossil fish. In this regard, however, the trials marked the beginning of an extended scientific controversy over the Albert mineral and coal.

CHAPTER 3

The American Sciences of Coal

Perhaps in no country have more frequent inquiries been made
in relation to COAL; to its infinite varieties, adaptations and
modifications; its innumerable depositories and its geographical
distribution, than in the United States of America.
—Richard Cowling Taylor, *Statistics of Coal* (1848)

T HE DECISIONS HANDED DOWN by the supreme courts of Nova Scotia and
New Brunswick brought closure to the legal and commercial questions surrounding the Albert mineral, but they did not resolve the question of fact. On
the contrary, the Albert trials reinvigorated a long-standing and difficult scientific problem: what was coal?

To anyone not familiar with the story of the Albert mineral, it might seem
astonishing to learn that nineteenth-century men of science could not answer
that straightforward question. Was coal not the most common mineral in the
world? Yes, according to Richard Cowling Taylor's *Statistics of Coal*. Taylor's big
book was loaded with facts from every place that had coal, from local weights
and measures, through regional prices and production, to national imports and
exports; it also displayed the "general outlines" of the science of coal, which, Taylor boasted, was his generation's greatest achievement. It was no small irony,
then, that the Albert mineral did not fit those outlines. Indeed, the controversy
revealed how far from established they were. To get a glimpse of the trouble
caused by coal, readers of Taylor's tome need go no further than the first page
where he listed "mineral combustibles," "bituminous substances," and "fossil
fuel" as synonyms for coal. Nowhere did he define these terms or explain how
they were related to one another. For Taylor, coal was what coal did; it fueled industrial expansion and stimulated science.

Coal does indeed have a rich scientific history, one that Taylor captured succinctly, yet unwittingly, in his list of synonyms. Coal, however, has not figured
very largely in histories of science.[1] As an object of inquiry, nineteenth-century

men of science treated coal as both a mineral and a rock. Its dual nature corresponded roughly to the two disciplines—chemistry and geology—that lay claim to it. Chemists studied coal in laboratories and tried to figure out what it was made of and what it could be made into, whereas geologists worked in the field, where they investigated the age, origin, and structure of coal deposits. Coal was thus a boundary object binding together two disciplines and at the same time separating them. In a sense, the Albert trials simply brought inside the courtroom a disciplinary division already well established outside it.

Mineral Coal

One curious feature of the Albert trials was the noticeable absence of any mineralogist. There were certainly men of science who studied minerals—Benjamin Silliman Jr. and Augustus A. Hayes, for example—but they identified themselves as chemists and thereby signaled to the judges and juries that the authority for classifying coal or asphaltum lay with a science other than mineralogy. This shift in the interpretation of minerals, from natural kinds to chemical compounds, is a story beyond the scope of this book;[2] however, it can be recapitulated and brought to bear on the scientific history of coal and the Albert mineral through a discussion of the career of James Dwight Dana.[3]

By temperament and training, Dana was a natural historian, an arrant compiler, sorter, and systematizer of facts. He had gone to Yale in 1830 to study with Benjamin Silliman Sr. (1779–1864), professor of natural history and chemistry, but it was Charles U. Shepard (1804–1886), professor of mineralogy, who taught him that the mineral kingdom was a vital branch of natural history.[4] Dana learned that minerals had to be studied in their totality to reveal their natural affinities. Totality meant other minerals, not geographical or geological aspects. A "natural system," Dana explained, was "a transcript of nature" in so much as it consisted of "those family groupings into which the species naturally fall."[5]

The natural system had been introduced by the famous Swedish systematist Carl Linnaeus. In his *Systema naturae* (1768), Linnaeus had classified coal and asphaltum as bituminous substances, a family of minerals characterized by their inflammability and pungent odor. "Bitumen" was what made these minerals burn and smell so bad, and it was thought to be of organic (meaning vegetable) origin. Bituminous substances comprised a continuous spectrum ranging from liquid petroleum through semisolid asphaltum to solid coal.[6] At the Albert trial, James Robb espoused this understanding of bituminous substances when he described the difficulty in trying to fix the point on an idealized line where asphaltum stopped and coal started.

During the 1820s, Linnaeus's natural system was revised by the German min-

Table 3.1. James Dwight Dana's Natural-Historical System of Mineralogy (1837 and 1844)

Class I
EPIGÆA (epi = upon, gaia = Earth)
    Specific Gravity under 3.8
    Fluid or soluble
    No bituminous odor
    Taste of solid individuals, acid, alkaline, or saline

Class II
ENTOGÆA (ento = within, gaia = Earth)
    Specific Gravity above 1.8
    Insoluble

Class III
HYPOGÆA (hypo = beneath, gaia = Earth)
    Specific Gravity under 1.8
    Resinous or carbonaceous
    Combustible

---

eralogist Friedrick Mohs, Abraham Werner's successor at Freiberg, and it was the "illustrious" Mohs whom Dana tried to emulate. In 1837 Dana published *A System of Mineralogy,* an ambitious attempt by a twenty-four-year-old to classify all minerals according to a natural-historical system, and in 1844 he updated his system. In both editions, Dana divided minerals into three classes: Epigæa, Entogæa, and Hypogæa. In distinguishing these classes, the character "bituminous" was critical; its absence characterized the first class, while its presence defined the third (table 3.1).

Dana coined the term hypogæa to describe how minerals in his third class were formed from the burial and subsequent alteration of animal and vegetable remains. But, unlike the traditional family of bituminous substances, Dana's Hypogæa minerals did not constitute a series. Pittinea (Pitch) was distinct from Anthracinea (Coal); the former was fusible, the latter was not (table 3.2). At the trial, the Albert mineral's fusibility was one of the most contentious points.

According to Dana, naphtha and asphaltum were two varieties of the species, *Bitumen communis.* The varieties described the end points of a natural transition; liquid petroleum changed into semisolid asphaltum by exposure to air. It was this presumed transformation that prompted several questions at the trial about the proximity of petroleum springs to the Albert mine. James Robb, for instance, said the Albert mineral was "accompanied by springs of naphtha or petroleum." Charles Jackson, however, saw no springs near the mine.[7]

Table 3.2. James Dwight Dana's Class III: HYPOGÆA

Order I. Pittinea (Pitch)

Genus I. Succinum (H = 2–2.5; Sp.G. = 1–1.1; Transparent-translucent; Color light)
　　1. S. Electrum (Amber)
　　2. S. Copallinum (Fossil Copal)

Genus II. Steatus (Sp.G. = 0.65–1.1; Whitish; Crystalline)
　　1. S. acicularis (Scheerite)
　　2. S. obliquus (Hartite)
　　3. S. sebaceus (Hatcherite)

Genus III. Bitumen (H = 0–2.5; Sp.G. = 0.8–1.2; Amorphous; Solids opaque or subtranslucent)
　　1. B. flexile (Mineral Caoutchouc)
　　2. B. fragans (Retinite)
　　3. B. amarum (Guyaquillite)
　　4. B. communis (Naphtha and Asphaltum)

Order II. Anthracinea (Coal)

Genus I. Anthrax (Lustre unmetallic)
　　1. A. bituminosus (Bituminous coal)
　　2. A. lapideus (Anthracite)

Genus II. Plumbago (Luster metallic)
　　1. P. scriptoria (Graphite)

As for coal, Dana split it into two species, hard (anthracite) and soft (bituminous), a distinction that went back to Linnaeus, although the term anthracite had been introduced by the French mineralogist René-Just Haüy in his *Traité de minéralogie* (1801) to underscore its crystalline form in contrast to the amorphous bituminous coal. Dana thought neither coal showed a regular structure. They were different because of hardness and specific gravity.

A unique feature of Dana's mineralogy was his nomenclature, a full Latin binomial system following Linnaeas's rules. Dana thought that Latin names would be more transparent, consistent, and useful. But he was wrong. By the late 1840s, he realized that his nomenclature was "out of place."[8] There were only a small number of minerals, "far less than a thousand," and common names could be used without confusion.[9] No mineralogist who published in the *American Journal of Science,* of which he was coeditor, employed Latin names, and no one at the Albert trial referred to asphaltum as *Bitumen communis.*

Another troublesome feature of Dana's natural-historical system was its impracticality for students, teachers, and artisans. Mineralogy had to be useful; and this, above all else, propelled a complete conversion in Dana's method. In 1850 he published a third edition of his mineralogy—"rewritten, rearranged and enlarged." During the preceding six years, chemical composition had proved to be the more certain and generally accepted basis for classification. Dana therefore abandoned the natural-historical system and, in a dramatic preface, acknowledged the sea change in his science.

> To change is always seeming fickleness. But not to change with the advance of science, is worse; it is persistence in error; and, therefore, notwithstanding the former adoption of what has been called the Natural History System, and the pledge to its support given by the author in supplying it with a Latin nomenclature, *the whole system, its classes, orders, genera, and Latin names, have been rejected.*[10]

Dana embraced the chemical system because science had progressed. He was not the instigator of this, merely a recorder of the most recent discoveries. His new system was therefore a reflection, not a source, of scientific change. As a comprehensive compilation of consensual science, not a contender for controversial ideas, *Dana's Mineralogy* continues to be updated and published to the present day.[11]

In changing systems, Dana decided that the natural-historical method was "false to nature" and that chemistry was the road to a more perfect arrangement.[12] Minerals could no longer be separated from other chemical compounds. "The distinction between *natural* and *artificial* inorganic products," he explained, was "contrary to strict science."[13] By dissolving the difference between the natural and the artificial, Dana made mineralogy redundant. Chemistry was the science that treated the full range of inorganic substances, regardless of whether they "proceed from the laboratory" or from "a running brook, an exhaling spring, or volcanic fumarole."[14] "The term Mineral Kingdom," he declared, "is fundamentally erroneous."[15] Nonetheless, Dana warned readers about assigning too much certainty to his new system. "Chemistry is slowly preparing the way for a perfect classification of inorganic substances, yet it is at present far from this result. Never perhaps in the history of the science have opinions been more widely different than now among leaders in this department."[16]

Chemistry also became Dana's new route to Providence. Since 1838, the year of his religious awakening, Dana had been deeply committed to the search for God's order in the natural world. He firmly believed that America would display the most useful evidence of God's plan. For Dana, *American* science was exceptional by virtue of its truth. That was one reason "a thorough American work on Mineralogy" was needed.[17]

Table 3.3. James Dwight Dana's Chemical System of Mineralogy (1850)

I. Nitrogen, Hydrogen

II. Carbon, Boron

III. Sulphur, Selenium

IV. Haloid Minerals; or the Alkalies and Earths, and their compounds with Water or the Soluble Acids, (Carbonic, Sulphuric, Phosphoric, Arsenic, Boracic), or their metallic bases with Chlorine or Flourine

V. Earthy Minerals; or Silica and Siliceous or Aluminous compounds of the Alkalies and Earths, and substances Isomorphous

VI. Metals and Metallic Ores, (exclusive of the Alkalies and Earths)

VII. Resins

Piety and patriotism notwithstanding, Dana remained practical and so hedged his conversion. Rather than conform to a pure chemical system, he chose "one less strict to science."[18] Thus, he did not consider "every separate chemical compound . . . a separate species."[19] In trying to preserve some of the natural-historical family groups, he presented mineralogy "in its economical bearings."[20] For science to be useful, it had to be useable. This was the demonstrably American side of Dana's new mineralogy (table 3.3).

In Dana's chemical system, coal, diamond, and graphite were placed in Class II (Carbon). Dana considered diamond and graphite different crystalline forms (dimorphs) of pure carbon. Coal was the third (amorphous) form of carbon. By classifying coal as carbon, Dana elevated it from an organic bituminous substance to an inorganic carbon mineral. He stuck the other former Hypogæa minerals in Class VII (Resins), organic compounds of mixed chemical composition. By this classification, Dana made the distinction between inorganic coal and organic asphaltum more significant; they were two distinct mineral species from two different *classes*. In the courtroom, Hayes made exactly that point, and Dana was surely the authority backing his statement. Dana's chemical conversion would thus seem to have been crucial to the outcome of the trial.

On the contrary, the Albert mineral upended more than it upheld Dana's new classification system. The problem, as Jackson's first analysis showed, was that Albert "coal" was not composed primarily of carbon. Jackson found only 40 percent carbon and 60 percent bitumen. Dana's chemical arrangement did not have a place for such a bitumen-rich mineral. Hence Dana, not Jackson, coined the term "asphaltic coal." But a neologism could not cover an anomaly. In short, the apparent failure of Dana's new chemical system to accommodate the Albert mineral might have helped to precipitate the controversy in the first place.

Only after the Albert trial, in the fourth edition of his *System of Mineralogy* (1854), did Dana deal with the percentage problem. Mineral coal could now contain less than 50 percent carbon. To justify this expansion of the Carbon class, Dana relied on three pieces of evidence. The first, of course, was the Albert mineral. The second was a coal recently discovered in Breckenridge, Kentucky, which Silliman, who had just returned from a consulting engagement there, had analyzed and found to be very much like the Albert mineral. The third was the Torbanehill mineral (or Boghead cannel). Discovered near Glasgow, Scotland, it too contained 60 percent bitumen, and it too had become the subject of an extended legal trial. In the Torbanehill case, an Edinburgh jury had decided that the mineral was to be defined, for legal and commercial purposes (gas lighting), as coal.[21] At the Albert trial, both Silliman and Hayes referred to the Torbanehill decision. Dana presented all this evidence to justify his new definition of mineral coal (table 3.4).

The scientific impact of the Albert trial on American mineralogy was thus profound. If the Albert mineral were to be classified as a coal—as the courts had decided—then Dana had to change the Carbon class to include it. Coal, by the mid-1850s, was mineral carbon. Asphaltum was resin, an amorphous, organic compound. On these points, practically everyone agreed, except Charles Wetherill.

Drawing on the new organic chemistry he had learned in Germany,[22] Wetherill classified coal and asphaltum as hydrocarbons because both contained carbon, hydrogen, oxygen, and nitrogen, just in different proportions. At the trial, he had put it succinctly: "Bituminous coal, so-called, does not contain bitumen."[23] To prove his point, Wetherill conducted further experiments on the Albert mineral. In his laboratory, Wetherill once again melted a sample, which seemed "an '*instantia crucis*' on the subject." In addition, he distilled naphtha from the mineral, a liquid that was usually distilled only from asphaltum. Coal, it was generally thought, could not produce naphtha. When heated, coal produced illuminating gas. Wetherill thought the naphtha was a remarkable distillate, one that seemed to raise a number of commercial and chemical questions. "I think an examination of this subject with regard to its products of decomposition," he suggested in a letter to Silliman, "would be very interesting to science & would yield some new substances."[24] Such suggestions about oil from coal would soon prove prophetic.

In broad perspective, the chemical turn in American mineralogy explains the paradox of the Albert trial: the absence of mineralogists, per se, as expert witnesses. There were no mineralogists because by the early 1850s there was no distinct science of mineralogy; it had been subsumed by chemistry.[25] "The

Table 3.4. James Dwight Dana's Analyses of Carbon Coal

	Moisture	Vol. combustible matter	Fixed carbon	Ash and clinkers
1. Pennsylvania anthracites	1.34	3.84	87.45	7.37
2. Maryland free-burning bituminous coal	1.25	15.80	73.01	9.74
3. Pennsylvania free-burning bituminous coal	0.82	17.01	68.82	13.35
4. Virginia bituminous	1.64	36.63	50.99	10.74
5. Cannelton, Indiana, bituminous	2.20	33.99	58.44	4.97

The following are other analyses: 6–13, by B. Silliman, Jr.; 14, Frazer, (Am. J. Sci. [2], xi, 301).

	Vol. matter	Fixed carbon	Ash
6. Grayson Co., Jet Cannel	61.95	30.07	7.98 = 100.08
7. " "	65.59	27.22	7.27 = 99.91
8. Breckenridge Co., Jet Cannel	64.30	27.16	8.48 = 99.93
9. Grayson Cannel, G. = 1.371	62.03	14.36	23.62 = 100.00
10. Grayson bitum.	41.06	54.94	4.11 = 100.11
11. Albert Coal	61.74	36.04	2.22 = 100.00
12. Boghead Cannel (Scotch)	66.35	30.88	2.77 = 100.00
13. Pittsburgh bitum.	32.95	64.72	2.31 = 99.98
14. Cowlitz, Oregon (Brown Coal)	49.5	42.9	2.7, water 4.9 = 100

Source: James Dwight Dana, *A System of Mineralogy*, 2 vols. (New York: George P. Putnam, 1854), 2:27.

mind uneducated in Science may revolt," Dana pontificated in the 1854 edition of his *Mineralogy*, "[b]ut it is one of the sublime lessons taught in the very portals of Chemistry, that nature rests no grand distinctions on lustre, hardness, or color, which are mere externals, and this truth should be acknowledged by the Mineralogist rather than defied."[26] This was a sober fact that Gesner understood. Looking back on the trial a decade later, he decided that the reasons for his loss lay in the weakness of his chemistry.[27]

The Carboniferous

The strength of Gesner's case had been the geology. But geologists, too, had had difficulty classifying coal since at least the late eighteenth century. In Abraham Werner's famous classification of rocks, Coal was a formation belonging to the Secondary. All formations, Werner assumed, were *universal,* insofar as they had been deposited in a global ocean. Coal was the exception. It did not have universal extent or continuity. As Robert Jameson (1774–1854), the Regius Professor of Natural History at Edinburgh University and Werner's student, explained, coal "shews a very peculiar character."[28] It seemed to occur in independent basins, and Werner thought it must have originally been deposited there. Water had carried plants to the basins, where the organic material was converted, without heat, into bituminous coal. Because such independent deposits were found everywhere, Werner argued that coal must be a universal formation, hence the *Independent* Coal formation.[29]

Besides the Secondary, Werner also found coal in the Transition and Alluvial. But among these rocks, coal was a *subordinate,* rather than the *predominant,* bed. One way to distinguish predominant coal from subordinate coal was by mineralogy. The coal of the Transition was anthracite, and Werner assumed that it was a chemical (mostly carbon) precipitate of the universal ocean. Anthracite contained no fossils.[30] The Independent Coal formation was composed "exclusively" of black bituminous coal and contained many fossils. The Alluvial contained "exclusively" lignite, a brown coal of very visibly decomposing plants.[31] Another way to distinguish these different coals was by age and relative position. Anthracite was the oldest and lay below bituminous coal, which was older than and lay below lignite. In theory, then, geology and mineralogy could be correlated.

In the field, they did not always agree. Anthracite was sometimes found among fossiliferous beds that seemed to belong to the Secondary, whereas bituminous coal was often found in beds that were highly bent and contorted, an orientation that did not agree at all with Werner's theory that Secondary formations were flat-lying (i.e., Floetz). Jameson warned that Transition anthracite "must not be confounded with the independent or true coal formation."[32] But the age and relative position of Werner's Independent Coal formation was confusing, according to the English geologists William Conybeare (1787–1857) and William Phillips (1775–1828). They, therefore, decided to create a new class of formations—the Carboniferous (table 3.5). Situated between the Transition and Secondary, the Carboniferous was characterized by thick beds of bituminous coal. (The *iferous* literally meant "containing," and carbon was the characteristic constituent of coal.)[33]

Conybeare and Phillips introduced the Carboniferous in their *Outlines of the Geology of England and Wales* (1822).[34] In this handbook, they described all the stratified and fossiliferous formations found in England and Wales, in detail and in order, beginning with the most recent and ending with the oldest. That organization (an inversion of the Wernerian system of oldest to youngest)[35] had the effect of saving the most innovative part of their book till last, where they presented for the first time "a regular and connected account of the Coal-districts of this country."[36] Much like Werner had done, Conybeare and Phillips grouped all the disconnected coal basins into one formation called the Coal

Table 3.5. William Conybeare and William Phillips's Classes of Formations (1822)

Primary
Transition
Carboniferous
Secondary
Tertiary

Measures.[37] The Coal Measures sat atop the Millstone Grit, Mountain Limestone, and Old Red Sandstone (table 3.6). This stack of formations could not be separated; otherwise, "inextricable confusion will result."[38] Conybeare and Phillips thus grouped them all together as the new Carboniferous.

According to Conybeare and Phillips, the Carboniferous deserved to be a class unto itself because of "its proportional importance in the geological scale [and] its peculiar characters," especially its structural complexity.[39] Coal beds were "most frequently much inclined" and exhibited "contortion and disturbance." It was comparatively "easy," Conybeare and Phillips thought, "to trace the position of the horizontal strata above the coal," which had been done in the 1810s by William Smith and the gentlemanly geologist George Greenough.[40] "But to reduce our description of the coal-fields, scattered as they are in unconnected basins and exhibiting every possible mode of disorder and derangement, into a regular systematic form, is a task of much more difficult accomplishment."[41] Delineating complex structures was the virtuosity of coal geologists, and the development of structural geology can be traced directly to coal.

Another peculiar character of the Carboniferous was, of course, the coal. The Coal Measures, Conybeare and Phillips emphasized, referred "to the great and principal deposit of that mineral."[42] These were the same thick beds that characterized Werner's Independent Coal formation, and Conybeare and Phillips cautioned geologists "carefully to guard against the error" of mistaking other beds for true coal.[43] Like Werner, Conybeare and Phillips located lignite in the Tertiary (the youngest formations) and anthracite in the Transition. They were, however, hesitant to draw "a line of distinction" between anthracite and bituminous coal. Mineralogically, both were composed of carbon, and geologically they sometimes were found intermixed. "All that can be done, therefore, in the present state of science is, to state the difficulty and leave it for solution to that more advanced period towards which we are now only securing the approaches, by preparing a firm ground-work of induction from facts."[44]

While anthracite remained a mystery, the origin of bituminous coal was not.

Table 3.6. Formations of the Carboniferous Class (1822)

Coal Measures
Millstone Grit
Carboniferous or Mountain Limestone
Old Red Sandstone

There were a heap of facts, namely, fossils. Conybeare and Phillips estimated 400 fossil plants had been identified.[45] Coal was unequivocally of vegetable origin, and Conybeare and Phillips felt confident in identifying and naming the Coal Measures as *the* great and principal deposit of "fossil combustibles from the vegetable kingdom."[46]

Putting the paleontological, structural, and stratigraphical information together, Conybeare and Phillips theorized about the "original habits" of the Carboniferous. The plants, mostly ferns, firs, and palms, indicated a hot, moist climate, not a dry, temperate one, as then existed in northern Europe. The few shells that had been found came from marine species. Conybeare and Phillips thus pictured a scattered group of islands covered by vegetation where torrential rivers would frequently tear up the plants and carry them to adjacent estuaries or arms of the sea. These uprooted plants then settled to the bottom and over a long time were chemically converted, without heat, into coal. In short, Conybeare and Phillips deduced that coal was formed by the bitumenization of land plants within a marine environment. The "estuary theory," as it became known, was the dominant explanation for the origin of coal during the 1820s and 1830s.[47]

Searching for American Coal

According to one English geologist, the Carboniferous was the marker for "a most important and extensive change in the condition of the globe," from a world covered in ocean to one with extensive tracts of dry land.[48] That the Carboniferous represented such a dramatic moment in the earth's history required more than an understanding of English or German coal fields. It needed global correlation. Coal was thus an ideal starting point for American geologists. They could test European theories, modify them, and/or develop their own to fit American discoveries.[49]

Among the first to set out in search of coal was Edward Hitchcock (1793–1864), professor of natural history and chemistry at Amherst College. Hitchcock had studied at Yale, where Silliman Sr. had introduced him to Werner's theories.[50] But after reading Conybeare and Phillips, Hitchcock opted for their stratigraphical system and began his coal geology along the Connecticut River, where small outcrops of bituminous matter had convinced him of the existence of the Coal Measures. Hitchcock's convictions also helped sway the Massachusetts governor. Calling coal an "object of inquiry essential to internal improvements, and the advancement of domestic prosperity," Governor Levi Lincoln proposed a geological survey. The legislature agreed and authorized $2,000, over three years (1830–1833), for Hitchcock to examine the Commonwealth.[51]

The Massachusetts survey was a watershed—the first to be publicly funded, to complete a geological map, and to issue a final report. Hitchcock, however, released the most practical part of that 1833 report a year earlier as *Economical Geology* (1832), wherein he described many useful minerals, but not coal.[52] Contrary to his first impressions, "further investigation has brought me, unwillingly, to the conclusion that no such formation exists along the Connecticut." What Hitchcock had thought were "real coal measures" turned out to be thin seams belonging to the New Red Sandstone. The rocks along the Connecticut River were too young to contain "genuine bituminous coal."[53] This conclusion was seconded by James Percival, the state geologist of Connecticut. In 1835 Percival began searching for coal along the lower Connecticut River, but he too discovered that the rocks belonged to the New Red Sandstone. There was no true coal in New England.[54]

There was some anthracite, though. Hitchcock investigated a seven-foot-thick bed at Worcester, which was, possibly, "exceedingly valuable"; however, Hitchcock discovered it burned poorly, and mining operations had been suspended. Hitchcock reported that anthracite mines in Portsmouth, Rhode Island, had also been abandoned because of "the prejudice against it in the market."[55] Pennsylvania anthracite was better fuel. According to Hitchcock, New England's mining failures could be put down to geology. Massachusetts's and Rhode Island's anthracites were older than Pennsylvania's variety. Although Hitchcock put all three in the Transition, he wondered whether "all these varieties of carbon have the same . . . vegetable origin?" Perhaps the difference among the anthracites, as well as between them and bituminous coal, was simply a matter of the amount of organic remains and the degree of heat. Hitchcock thought he had hit upon "an easy and satisfactory explanation of the mode in which all the varieties of carbon were produced, except perhaps the diamond."[56]

Rhode Island's anthracite got further attention in 1839 when the legislature commissioned Charles Jackson to conduct a geological survey. Rhode Island did not present a great diversity of rocks (Jackson found no Secondary formations, for example), so Jackson spent most of his time studying the abandoned Portsmouth mines. He found a few plants, which had a "remarkable resemblance" to fossils from the coal fields of New Brunswick and Nova Scotia. Nevertheless, Jackson, like Hitchcock, placed Rhode Island anthracite in the Transition because of its association with highly disturbed rocks. But unlike Hitchcock, Jackson thought Rhode Island's anthracite "will answer for furnaces and for cylinder stoves." There existed 37.8 million tons of it, he declared, "an ample supply for the citizens of Rhode Island for many years."[57] Years later, Taylor remarked on Jackson's mistaken calculations: "we should not like to be the purchasers [of Rhode Island mines] on the basis of [Dr. Jackson's estimate]."[58]

Optimistic advice was not unusual for Jackson or for most surveyors. Hitchcock and Percival, for example, did not reject, totally, the possibility that marketable coal could be discovered somewhere in their states. Percival noted that "perhaps, however, beds of coal may be found at a greater depth, than explorations have yet reached."[59] Such hope had some basis in stratigraphy; the Coal Measures did underlay the New Red Sandstone. Hitchcock gave a different rationale for why Massachusetts should keep the coal fires burning. First, he pointed out that geologists were still uncertain whether American bituminous coal correlated exactly with the English Coal Measures. Second, it was unclear why *thick* seams of New Red coal might not exist if there existed *thin* ones. Under the right conditions, plants might have converted to coal at any time in the earth's past. "All these facts prove," Hitchcock decided, "that it was a hasty generalization which limited workable coal to the coal measures."[60]

Coal was a process, not a product. That was a theory shared by the New York geologist Amos Eaton (1776–1842). Outspoken, independent-minded, and irascible, Eaton was a lawyer and land surveyor, who spent some time in prison accused of forgery. Upon being pardoned by the New York governor, he dedicated his life to the pursuit of science and went to study with Silliman Sr. at Yale. In 1818 Eaton began his geologizing with a tour along the Boston–Albany stage route, and in the following two years, he completed agricultural-geological surveys of Albany and Rensselaer counties along the Hudson. The public and scientific success of these undertakings induced the patroon Stephen Van Rensselaer III (1764–1839) to sponsor an ambitious project to survey the entire region bordering the newly completed, 363-mile-long Erie Canal. Surveying this "Grand Canal," the world's longest, featured prominently in all of Eaton's work, both substantially and symbolically. It contributed a wealth of facts and covered a lot of ground. Eaton repeatedly claimed to have surveyed more land, on the order of tens of thousands of miles, than any other geologist, American or European, of his time.[61]

The scale of American geology convinced Eaton that no European system would suffice. But figuring out an American one took time. Eaton produced no fewer than nine between his 1818 *Index to the Geology of the Northern States* and his 1830 *Geological Text-book*. The reasons for Eaton's "fickleness" had much to do with coal, and to get an idea of Eaton's coal troubles, it is useful to look at his 1828 system, *A Geological Nomenclature for North America*.[62]

GEOLOGICAL NOMENCLATURE

CASE OF SPECIMENS. CLASSES 2 & 1.	GENERAL STRATA and SUBDIVISIONS.	VARIETIES.	IMBEDDED and DISSEMINATED.
12	SECOND GRAY-WACKE. B. Rubble. A. Compact.	Red sandy, (old red sand?) Hone-slate. Grind-stone.	Manganese. Anthracite.
11	METALLIFEROUS LIMEROCK. B. Shelly. A. Compact.	Birdseye marble.	
10	CALCIFEROUS SANDROCK. B. Geodiferous. A. Compact.	Quartzose. Sparry. Oolitic.	Semi-opal. Anthracite. Barytes. Concentric concretions.
9.	SPARRY LIMEROCK. B. Slaty. A. Compact.	Checkered rock.	Chlorite. Calc spar.
8.	FIRST GRAY-WACKE. B. Rubble. A. Compact.	Chloritic.	Milky quartz. Calc spar. Anthracite.
7.	ARGILLITE. B. Wacke Slate. A. Clay Slate.	Chloritic. Glazed. Roof-slate. Red. Purple.	Flinty slate. Anthracite. Striated quartz. Milky quartz. Chlorite.
6.	GRANULAR LIMEROCK. B. Sandy. A. Compact.	Verd-antique. Dolomite. Statuary marble.	Tremolite. Serpentine. Chromate of iron
5	GRANULAR QUARTZ. B. Sandy. A. Compact.	Ferruginous. Yellowish. Translucent.	Manganese. Hematite.
4	TALCOSE SLATE. B. Fissile. A. Compact.	Chloritic.	Octahedral crystals of iron ore. Chlorite.
3	HORNBLENDE ROCK. B. Slaty. A. Granitic.	Greenstone. Gneissoid. Porphyritic. Sienitic.	Granite. Actynolite. Augite.
2	MICA-SLATE. B. Fissile. A. Compact.		Staurotide. Sappare. Garnet.
1	GRANITE. B. Slaty, (gneiss.) A. Chrystalline.	Sandy. Porphyritic. Graphic.	Shorl. Plumbago. Steatite. Diallage.

OF ROCKS IN PLACE.

CASE OF SPECIMENS. CLASSES 4 & 3.	GENERAL STRATA and SUBDIVISIONS.	VARIETIES.	IMBEDDED and DISSEMINATED.
20.	BASALT. B. *Greenstone trap* (columnar.) A. *Amygdaloid,* (cellular.)	Granular. Compact. Toadstone.	Amathyst. Calcedony. Prehnite. Zeolite. Opal.
19.	THIRD GRAY-WACKE. B. *Pyritiferous grit.* A. *Pyritiferous slate.*	Conglomerate, (breccia.) Calcareous grit. Red sandstone, (old red sandstone?) Red-wacke. Argillaceous.	Grindstone. Hornstone? Honeslate. Bituminous shale and coal. Fibrous barytes.
18.	CORNITIFEROUS LIMEROCK. B. *Shelly.* A. *Compact.*		Hornstone.
17.	GEODIFEROUS LIMEROCK. B. *Sandy.* A. *Swinestone.*	Foetid.	Snow-gypsum. Strontian. Zinc. Fluor spar.
16.	LIAS. B. *Calciferous grit.* A. *Calciferous Slate.*	Shell grit. Argillaceous. Conchoidal.	Shell limestone. Vermicular. Water cement. Gypsum.
15.	FERRIFEROUS ROCK. B. *Sandy.* A. *Slaty.*	Conglomerate. Green. Blue.	Argillaceous iron ore, (reddle.)
14.	SALIFEROUS ROCK. B. *Sandy.* A. *Marle-slate.*	Conglomerate. Grey-band. Red-sandy. Grey slate. Red slate.	Salt, or salt springs.
13.	MILLSTONE GRIT. B. *Conglomerate.* A. *Sandy.*		Coal?

Amos Eaton's geological nomenclature for North American rocks (1828). Eaton's order (from oldest to youngest) and descriptions (lithological) corresponded, for the most part, with Abraham Werner's system. Eaton's formations 1–6 belonged to the Primary Class, 7–12 were Transition, 13–19 were Secondary, and 20 was Volcanic. Coal was *not* a separate formation but a subordinate bed of the Millstone Grit (13). Anthracite was a characteristic bed of the Transition Class. Eaton identified these rocks along the route of the 363-mile Erie Canal. *Source:* Amos Eaton, "Geological Nomenclature, Classes of Rocks," *American Journal of Science,* 14 (1828):145–159, 359–368, facing page 145.

Eaton identified, ordered, numbered, and named twenty formations within four classes: Primitive (1–6), Transition (7–12), Secondary (13–19), and Super-incumbent (20).[63] Coal was not among them. Eaton rejected Werner's Independent Coal formation along with Conybeare and Phillips's Coal Measures and their Carboniferous class. He relegated bituminous coal to a subordinate, yet characteristic, bed of two Secondary formations—Millstone Grit (13) and Third Graywacke (19). Likewise, anthracite was a subordinate, yet characteristic, deposit of four Transition formations—Argillite (7), First Graywacke (8), Calciferous Sandrock (10), and Second Graywacke (12). Transition formations, Eaton emphasized, "never contain bituminous coal."[64]

Over the next two years, Eaton extended his surveys into adjoining parts of New England, New Jersey, and Pennsylvania. By 1830 he had mapped out the geography of American coal.[65] Lignite was found in New Jersey in Tertiary rocks. Bituminous coal occurred in western Pennsylvania at Tioga and Lycoming in "the lowest series of upper secondary rocks," Third Graywacke (19). "Genuine" anthracite was found in Transition Argillite (7) near Worcester and Portsmouth. Eaton then introduced a fourth variety, *anasphaltic coal.* Mineralogically, this new coal was much like anthracite; geologically, however, anasphaltic coal was younger than genuine anthracite and older than true bituminous coal. Anasphaltic coal characterized the "lowest of the lower secondary series of rocks," Millstone Grit (13). Eaton found thick beds of anasphaltic coal in eastern Pennsylvania at Carbondale, Lehigh, Lackawanna, and Wilkes-Barre, precisely where Hitchcock and others had identified genuine anthracite. Eaton thought these deposits were *not* anthracite because they contained fossil plants similar to those in bituminous coal. The Worcester and Portsmouth deposits, by contrast, had no fossils. Eaton then correlated his anasphaltic coal with "all the great coal measures of Europe."[66]

Coal thus shaped Eaton's classification of American rocks without ever being identified as a separate formation. This stratigraphic absence, oddly enough, boosted his confidence that profitable deposits would be found in New York. "[A]ll our hopes of discovering valuable coal beds," Eaton told state legislators in 1830, "are limited to the [anasphaltic] coal."[67] Eaton claimed to have traced these great beds from eastern Pennsylvania to the foothills of the Catskills, where they passed underneath his Saliferous Rock (14), a formation that extended more than 200 miles over most of western New York. All that New Yorkers need do was bore to a depth of 600 feet. "[I]f coal is not found beneath the saliferous rock," Eaton assured the legislators, who were debating a bill on coal boring, "it will be truly a geological curiosity."[68]

Curious it was. Extensive and expensive borings did not find coal. And so in 1836, the legislature established the New York Natural History Survey to redirect the search.[69] The progress and results of that survey were of great interest to contemporary geologists (and later historians), not least because the survey established a new class of formations (now referred to as a system)—the aptly entitled New York system—which described, ordered, and named the oldest fossiliferous rocks in North America. Lying between the Primary and the Carboniferous, the New York system replaced the old Transition class. Eaton's stratigraphy and the search for coal were dashed. "It is doubtful whether a more perfect series of Transition strata than that of New-York can be found in any part of the world. . . . Nine distinct [formations] occur within the limits of the State, all *below* the great coal formation and Carboniferous . . . which lie in the bordering counties of Pennsylvania."[70] New York rocks were too old to contain bituminous coal or Eaton's anasphaltic coal. Moreover, no anthracite was discovered. All coal, it seemed, had stopped at the Pennsylvania border.

This finding was not particularly welcome to politicians and entrepreneurs who had clung to Eaton's interpretation. On the contrary, here was cause for complaint. To geologists and their political allies, here was cause for concern. The New York survey thus provides another illustration of the dilemma brought on by unrequited science. A negative pronouncement needed explanation (what might be called "spin" today). Hence the geologists and their supporters spoke of savings (no more money wasted in coal borings) and the prevention of speculation (no more worthless mining ventures). More generally, the need for such an explanation (indeed, the whole dilemma itself) might be regarded as characteristic of American geology, or of surveys, at least. Charles Lyell, a keen observer of American habits and beliefs, recognized it as such. In his opening address at the Lowell Institute lectures in October 1841, Lyell acknowledged and subtly criticized Americans' need for a scientific payoff. "This is a kind of advantage which is never easily appreciated: because to prevent mischief is never so clear and palpable a benefit to the multitudes as to find mineral wealth."[71]

The Great Appalachian Coal Field

Compared to conditions in New England and New York, finding coal in Pennsylvania was easy. What was at issue by the mid-1830s was the extent and feasibility of its extraction. Capitalists, especially in Philadelphia, were keen on mining anthracite, and legislators, naturally responsive to such interests, wanted information on its suitability for heating and manufacturing. In 1836 this confluence of interests resulted in the establishment of the Geological Survey of Pennsylvania, the largest, most expensive, and scientifically significant study of coal geology in antebellum America, if not the world.[72]

The Pennsylvania survey was put under the direction of Henry Darwin Rogers (1808–1866).[73] Born in Philadelphia, Rogers had learned chemistry and natural history from his father, a physician and professor at the College of William and Mary. In 1829 Rogers became professor of mathematics and natural philosophy at Dickinson College, in Carlisle, Pennsylvania, but after a disagreement over educational reform, he left Dickinson and traveled to England. In London, at meetings of the Geological Society, Rogers found a new calling in geology. Returning to Philadelphia in 1833, he became friends, albeit temporarily, with the well-connected Alexander Dallas Bache (1806–1867), grandson of Benjamin Franklin, professor in the Franklin Institute, and a powerful force for reform in American science. Bache helped Rogers secure a lectureship in the Franklin Institute and then an appointment as professor of geology and mineralogy at the University of Pennsylvania.[74] In 1835 Rogers became director of the New Jersey survey. Thus, when Pennsylvania went looking for a geologist, Rogers seemed the obvious choice.

Rogers could not conduct the survey in Pennsylvania alone as he had in New Jersey. He was assisted by a skillful and hardworking group of twenty-three young men.[75] Rogers sent them into the field, over the mountains, and down the mine shafts, while he pursued his own related investigations, often out of state and with his older brother William Barton Rogers (1804–1882), who was directing the Virginia survey at the time. Rogers then gathered up his assistants' field notes and drawings and compiled them into annual reports, short narratives of the year's operations filled with general, practical results. Rogers did not think annual reports were the proper place to give full accounts of the facts or theoretical explanations of them. That would be done in a final report. But all is not well that does not end. Rogers high-handed manner had the predictable effect of driving away assistants, thus leaving him with a mass of notes and no one to interpret them. Such disorganization made it impossible to write up a final report. The survey's scientific achievements thus had to be taken, unpolished and incomplete, from the annual reports.

During six years of fieldwork (1836–1842), Rogers and his assistants had established the geographical distribution, lithology, and order of thirteen large formations. Following the Wernerian custom, Rogers numbered these formations from oldest (Formation I) to youngest (Formation XIII). Roman numerals were chosen as a temporary expedient in 1837 in order to avoid premature correlation.[76] Rogers, however, was willing to make one early and influential correlation. Formation XIII (the Great Coal formation) was the same as Conybeare and Phillips's Coal Measures. Rogers considered the possibility of correlating Formation XII (the Great Conglomerate) with the Millstone Grit and Formation XI with the Mountain Limestone, but he rejected Conybeare and Phillips's Carboniferous class on the grounds that Formation XI was a marine deposit, whereas Formations XII and XIII were terrestrial. Another reason Rogers adopted Roman numerals was to underscore the continuity of the formations. There was no major break in the sequence; Pennsylvania rocks lay conformably one on top of the other. Hence Rogers classified *all* his formations (I–XIII) as Secondary. Years later, one of Rogers's assistants recalled that the survey had "put a stop to all talk about 'Transition Rocks'" as early as 1837.[77]

If there were no Transition formations in Pennsylvania, then what was the anthracite in the eastern part of the state? From early on, Rogers entertained the possibility that anthracite and bituminous coal were not separated by any theoretical line of distinction. Both belonged to Formation XIII. Despite its importance, Rogers's idea might have remained unknown (at least until he published his final report) had not Lyell spurred him to act.

Lyell had spent the months before his Lowell Institute lectures (scheduled for October 1841) touring New England and New York.[78] At one point, he broke off from the geologist James Hall (1811–1898), his New York guide, to examine the bituminous coal mines at Blossburg, Pennsylvania. A few weeks later, Rogers led Lyell on an eleven-day tour of the anthracite regions. Lyell subsequently sent a letter to the Geological Society of London discussing the relations among bituminous coal, anthracite, and mountain building. The letter troubled Rogers. Not only did he risk being scooped, but worse, his final report might be forfeit should Pennsylvania legislators think a few days' observation by Lyell a substitute for years of surveying.[79] Rogers decided to present his theoretical findings at the third annual meeting of the Association of American Geologists and Naturalists to be held in Boston in April 1842.

The Boston meeting was the setting for Rogers and his brother William's famous paper on the origin of the Appalachians. According to the Rogers brothers, the mountains were uplifted by a series of parallel earth waves, generated by the shock of a catastrophic earthquake and spread out from the southeast toward the northwest. While original and bold, the Rogerses' theory was not widely accepted in America or Europe, a failure that overshadowed the success of Rogers's other paper on the origin of anthracite and bituminous coal.[80]

According to Rogers, the most distinctive feature of the Appalachian bituminous coal field was its size, nearly 900 miles long, 200 miles wide, and covering parts of seven states—Pennsylvania, Maryland, Virginia, Ohio, Kentucky, Tennessee, and Alabama. "Here, then," Rogers announced, "we have a coal formation." Rogers's formation actually comprised several detached coal fields mapped by several state surveys besides his own. Basically, Rogers did the same thing as Werner and Conybeare and Phillips had done; he connected them all into "one great formation."[81]

Rogers and his assistants did this through a method of lithological and structural comparison. Carefully tracing the coal seams and the intermixed sandstones, conglomerates, and limestones from basin to basin, they showed how the beds maintained the same order and, over a remarkable extent, the same thickness. Rogers described "uniform sheets of [carbonaceous] material" regu-

Idealized section of Paleozoic formations (circa 1860) showing the correspondence of the English system with three American systems: New York, Pennsylvania (Henry Darwin Rogers's final nomenclature), and Roman Numerals (Henry Darwin Rogers's first nomenclature for Pennsylvania rocks). *Source:* Samuel Harris Daddow and Benjamin Bannan, *Coal, Iron, and Oil; or, The Practical American Miner* (Pottsville, PA: Benjamin Bannan, 1866), 35–36.

FEET.		PENNSYLVANIA.	NEW YORK.		ENGLISH.
500 to 3000	XIII	Seral Coal Measures.	Absent in New York.	Carboniferous.	Coal Measures.
10 to 1000	XII	Seral, Conglomerate.	Absent in New York.		Millstone Grit.
0 to 2000	XI	Umbral, Red Shales.	Carboniferous Limestone.		Carb. Limestone.
200 to 2000	X	Vespertine, White Sandstones.	Gray and Yellow Sandstone.		Subcarboniferous.
0 to 5000	IX	Ponent, Red Sandstones.	Catskill.	Devonian.	Old Red Sandstone.
1700 to 5000	VIII	Vergent, Shales and Sandstones.	Chemung Group. Portage Group.		Eifel.
1000 to 2000		Cadent, Black Slates, &c.	Genesee Slate. Hamilton Slate. Marcellus Shale.		Eifel.
	VII	Post Meridian.	Corniferous Lime. Onondaga Lime. Schoharie Grit, &c.		
600 to 1500	VI	Meridian, Sandstone. Per Meridian Limestone.	Oriskany Sandstone. Lower Helderberg.		Ludlow.
1000 to 2000		Scalent.	Saliferous. Niagara.	Upper Silurian.	Wenlock.
1600 to 2000	V	Surgent, Red Shales.	Clinton Group.		Wenlock.
200 to 2500	IV	Levant, White Sandstones.	Medina Sandstones. Oneida Conglomerate.		Caradoc.
500 to 2500	III	Matinal, Slates and Limestones.	Hudson River Slates. Utica Slates. Galena (lead) Limestones	Lower Silurian, or Cambrian.	Bala Rocks.
1000 to 6000	II	Auroral, Limestones.	Trenton Limestones, &c. Chazy Limestones.		Festiniog Group.
1000 to 4000	I	Primal, Sandstone.	Calciferous Sandstones. Potsdam Sandstones.		Lingula Flags.
5000 to 10,000		Gneissic.	Gneiss.		

larly and widely extended across the North American continent.[82] The best example was located near Pittsburgh, where the "great" Pittsburg coal seam, sometimes 12–14 feet thick, outcropped. Rogers himself traced the Pittsburg coal over most of western Pennsylvania and into Virginia and Ohio, a total of 34,000 square miles, "a superficial extent greater than that of Scotland or Ireland. . . . It is still by far the most extensive coal-bed yet explored in any country, and the mere fact of its great extent must exert an influence on our views concerning the conditions under which the whole coal-formation originated."[83] Coal was big science, and not just in terms of geography, but of theory, too.

Rogers then turned to the question of the origin of coal. The Appalachian coal field's enormous size was evidence, "strongly adverse," to the estuary theory. He could not imagine any mechanism whereby plants were floated across such a vast region and deposited in such a uniform way. The absurdity of the estuary theory, Rogers noted in an aside, invited harsh criticism on its supporters, namely, English geologists. He mused, in heavily ironical tones, how the peat-bog theory could have "escaped so generally [their] attention."[84]

Rogers did mention "the exception of Lyell," who happened to be in the audience. Lyell, however, was not yet a convert to the peat-bog theory. In his *Principles of Geology* (1830–1833), Lyell expounded the estuary theory, which fitted very well with his uniformitarian ideas. He even relied on American evidence. "The prodigious quantity of wood," annually drifted down the Mississippi River, Lyell explained, "illustrat[es] the manner in which [an] abundance of vegetable matter becomes, in the ordinary course of Nature, imbedded in submarine and estuary deposits."[85] In his *Elements of Geology* (1841), published just before embarking for America, Lyell again endorsed the estuary theory, although modified by the concession that some coal might have formed in situ.[86]

The peat-bog or in situ theory had been championed early in the nineteenth century by the English geologist Robert Bakewell (1768–1843). In 1828 Bakewell had published the third edition of his popular and practical textbook, *An Introduction to Geology,* wherein he explained why the prevailing estuary theory was not "probable."[87] Bakewell pointed to a layer of fireclay beneath every coal seam and to upright trees found in coal fields. The fireclay was the soil in which the freshwater coal plants originally grew.[88] As to the trees, Bakewell thought it impossible for long trunks to remain vertical after being uprooted by torrential rivers. "We are therefore certain," Bakewell argued, that the plants composing coal had "grown in the situation where [they] stood."[89] This explanation came to be known as the "*in situ* or peat-bog theory," because peat bogs were considered to be extant analogies to ancient coal marshes. The theory was largely discarded by English geologists, but it was supported by the French geologist Alexandre Brongniart and was well known to Americans because Silliman Sr. chose Bakewell's *Geology* as his textbook.

Rogers called his explanation of the origin of coal the "growth on the spot" theory and he drew heavily on Bakewell.[90] Appalachian coal came from beds immediately beneath it, the fireclay belonging to Formation XII, which contained the distinctive fossil plant, *Stigmaria*. In describing the role of *Stigmaria*, Rogers relied on the work of the Canadian geologist William Logan, who in 1840 reported to the Geological Society of London his discovery of nearly 100 coal seams in South Wales underlain by fireclay containing innumerable *Stigmaria*.[91] In his letter to the Geological Society, Lyell had noted the *Stigmaria* in the fireclay at Blossburg, so Rogers wanted to make plain his priority in finding *Stigmaria* beneath all the coal seams in the Appalachian field.[92]

Another crucial piece of evidence in support of the growth-on-the-spot theory was Rogers's discovery of the uniform *condition* of the Appalachian coal. The purity of each seam (no intermixing with sand or clay) suggested a very stable and quiet environment of deposition, not the turbid waters of an estuary. "I would ask, is it conceivable, that any lake, bay, or estuary, could have been the receptacle of a deposit so extended, or that any river or rivers could have possessed a delta so vast?"[93] The answer was obvious; American coal meant a massive continental-sized marsh.

Rogers then sketched "an amazing picture" of this ancient marsh. In a flat-lying region about 150–200 miles wide, bordered on one side by a shallow, open sea and on the other by a hot, wet forest of tall trees of various species, especially *Sigillaria*, grew an extensive morass or marshy savannah clothed in a spongy matting of *Stigmaria*. The *Stigmaria* flats experienced a general and slow subsidence, hence the thick, pure coal beds. There were interruptions in this subsidence and even times of gradual upward movements, which were indicated by coarse sandstones sometimes found between coal seams. Sudden drops, by con-

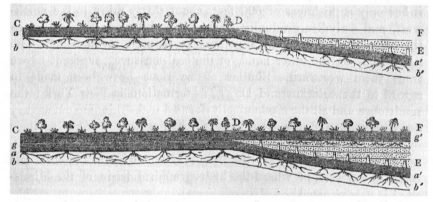

Charles Lyell's interpretation of gradual subsidence and the formation of coal beds (a′) following Henry Darwin Rogers's growth-on-the-spot theory of the origin of coal. *Source:* Charles Lyell, *A Manual of Elementary Geology,* 4th ed. (London: John Murray, 1852), 330.

trast, were indicated by the presence of upright trees. The trees, though, contributed little material to the coal; trunks were rarely, if ever, found standing in coal seams.[94]

Finally, Rogers turned to anthracite. It belonged to Formation XIII because the beds were on the same horizon as bituminous coal. In other words, anthracite and bituminous coal were parts "of a once continuous deposit."[95] What now separated eastern anthracite from western bituminous coal was a series of mountainous folds in the middle of Pennsylvania, on top of which no coal existed. Pennsylvania's barren center had led many geologists to the erroneous conclusion that anthracite and bituminous coal were of different ages and origins. Rogers thought the coal had simply been eroded. "This distribution of coal . . . is a direct result of the system of vast flexures, into which the whole of the Appalachian rocks have been bent, by the undulatory movements that accompanied the final elevation of the strata, and terminated the era of the coal."[96] The deposition of continental coal came to an end with the rise of the Appalachians.

The process of mountain building did more than disrupt the vast savannahs, however; it transformed the coal. During the catastrophic uplift, a "prodigious quantity of intensely heated steam and gaseous matter" was emitted through the crust. The "hot vapors" concentrated in eastern Pennsylvania, where the paroxysmal force was most intense and where the Appalachians were most tightly folded, faulted, and thrusted up. There, the coal was "steamed," thus releasing its bitumen or volatile matter. In the west, where the paroxysmal force was less intense, the unsteamed coal retained its bitumen. In a word, anthracite was bituminous coal that had been "de-bitumenized."[97]

According to Rogers's "law of gradation," coal functioned as a natural thermometer and pressure gauge; its varieties could be mapped across Pennsylvania from east to west in direct relation to the intensity of mountain building. Eastern anthracite contained 6 percent to 12–14 percent bitumen; Allegheny Mountains coal contained 16 to 22 percent bitumen; and the "great Appalachian basin" of bituminous coal contained 30 to 50 percent bitumen. The spectrum from anthracite to bituminous coal was thus an index of deformation, not of age.

The linchpin between stratigraphy and structure was the age equivalence of anthracite and bituminous coal. And to that extent, Rogers's theory might be considered geographically determined. Pennsylvania was the only state (and one of the few polities on the globe, the other being Britain) that contained within its borders *both* anthracite and bituminous coal. But the theory also depended upon a specific geological method practiced by the Pennsylvania survey, what was called topographical geology.

Rogers's theory of the origin of coal was an impressive achievement.[98] Lyell,

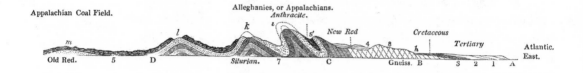

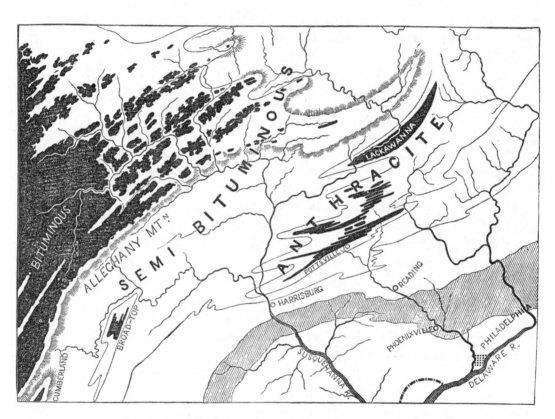

Top, Charles Lyell's ideal geological section from the Atlantic coast to the great Appalachian coal field of the Midwest. Length from east to west approximately 400 miles. *Source:* Charles Lyell, *A Manual of Elementary Geology*, 4th ed. (London: John Murray, 1852), 327. *Bottom*, Map of the coal regions of Pennsylvania exhibiting Henry Darwin Rogers's law of gradation: anthracite in the east, semibituminous coal in the center, and bituminous coal in the western part of the state. The coal fields are in black; the Mesozoic Red Sandstone formation is obliquely lined across. *Source:* James D. Dana, *Manual of Geology* (Philadelphia: Theodore Bliss, 1863), 323.

for instance, admitted in his Lowell lectures that "contrary to my early impressions" the greater part of coal had grown on the spot.[99] "I have been asked by several persons," Lyell continued, "if I could explain the difference between Coal and Anthracite." "It is certainly a good question," and Lyell proceeded to explain that "*Anthracite has once been Bituminous Coal,* but has lost its bituminous matter." By way of illustration, Lyell drew the audience's attention to the gas used to light the auditorium. "You are aware that this gas cannot be produced from anthracite because in that [coal] the operation has already taken place; [gas] is manufactured entirely from bituminous coal."[100] A gasworks replicated the debitumenization of coal in the earth. The artificial reproduced the natural. The cause of natural debitumenization, however, was debatable. Lyell rejected Rogers's catastrophism. He suspected hot igneous intrusions, not paroxysmal earth waves, cooked the coal.

Taylor was also impressed by Rogers's theory. "[I] entered upon the investigation of American geology," he reminisced in 1848, "imbued with the prevailing impression at that time [1830] generally advocated in Europe, and taught by nearly every geologist of eminence, that the anthracite deposits were older than those of the bituminous coal. In fact, the presence of anthracite was at one time thought to be conclusive evidence of a transition . . . period, in contradistinction to the bituminous coal of the secondary formations."[101] Now, however, "the doctrine of the supposed antiquity of the Pennsylvania anthracites had been abandoned, by common consent. It seemed no longer debatable in the United States."[102] There was also no doubt about the vegetable origin of coal and anthracite. It was generally admitted, Taylor reported, "that nearly every coal seam in the world is embedded upon an argillaceous stratum [fireclay]. These argillaceous beds are characterized by the abundant traces of the fossil vegetable, Stigmaria." The growth-on-the-spot theory, Taylor concluded, "seems now most generally believed."[103]

To support this conclusion, Taylor, like Bakewell, pointed to upright trees, especially *Sigillaria* (tree ferns). The "most important discovery" of the 1840s, Taylor thought, was made by the French paleontologist Adolphe Brongniart, Alexandre's son, who had determined that *Stigmaria* were "*but the roots of Sigillaria.*"[104] This connection had subsequently been confirmed by the English geologist Edward William Binney (1812–1881). In the St. Helen's coal mine in Lancashire, Binney had discovered an upright trunk of *Sigillaria,* nine feet high, with its *Stigmaria* roots, eight or nine feet in length, still attached.[105]

Sigillaria and *Stigmaria* were the most sensational coal fossils, but they were by no means the only ones. By the mid-nineteenth century there were more than 800; coal accounted for half of all known fossil plants.[106] These discoveries, combined with Rogers's theory, led Taylor to conclude that the "general outlines" of the science of coal were settled. Lyell, too, was sure that coal was a problem already solved. As he explained in his Lowell Institute lectures, if one started looking in the rocks below the Coal Measures, "we should certainly never reach the coal until we had bored through the whole earth."[107] At least that was theory before the Albert mineral surfaced.

Paleontology versus Structural Geology

Rogers was confident his theory covered everything west of New Jersey, including the recently explored coal fields in Indiana, Illinois, Michigan, and Missouri. He did not extend his theory eastward. There was no mention of Rhode Island, Massachusetts, New Brunswick, or Nova Scotia. Lyell, though, was very interested in those places. And so, after examining Pennsylvania anthracite with Rogers, he went to New England. In Massachusetts and Rhode Island, Lyell found a "series of transmutations" by which the anthracite had been cooked more thoroughly than the Pennsylvania variety. "The progressive debitumenization of the coal of the United States, as pointed out by Professor H. D. Rogers, lends support to this conjecture," Lyell explained.[108] Lyell then traveled to New Brunswick and Nova Scotia, where he met Gesner and together they visited the Joggins. They counted seventeen upright trees, six to twenty feet tall and one to four feet wide.[109] "This subterranean forest," Lyell remarked, "exceeds in extent and *quantity of timber* all that have been discovered in Europe put together."[110] Here was incontrovertible evidence for the growth-on-the-spot theory.[111]

After inspecting the Joggins, Lyell visited Pictou, where John William Dawson served as his guide. The meeting between Lyell and Dawson had a profound and propitious impact on both.[112] Lyell became Dawson's mentor and patron.[113] In matters of theory, Dawson adopted Lyell's principles of uniformitarianism right down to the belief in the nontransformation of species, a position Dawson held till the end of his life. In matters of practice, Dawson concentrated on paleontology, especially the fossil plants of the Carboniferous. Lyell, in turn, capitalized on a continuous supply of information from Dawson. Dawson wrote or coauthored with Lyell ten articles for the Geological Society of London detailing Nova Scotia geology.[114]

Dawson and Lyell divided the Carboniferous of Nova Scotia into three formations. The Upper Carboniferous comprised a series of sandstone beds barren of coal. The Middle Carboniferous contained the thick coal beds exposed at Pictou and the Joggins. The Lower Carboniferous was the new name Dawson and Lyell gave to the Gypsiferous formation, which contained gypsum and limestone beds along with some thin seams of impure coal.

In August 1852, when Lyell made his second visit to British North America, he joined up with Dawson again to study the fossil trees of the Joggins.[115] During a week-long investigation, they uncovered inside one of the trees the re-mains of a reptile, *Dendrerpeton acadianum*. This was not the first reptile to be found in Carboniferous rocks (other examples came from Germany), but it was significant for being the only one discovered in North America.[116] After the Jog-gins, they went to the Albert mine.

Lyell's interest in the Albert controversy had been spurred by Gesner, who had consulted him about a possible role as an expert witness. Dawson, on the other hand, had been contacted by the Halifax Gas-Light Company. Both Daw-son and Lyell, however, "[took] no part in the litigation."[117] Dawson was pre-occupied with his duties as superintendent of Nova Scotia schools, a position he had taken in 1850 at the request of Joseph Howe, a leading Reformer and po-litical ally of James Uniacke, one of the principals of Halifax Gas. Dawson also did not approve of courtroom controversies. He thought the examinations un-dertaken by "eminent" geologists and chemists "at the instance of the parties interested, [were] unfortunately . . . of a remarkably conflicting character."[118] Dawson cast doubt on the impartiality of the "scientific gentlemen summoned from the United States as witnesses experts," and he only darkened that im-pression with the somewhat disingenuous qualification that the gentlemen be-lieved "in all sincerity" in their own opinions. Dawson's opinion of the whole spectacle was succinct: "This was not wonderful."[119]

In contrast to the expert witnesses, Dawson, and presumably Lyell, "enjoyed" the chance to examine the Albert mine, thereby implying that the absence of le-gal and/or financial pressures produced salubrious scientific effects. Dawson's goal, nonetheless, was the same as the other experts: to classify the mineral. The "true place" of the deposit, Dawson decided, lay in "*the lower part of the Lower Carboniferous.*"[120] He figured out the stratigraphy through correlations with Nova Scotia rocks, especially the Gypsiferous formation at the Joggins.[121] Daw-son placed the Albert deposit on the same geological horizon as the "pseudo-coal-measures" of Nova Scotia, a band of thin, impure coal in the Lower Car-boniferous *below* the gypsum and limestone beds (table 3.7). The Albert mineral thus occupied the very base of the Carboniferous system in Nova Scotia and New Brunswick. According to Dawson, it marked "the dawn" of coal.

To the question of the mineral's origin, Dawson entertained three possibili-ties: hardened petroleum, altered bituminous shales, and bitumenized plants. Regarding the first, Dawson was uncertain about the process by which petro-leum changed into coal, or if nearby springs could supply enough petroleum for the large Albert deposit. As to the second, Dawson thought petroleum might be distilled from underlying bituminous shales, but distillation required heat, and he found "no indications of metamorphism or of the passage of heated vapours."[122] He did agree with Taylor and Robb that there existed an anticlinal bend to the strata, evidence of lateral pressure that could have produced a "pasty

Top, Erect *Sigillaria* containing the fossil reptile, *Dendrerpeton acadianum,* discovered by Charles Lyell and John Dawson at the Joggins, Nova Scotia, in 1850. *Source:* John William Dawson, *Acadian Geology,* 2nd ed. (Edinburgh: Oliver and Boyd, 1868), 192. *Bottom,* Section of the Joggins, Nova Scotia, showing erect *Sigillaria, Calamites,* main coal bed, and underclay. *Source:* John William Dawson, *Acadian Geology* (Edinburgh: Oliver and Boyd, 1855), 172.

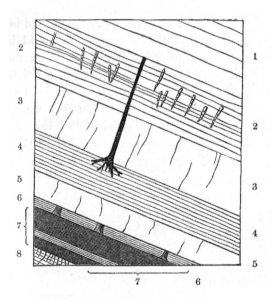

1. Shale and sandstone. Plants with Spirorbis attached; Rain-marks (?)
2. Sandstone and shale, 8 feet. Erect Calamites.
3. Gray sandstone, 7 feet.
4. Gray shale, 4 feet.
5. Gray sandstone, 4 feet.

{ An erect coniferous (?) tree, rooted on the shale, passes up through 15 feet of the sandstones and shale.

6. Gray shale, 6 inches. Prostrate and erect trees, with rootlets; leaves; *Naiadites;* and *Spirorbis* on the plants.
7. Main coal-seam, 5 feet coal in two seams.
8. Underclay, with rootlets.

Table 3.7. John William Dawson's Carboniferous
System of Nova Scotia and New Brunswick (1855)

1. Upper Coal
2. Middle Coal (Coal Measures)
3. Lower Coal or Gypsiferous
   a. Grey and Reddish Sandstones
   b. Limestone and Gypsum
   c. Conglomerates
   d. Albert Deposit

or fluid condition."[123] But pressure could not generate enough heat to produce distillation. The third hypothesis seemed most likely, not so much by default as by circular reasoning; if the Albert mineral were a type of coal, it must have been produced by the same process as other coals, namely, the bitumenization of plants. This theory, however, was "accompanied with serious difficulties." Dawson found no evidence of growth in situ. "I could not," he explained, "in any part of the mine, find beds corresponding to the *Stigmaria* underclay of ordinary coal-seams."[124] Neither could he detect, under the microscope, any organic structure in the mineral (a point Joseph Leidy had made at the trial).[125] This was not such an insurmountable hurdle because other coals, such as cannel, showed little structure. But combined with the absence of *Stigmaria,* it raised a big question about the supposed plants from which the Albert mineral was composed. The answer, Dawson suggested, might lie with Jackson's fish, although he did not know how fish transformed into coal. In the end, Dawson reasoned that a *lack* of evidence should not be construed as "evidence *against* [the mineral's] accumulation in the manner of ordinary coal."[126] Dawson thus envisioned the deposition of some kind of woody matter in a calm freshwater swamp during the Lower Carboniferous period.

In a decidedly tepid conclusion, Dawson dubbed the Albert mineral a "formation of a very peculiar character"[127]—so peculiar, in fact, that he vacillated on naming it. Initially, and "for convenience," he called it "coal."[128] Then he tossed around "Pitch Coal," a term little different from Jackson's and Dana's "asphaltic coal." Eventually, he settled on "Albertite"—the name, used "by persons desirous of not committing themselves," for "a distinct mineral species, intermediate between coals and asphalts."[129] So Albertite was "a really new material."[130] That explained why its classification caused so much controversy.

Dawson's account of the Albert mineral found its place at the heart of his magnum opus *Acadian Geology* (1855), a book designed "to tell a very intelligible tale" about coal.[131] It was a story about fossils, the dominant method for identifying, ordering, and correlating formations. But there was another story about coal, one that dealt with structure.

Structural geology was developed by Rogers and his assistants. In Pennsylvania, J. Peter Lesley (1819–1903) explained, "paleontology [was] impossible."

> We had an empire to survey and little time to do it. The country . . . was a waste of sand, mud, iron, and limestone strata of various texture, and color, in endless repetition . . . to know it here was to know it everywhere. Nothing remained but dynamic forms; and these so grand, so variously grouped that they excited perpetual enthusiasm. . . . [T]hus [we] became not mineralogists, not miners, not learned in fossils, not geologists in the full sense of the term, but topographers, and topography became a science and was returned to Europe and presented to geology there as an American invention.[132]

Topography fit the American idiom. Lesley regarded it as "new and large," in stark contrast to "that customary European local research," which was "tedious and puerile." "The fossils," Lesley carped, "distracted study from the topography."[133]

> It is not unfair to the geologists of the Old World to say that Topographical Geology was born in Pennsylvania. . . . As there is but one such Anthracite Coal Field known in the world, so there is no field of investigation for the topographical geologist so perfectly adapted in all respects for suggesting, at sight, the principles of his branch of science, and for testing their application to all accidents of the earth's surface.[134]

Lesley was proud to be called a topographical geologist, for this title carried a special identification with Pennsylvania and coal. He had graduated from the University of Pennsylvania in 1838 and had planned to enter the Presbyterian ministry, but his father thought some outdoor work on Rogers's survey would be good for his health. Rogers assigned Lesley to the anthracite regions, but he soon worked on all the major coal fields and became Rogers's principal workhorse in the field and office. In 1842, when the survey's funding was cut, Lesley "indulged in the luxury" of studies at the Princeton Theological Seminary.[135] He then spent a year, 1844–1845, in Europe, where he became acquainted with the Jura and the Alps and the theories of the French geologist Léonce Élie de Beaumont (1798–1874). Returning home, Lesley joined the Congregational Church, whose form of government he preferred, and took to the pulpit in Milton, Massachusetts, where he met Susan Lyman, whom he married in 1849. But his calling as a minister was brief.[136] By 1851 he was back with Rogers and working on the final report of the Pennsylvania survey. But within a year, they had an acrimonious split.[137] Rogers went to the University of Glasgow as a Regius Professor to finish *The Geology of Pennsylvania* (1858), and Lesley stayed in

Philadelphia pursuing a career as a professional geologist and writing his own book, *Manual of Coal and its Topography* (1856).

In some ways, Lesley's *Manual* scooped Rogers's *Geology* for it included both an explanation of the Pennsylvania survey's structural methods and its principal results. The book also marked the beginning of Lesley's rise to prominence as the authority on American coal. For Lesley, topography was the science of geological dynamics dealing with the great laws of mechanical force—in other words, the uplift and erosion of mountains. And Lesley considered the surface of the earth "as a congeries of mountains."[138] Topography also encompassed the art of map making. It was Lesley (not Rogers) who explained how to do a survey in the field and how to translate all those precise observations, plottings, and measurements onto a map.

The most innovative part of Lesley's *Manual*, and from a historical perspective one of the most important contributions of Lesley's scientific career, was his construction of contour maps. Contours "describe themselves" by points of the same given elevation above the sea. They were imaginary lines that displayed on paper a three-dimensional relief. "[A] perfect map in contour lines" revealed geological structure—the steepness of hills and valleys, the directions of cliffs and terraces, the grades of rivers, and the spread of plains. Upon contour maps, geologists could trace the outcrops of beds and calculate the geometric curves of entire formations.[139] Lesley's *Manual* contained several examples of contour maps, which he called topographic maps. Lesley dated the actual "invention" of the terms "contour-curves" (showing equal elevation above sea level) and "underground contour-curves" (showing equal depth below the surface) to around 1860.[140] Since then, contour maps have been a standard tool of geology the world over.

Topographical geology and contour maps constituted only half of Lesley's *Manual;* the other was devoted to coal. Lesley provided the first systematic classification of Appalachian coal (bituminous and anthracite). He identified, ordered, numbered, and named all the coal beds in detail (and he meant it, "*detail* is truth").[141] The key was the great Pittsburg coal bed. By Lesley's reckoning, it covered an enormous 200,000 square miles (three times what Rogers had calculated), from eastern Pennsylvania into Ohio and parts of Maryland and Virginia, and it maintained a nearly uniform thickness of eight feet.

Taking into account the immense area over which it must originally have stretched . . . , [the Pittsburg bed] presents itself as one of the most remarkable and significant facts of science, a starting point for many theoretical speculations, and, practically, a sure horizon of observation from which to measure and identify the rocks above and below it. It has, in fact, been made the base line of our carboniferous geology.[142]

As *the* stratigraphical referent, the Pittsburg bed divided Formation XIII (the Great Coal) into two "systems": Upper and Lower. The Upper Coal contained the largest and most workable seams, including the Pittsburg bed; the Lower had thinner and fewer seams along with cannel coal. The Lower Coal rested conformably on Formation XII (the Great Conglomerate) and below that lay Formation XI, which Lesley variously styled as the "Proto-Carboniferous Formation," the "Subcarboniferous Formation," and the "False Coal Measures." "The coal of No. XI," Lesley warned, was not "real coal." "[I]t requires the strongest assurances of correctness," he declared, to distinguish "the lowest true coal" from false coal.[143]

Lesley's ability to distinguish true from false coal rested on his knowledge of all the Pennsylvania seams. He believed that coal beds across North America could likewise be identified.

> There can be no doubt that the time is not far distant when every persistent coal will be known wherever it outcrops, from [Pennsylvania], to the southern limits of Tennessee. And if so, it is not too much to expect the same success with the scattered fields of Illinois. The coals of . . . St. Clair [Pennsylvania] are evidently the beds of Iowa. Even some of the innumerable layers of Nova Scotia coal may, in time, be found to be the same with some which crown the Pottsville basin.[144]

Lesley's coal vision was truly continental.

"Here, however, we must stop." The idea of matching individual coal seams in England and Pennsylvania, Lesley thought, was simply "absurd." For transoceanic correlation, "fossil plants must be our guide at last."[145] But even then there was the problem of definitions and national styles. To illustrate the point, Lesley turned to "the remarkable instance" of a litigated coal.

The coal Lesley chose was not the Albert mineral but the Torbanehill mineral. Lesley used the story of its lawsuit to show the beguiling nature of false coals and "the opposite genius" of the English and Germans. English geologists were "practical"; whatever "lay in the usual form, position, and general relationship of coal was *practically* nothing else than coal." (That had been Jackson's and James Percival's geological opinion at the Albert trial.) The Germans were "ideal"; whatever "has not the essential traits of coal has no right to its name." Essential traits were determined chemically and microscopically. (That had been Charles Wetherill's and Joseph Leidy's opinion.) Because the Torbanehill mineral contained too much bitumen and too little "carbon," it could not be classified according to "true scientific nomenclature" as coal.[146] Presumably, Lesley thought the Albert mineral was also a false coal. But he did not say so explicitly. He tried to explain how different coals (true and false) resulted from different plants, but because he was no paleontologist, his brief discussion of

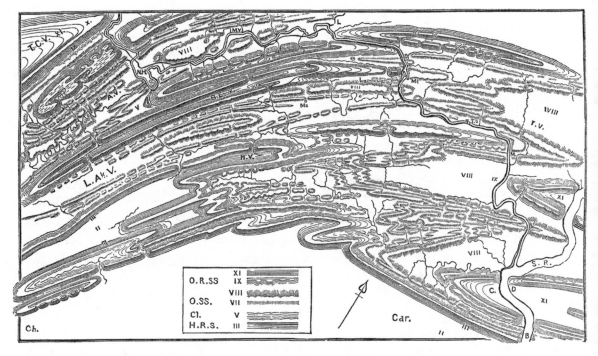

One of the earliest examples of geological structural contouring. J. Peter Lesley's "zigzag" topography of the area around Chambersburg (Ch.) and Carlisle (Car.), Pennsylvania. The letters on the map read as follows: Trough Creek Valley (T.C.V.), Newton Hamilton (N.H.), McVeytown (M.V.), Lewistown (L.), Mifflintown (Mf.), Thompsontown (T.V.), Awkwick Valley (A.V.), Little Awkwick Valley (L.Ak. V.), Horse Valley (H.V.), and Black Log Valley (B.L. V.). The Susquehanna River (S.R.) is crossed by the Pennsylvania Railroad bridge at B. and the coal mines were located at C. & D. In the square, the letters read: Old Red Sandstone (O.R.SS.), Oriskany Sandstone (O.SS.), Clinton Group (Cl.), and Hudson River Slates (H.R.S.). These were formations of the New York system; the Pennsylvania system equivalents were shown with Roman numerals. *Source:* J. Peter Lesley, *Manual of Coal and Its Topography* (Philadelphia: J. B. Lippincott, 1856), 137.

fossils was reduced to vague generalities about the decomposition of firs, ferns, mosses, or especially resinous pine trees. In the cases of litigated coals, Lesley conceded that topography could not resolve the "inherent difficulties involved in the question: What is coal?"[147]

Rogers thought he could. In 1858 *The Geology of Pennsylvania* finally appeared in two thumping volumes totaling more than 1,600 pages. The wealth of facts was new and interesting, but Rogers's geology was deeply flawed by an unacceptable theory of mountain building and an idiosyncratic nomenclature.[148] These flaws undermined the strengths, of which two were scientifically important. The first was Rogers's structural geology. In far greater detail than Lesley, Rogers developed a typography for the folds and flexures found in the Appalachians—the source for the term Appalachian-type mountains—as well as for the features of denudation. What Rogers did not present, although he indicated he would do so in the preface, was an explanation of methodology. Lesley's *Manual* remained the standard reference on contour maps.

The other strength was coal, which took up almost the entire second volume. Rogers provided a "full and minute" description of the Pennsylvania bituminous and anthracite fields. More importantly, he made a sweeping survey of North American coal geology, from Nova Scotia to Alabama and from Rhode Island to Missouri. Naturally, in his discussion of New Brunswick, he took up "the most important, certainly the most curious" mineral. "This substance," Rogers explained, "occasioned much discussion in connection with a legal definition [of coal]." Rogers characterized the Albert mineral's chemical nature, geological relations, and physical appearance as "abnormal," which meant it contained "an unusual proportion" of bitumen, did not "constitute a true bed or layer conformably embedded between the strata," and did not "wear the aspect of ordinary coal." Therefore the key to classifying it lay not in chemistry, geology, or mineralogy, but in dynamics.[149]

To illustrate how faulty knowledge of dynamics led to erroneous interpretations of the Albert mineral, Rogers pounced on Dawson's *Acadian Geology*. Dawson had assumed that "mechanical disturbances" had forced the Albert mineral into a fissure and, in the process, somehow "very singularly distorted" it.[150] For Rogers, it was "difficult to conceive" how the mineral's peculiar chemical and physical properties arose from "mechanical pressure, however energetic, upon an ordinary bed of coal."[151] "[T]he history of coal mining has hitherto failed to exhibit any instances of such exudation of the bituminous portions of coal-seams," Rogers asserted, "or indeed any change of chemical composition from mere compression." The Albert mineral was *not* "a metamorphosed condition of true coal, or a derivative part of a genuine coal-seam."[152]

The mineral's "peculiarities" therefore had to come from "an original product or formation." Rogers, like Lesley, tried to envision the "conditions under which the deposit was originally collected and elaborated."[153] But, ironically, he too found that dynamics did not help and so retreated to fuzzy notions of chemical action on vegetable matter. Nevertheless, Rogers arrived at a categorical conclusion.

> Regarded from a geological point of view, this much-discussed hydro-carbon must rank therefore as a genuine coal, using this word to designate not a specific mineral or chemical compound, but a genus of substances having common or similar geological history, possessing carbon for their base or principal constituent, traceable to a similar origin in vegetable matter, and applicable to similar uses in the arts.[154]

So the Albert mineral was coal, in particular a hydrogenous coal (table 3.8).

Table 3.8. Henry Darwin Rogers's Coal Classification (1858)

I. Anthracites (Mostly carbon with little bitumen)
   1. True (i.e., Hard or Dry)
   2. Semi (i.e., Gaseous)
II. Bituminous (Balance of carbon and bitumen)
   1. Semi (i.e., Dry, more carbon)
   2. True (i.e., Soft or Fat, more bitumen)
III. Hydrogenous (Mostly bitumen with little carbon)
   1. Cannel
   2. Gas

Rogers concluded his grand geology with a new classification system for American coal. He defined three species—anthracites, bituminous, and hydrogenous. Bituminous coal was the common form, best exemplified by the great Appalachian coal field. Anthracites were "the metamorphic form of ordinary bituminous coal" and occurred in regions once subjected to a "high distilling temperature" and "great pressure"—for example, eastern Pennsylvania or Rhode Island, home of the most highly metamorphosed coal in North America. Hydrogenous coals were subject to the least degree of heat and pressure and were found in Ohio or Breckenridge, Kentucky, on the western edge of the great Appalachian coal field. The Albert mineral, oddly enough, did not fit Rogers's classification. It occurred in a region that everyone, including Rogers, thought had been subjected to an elevated degree of heat and mechanical compression, yet the mineral seemed to gain, not lose, "hydrogenousness." Understandably, Rogers did not dwell on the anomaly beyond mentioning that the Albert mineral was good for making "highly-illuminating gas" and a "peculiar oil."[155]

> It will be the serious business of geologists to discuss and describe all these [coal] varieties until they are understood and published to the business world, for every change in the character of coal provides a new and more efficient agent to some branch of the arts which has been waiting for it to complete its own development.
> —J. Peter Lesley, *Manual of Coal and its Topography* (1856)

THE CONNECTIONS among coal, commerce, and science could not have been made more explicit. Coal was "serious business," and not just because it required lots of money and politics or involved the complex organization of mining companies or forced the intransigent coordination of management, labor, and technology, but because coal meant science. No country on earth, Taylor declared, had experienced such a "rapid expansion of the field of industrial operations" as had the United States in the second quarter of the nineteenth century. Americans had thus created an unparalleled demand for science "on the subject of coal."[156]

Coal geology was basically of two sorts: knowing where coal was and knowing where it was not. The former was obviously preferable; the latter often required explanation. Inevitably, both sorts were qualified by the terms *true* and *false.* On one level, this might seem self-evident—true meant those beds that could be mined for a good quality and quantity of coal; false referred to unprofitable seams that were irregular, limited in extent, or poorly combustible. In these instances, true and false carried tangible economic import, which any miner, manufacturer, entrepreneur, or householder understood. On another level, the terms implied an understanding of *why* a particular coal was true or false. Perhaps it was a question of composition; false coal might contain too much sand or too little bitumen. Such deposits, Lesley warned, were "often mistaken for coal."[157] Composition was no doubt important, but geology, not chemistry, was the more common framework for understanding true and false. True referred to the thick beds of anthracite and bituminous coal of the Carboniferous system; false referred to the thin seams of the Subcarboniferous system. The Albert mineral occurred in the Subcarboniferous; in geological terms, it was a false coal.

True and false also characterized the trustworthiness of the geologists themselves. True meant a man of science knowledgeable and experienced in unraveling the structure of coal fields, the order of strata, and the age of formations. False meant either outright dishonesty (a rare charge) or, more likely, mistaken identity. But even in cases of misjudgment, such as Gesner in New Brunswick or Eaton in New York, geologists were usually not regarded by the public, politicians, or private enterprise as false prophets, but rather as overly optimistic. Hope was hardly an uncommon attribute among coal geologists (viz. Hitchcock, Percival, Jackson).

Still, there was no truer test of a geologist than the ability to expose a believed-to-be-true coal as false. In one episode in Pennsylvania, Lesley recounted how "extensive arrangements [had been made] for an eastern trade." "[A] prosperous adventure" was advertised by land agents who believed that certain beds belonged to "the lowest of the true coal beds." But Lesley showed that the beds were "six hundred feet beneath the base of the true coal measures." As the true geologist, Lesley predicted the outcome: the mines "never paid."[158]

The high cost of misguided adventure was a recurrent theme among coal geologists.[159] In *The Geology of Pennsylvania,* Rogers included an entire section on "Searching for Coal." He warned so-called coal seekers, coal hunters, and coal diggers against fallacy, flagrant frauds, serious mistakes, self-deception, and, of all things, exaggeration. "A sagacious use of geological knowledge," Rogers recommended, was the antidote to the ills that bedeviled the "uninitiated explorer."[160]

Without geology, an equally concerned Dana warned, "[d]isastrous errors are often made."[161] Dana's admonition appeared in his 1863 *Manual of Geology*. Like his mineralogy textbooks, Dana's *Geology* was a synthesis of the best work by Americans on America, and as usual, he was unabashedly chauvinistic. "[O]n account of a peculiar simplicity and unity," he announced, "American Geological History affords the best basis for a text-book of the science."[162] The part of that history that best illustrated the "American character" of geology was the Carboniferous Age.[163]

Coal shaped the theories and practices of American geology, and Americans became the foremost coal geologists. "The special history of the Coal period of Europe and Britain might be followed out, as has been done for North America," Dana explained, "[b]ut it would illustrate no new principles."[164] By the 1860s, Americans had reestablished the general outlines of the Carboniferous Age and were routinely dividing it into two systems: the Carboniferous system,

The Carboniferous Age. *Source:* James D. Dana, *Manual of Geology* (Philadelphia: Theodore Bliss, 1863), frontispiece.

a repeating series of sandstone, coal, clay, shale, and conglomerate beds (Formations XIII and XII in Pennsylvania or the Coal Measures and Millstone Grit in England);[165] and the Subcarboniferous system, mostly thick limestone beds (Formation XI or the Mountain Limestone in England).[166] In the late nineteenth century, Americans would rename these systems. The Carboniferous became the Pennsylvanian, and the Subcarboniferous, the Mississippian. In the modern geological time scale, the Pennsylvania and the Mississippian are the only two systems named by Americans.[167]

But beyond taxonomy or methodology (i.e., contour maps), American geologists thought their most significant contribution lay in the very discovery of the coal fields themselves. According to Taylor, the coal surveys were a "remarkable epoch, wherein was accomplished one of the most rapid and successful geological developments, that has occurred in the history of science."[168] Rogers estimated that Americans had discovered more coal than any other geologists in the world, "twenty times the area [of] all known coal-deposits of Europe."[169] Dana figured that North America (the United States, New Brunswick, and Nova Scotia) had twelve times as much coal as Great Britain. "The contrast is striking," he concluded, "in its bearing on the earth's future and has a profound historical interest."[170]

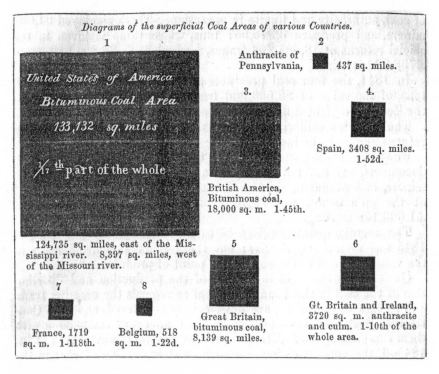

Diagrams of the superficial coal areas of various countries. *Source:* Richard Cowling Taylor, *Statistics of Coal* (Philadelphia: J. W. Moore, 1848), xvi.

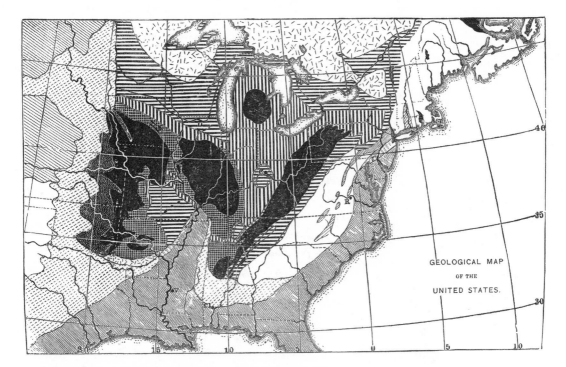

Geological map of the United States (1860). The black shaded regions represented the Carboniferous and corresponded to the six largest and best-studied coal fields, beginning in the West: the Iowa-Plains, Illinois-Missouri, Michigan, the great Appalachian (60,000 square miles), Rhode Island-Massachusetts (only 1,000 square miles), and Acadian. Dana, however, did not color in much of the Acadian or New Brunswick–Nova Scotia coal field, which he estimated at 18,000 square miles, a figure he took from Abraham Gesner. *Source:* James D. Dana, *Manual of Geology* (Philadelphia: Theodore Bliss, 1863), 133.

CHAPTER 4

Mining Science

In our Country the business of exploring & surveying mines is given to Geologists who are supposed to know all that the earth contains from the superficial soil to the molten central regions & the slightest smattering in the elements of that science is regarded as a sufficient claim to a commission or employment as a Mining Engineer.

—Charles T. Jackson, "Remarks on Mining Operations" (1846)

ACCORDING TO CHARLES JACKSON, American geologists' ability to discover, describe, and develop mineral resources had created a problem. Smatterers in science were capitalizing on their success by soliciting commissions as mining engineers. These so-called surveyors, who "know nothing about mines or minerals or the art of tracing a vein over irregular ground," contributed directly, and unfortunately often, to the disappointment and loss attendant to so many mining ventures. For Jackson, mining was the foundation of American agriculture, manufacturing, and commerce. There was no profession more important than mining engineer, and no science more important than geology.[1]

Geology-based mining was a practical necessity *and* a business imperative. For most interests and purposes, geologists were the most knowledgeable and experienced experts in America. In large part, this was due to the absence of trained mining engineers, for unlike continental Europe, no mining schools existed in the United States or British North America until the 1860s. The handful of emigrant European-trained engineers available were uninformed or misinformed, or so Jackson claimed. The same applied to practical men. Jackson contrasted the experienced mine viewers of Britain, the other country without a mining school in the first half of the nineteenth century, with their American counterparts. British viewers worked in one region for years (even generations), whereas Americans lacked the required long-term familiarity with the land and its resources. All this combined to put geologists in a vital commercial position.

Mining was also about science. Most geologists believed in the complementary relations of private and public surveying, of science paid for by mining companies and sponsored by state or federal government. Granted there were important differences in resources (equipment, personnel, and funding) and in scope (an entire state versus one property), but both activities provided geologists with opportunities to do fieldwork, the essence of their science. Mining, in other words, involved more than the routine application of elementary principles; it meant research.

Such research was valuable to both business and science, and in some cases, it was published in special consulting reports. These reports provide historians with evidence for evaluating the impact of mining on the development of American geology and, conversely, allow insight into the commercial influence of men of science. The reports contain not only what consultants did, but often how and why they did it, and to what purpose their expertise was put. In short, these reports represent a unique genre of scientific-commercial literature.

American geologists generally conducted the business of exploring and surveying mines smoothly. There was nothing especially controversial or novel in Jackson's demand for greater commercial involvement by geologists and for a more discriminating selection by those who hired them. The development of consulting is a story of interests more in common than in conflict. In this analysis, the practice unfolds as a series of engagements undertaken by a small group of geologists beginning in the 1820s and gathering steam by the 1850s. Although it is impossible to examine every geologist, conclusions can be drawn from select episodes (both common and uncommon) from the professional careers of the most prominent consultants during the first half of the nineteenth century. In this way, consulting and its contributions to geology (and to the relations between science and industry) can be explored and explained.[2]

Origins

The origins of American consulting go back to the careers of Amos Eaton and Benjamin Silliman Sr., whose most active years (the 1820s and early 1830s) predated the era of the first geological surveys (the late 1830s and early 1840s).

Eaton was well known for his geology and for founding the Rensselaer School in Troy, New York.[3] His surveys (and to some extent his school) combined scientific and commercial interests. Both endeavors were carried out "under the direction and at the expense" of Stephen Van Rensselaer III. This would seem to place the surveys under the broad rubric of consulting. After all, Rensselaer owned some of the land Eaton covered.[4] But Eaton's personal relationship with Rensselaer should be regarded differently. It was based on very unequal degrees of social status, political power, and wealth. Unlike later consulting geologists,

Eaton worked only for Rensselaer; he discharged a "duty" to his "patron."[5] Literally born to the manor, Rensselaer considered his sponsorship of Eaton's science in line with the obligations of his patrician class. Between 1821 and 1828, Rensselaer expended more than $18,000 in the cause of "improving and extending the natural sciences."[6] Rensselaer assumed that men of the moneyed gentry would take charge of civic affairs and cultural activities. He himself served in the New York legislature and in Congress. In short, Rensselaer's kind of patronage belonged more to an aristocratic eighteenth century than to a market-oriented democracy of the nineteenth.

Silliman Sr. had no patroon and thus had to look elsewhere for scientific support. He could, of course, rely on Yale College to supply a room and apparatus for chemical experiments, although he had little time to pursue them.[7] Outside the College, New Haven capitalists and companies submitted questions (e.g., minerals to be analyzed) and additional funds. From early on, Silliman Sr. displayed a keen interest in the useful application of science. In May 1810, for example, "at the request of the proprietors" he examined and wrote a report on a lead mine near Northampton, Massachusetts.[8] By the 1830s his services as a mining engineer were in high demand. He accepted a number of commissions, some of which involved large amounts of time, travel, and toil and were accordingly remunerative. He received as much $1,400, the equivalent of a professor's salary, plus expenses for traveling and assistants (usually a student or his son Benjamin Jr.), to examine properties as far away as Virginia, where he went in August and September 1836 to investigate gold mines.[9] His expertise was equally valuable to coal interests in Virginia, Maryland, and Pennsylvania, and his 1830 study of Mauch Chunk and Pennsylvania anthracite was among the first scientific-commercial reports on the area. It was subsequently published, as were his gold surveys, in slightly different forms, in the pages of the *American Journal of Science*, which he founded in 1818.[10]

Because Silliman Sr. brought the results of his private surveys to the attention of other geologists and the public, his entrepreneurship conformed to the expectations of Yale and the scientific community. From a cultural perspective, Silliman Sr.'s consulting reflected and reinforced a strong and widely held moral belief in service to others. Ideally, consulting served both selfish and selfless ends: the acquisition of knowledge and money complemented a duty to advance science and a patriotic obligation to promote industry and prosperity.[11] Beyond these ideals, the ability to balance private and public interests, practical and theoretical concerns, was regarded by Americans as characteristic of respectable science.[12] Silliman Sr.'s prominence popularized consulting and made it a practice worthy of professors and men of science in general.

Professionals

Silliman Sr.'s success, ironically, helped bring to an end the initial flurry of consulting. In the mid-1830s several states, including Virginia, New Jersey, Pennsylvania, New York, Massachusetts, and Maine (and the province of New Brunswick), commissioned surveys to bring the knowledge of mineral resources to the attention of the public and private interests alike. These surveys were crucial to the development of American geology. They provided training and experience along with the opportunity and resources (material, cultural, and factual) for crafting distinctively American practices and theories.[13]

"Most of the State Surveys were in the popular phrase 'finished'" by the early 1840s, according to J. Peter Lesley.[14] Public works projects were increasingly discredited as various states faced financial crises, the most spectacular of which was Pennsylvania, which defaulted on its loans in 1842. The American economy suffered a downturn, and "working geology," Lesley remarked, "received a death-blow."[15] But geologists still needed to work. Here then was the impetus to scientific entrepreneurship.

Possible options such as other survey positions, teaching, lecturing, and writing were limited. Private mining engagements, by contrast, seemed plentiful and profitable. The two decades before the Civil War can be described as the boom period (or at least the first boom) in consulting. Between 1843 and the Panic of 1857 the economy grew rapidly, especially coal mining, and the territory of the United States expanded across the Mississippi River to the Plains. The period also marked a shift away from public expenditure on improvements toward private financing. Instead of governments constructing canals or harbors, private companies began to build railroads or open mines. The rise of private corporations with their attendant stocks, exchanges, and banks (not to mention jobbers, investors, and capitalists) created and coincided with an increasing demand for expertise of all sorts—legal, commercial, and scientific.[16] In a sense, geologists followed the money by switching from public to private surveying.

Survey geologists' qualifications perfectly suited them for engagements as mining engineers. According to Jackson,

> 1st He must be well versed in Geology & practically acquainted with every variety of rock & know how to trace out the connections of one formation with another
>
> 2d He must be perfectly familiar with the nature of minerals, for something more than a knowledge of *ores* is required in tracing out mineral associations
>
> 3d He must be familiar with the Science & practice of chemical analysis so as to be able to examine any mineral he may meet with & determine its nature
>
> 4th He should be well versed in Geometry & Trigonometry as applied to Sur-

veying or dialing a vein or bed & should be practically acquainted with the use of surveying instruments.[17]

The list conformed, interestingly enough, to Jackson's strengths, geology and chemistry, and omitted other skills such as paleontology.

Surveys supplied more than a corps of geologists and body of knowledge; they fostered connections between consultants and their commercial patrons. This cozy relationship was not altogether unexpected because surveys were designed to provide expertise on mineral resources. Contacts made with mine operators, landowners, capitalists, and politicians during public surveys became sources for private engagements afterward.

The New York State geologist James Hall understood, better than most, the politics of survey science.[18] A student of Eaton at Rensselaer, Hall began as an assistant on the New York State Natural History Survey in 1836. The next year he took charge of the western part of the state. In the summer of 1837, Hall allowed Archibald McIntyre, a prominent iron manufacturer and business partner of Governor William Seward, to join him on a search for ore deposits in the Adirondacks. In 1839 Ebenezer Emmons, another New York State geologist, prepared a special study of McIntyre's properties. Emmons, a man of strict religious observance, was troubled by this blatant, though not illegal, favoritism. Hall, however, had few ethical or religious reservations about the need to appease powerful constituents. He willingly used his survey position to gain introductions to politicians and capitalists who might be interested in his services.[19] For more than forty years, from the late 1830s through the 1870s, Hall earned part, sometimes most, of his livelihood as a professional geologist.

The other part of his income came from the state, for Hall had learned how to play politics, too. When the survey officially ended in 1842, he contrived to remain on the payroll as the state paleontologist. (Emmons also extended his employment as state agriculturalist.) Initially, Hall's job was to finish the fossil project already underway, but he dunned legislators into funding thirteen massive volumes on paleontology over the next fifty-odd years. The science was of the highest standard. Hall's peers regarded him as a leader of American geology whose work brought national and international recognition. Even so, Hall's government position was never secure, either politically or financially. He constantly had to lobby, lecture, and even bully the New York legislature.

Besides the necessary income, Hall was quite attuned to the scientific value of consulting. He frequently accepted engagements to visit places not previously surveyed. In 1845, for example, two years prior to Jackson's federal survey, Hall examined the Lake Superior copper regions for some Albany capitalists.[20] In 1865 Hall cobbled together several commissions to explore regions of Minnesota (where the first state survey had not yet begun) and northern Georgia (just recovering from the Civil War).[21] In short, Hall used these engagements as another means to increase his fossil collection (for his New York paleontology) and his understanding of North American geology.

In Pennsylvania, the same commercial-political dynamic was at work. Henry Rogers, the survey director, made contacts with many influential Philadelphians, including Nicholas Biddle, the president of the Bank of the United States (1822–1836). Biddle might even have invested in Rogers's iron furnace, opened in 1838 on Buffalo Creek, just north of Pittsburgh. When the furnace failed in the early 1840s, Rogers suffered a financial blow.[22] He also lost his state funding, and so had to support himself entirely through consulting. Rogers worked mainly for large Pennsylvania coal and anthracite firms, although he did take a commission to study Michigan copper. Rogers's consulting, like Hall's, was a way to make a living and the necessary means to continue his geology. According to Lesley, "the knowledge thus acquired" was of great value to Rogers's eventual publication of the Pennsylvania survey.[23] Moreover, Rogers funneled his consulting fees back to his assistants, like Lesley, whom he hired to help finish the survey. In this case, private and public geology were definitely not in conflict, for without the former Rogers would not have finished the latter.

Despite his commitment to Pennsylvania geology, Rogers lost most of his assistants. Many went to work for companies in the areas they had surveyed, as much for the lucrative offers as because they could not stand Rogers. Thus, in 1839, Peter W. Sheafer (1819–1891) went to Pottsville, the heart of the anthracite region, to pursue a career as a mining engineer.[24] James T. Hodge (1816–1871) likewise became a professional geologist and gained a national reputation by the 1850s. In 1852 Henry Poole left Rogers to become the superintendent of the General Mining Association's coal mines in Pictou, Nova Scotia. And that same year, Lesley, Rogers's protégé, departed in a huff.[25]

Lesley opened a professional office in Philadelphia and by the mid-1850s was the most sought after coal geologist in the country.[26] An excellent draftsman, as well as a perfectionist, his consulting reports were remarkable for their sumptuous detail. He was extraordinarily able at unraveling the structure and stratigraphy of coal fields, skills he clearly advertised on a professional letterhead.

Geology and Topography

Geological and other Maps constructed; Surveys of Coal Lands made; Mineral Deposits examined; Geological Opinions given to guide purchaser, and Reports made to Owners and Agents.

Orders for elaborate Topographical Surveys from Rail-road and other companies will be executed on scientific principles, and in the highest style of art.[27]

Lesley's commercial network spread throughout the Philadelphia and New York City areas, business communities he had gotten to know during his time with Rogers. Lesley usually worked in the Pennsylvania anthracite districts, where he had the added advantage of continued friendship with Rogers's former assistants, most importantly Sheafer. Lesley also had the help of his brother, Joseph, an engineer with the Pennsylvania Railroad, who rose to the position of vice-president. In 1853 and 1854, Lesley himself was employed by the Pennsylvania Railroad (at a salary of $100 a month) to draw maps and explore for coal. In 1856 Lesley was hired, at a salary of $1,000 per year, as secretary of the American Iron Association. Part of his duties entailed compiling a guide to *all* the furnaces, forges, and rolling mills in the United States. Writing *The Iron Manufacturer's Guide* (1859) was the best introduction to companies and capitalists that a professional geologist could hope for.[28] *The Iron Manufacturer's Guide* was also a scientific work, in which Lesley discussed the geology and mineralogy of coal and iron. He always looked to promote *both* science and industry. Much like Silliman Sr., he strove to apply what he studied. Whether as a public surveyor or private consultant, Lesley balanced practical and theoretical work. He epitomized the professional geologist.

A similar sense of obligation guided Jackson's career. Contemporaries praised and criticized him for his emphasis on economic geology. Jackson's interests and aptitude were understandably attractive to politicians, and it is no wonder that he was appointed to direct three state surveys—those of Maine, Rhode Island, and New Hampshire. But Jackson did not make new commercial contacts during these surveys so much as reinforce and extend those he already had. Jackson held very strong opinions about *who* should benefit from his knowledge. In Rhode Island, for example, he made sure to survey the estates of the governor, former governors, justices of the state courts, and members of the General Assembly.[29] While such attention might have appeared as due recompense (and in this regard Jackson seemed to have taken the Nova Scotia kerfuffle with Abraham Gesner to heart), Jackson rejected any accusation of a quid pro quo. His fieldwork merely reflected the obvious fact that the politicians who appointed him happened to be among the largest landowners, nothing unusual for a state like Rhode Island where political office (and the franchise) was restricted to landowners. More importantly, Jackson approved of that class of wealthy New Englanders trying to improve their property. He thought these upstanding gentlemen were the right kind of capitalists.

Jackson, more than most, harped on the moral economy of science. Whether it was the proper behavior of mining engineers or the credit due to original investigators, he knew what the obligations of men of science and the responsibilities of those with whom they dealt should be.[30] But it was such sanctimony that got him into trouble. During his appointment as federal surveyor of the Michigan copper district, Jackson apparently withheld his geology from local landowners, miners, and others whom he regarded as mere speculators. He had no scruples, however, about passing along intelligence to Boston capitalists whom he deemed "legitimate." Such preferential science, understandably, angered more democratically inclined politicians and the Michigan electorate. Jackson subsequently lost his federal position, but not his Boston patrons.

Jackson's position on the copper survey was taken over by Josiah Dwight Whitney (1819–1896), his assistant and in many ways his protégé. Whitney had trained in chemistry in Jackson's private laboratory before going to Europe to study with the French geologist Élie de Beaumont (Jackson's old teacher) and the German chemist Justus Liebig. Upon his return, Whitney rejoined Jackson as an assistant on the New Hampshire survey and then on Michigan's. After he finished the copper survey, Whitney turned professional. For at least six years, between 1850 and 1855, he made his living from consulting. He had more job offers than he could handle and turned down a lower-paying academic post at the Yale scientific school.[31] In 1855 Whitney accepted the not-so-very-lucrative position of state chemist and mineralogist on the Iowa survey; Hall was the official state geologist. This return to government work marked the beginning of the end of Whitney's consulting career. By the late 1850s, the poacher had turned gamekeeper; Whitney began to despise mining. He had always been prejudiced against stock and land speculators, although, like Jackson, he had favored and been favored by Boston capitalists. Whitney's new attitude, however, was harsh and unforgiving. He regarded most mining ventures as swindles and most professionals as mercenaries. This opinion, not surprisingly, put him at odds with many consulting geologists, including his mentor and, most especially, Benjamin Silliman Jr.

Silliman was the exception to the generalization that the professional practice of consulting grew out of government employment. He had no training or experience on state or federal surveys. Instead, he had followed his father's footsteps by launching his consulting practice from an academic base, namely, Yale, where since 1846 he had been the "Professor of Chemistry and the Kindred Sciences as Applied to the Arts." Like his father, he used that reputable institution to back the trustworthiness of his character and the truth of his opinion. He made sure to write his affiliation below his name on all his consulting reports, a custom that became commonplace by the 1850s. The Silliman name carried authority, not least because of the *American Journal of Science,* the oldest and most distinguished scientific forum in the country (Benjamin Silliman Sr. had founded it in 1818). It was often referred to simply as Silliman's *Journal,* and Silliman had been coeditor since 1838. Of course, the Silliman name brought its own connections. Silliman accompanied his father on mining surveys, and it was through his father's network that he made his own commercial contacts. In short, given his name, institutional identity, and family history, Silliman could offer businesses advantages that no other consultant of his generation could match.

He could also supply a few intangibles of his own. His personal charm and optimistic temperament were very appealing. Like Hall, with whom he often worked, Silliman was not burdened by ethics. But whereas Hall was more of a coal canoodler, Silliman was a committed scientific capitalist. He was patently interested in money and willing to engage in ventures, more speculative than most, to get it. To be fair and balanced, Silliman wanted the means for noble ends; money would free him from engagements and let him pursue science. But in the 1850s, he found himself caught in commercial commissions at the cost of his research and possibly his standing in the scientific community.

Engagements

In mid-nineteenth-century America, men of business and men of science made an exchange.[32] In its crude form, consultants "sold" and businessmen "bought" science. In a more refined manner, consultants offered expertise in return for professional fees. In either formulation, the transaction would not have been regarded as inappropriate for it rested on the common cultural assumption that science was useful and therefore should be used.

The business of contacting and contracting with a consultant was conducted largely through correspondence, or at least the evidence of the practice has survived mostly in the form of letters between consultants and their commercial patrons. There were few, if any, formal, legal contracts drawn up between geologists and companies (or their respective lawyers). Letters served as the basis for engagements.[33] And consultants referred to their business affairs as "engagements." The word meant both an occupation and a pledge, an honorable agreement between equal parties. Consultants never used the term *client* either to describe themselves (as in the aristocratic sense of patron and client) or the individuals engaging them (as in the modern sense of professional and client).

Engaging a consultant was straightforward. A capitalist, such as the president or secretary of a company, addressed a short letter to a professional geologist or chemist requesting his services. It was not necessary for the two parties to know each other, although they often did, but when they did not, the capitalist did not forward a letter of introduction. Such anonymity could, on occasion, result in trouble, both financial and ethical, for the consultant. In their letters of request, capitalists did not specify what services they required beyond a desire to have a mine or property examined. "I wish to have a Geological Survey made of some Coal (Cannel) Lands in Cambria Co., Pa." one capitalist wrote to Lesley. "Will you please write me & name the time it would suit you to undertake it, if at all, your terms &c[?]" Lesley said he would do the survey, make a map, and write a report for $300 to $400. This, however, proved too expensive for the landowner.[34]

In rare instances, capitalists made detailed demands. Henry A. DuBois, president of the Virginia Cannel Coal Company, asked Hall to "furnish me with your freely expressed opinion in regard to the merits of our coal as derived from your own personal knowledge . . . stating what you know in regard to the probable extent of the deposit, ease of mining, proximity to the coal river, & the probable extent of the demand for fuel of this character."[35] Such information was included in all consulting reports. DuBois's explicit instructions suggest he was unfamiliar with consulting, and his request that the report be "signed professionally" underscores his unfamiliarity with Hall. DuBois went on to explain that he was trying to negotiate a large loan in London, so perhaps it was the British investors who did not know Hall. In the event, Hall made the survey and wrote a report, for which he received $200; the outcome of DuBois's loan is unknown.[36]

Not all engagements were so well defined. Hall's trip to Lake Superior between 28 July and 21 October 1845 took so long, in part, because he was *not* told where he was going or what he was supposed to do. "I was engaged to go to Lake Superior," he complained, "the plan was a secret only divulged partially by Mr. [Thomas] Olcott & Mr. [Thurlow] Weed just previous to my starting." Olcott was president of the Mechanics' and Farmers' Bank of Albany. With a capital of nearly $500,000, he was the most influential financier in America after Biddle and the Bank of the United States. Weed was publisher of the *Albany Evening Journal.* These Albany capitalists informed Hall, en route, that he was heading to Isle Royale "to examine ground which had been publicly noticed by the [War] department as prohibited from location [i.e., staking claims]." Olcott and Weed were also leaders of the New York Whig Party and promised to use their "influence" with William Marcy, former three-term New York governor (1833–1839) and then secretary of war, "and by covert mode of gaining a footing to secure all the valuable parts of Isle Royale." As it turned out, Hall encountered many "difficulties and prohibitions attending [the] visit and examination," and, in fact, "[I] periled my life and the lives of those with me." Hall's flair for melodrama highlighted the futility of the engagement. "This time was in effect wasted for no possession [of property] can be obtained [on Isle Royale]."[37]

Under ordinary conditions, engaging a consultant meant setting up an arrangement that fell somewhere between a secret rendezvous and a step-by-step instructional. Engagements were somewhat open-ended and more or less informal depending on the degree of trust (by both parties) in the whole transaction. An example of a "standard" request can be found in the brief note sent by William B. Hotchkiss, the secretary of a coal company, to Silliman.

> The St. Clair R[ailroad] & Coal Co. desire your services in making a scientific examination of their coal mines in the vicinity of St. Louis.

Will you give us an opportunity to confer with you on the subject by calling on me [in New York City] at as early a day as is practicable[?][38]

Silliman was "too urgently occupied" and declined the offer; however, he referred Hotchkiss to his friend Lesley. He then notified Lesley, and Hotchkiss subsequently contacted him.[39]

B. Silliman Jr. Esq. in an answer to an invitation to him to make a scientific examination of Coal Lands belonging to the St. Clair R.R. & Coal Company, has given your name, accompanying it with expressions of so favorable a character that we have been induced to communicate with you on the subject.

. . . Are you disposed to accept the invitation sent Mr. Silliman[?] If so, will you give us an opportunity to see you here [New York City], or shall we visit you at Phila.[?][40]

Consultants regularly conducted business meetings in lawyers' offices or in their own homes in order to sort out the details of an engagement. In this instance, Lesley accepted the invitation and visited the coal mine at the end of May 1854, for which he received $200. He then wrote a report for an additional, undisclosed amount, with which Hotchkiss was "well satisfied."[41]

Referrals were thus another way of contracting business. The circle of Silliman, Hall, and Lesley was quite close in this respect. Both Silliman and Hall tended to refer coal companies to Lesley or to request Lesley's assistance because of his acknowledged expertise. "I have been invited to go into Ohio for a few days to look at a coal tract," Hall wrote to Lesley in May 1856, "[I] inquire if you intend going to western Pa. or Ohio during June, and if so we might travel together and make the journey pleasant and advantageous. I would indeed prefer your opinion of the coal tract to my own."[42] A decade later he wrote to another capitalist: "Mr. J. P. Lesley of Philadelphia is, in my opinion the best person for your purposes, if that be to inform yourself of the real nature and value of your property. Mr. L. has had much experience, is a gentleman and a man of honour."[43] In sum, contacting and contracting with a consultant meant becoming involved in a scientific network designed to supply business with expertise and men of science with jobs.

Surveys

Once an engagement was agreed upon, the consultant made preparations to go into the field. As with any fieldwork, engagements could be very physically and mentally taxing; they often involved as much patience, fortitude, and organizational skill as they did scientific expertise. Hall's examination of the Lake Superior copper regions required travel by train, steamboat, coach, canoe, horseback, and foot. Hall not only had to arrange for all this, plus accommodations ("Huff's Hotel in Buffalo if you have to stay over night") and provisions ("two porter bottles full of Brandy" and "a few good yeast cakes, so that we can make a little bread"), he had to purchase all the equipment (hammer, drilling tools, life preserver, toothbrush, and "pistol if you think necessary") as well as organize field assistants (one of whom was his brother William) and other helpers.[44] Only then could Hall actually survey the property. He then had to get everything and everyone back home, where he could write up the report. Lesley, in contrast, tended to work closer to home, which explains in part how he was able to do a great many projects. He could make a quick visit (via train) to the Pennsylvania anthracite districts, where he could stay with friends, like Sheafer, and return to Philadelphia in a few days.

Much of a consultant's time was thus devoted to traveling. By the mid-1850s, the extensive railroad and steamboat network in the East and Midwest facilitated travel to distant locations, and the telegraph helped to speed arrangements for meeting businessmen, often on short notice, in widely scattered cities. Nevertheless, mines and undeveloped property were usually far removed from railroad connections and telegraph offices. Consultants had to keep records of their expenses, which were then charged to the company. Hall kept meticulous accounts, which is not surprising given his penchant for adventure. His trip to Lake Superior, for example, cost $780.52 (an "*exorbitant*" amount, he thought), while his professional fee was only $750.[45]

Once on location, the geologist could spend up to a week going over the property, mostly on foot, and only occasionally, where terrain permitted, on horseback. Like any survey it involved locating outcrops (often with the help of local guides), examining minerals and fossils, collecting samples, identifying rocks and formations, ordering these, tracing them over distances, working out structural features (anticlines, synclines, faults), and making notes, sketches, and diagrams. It was arduous work, done on short order, and so consultants frequently relied on assistants in the field. Hall called on his younger brother William, who later secured a position as an engineer on the Erie Canal. Lesley hired his nephews, Joseph Lesley Jr. and Benjamin Smith Lyman (1835–1920), who became an able consultant in his own right. Collaboration with other geologists was also quite common. Lesley liked to work with the Swiss émigré Leo Lesquereux (1806–1889), whose knowledge of fossils, especially coal plants complemented his own topographical skills. The two often discussed matters of theoretical importance (like the origin of coal) during engagements.

At some point during a survey, the geologist was accompanied by officers of the company, operators of the mine, or owners of the property. (Many letters and telegraph dispatches were dedicated to such arrangements.) These meetings were valuable to both parties. Companies learned the results or progress of the survey such as locations of mineral resources, places to start mining, or where to *stop* prospecting. Geologists got practical information (usually about earlier mining operations or mineral discoveries) or, more generally, an update on the latest developments in the area. There were, of course, meetings that did not go smoothly. Lesley explained.

> [T]he commonest miners from foreign and quite different geological regions . . . suddenly exchanged the character and position of hewers of coal and pumpers of water at home, for the character and position of mining engineers in America. Ignorant, undisciplined, obstinate, narrow minded and superstitious by nature and habit, and rendered presumptuous and dogmatic by their strange advancement, they were as unwilling to accept as they were unable to acquire a correct knowledge of our geology, so different from their own, and hated *professional geologists* because these had never lived in childhood, pick in hand, under ground,—because they taught new things hard to comprehend. . . . The jealousy of professional and "theoretical" interference with traditional and "practical" usages . . . was in 1842 in all its vigor; and is shared by the landed proprietors, the directors of companies and the general superintendents of collieries and mines.[46]

Lesley's generalization was rather rough and stylized in its dramatization of the conflict between truth and ignorance. But it revealed a real struggle for control over the mines. From the miners' perspective, professional geologists were not imparting information so much as stealing it. Coal geology was found under the miner's helmet.

On most occasions, geologists and capitalists were not in conflict. After all, the latter were paying for the expertise. If both parties agreed, the consultant would write a report. Most contained positive evaluations, or at least such useful information as the company or capitalist wanted to see on paper. The report could then be used to persuade potential investors or to convince superintendents (or miners) to change certain practices. In any event, the information in reports was considered by consultants and their business patrons as private. Companies, especially, were forbidden to publish reports without the consultant's approval.

It would be a mistake to assume that all written reports were glowing. Lesley, for example, dutifully explained to William E. S. Baker, the secretary of the Duncannon Iron Company of Pennsylvania, the reasons for his low estimation of a property.

> On careful consideration of my notes I am reluctantly obliged to repeat the opinion which I verbally expressed to you, last week, at Duncannon, respecting the iron ore, opened in the drifts & shafts, upon Cove mountain.

[L]ittle chance there is of the successful issue to any further prosecution of your trial works on the Cove mountain. I believe that *the lower ore of [Formation] XI* is quite worthless all round the inside rim of the Cove.[47]

For this advice Lesley received $208.50: a professional fee of $200 plus $8.50 for expenses in the field and office.[48]

In another case, Lesley wrote a detailed letter, practically the equivalent of a report, although, significantly, he did not get paid for it, in which he explained the value of negative assessments.

I have written out my reasons at some length, because, I take it, you wish an open & candid professional opinion of the state of the case; and because I look upon it as an important function of science—the most important in fact—to expose an error and prevent a mistake, which would cost much money and keep the mind anxious for a long time.[49]

Presumably, such peace of mind was worth Lesley's fee. From the company's perspective, saving money might not be as good as making it, but it was better than losing it down a barren mine.

In general, consultants did not write up reports or pen lengthy letters when the results were negative, in contrast to government geologists who were usually instructed by legislatures to make a *full* and *complete* survey. Consultants gave such assessments in person, and in this way, they saved time and the company cash, for reports were a separate charge from the investigation itself. Companies were always free to ignore any advice, positive or negative, verbal or written.[50]

Reports

Under special circumstances, consulting reports were published. Publication, however, was a delicate matter, to which both parties had to agree beforehand. Companies, understandably, did not want to broadcast valuable, private information to potential competitors, a situation that Hodge discovered when he worked for the McCullock Copper and Gold Mining Company of North Carolina. "I doubt whether the McCullock report will be published," Hodge groused to Hall, "though I should much like it should be. It is one of the most favorable reports I have ever written, but it is too contradictory of some others I suspect to [be] made public."[51] Conversely, consultants did not want to publicize unsound science or speculative judgments. The reports thus embody a fascinating tension between restraint and optimism, precise facts and hedged pronouncements.

Writing for publication was a serious professional undertaking. Consultants tried to be as comprehensive and objective as possible. For some, like Lesley, reports were the equivalent of short geological surveys and were prepared as such, carefully worded with well-labeled maps and vertical sections. When one company tried to publish an edited version of a report without his approval, Lesley fired off a sharp rebuke: "[H]ad you requested me courteously to modify the *form* of my report *for publication,* I should have been happy to do so. The report I sent you was a full recognisance statement for *yourself* & *the parties interested at present with you.*"[52]

If Lesley's report was to be printed, he made sure it was "put correctly through the press."[53] Such tight control was necessary because once a report was released it was free to be used, or misused, by anyone. Other companies and capitalists often edited or reprinted selections from published reports. This also happened with government surveys. Geologists were very attuned to such opportunistic commercial practices.

Reports were addressed to both scientific and commercial audiences. They contained practical advice written in nontechnical language along with detailed descriptions and precise observations. Published either as pamphlets or as part of a company prospectus, usually with incorporation acts and bylaws, reports were designed to attract attention, and because most companies were public joint-stock ventures, a good report could have an immediate impact on the price and sale of stock. It might also impress men of science. Geologists frequently read and cited consulting reports; they thus did double duty by circulating within two related but distinct communities. Among companies and capitalists, reports revealed investment prospects; among men of science, they contributed to geology. The coal geologist Richard Taylor commented on this symbiotic relationship between American science and commerce.

> Among the class of periodicals and occasional documents, not strictly scientific, yet comprising authentic communications of a business character, may be named the numerous reports of companies. . . . We advert to this temporary and commercial literature, because of its remarkable diffusion, its cheapness, its influence, and its employment, in this country, to an extent unknown in any other part of the world. It forms an economical substitute for books of a more expensive and pretending character, and may be found in every man's hand.[54]

Taylor tried to collect all the reports on Pennsylvania coal, but he had to give up. "So many reports have been published relating to the property of individual companies, and to the general interests and characters of this [anthracite] district, that we cannot undertake even to enumerate them."[55]

By the 1850s reports had assumed a standard form. They were ten to twenty pages long, although some were as short as five, while a few, such as Lesley's ninety-nine page report on iron ore for Lyon, Shorb & Company of Pittsburgh, seemed never-ending.[56] Typically, they were comparable to articles in the *American Journal of Science,* but with fewer references and less technical jargon. On the other hand, they contained a wealth of details about time, place, and participants generally excluded from scientific articles. Such information was locally specific and intended for commercial audiences not familiar with the property or company. At the same time, these facts enhanced the knowledgeable and impartial tone of the report. For historians, consulting reports present a trove of insights into how nineteenth-century geologists perceived their practice, their profession, and their discipline. Authors often explained terms, methods, and concepts that otherwise would have remained unarticulated.

A clear illustration of these points can be found in Jackson's consulting report on the Albert mineral. It was published, along with several shorter ones, in the pamphlet, *Reports on the Geological Relations, Chemical Analyses, and Microscopic Examination of the Coal of the Albert Coal Mining Co. situated in Hillsboro, Albert Co., New Brunswick* (1851). Jackson began his twenty-eight-page report, as did most consultants, in an epistolary fashion with "Dear Sir," referring to a director of the Halifax Gas-Light Company, followed by modest declarations of his thoroughness and diligence. Two paragraphs later, he explicitly addressed his other intended audience. "I shall be able to prove to the satisfaction of all scientific men, that the Albert coal is a true bed, included in . . . the true coal formation, as appears . . . from the nature of the numerous fossils, fishes and plants, characteristic of the coal formations."[57]

To justify this classification, Jackson presented six new species of fossil fish. Their significance was plain to other geologists but might have eluded non-trained readers. "It is generally conceded by experienced geologists," Jackson explained, "that the character of the organic remains found . . . between the strata . . . is the most reliable evidence of their geological age, and best determines their position in the scale of rock formations."[58] Jackson went on to relate how he made his discovery: "On the 5th of May, I obtained a slab of shale from the 8th level of the mine, which, upon being split open, disclosed an entire fish most beautifully preserved with its shining armour. It was an entire and perfect Paleoniscus, with its head, fins, and tail well preserved, and its scales shining with silvery brilliancy, though of a delicate brown color."[59]

Jackson named it *Paleoniscus cairnsii,* "after the highly intelligent superintendent of the Albert coal mine, William Cairns, to whose active and unremitting labors I am indebted for so many specimens of these interesting fossils."[60] Actually, Jackson had announced his fish findings at a meeting of the Boston Society of Natural History several months before the publication of his consulting report, but he did not provide such an extensive explanation to the society's members.[61] Significantly, geologists and paleontologists who later referred to these fossils cited Jackson's consulting report, not his announcement at the meeting. The report was even reviewed in the *American Journal of Science,* which was not unusual for reports containing important research on coal.[62]

Nor was it unusual that Jackson concluded his report with vague, yet promising, predictions about the commercial potential of the Albert mineral: "suitable for the production of gas, and flaming fires."[63] He signed off "respectively, your obedient servant" and made sure to identify himself as the "Assayer of the State of Massachusetts, &c. &c." Readers would have been in no doubt as to what the etcetera referred because Jackson placed on the title page a long list of his awards and achievements, the first of which was the grand-sounding French title "Knight of the National Order of the Legion of Honor."

Like all published consulting reports, Jackson's was a well-crafted act of scientific-commercial theater. It was no different than any other piece of self-promoting rhetoric. Designed to present the consultant as both humble and authoritative, busy but focused, thorough yet not entirely at leisure to pursue the requested study in such a manner as required by a more scientific audience, reports assured readers that consultants' conclusions were cautious yet confidant. When done well, consulting reports were assuredly convincing.

Science

They were also factually based, and sometimes argumentative in their scientific conclusions. Jackson's report was just one example of where a commercial engagement produced important findings or contributed to an ongoing debate— the classification and origin of coal. In other examples, a particular location or mine was interesting in and of itself, especially if it had not been studied before or displayed unusual features. Commercial questions thus shaped the discussions, theories, and research agendas of geologists. This is not to say that the geology was determined by a company or industry. Consulting concentrated research on practical problems.

On most engagements, the work was straightforward and focused on local phenomena (identify, count, and measure the thickness of workable coal seams) and on a limited area (a hundred acres or so) or just a single mine. The scientific value of consulting therefore lay not in these individual cases but in their aggregate. Innumerable engagements added up to a lot of fieldwork covering a good deal of ground. This was normal science in a commercial context.[64]

Taylor treated private surveys this way. In his *Statistics of Coal,* he cited numerous reports (including his own) and discussed their detailed observations and conclusions. He frequently compared consultants' work, and especially liked to contrast "the early reports, which were generally made by unscientific persons," with those of scientific geologists "who had acquired experience in unravelling the intricacies of districts of complicated structure."[65] He noted, for example, that the special report to the president of the Chesapeake and Ohio Canal Company (dated 22 October 1839) contained exaggerations about the quality of the Cumberland coal of Maryland. The "true value" of the coal was later established by Silliman Sr. in his consulting report. Likewise, in a report to the Maryland Mining Company, Silliman Sr. corrected a "miscalculation" of coal seams made in a report by an unscientific engineer to the Baltimore and Ohio Canal Company.[66] A basic, yet vital, contribution of consulting geologists was their ability to correct errors of repetition—"counting the same seams two, three, or more times over"—and it was another reason why companies, and governments, needed "true" geologists.[67]

In *The Geology of Pennsylvania*, Rogers also relied on consultants' findings, usually his own or his assistants. On the subject of igneous rocks, for example, Rogers discussed the mineral veins (mostly lead) of Montgomery and Chester counties near Phoenixville. Rogers had surveyed these lands for Charles M. Wheatley, a wealthy landowner, in the early 1850s, and then reprinted large selections of his consulting report.[68] The use of such reports, especially ones dealing with specific mines, was hardly unusual. The state geologist's job was to evaluate the productivity and future prospects of mines, which inevitably entailed collecting private surveys. Consulting reports consequently formed the patchwork of field studies that made up government surveys.

Private mining surveys were at the heart of Whitney's *The Metallic Wealth of the United States* (1854). Solid and statistical, the book dealt with hard-rock mining (gold, silver, mercury, tin, copper, zinc, lead, and iron). The *American Journal of Science* praised Whitney's work as "certainly a most important addition to scientific literature" and "a model of pure scientific style."[69] To modern ears, pure science might be an odd sounding description for a book treating the world's major mines and mining companies. But it reinforces the general theme of symbiosis.

Another way to describe Whitney's work would be as one big consulting report. His purpose was to make a comprehensive and general investigation of "[a]ll of our important mining regions on this side of the Rocky Mountains, and most of the prominent mines."[70] It included geology, mineralogy, mining methods, and Whitney's professional appraisal of certain mineral deposits. It was based on his own private surveys, government surveys, and "reliable published accounts," such as the consulting reports of Jackson (on Michigan copper), Silliman Sr. (on Virginia gold), and Silliman Jr. (on Virginia copper).[71] Whitney also leaned on his friend Hall. "If you have any old mining reports . . . which I can't get at here [Cambridge]," he wrote, "I wish you would make a note of them (when found) & bring [them] down."[72] Hall did, and Whitney compiled them into his book.

A similar plan was embodied in Lesley's *Manual of Coal and its Topography*

(1856), arguably the best example of the symbiotic relationship between the interests of industry and science.[73] The title was deceptive, as many reviewers noted. "[W]e have here an admirable treatise on the geology of the older rocks of North America," the *Journal of the Franklin Institute* observed, "plainly written . . . as to be both intelligible and interesting to the general reader . . . [and] as to give it a very high value as a strictly scientific treatise."[74] The *American Journal of Science* similarly applauded Lesley, but it focused more on the science. "The scope of this work is hardly indicated by its title," it declared. "It does not take up the subject of coal in its economical bearings, but rather in its lithological and topographical relations, as illustrated in the Appalachian Regions, especially in Pennsylvania."[75] The twentieth-century geomorphologist W. M. Davis referred to Lesley's book as "the most important matter of this decade, geologically considered."[76]

As for its commercial aspects, Lesley confided the book's true nature to his friend Hall. "I whisper a secret it is in fact a business affair more than a work of love for science and intend just now to be the vehicle of certain fresh facts which have been too long shut up & will not keep much longer."[77] Lesley's *Manual* was the first general treatise on coal geology. His intention, much like Whitney's, was to provide reliable information to mining interests. And his method, too, was much like Whitney's; collect the facts from his own surveys (both public and private) as well as from other consultants, such as his friends, Sheafer and Lesquereux, although he did not cite the reports. Pull it all together, and the reader had a professional geologist's guidebook to coal.

The practice of consulting was thus very much of a piece with the practice of geology. From small fossil discoveries such as Jackson's *Paleoniscus* to major works like Lesley's and Whitney's, engagements added facts and theories to American geology. On one level, this makes perfect sense. Private surveys were in many ways the same as public ones; new discoveries were as likely to be produced from one as the other. On another level, it is important to emphasize the commercial support of American science, for so much emphasis has been placed on government surveys. That is not say that the two were equivalent, far from it.

Money

Whether their reports were positive or negative, written or verbal, acted upon immediately or ignored entirely, consultants expected and insisted on being paid for their opinions, or what Hall called "services in exploration."[78] Professional fees could easily exceed what a geologist earned from teaching or government surveying, which by the mid-1850s averaged between $1,500 and $2,000 annually.[79] Lesley, for example, received $2,000 for an extensive survey of Cape

Breton coal fields in the summer of 1865.[80] Whitney made $500 a week during the early 1850s.[81] On a more modest scale, James Percival made $2,000 in 1853 for nine months' work for the American Mining Company searching for lead in Illinois and Wisconsin.[82] In general, detailed information about professional fees is sketchy, but some insights can be gleaned from consultants' correspondence about how fees were set and paid and the troubles both consultants and capitalists faced in getting their money's worth.

The first thing to note is the language of money. Consultants typically belittled professional fees in their private letters with such euphemisms as "the needful" or "bread & butter."[83] While the sums involved were far from negligible, the euphemisms suggest a degree of discomfort with the act of taking cash for scientific services. "You need no excuses in my eyes," the Swiss-born geologist Louis Agassiz assured Hall, "for [accepting] any engagement which is not strictly scientific."[84] Indeed, it is not too much to say that professional fees were necessary for survival.

Of all the consultants, the one who seems to have survived best, indeed flourished financially, was Lesley. He supported himself and his family entirely by professional fees for more than twenty years, from the early 1850s until 1872, when he became professor of geology and mining at the University of Pennsylvania. Even then Lesley continued to consult and only gave up the practice when he became director of the Second Geological Survey of Pennsylvania in 1874.[85] During the Civil War, a time of growing mining speculation and rising inflation, Lesley expected to make at least $10,000 a year from consulting.[86] When he was offered the chair in geology and paleontology at Columbia University in 1866, Lesley turned it down,[87] and again in 1868 when he was offered the chair in geology at Cornell, he chose professional over academic geology.[88] In both cases, Lesley thought the time required for teaching would take away from field-work, regardless of the drastic reduction in income.

Not all consultants were as prosperous as Lesley. Hall, the only other prominent consultant for whom extensive billing records exist, did not pocket wads of cash. By his own reckoning, he made only $400 a year between 1844 and 1855. But Hall's accounting is suspicious. The low figure translates into only two engagements annually, while Hall's correspondence reveals many more. He drew up the account of his income in response to questions from the New York State legislature, and given his tempestuous relations with that body, it might not be unreasonable to suspect he underreported his professional fees to demonstrate his need for continual government support.[89]

Professional fees depended on the consultant's status as well as the engagement's duration and importance, which included the prestige of the capitalists or the company and the potential value of the examination. Fees covered the consultant's services (visiting the site, surveying, and giving a verbal opinion) and the expenses for travel and special equipment. Preparing a written report (if requested), or a publishable one, if agreed upon, were additional charges. On the other hand, there were "reports" made *without* doing a survey; basically, this meant professional advice on a site unseen. Lesley charged $100 for what he called a "regular report."[90] If he visited the site, surveyed it, and wrote an unpublished report, Lesley charged $500 plus expenses.[91]

Like many consultants, Lesley's fees were negotiable. The Broad Top Railroad and Coal Company, for example, thought $200 for Lesley's "regular report" was too high, especially because he had the materials at hand. They asked if he would take less.[92] Similarly, in 1864, after a negative appraisal of a location Lesley reduced his normal fee of $500 to $200.[93] At other times Lesley raised his fee, especially if he spent more time than usual on an examination or if he had a tight schedule. Such was the case in the spring of 1865 when Alexander Agassiz asked if he would advise on the Princess Alexandra mine, which was adjacent to the famous Albert mine. Lesley boosted his fee to $2,000 because he was very busy and would have had to drop or postpone a planned trip to the Cape Breton coal fields. If Agassiz waited till summer, Lesley would charge the usual $500.[94]

Part of Lesley's fee had to be paid up-front in order to defer costs for travel, preparations, and provisions. Some capitalists, however, objected to such an "unusual demand." "The imputation inferable, from your demand *in advance*," huffed General A. P. Wilson of Huntingdon, Pennsylvania, "was uncalled for."[95] Lesley did not mean to cast aspersions on the general's character, as he tried to explain.

> [Y]ou are mistaken in thinking my call for $50 in advance *unusual,* on the contrary it is the *usage* & a very good usage, as a little reflexion will convince you! I believe gentlemen of your profession [i.e., law] practise the same usage with entire satisfaction to yourselves and to laymen also. It infers no imputation but relieves the geologist who is never rich of the principal burden of his work.[96]

Actually, Lesley did think Wilson and his partners were "rascals." Such companies were always trying to squeeze consultants for information before paying.[97]

Many consultants adopted a per diem fee schedule rather than flat rates. At the Albert mineral trial, Jackson said he charged Halifax Gas $10 per day for his services, which probably excluded expenses.[98] Hall, likewise, worked for a daily rate, which he explained to a clueless Charles Lyell. "I charge always $20 or $25 per day when I charge professionally for explorations and examinations, and the Report is an additional charge, in accordance with the magnitude or importance of the subject or the value of the scientific opinion given."[99] Lyell, the consummate gentlemanly geologist, had consulted Hall about the Crystal Palace Exhibition in New York City in 1854. As a commissioner for the British government, Lyell was supposed to have examined the minerals, but he did not have time before returning home. So Hall wrote the commission's thirty-two-page report, but he spent four months doing it, traveling back and forth from the exhibition to his home in Albany. Hall thought $1,000 was reasonable compensation for such effort. "[H]ad I been employed by a corporation or by our State government, I should have charged more than the present amount."[100] As a personal favor to Lyell, Hall was willing to accept *only* $1,000, but he had to submit a formal "bill" to the British government. After some prodding from Lyell, it eventually paid the bill.[101]

Payment was at the heart of most disagreements between consultants and capitalists, and sometimes capitalists refused to pay. Dr. McCleod of Cape Breton, for instance, sent back Lesley's "useless" report because he had already sold his property. Lesley denounced McCleod as "unmanly and dishonest," but there was little he could do.[102] Lesley had a similar run-in with the Wood River Mining Company. He responded to its inquiry with a "regular report" on the company's property near St. Louis. He sent this to the directors along with a bill for $100 and a proposal for a thorough on-site survey. The company rejected the proposal, returned the report, and enclosed a copy of a survey by the geologist T. S. Ridgeway as an example to Lesley of what a proper consulting report should look like![103]

Nonpaying companies were just as familiar to Hall. After he finished surveying coal lands in Jefferson County, Ohio, he received only $150 of the $250 he had agreed upon. The owners made weak promises about the remainder, but only *after* the coal mines were established. Two years later, the company was thriving, and Hall demanded the additional $100. He never received it.[104] Such financial losses were part of the business of consulting.

Sometimes, instead of grumbling resignation, consultants took legal action. When the Boston and Pictou Coal Company refused to pay a $400 bill for services and a report, Lesley threatened to sue.[105] And when Bell, Garretson & Company paid only $100 of his $150 fee, Lesley sent his attorney after the company.[106] Its attorney promptly responded: "I cannot comprehend why there existed a necessity of sending the bill for collection to an Attorney, it is certainly out of the usual order [of] business, accounts are not usually sent to Attorneys for collection, until there is at least notice that the money is wanted."[107] Lesley wanted his fee, and he got it.[108]

Besides legal action there were more subtle ways of dealing with difficult capitalists. Silliman and Hall refused to write a report for Stephen F. Headley, a director of a Kentucky coal company, if Headley did not agree to pay their $1,000 fee, which he eventually did.[109] Silliman adopted this technique of withholding reports on several occasions, but Hall was less savvy. When confronted with another case of nonpayment, he asked Whitney for help. In typical histrionic style, Whitney advised:

I have adopted one rule in all such cases:

Pay in advance!

That is the only rule which will work. It may be safely taken for granted that all getters up of mining speculations are

Swindlers!

and must be treated as such.[110]

Whitney railed against the all too common practice, from his perspective, of payment *after* a company was established; he called it being paid "by-the-by." Contingency placed the consultant at the company's mercy. He became a *dependent* partner rather than an *independent* professional. Contingency was the avenue toward undue interest, and most consultants could readily identify such dependency as a source of ethical troubles.[111]

Interests

Objectivity and usefulness, the professional geologist's twin virtues, were certain to be undermined by any undue interest taken in mines, minerals, or lands. According to Jackson, a mining engineer served as the conscience of good business, a moral warrior in the battle against reckless speculation and unscrupulous greed.

> When a man devotes himself to the duties of a Mining Engineer he should consider himself as standing between the Company employing him & the public and ought to hold no interest directly or indirectly in the mine he Reports upon. His pay should always be agreed upon before he commences his work & he should never take any reward beyond his stipulated salary which should be the same [whether] his report be favorable or unfavorable.[112]

Both business and the public should thus heed the expertise and example of the professional geologist. Above all else, the consultant must remain disinterested.

Such pronouncements sit unconformably with Jackson's actions on the Michigan copper survey. He was caught giving aid and advice to "legitimate" mining interests (meaning Boston-based), which not everyone could see as distinctly different from "illegitimate" (meaning Michigan-based) interests. Instead of standing between companies and the public, Jackson sided *with* certain companies *against* the public. Jackson vehemently denied any "unworthy motives" and proclaimed his innocence on all charges of having taken any "interest" in copper firms for whom he consulted. In high dudgeon, Jackson reiterated his guiding principle; when it came to consulting, he held no other object in view "than to advance the interests of science, and to aid in promotion of the *legitimate* mining operations of this country."[113]

From a historical perspective, there are grounds for treating Jackson sympathetically. Rather than read his Michigan dismissal as a tale of rhetoric versus reality, it might be treated as a pitiful and severe experience, which, according to one observer, nearly drove him insane.[114] It was bitterly ironic that Jackson, the most outspoken consultant on ethics, should serve as an example of the interested expert. His behavior was certainly no worse than most and in some ways remarkably restrained.[115]

Hall, for example, was very willing to take part in a Lake Superior copper scheme. "[T]hey intended to give me an interest," was how he described the terms of his covert engagement. "What this interest should be was never mentioned."[116] After he completed his examination, he received 200 shares, valued at $95 per share, in the Pittsburgh Copper Harbor Mining Company, on top of his professional fee and expenses.[117] In addition, Hall took out a lease for himself and one for his brother on one square mile of mineral land. "I find that those in the country exploring have the privilege of making applications for locations," he explained, and so he had, as if it were the only natural thing to do.[118]

And it might have been. Land purchases were often regarded by consultants as nothing more than well-informed investments. According to the Philadelphia-based chemist Frederick Genth, "there is no doubt that *Coal* mining and *Smelting* are the two best speculations which can be made at present."[119] Genth had written to Silliman in April 1854 to congratulate him on the pending success of his "western scheme," the Kentucky Mining and Manufacturing Company, which owned 12,000 acres in the eastern part of the state along the Sandy River.[120] It looked like good coal country to Silliman, and he agreed to become a director and wrote to Hall in June 1854 brimming with enthusiasm. "The time to make our fortunes in coal has arrived. The plans which for many months I have been maturing are now ready for fulfillment, and if you are disposed to join me I think I can show you exactly how it will happen out."[121]

Silliman hoped to raise $100,000 by selling shares; for an investment of $10,000, Hall could have two shares, which he decided to take.[122] The next part of the scheme involved examining the property. Because they were coal lands, Silliman thought Lesley might want a share and so wrote to him, "in confidence."

> Hall & I do not mean to become coal miners, but we feel it is high time we did something for ourselves, while we do so much for others. . . . I believe it is quite worth our while to spend some time & effort & careful thought on a plan which will be if well conceived & carried out the means of giving us plenty of sea room hereafter for pure science. What do *you* say[?][123]

An investment toward research sounded appealing, but Lesley was doubtful and declined to go along. The survey went very well without him; Silliman and Hall found five coal beds, "two of them worthy of special attention."[124] Silliman again tried to interest Lesley. "[Hall and I] are both well aware of the folly of a misstep in such a matter and shall take good care that it does not occur."[125]

They decided to step forward with the scheme, and once more Silliman solicited Lesley. "You know very well that Hall & I would not risk our reputations & our money & that of our friends in any doubtful enterprise and we are still ready to abandon it entirely if it appears on more minute investigation that there is any essential flaw in it."[126] They were planning to sell the company's stock in London, where they knew capitalists who would buy it. Silliman assured Lesley that their names would not "appear in any subscription paper." For Silliman knew that "a man of science loses caste by having his name connected with anything of the sort in a public way."[127] Taking an interest in a company required discretion or, perhaps, secrecy. But a lack of full disclosure could cut both ways. Within a week, "things [had] taken an entirely new and unexpected turn." Silliman and Hall discovered a "serious moral dereliction" on the part of one of the company's directors.[128] The exact nature of this "flagrant outrage against morality" is unknown, but for Silliman and Hall, it was more than sufficient reason to abandon the Kentucky Mining and Manufacturing Company.[129] "Let us never say or think any thing more of [this] scheme," Silliman counseled Hall, and then he went on unabashed. "I know of no mode in which I can be so likely to make you amends as to tell you of something now in hand which has the merit of present realization to it." Silliman had another coal scheme in the works, this time in Ohio.[130] "I am not the less determined than I was 2 years ago to make mother Earth give me a competent support," Silliman wrote to Hall in 1856. He was now negotiating to take hold of $50,000 worth of stock in the Ohio Diamond Coal Company. "If we can make it all right, would you like to join our party in the sum of say $5000 or $10,000[?]" Charles Wheatley, the Pennsylvania land developer, along with some "New Haven men," were interested. And so, Silliman confided to Hall, "there was no hocus pocus about it."[131]

Magic was an apt metaphor. Some consultants were keen on transmuting their science into gold through judicious purchases of lands and mines. Hall, more than Silliman, was the master of such manipulations. He had begun looking into coal lands as early as 1837, his first year on the New York survey. With the encouragement of his friend and Rensselaer schoolmate Caleb Briggs (1812–1884), Hall had become interested in Ohio property. Briggs was an assistant geologist on the Ohio survey under the direction of William W. Mather (1804–1859), and he thought he could buy land at a low price. "I can enter lands at the government price [$1.25 per acre in cash] which will double in value for several years. The lands to which I have reference embrace valuable deposits of coal & iron ore. Investments made in them can not fail to be successful." Briggs could not make purchases in his own name, unless he were "disconnected with the Survey." Nor did he have the money. He wanted Hall "to induce some Capitalist to embark with me in the purchase of government lands, with a stipulation that I shall receive a certain share of the profits equivalent for my local knowledge."[132] Briggs suggested the seemingly ubiquitous Archibald McIntyre. Hall, however, found a "monied man" in Philadelphia who commissioned *him* to go into Ohio in early 1838 and select the lands, on his "judgement and prudence." For his effort and expertise, Hall received a quarter, undivided, of the lands and the warm appreciation of a "profitable speculation."[133]

Briggs had to wait for his share until he left the survey, which was not long. The Ohio legislature learned of the land speculations, despite Briggs's best efforts to keep the deals secret, and called Mather and the rest of the geological corps to testify. Mather admitted that 700 acres had been purchased, 500 of which belonged to him as a residence, but the land's mineral resources were unknown and undeveloped, thus the geologists were innocent of any charges of using supposedly public information for private gain. The legislature disagreed and discontinued the survey in 1838.[134] Hall, Mather, and Briggs, and possibly the other assistant Charles W. Whittlesey (1808–1886), kept their land shares, which contrary to Mather's testimony amounted to more than 2,000 acres in Hall's case.[135] Mather later formed his own company, the Ohio Iron Manufacturing Company, based on the iron beds on his property. Briggs became a director of the Ohio Iron and Coal Company in 1849.[136]

As the Michigan and Ohio episodes reveal, survey geologists had a unique advantage when it came to land deals. Some regarded their positions as golden opportunities; others saw a corrupt spoils system. With any land deal, geologists ran the risk of being dismissed or denounced. Nevertheless, they appear to have been enormously skillful at *not* getting caught in compromising positions or even appearing to compromise the public's trust. It was only in 1853 that Missouri wrote into its organic act for a state survey the first explicit prohibition against geologists partaking of "pecuniary speculations."[137] The following year the Kentucky legislature wrote an even more detailed proscription.

> The principal geologist and each of his assistants, before entering upon the duties of their offices, shall take an oath faithfully to perform all the services required of them under this act, and to abstain from all pecuniary speculations during their progress, and that they will not conceal any valuable discovery or information from the owner or owners of the land on which such discovery is made.[138]

The point about concealment could easily have been inserted in response to Jackson's behavior in Michigan. The clause about speculations might have targeted Silliman. The Kentucky Mining and Manufacturing Company had been organized after Silliman was appointed by the governor as a special commissioner to investigate the company's coal.

The other partner in the Kentucky coal scheme, of course, was the wily Hall. In the early 1850s, after he had pocketed the profits (between $10,000 and $15,000) from the sale of his Ohio property, Hall had been nosing around for another good deal.[139] "I have been intending to go into Iowa to 'locate' some coal lands," Hall's former student John Strong Newberry (1822–1892) wrote in April 1854, "and should delight to have your company."[140] Hall could not go (he went with Silliman to Kentucky instead), so Newberry ended up "prospecting" by himself in the Ohio coal regions.[141] "I have been indulging in a little speculation in mineral lands which seems to promise very well," Newberry later confided to Hall, and "I hope to acquire some interesting geological facts and something handsome in the way of the 'needful.'"[142]

Hall was not put out for long. In May 1854 the chemist Eben Horsford (1818–1893), Rumford Professor of the Application of Science to the Useful Arts (1846–1863) at Harvard and another close friend from Rensselaer days, wrote to tell him "of an opportunity where I think you may be certain of 20 percent." Horsford knew some capitalists willing to invest $30,000 in the West Columbia Mining and Manufacturing Company of Virginia (across the Ohio River from Pomeroy, Ohio).[143] Hall, unfortunately, missed out; the stock was bought up before he could get some. Horsford did set him up with a consulting engagement in western Pennsylvania, from which Hall made $950.[144] But it was Virginia wine that was on Hall's mind. A few months earlier, he had toyed with a fantastic land scheme (10,000 acres and $20,000 to $30,000 capital). The company would produce coal ("the main object of the Enterprise eventually"), but in the meantime it would sell lumber and "Sparkling Catawba" at $1 a bottle. The plan was to "select" German families from the Rhine region, "the most famous wine producing section," and settle them on Virginia vineyards.[145]

Besides lands and shares, geologists sometimes took an interest in the management positions offered by the companies for whom they consulted. Joseph Lesley wrote to inform, and reassure, his brother Peter about an offer he had just received from a Kentucky coal company. "I have concluded to accept the Presidency of the Company (they wanting my name & not my time) and will see the thing started."[146] Such a stake might seem to be a clear conflict of interest and a threat to disinterestedness, which was why Jackson, for one, opposed such positions. On the other hand, what better endorsement for a mining company than having a geologist as president? Conversely, what better endorsement of a consultant than a position with a profitable company? Many reputable companies and upstanding capitalists wanted the name of a distinguished man of science on their board of directors.

Silliman seemed especially willing to oblige. During the mid-1850s, he was the president of no less than three different companies: the New Haven Gas-Light Company, the Pennsylvania Rock Oil Company, and the Bristol Mining Company of Connecticut. Eliphalet Nott (1773–1866), the president of Union College in Schenectady, New York, was the sole owner of the Bristol copper mine. He had consulted Silliman Sr. back in 1839, but in 1855 Nott wanted to convert the mine into a joint-stock company. He engaged Silliman Jr. and Whitney to examine his mine and write a report. Silliman and Whitney presented a very favorable prospect noting that the ore in sight would be worth $854,000. Silliman was elected company president, and Silliman Sr. became a shareholder.[147] Whitney did not take a company office, although he may have received stock. He was certainly interested in lands and mines; after all, he supported Hall's Sparkling Catawba scheme.

Other consultants actually established their own companies. Rogers started the iron furnace on Buffalo Creek. Lesley became a director of a Pennsylvania iron and coal company. But he, like Rogers, went broke. Bankruptcy, in fact, was a distressingly common occurrence among professional geologists. Leo Lesquereux lost all his money in mining, which served as a source of commiseration with his friend Lesley. In the end, the only ones who seemed to have made any money were Mather and Hall, and they did so by selling mineral lands.[148]

"Mining at the best is a lottery," warned Hodge, "& science, I fear, will never reach that point to judge with safety of the value of a mineral vein, any more than foretell what weather will be next week."[149] Hodge had just lost more than $1,000 ("a large interest") when his stock in an iron mine and furnace located in Ulster County, New York, had plummeted from $2.50 to $1.50 a share. "I have been obliged consequently," Hodge moaned to Hall, "to seriously reduce my interest at the worst time." Hodge wanted to know if Hall would examine the mine in order "to avail myself of the best advice I could obtain, as well as for the interest of the Comp[an]y, as my own."[150] Hall agreed and wrote a report for $100, but it seemed to do little good for the company.

Bristol Mining Company stock certificate, owned by Benjamin Silliman Sr. and signed by Benjamin Silliman Jr., as president. *Source:* Benjamin Silliman Jr. Mss, Smithsonian Institution, National Museum of American History, Washington, DC.

Hodge decided to go on a ten-week tour of the gold mining regions of North Carolina. "I met Dr. Jackson out there," he reported to Hall, somewhat surprised. Jackson was telling everyone how Whitney and John W. Foster had connived in his dismissal from the Michigan survey. Jackson was also getting "a salary of $500 & expenses, & if underlet by his employers, he receives $10 more per day [above] the *$30* charged!"[151] Jackson's dismissal had obviously not hurt his earning power.

For all the interest that consultants took in companies, it is important to emphasize that it did not work the other way round. Consultants never allowed companies or capitalists to take an interest in them, especially in their published reports. That smacked of interference. In a letter to a particularly difficult capitalist, Lesley made the point clearly.

> I sent you a recognisance report, which was what you ordered; carefully written, elaborately illustrated & as truthful as it could be made. I permit, of course, no interference with my methods. I guarantee the correctness of my work, & its completeness & intelligibility. I know nothing about your financial interests, & care nothing about your jealousies & rivalries, if you have any. I acted in good faith with you as I do with every body else.[152]

Lesley had been asked to make his report more positive, and he would not do so.

Whitney, too, bridled at outside pressure. There was nothing more important to him than "no interference." He complained to Hall that "[s]ome of the big boys at Wash[n], owners of mining stock on Lake Sup[r]," were willing to buy copies of his *Metallic Wealth* "by the hundreds," if he would only "crack up" or praise their mines. It was enough to "disgust" him.[153]

Consultants worked diligently to maintain their credibility in the face of what they took to be continual efforts on the part of capitalists to exploit them. Reports tailored to the wishes of a company (as Jackson had been accused of doing for Halifax Gas), or even the perception of such alterations, could unravel a consultant's reputation. What was worse, that kind of control meant corrupting science in pursuit of the almighty dollar. But was there a meaningful difference between a consultant taking stock in a company and a company taking stock in a consultant? Yes. Professional geologists believed their science was objective and accurate, so when they bought mineral lands or invested in companies they were acting on that belief. They did not compromise the science. Their behavior endorsed their trustworthiness.

The facility with which the public allows itself to be deceived, in regard to everything connected with mining is . . . remarkable, [and] the machinery [of] . . . the swindling speculation is . . . simple. The locality is selected, and visited by some very distinguished scientific geologist, who for a sufficient consideration will write a sufficiently flattering report, and demonstrate the absolute certainty of the success. The value of the mine is fixed at an enormous sum, and divided into one or even two hundred thousand shares; the company is organized, and the stock brought into the market. Every means possible is then taken to inflate its value. . . . As soon as a sufficient quantity of the stock had been thus disposed of, and the getters-up of the scheme have pocketed the proceeds of their skilful manoeuvering, the natural results follow: . . . The property which a few days before was quoted at hundreds of thousands can now hardly be given away; the unfortunate victims having nothing left as the tangible evidence of the brilliant dividends promised but the elegantly engraved stock certificates, and the equally valuable reports by which they were deluded.

—J. D. Whitney, *The Metallic Wealth of the United States* (1854)

MID-NINETEENTH-CENTURY Americans saw corruption everywhere—politics, business, journalism—and there was good deal of it around.[154] As Whitney made clear, mining frauds were abundant and apparently easy. The public was gullible and greedy. Companies were cunning and greedy. And the "very distinguished scientific geologists" were especially devious and greedy.

This characterization of swindling was too simplistic. Jackson, from whom Whitney had learned the swindling refrain, was more sensitive to its intricacies. "It is difficult," Jackson observed, "to *prove* fraudulent intentions though they may be fairly informed." There were certainly "cases where men operate with imaginary mines and extract money from the pockets of deluded people by the sale of fancy stocks as they are called." And these "downright frauds" should be "punished by law as Swindling." But even "the best mining prospects" can fail through no fault of their "honest & industrious adventurers." Moreover, the obsession with stock prices, their rapid inflation and their subsequent crash, was something akin to a national foible. "[O]ur adventurers," Jackson explained, "are too impatient." "[T]he American rushes into the market with his stocks and operates mostly on Change." The error, as Jackson called it, was this: Americans were disposed to make "a Capital from an income instead of an income from a Capital."[155]

Americans did seem more interested in money than in mines, but this did not necessarily make them dishonest. Distinguishing hard luck from downright fraud, or hastiness from cheating, could be very difficult, which was precisely why men of science had to be constantly on guard. When Silliman assured Hall of the absence of any "hocus pocus" in a land scheme, he spoke directly to the possible deceptions perpetrated by overly enthusiastic capitalists. The greatest danger of all to a consultant was to be the unwitting tool of a fancy stock swindle. That is why Jackson advocated fixed payments for stated services, and why Whitney warned Hall against contingency fees. It was also the reason why consulting reports (whether published or not) were often guarded in their language and predictions.

Companies were likewise wary of consultants. They too might be deceptive, impatient, or incompetent. Companies could easily be disgusted by "very distinguished scientific geologists" who might try to pass off "interested" reports. The St. Clair Railroad and Coal Company, the one that Silliman had referred to Lesley, sent a copy of Lesley's report *to* Silliman for his endorsement. Of course, he gave it "cheerfully."[156] Nonetheless, the company's actions revealed a greater trust in Silliman (a Yale chemist) than in Lesley (a professional geologist). In a similar vein, Hall received one of Taylor's consulting reports on Pennsylvania anthracite, and he too was asked for an endorsement. "I have too much regard for the labors of such a man," Hall responded, "to report against his views."[157] This was especially true when Hall had not examined the property. In another case, Hall received one of his own reports!

> You will please do me the favor to examine this report and if *yours* and correct, give it your approval as indorsement [*sic*].
>
> In times like these, one cannot observe too much caution relative to reports or statements put forth under the signature and sustained by the reputation of distinguished men.[158]

Unsound advice could prove costly, not only in money, but in public confidence. Honest companies and upstanding capitalists did not want a reputation for passing biased reports.

It was in the interest of men of business and men of science and the public to preserve the disinterestedness of consultants. But disinterestedness was not easy to maintain. "There are people," Louis Agassiz observed in 1855, "whose motives must be habitually improper that they cannot admit of the possibility of disinterestedness." Agassiz was complaining to Hall about Americans' tendency to see corruption everywhere. He himself had been accused of backing Joseph Henry and the Smithsonian Institution *because* Henry had paid him to lecture there. The implication of a quid pro quo was insulting. "I have never seen any thing so mean," snorted Agassiz, "as the weapons used in discussions in this country."[159]

Most consultants, and most companies and capitalists, had as much experience with pointed accusations as with sharp dealing. Jackson thought "[t]he evil is one to be cured by an improvement of the morals of the community."[160] Barring such an unlikely event, the next best thing was to encourage "legitimate mining." In theory, this meant reputable consultants working for reputable companies. In practice, it meant mining enterprises exercising caution commensurate with their willingness to tolerate risk, whether that was in the form of profits lost or reputations soiled. Consulting geologists lowered the risk of speculation and thereby increased the likelihood of return on investment. Consultants could not guarantee profits, as Hodge knew very well, although Jackson remained hopeful. "[M]ining is not so certain an art as to preclude losses," he said, "though it is not altogether conjectural in its estimates."[161]

Risk management was a positive feedback mechanism of the consultant-capitalist relation. That most prominent consultants knew one another (and many capitalists) helped to keep the practice within certain bounds, the moral latitudes set by a pattern of behavior. As a collection of individual engagements, consulting amounted to agreements among men of science and men of business as to what constituted acceptable conduct. But it was impossible to dictate a specific code of ethics. As is often the case, the boundaries of unarticulated norms were only discovered after they had been, or were perceived to have been, transgressed.[162] The most egregious transgression was, obviously, swindling.

It might be possible, on the other hand, to argue that familiarity of the sort that led to clandestine explorations and secret land deals could just as easily have bred swindling. But it does not seem to have done so. Hall did not like being kept in the dark by the Albany capitalists, Olcott and Weed, and Briggs and his fellow Ohio geologists did not think they had compromised the public's trust. By the mid-1850s, descriptions of proper behavior for survey geologists were beginning to find their way into legislative acts. For professional geologists, correspondence seemed to serve a similar function. That engagements were down on paper (although not yet formalized as contracts) helped consultants and capitalists keep the transactions business-like. Many consultants, like companies and capitalists, had their own attorneys.

Legal actions were rare, however. Not just because of social contacts or paper contracts, but because of trust. Companies trusted consultants to give their disinterested, scientific opinion of a prospect. At the same time, consultants trusted companies to be honest in their dealings. All getters-up of mining speculations were *not* swindlers. Nor were they trying to interfere with consultants and their work. Trust was a delicate and precious commodity, to be sure, especially because there were no guidelines or rules governing the practice of consulting. Swindling was most egregious because it broke the bonds of trust.

But trust could be restored, perhaps even made unbreakable, by science. Science was the most powerful antidote to swindling. It was the guarantee that a consultant's advice was open, honest, and objective. Companies could make decisions based on such sound advice, and the public could rely on these scientific opinions. To say that consulting was scientific meant more than the obvious fact that consultants were men of science. Consulting meant doing science, just under commercial sponsorship. Science was science, no matter where it was done. That, at least, was the argument for why consulting was good for science and good for business.[163] Trustworthy consultants—honest and scientific—could thwart fraudulent stock and land schemes. As Taylor explained, "the test of science restores all things to true value."[164]

The practice of consulting rested on the widespread support for science in American culture. For the most part, the reality of consultant and capitalist relations was harmonious and based on a shared optimism about the possibility of useful knowledge and a sense of mutual moral responsibility. Such was the ideal of respectable consulting: to act professionally and at the same time to prosper from it. In this way, the interests of industry and the disinterestedness of science came together in a symbiotic fashion in mid-nineteenth-century America in the form of consulting. And that is why charges of fraud involving a man of science could imperil not just the individual but the community of American science.

PART 2
KEROSENE

CHAPTER 5

The Technological Science of Kerosene

The business of manufacturing coal oils and the various products of the distillation of coals, is one of great magnitude in this country, as well as in other parts of the world. Although, comparatively speaking, it is a new branch of enterprise, much capital has been invested in it, and we regret to say, many have unwisely entered into the business in entire ignorance of its first principles, and thus large sums of money have been irretrievably lost, and their unfortunate owners ruined. This work of Dr. Gesner's is a valuable contribution to technological science.

—*American Gas-Light Journal*, 15 December 1860

IN A REVIEW OF Abraham Gesner's latest book, *Practical Treatise on Coal, Petroleum, and Other Distilled Oils* (1860),[1] the use of the term "technological science" was both apt and revealing, for it described Gesner's ability to explain the chemistry and geology of coal as well as the processes for manufacturing coal oil. Gesner had firsthand experience of this intricate working relationship; he was a founder of the North American Kerosene and Gas-Light Company, the consulting chemist at the oil works, and the patentee of Kerosene, the most popular coal oil in America in 1860. Gesner's book thus offered an insider's account of a rapidly growing business and an autobiography of a successful scientific entrepreneur.

Coal oil was a new midcentury industry relying on the knowledge of men of science and the know-how of practical men. It exemplified the raveling of science and technology. Historians, however, have struggled for years to disentangle the two, and in ingenious ways they have attempted to define or distinguish the one from the other. At the same time, they have also argued against trying to fix such categories. Technology and science need to be treated together in their particular times and places.[2] A study of coal oil can shed light on this enduring scholarly debate by pursuing the multiple meanings of "technological science" in mid-nineteenth-century America.

As the first manufactured mineral oil, "oil-from-coal" was both a new commercial product and a well-tested result of ongoing scientific research. Geologists and chemists were actively involved in locating and evaluating the best kinds of coal, in developing techniques for distilling and refining oils, and in testing their quality and safety. Men of science helped build competitors to Gesner's Kerosene Oil Company and, in the process, extended the practice of consulting. But they were not the only coal oil experts. Companies also relied on "practical men," most often chemists or engineers with experience in gasworks or chemical firms, who for specific reasons (usually lack of publication and research) did not have reputations within the self-defined scientific community. Still, other businesses took to (or were taken in by) less-than-honest and not-so-knowledgeable "coal-oil men." These charlatans exploited the very newness of coal oil and the unfamiliarity of its technological science. In effect, a marketplace of expertise sprang up alongside the burgeoning business in coal oils.

By the time Gesner published his book, coal oil manufacturing and marketing were national in scale and immensely profitable.[3] Coal oils replaced whale oil (and all other animal and vegetable oils) as the dominant lamp oil in America. They also served as durable and reliable lubricants that literally greased the wheels (and other metal parts) of American mechanization. How coal oils "increased with such rapidity" was a subject *Scientific American* thought worthwhile of investigation, for "their development appears to be something like a phenomenon."[4] Others referred to the progress and promise of coal oils as a "mania," "fever," or "El Dorado."[5] Taking a glance at this Kerosene boom is one way of tracing the cutting edge of midcentury technological science.

Inventing Kerosene

The supreme courts of Nova Scotia and New Brunswick had quashed Gesner's commercial aspirations in British North America. So he decided to take his plans and patents for Kerosene gas and move to New York City, where he arrived with his wife and five sons in early 1853.[6] Besides its familiarity, the city had an attractive market in gas lighting. Philadelphia, the other city Gesner visited in 1850, had a gasworks that was owned and managed by the city (a municipal monopoly much like that in Halifax). In the New York City area, private joint-stock companies supplied the gas.[7] Of these, the Manhattan Gas-Light Company was by far the biggest; by 1860 it was the fourth largest in the world.[8] Still, there was some chance of starting a gas company in the growing communities across the

East River on Long Island. Gesner and his financial backers decided to introduce Kerosene gas there.[9]

In March 1853 Horatio Eagle, a young partner in the firm Eagle & Hazard, ship's agents and brokers, issued an eight-page prospectus entitled *Project for the Formation of a Company to Work the Combined Patents (for the State of New York) of Dr. Abraham Gesner, of Halifax, N.S., and the Right Hon. The Earl of Dundonald, of Middlesex, England.* That Gesner wrote the prospectus is clear from the repetition of language taken directly from his lecture before the Academy of Natural Sciences in 1850. The only difference between the prospectus and the lecture was the raw material to be used in the gasworks. Because Gesner no longer had access to the Albert mineral, he praised the qualities of an "entirely new" substance he called "Asphalte Rock." Like its famous cousin, this "Rock" was found in "inexhaustible quantities" in New Brunswick, probably a mine in Dorchester, located across the Petitcodiac River from Albert Mines. Arrangements were being made to insure a constant supply of "Asphalte Rock."[10]

Gesner estimated that one ton of "Asphalte Rock" would furnish 15 gallons of naphtha (for paints and india rubber manufacture), 5 gallons of railway grease, 880 pounds of hydraulic concrete (for water proofing and paving), 200 pounds of mineral pitch (for varnishing and caulking), and undetermined amounts of coke, gas, paraffin for candles, and ammonia ashes for fertilizer. The "Rock" could also be distilled to produce 15 gallons of "Kerosine, or Burning Fluid." This was the first mention of any lamp oil, although Gesner did not elaborate.[11] According to his calculations, the profits from this new venture would be reassuring to any investor. A ton of "Asphalte Rock" cost only $5.00; products amounted to $43.00.[12] Such financial projections and long lists of wonderful products were common to all business prospectuses. Gesner was following a well-rehearsed script.

In two respects, however, the prospectus was distinctive. First, it announced that Gesner's "services" as a chemist were "secured to the company for a term of years, at a moderate salary."[13] Here was both a variation on Gesner's old relationship with Dundonald and a new twist in the science-technology relation— an in-house, salaried man of science. Second, the prospectus contained a short one-page biography of Gesner. Rarely were such biographies part of prospectuses, for companies, as a general rule, relied on well-known men of science. The biography thus served to instill confidence in potential investors that Gesner, although newly arrived from Nova Scotia, was competent to take charge of the New York operations.

The new company was called the Asphalt Mining and Kerosene Gas Company, a small joint-stock venture capitalized at $100,000 and divided into 1,000 shares of $100 each. The stock sold well enough such that by April 1853 the company had purchased, for $17,500, a seven-acre tract on the east bank of New-town Creek in Brooklyn and began construction of a manufactory.[14] By March the company was reorganized; the North American Kerosene and Gas-Light Company would be the legal name under which it would operate for the rest of the decade, even after it had abandoned all interest in gas.

The shift to oil came in June 1854, when Gesner received three patents for three oils.[15] The most important one was "a new liquid hydrocarbon," which he dubbed "Kerosene," the same name he coined in 1850 for the gas manufactured from asphaltum (and later the Albert mineral).[16] Kerosene was now an oil, essentially a liquid candle. Gesner assigned his patents to the company for $12,000, although he probably received stock instead of cash, which was needed to purchase new equipment and land.[17] Gesner's patents thus formed the basis on which the Kerosene Oil Company (as it was commonly known) would operate. They were also the opening act in what became known as the "coal oil controversy," a legal contest between rival producers over the ownership of the first coal oil patent.

There were three steps to manufacturing Kerosene oil: choosing a raw material, distilling it, and refining the distillate. In his patent, Gesner listed many raw materials—"petroleum, maltha, or soft mineral pitch, asphaltum, or bitumen, wherever found"—but not coal. Two years earlier, in March 1852, James Young (1811–1883), a Scottish chemist and entrepreneur, had received an American patent for the manufacture of "paraffine oils" distilled from "bituminous coals."[18] The missing coal in Gesner's patent suggests that he or his lawyers were well aware of Young's rights and tried to avoid infringing them.[19]

Whatever material was chosen, it had to be loaded into a retort and heated. Gesner did not lay claim to any invention regarding retort design, as he had in his gas patent, and he even blasted those "tyros in the art" who fancied wasteful novelties.[20] (He used a large, horizontal, revolving retort because it heated the raw material evenly and thoroughly.) Most of the equipment (retorts, worms, stills, and washers) and the chemicals used in refining coal oils were commercially available. The next step, distilling the raw material, was the crucial one. "There is, perhaps, no question of so much moment to the manufacturer of photogenic oils," declared the chemist Thomas Antisell, who, it will be recalled, had testified as an expert witness for Gesner during the Albert mineral trial and, subsequently, had become a chemical examiner in the U.S. Patent Office and the author of *The Manufacture of Photogenic or Hydro-Carbon Oils* (1859). "[T]he problem," Antisell explained, "[was] how to obtain, from a given weight of bituminous mineral, all of the volatile and heavy oils which it is susceptible of yielding under the most suitable application of heat."[21] Gesner conducted more than 2,000 experiments trying to solve the problem.[22] He and other chemists discovered that the retort had to be heated steadily and then held at the lowest possible temperature at which the raw material began to volatilize. Rapid heating or fluctuating temperatures resulted in different products being evolved. Too high temperatures, for example, produced illuminating gas, a common mistake, according to Gesner, which explained why so many gasworks went bankrupt in trying to move into coal oils. In his patents, Gesner specified that "the heat must not in any case be raised above 800° Fahrenheit." In Young's patent, it was described as "low red heat." How specific a patentee needed to be would become a much disputed point in the coal oil controversy.

Once the raw material began to volatilize, the gases were drawn off through a worm and condensed into a liquid usually called "crude coal-oil" or simply "crude oil," an effective reminder of its offensive odor, muddy color, and smoky combustion.[23] Crude oil needed to be refined by treating it with acids and alkalies and then washing it with water. The object was to remove the smell and change the color to white or lemon yellow. In his patents, Gesner specified a process for redistilling crude coal oil into three fractions. To distinguish them by their specific gravity, boiling point, and inflammability, Gesner labeled them Kerosene A, B, and C. Kerosene C, he believed, offered the best commercial possibilities. It burned "with a brilliant white light [and] without smoke or the naphthalous odor so offensive in many hydrocarbons having some resemblance to this but possessing very different properties." It was also "very good as a lubricant for machinery" and "a good solvent of gums."[24] Kerosene C was going to be *the* marketable product of the Kerosene Oil Company.

To manufacture it, Gesner designed and supervised the construction of the oil works. The technical drawings in his *Practical Treatise* were likely copies of the company's plans, and they remain to this day the most detailed diagrams of a coal oil manufactory. By the summer of 1854, the Kerosene Oil Company was in operation.

The public's willingness to buy Kerosene was another matter. The company's agents, John H. Austen and George W. Austen, had difficulty, initially, selling it because of the strong resistance put up by manufacturers of camphene and burning fluid, the most inexpensive lamp oils at the time. In addition, Kerosene had an offensive odor, despite the claims in Gesner's original patents. Gesner subsequently patented methods for further refining Kerosene, which apparently removed it.[25] But the most serious drawback to Kerosene was the lack of an appropriate lamp. John Austen eventually solved this problem with a model he found in Vienna. Designed for light oils, the "Vienna burner" had a flat wick, glass chimney, and, most importantly, an affordable price. In it, Kerosene burned cleanly, without smoke or odor, and soon the burner was marketed as the "Kerosene lamp."[26]

The Kerosene Oil Company could now offer a complete package—a safe, odorless lamp oil that burned with a steady, bright light in a Kerosene lamp, all

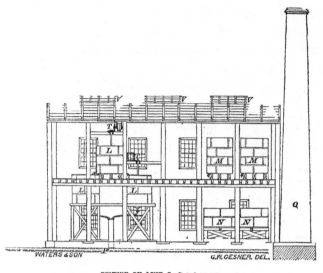

SECTION ON LINE C—D OF PLAN.

Coal oil plant. Based on the plans for Abraham Gesner's Kerosene oil works in Brooklyn, New York. *Source:* Abraham Gesner, *Practical Treatise on Coal, Petroleum, and Other Distilled Oils* (New York: Baillière Brothers, 1860), 110, 111.

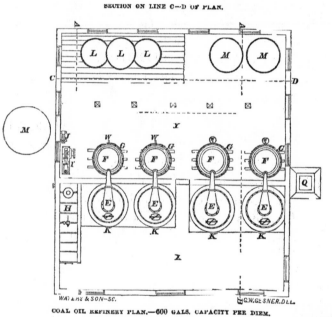

COAL OIL REFINERY PLAN.—600 GALS. CAPACITY PER DIEM.

SECTION ON BROKEN LINE A—B OF PLAN.

REFERENCES.

E. Stills.	P. Drain.
F. Worms.	Q. Chimney.
G. Worm tanks.	R. Water pipe.
H. Boiler.	S. Steam pipe.
I. Engine.	T. Washer gearing.
J. Steam pump.	U. Pipe from agitators to stills.
K. Still furnace.	V. Ventilators.
L. Washers, or agitators.	W. Tail pipes.
M. Receivers.	X. Still house.
N. Market tank.	Y. Refinery.
O. Syphon of still pipe.	

at a competitive price. In the fall of 1857, Gesner presented his Kerosene at the twenty-ninth annual fair of the American Institute in New York City. It was received as one of the most "important novelties" at the fair.

Kerosene's Competitors

As Gesner's Kerosene gained a commercial foothold, it attracted the attention of other oil entrepreneurs. Two companies in particular rivaled Gesner's: the Breckenridge Company of Cloverport, Kentucky, and the Downer Company of Boston.

The Breckenridge Cannel Coal Company was formed in February 1854 when the Kentucky legislature granted its incorporators a charter to set up a joint-stock venture to mine coal. The company owned 7,500 acres in Breckenridge and Hancock counties in central Kentucky along the Ohio River, about sixty miles west of Louisville. The legislature stipulated that an appraisal of the lands and an evaluation of the coal were to be made by "three discreet and disinterested persons." Governor Lazarus Powell chose to repose "especial trust and confidence in the integrity, diligence, and ability" of Benjamin Silliman Jr.[27]

Silliman was teaching in the Medical Department at the University of Louisville. He had left Yale in 1849, after much soul-searching, although he would return to Yale later in 1854. Silliman surveyed the company's land, analyzed the coal, and reported to the governor on 9 February 1854. He found plenty of coal, enough to last more than 500 years, and much of scientific interest; the coal was "entirely peculiar and unlike any other coal." According to Silliman's analysis, the Breckenridge coal contained over 60 percent bitumen.[28]

The commercial implications of Silliman's study were not lost on the directors of the Breckenridge Company. "The only coals yet discovered peculiarly adapted for the production of oils," they explained, "are the Boghead coal of Scotland, the Albert coal of the province of New Brunswick, and the Breckenridge coal."[29] In the fall of 1855 construction began on an oil manufactory. By April 1856 the Breckenridge Coal and Oil Company had been established, and in June of that year it was merged with the mining company. Capitalized at $4 million, it was much larger than Gesner's Kerosene Oil Company. The Breckenridge Company had twelve retorts producing 600 to 700 gallons of "a very rich crude oil" daily; this was then distilled and purified on site.[30] The role Silliman played in the reorganization or operations of the Breckenridge Company is unknown; the company's records have been lost. But as late as 1857, the company's annual report still included his analysis.[31]

The importance of the Breckenridge Company to the coal oil industry centered on its geographical location and internal organization. Unlike Gesner's Kerosene Oil Company, which had to import "Asphalte Rock," the Breckenridge

Company located its manufactory near its mines. Transportation costs were minimized, and effectively a vertically integrated firm was in place. Moreover, coal oil, in contrast to coal gas, could be packaged and shipped along the Ohio River or by rail to locations far removed from the manufactory, or at least to places that gas could never be piped. The Breckenridge Company soon came to dominate the Midwest market in lamp oils and lubricants.

In the Northeast, the Downer Company emerged as the market leader. Samuel Downer Jr. (1807–1881) had started in the whale oil business alongside his father in 1830. The Boston-based company called Samuel Downer and Son soon expanded by adding another partner, William R. Austin, as well as another line of products, lard oil. By the mid-1840s both senior members had retired, and Samuel Jr. ran the firm. Sometime in 1854 his attention was drawn to mineral oils by Joshua Merrill (1820–1904), the selling agent for the United States Chemical Company.[32] Merrill had been visiting a number of railroads, cotton and wool mills, and oil distributors throughout New England trying to introduce a new lubricant called Coup oil. Manufactured from the distillation of coal tar, the residuum left in retorts after distilling coal gas, Coup oil had been patented by Luther Atwood in March 1853. Atwood, along with his brother William, and Samuel R. Philbrick, had organized the United States Chemical Manufacturing Company to manufacture and market it.[33] All three were practical chemists with experience in the Boston pharmaceutical industry; however, their attempt to move into lubricating oils got stuck. Merrill thus began negotiations with Downer, who agreed to buy some stock in the company. Soon thereafter, Downer discovered that the company's finances were in grievous disorder so he decided to pay up the debts and take over ownership.[34]

The Downer Company (as it became known) continued to manufacture and market Coup oil alongside its whale and lard oils. Initially sales were slow; according to Merrill, 90 percent of first-time customers refused to buy it again.[35] The problem with Coup oil, as with early Kerosene, was its trenchant odor. Merrill, Philbrick, and the Atwood brothers began experimenting with substances other than coal tar. At the plant in Waltham, Massachusetts (just outside Boston), they tried "bituminous coals, bituminous shales, asphaltum, and petroleums" until they succeeded in making "a good lubricating oil from each of these sources." The new Coup oil was a mixture of these oils with cheaper animal and vegetable oils; and it did not smell bad.[36]

By 1855 the new Coup oil began to sell. This confirmed an early impression at the Downer Company that the most profitable market for mineral oils was lubrication, which was the case in Britain. In Glasgow, Scotland, George Miller and Company thought money could be made from marketing Coup oil and so wrote to Downer asking for assistance. Downer dispatched Luther Atwood in mid-1855, and Merrill arrived six months later to help install equipment. While in Scotland, Atwood invented a lamp oil. By October 1856, when he and Merrill returned to America, they were convinced that illuminants, not lubricants, would be the future for the Downer Company.[37]

Downer was willing to make the shift. As Merrill described it, "a great change in the direction of energy took place."[38] And Gesner played a key role. Merrill described Downer and Gesner as "being in close affiliation," and sometime in late 1856 or early 1857, Downer became a licensee of Gesner's patents and began producing Kerosene in his Waltham manufactory.[39] For about a year, Downer operated under Gesner's patents, but toward the end of 1857, Merrill, Atwood, and Philbrick began another search for a more economical raw material for manufacturing Kerosene. And they found it—the Albert mineral.

In 1858 Downer reached an agreement with the Albert Coal Company of New Brunswick (William Cairns's renamed and reorganized mining company) to supply his Waltham plant and a newly acquired manufactory in Portland, Maine. Gesner might have introduced Downer to the idea of using the Albert mineral, but it seems unlikely, because he would not have wanted to bring business to his arch enemy Cairns. The only surviving records discussing Downer's decision point to a connection between Downer and the Boston Gas-Light Company, for whom Charles Jackson and Augustus A. Hayes were the consulting chemists.[40] Another possible connection was Silliman. Downer was the secretary of the Bristol Mining Company of Connecticut, of which Silliman was the president. In the event, Cairns agreed to ship 3,000 tons per year for $25 in gold per ton on the Boston wharf. Downer thus became the sole American coal oil manufacturer to use the Albert mineral. By the end of the year, the Downer Company had increased its imports to 7,500 tons per year and had fifty retorts in operation. A Kerosene boom was underway.

Kerosene Fever

In January 1858 *Scientific American* noted that Kerosene was "fast coming into popular favor."[41] Merrill, now superintendent of the Downer Kerosene oil works in Waltham, said "people were hungry to get it." He had had 200,000 gallons of Kerosene in stock in the spring, but by the fall it was all gone. Tin cans, labeled and dated, were waiting for the next batch.[42] By December 1858, an astonished *Scientific American* reported that Kerosene was "very extensively consumed, and [its] manufacture has become the most extensive in our country."[43] The country had Kerosene fever.

The coal oil business displayed all the characteristics of a speculation bubble: commercial and popular excitement, innumerable writings on the subject, inflated prices, confidence schemes, and eventual bust. During the boom, coal oil came to control the American market in lamp oils and lubricants. As one observer declared: "More oils are made from coals in one week, in our country, than ever was obtained by our whale-fishers in the best year's fishing they ever enjoyed."[44]

Whale oil was indeed the primary competitor to Kerosene. For at least 200 years, sperm whale oil was the best lamp oil. When burned in a circular-wick Argand lamp, it produced a bright, clean light. After 1830 whale oils were also being used as lubricants, especially sperm whale oil; it worked well in fine machinery requiring thin oils with little or no wax, which tended to congeal in cold weather and thus "gum up" the works. It could also withstand the high temperatures generated by heavy machinery, like locomotives, and by high-speed power looms and steam printing presses. By the mid-1850s, sperm whale oil commanded $2.00 to $2.50 per gallon; inferior whale oil (from baleen or right whales) cost $.75 to $1.00 per gallon.[45]

The increasing mechanization of the American economy combined with a growing population thus placed ever greater demands for whales. Historians and economists debate whether this increase in demand or a decrease in supply of whales was responsible for rising whale oil prices.[46] Commentators at the time did not attribute higher oil prices to a diminishing stock of whales, although they did think that coal oils might save the whales. Contemporaries thought the oil market was undersupplied; hence, prices would continue to rise. In the so-called golden age of whaling, higher prices for lamp oils and lubricants gave manufacturers, inventors, and even men of science an incentive to concentrate their efforts on finding cheaper substitutes.

As early as 1825 the first alternatives to whale oil had appeared in the American market. Most were derived from plants whose seeds contained oil, such as cotton, sunflowers, and flax (known as linseed oil). By simply squeezing the seeds, a usable oil could be extracted. These oils, however, usually contained large quantities of gum and resinous matter that tended to clog machinery and the meshes of lamp wicks. Plant oils generally did not burn well in "common lamps" (those designed for whale oils). They tended to give off a dull reddish and smoky light.[47]

Two exceptions were olives and rapeseed (*brassica napus*), a member of the cabbage family.[48] Olive oil rivaled sperm whale oil for its brilliant light, but olives were not widely cultivated in the United States. Rapeseed, extensively and profitably cultivated in France and Germany, was introduced to America in the early 1850s on the recommendation of the Lighthouse Board, a government commission (Silliman Jr. was a member) set up to find the best lamp oil for United States lighthouses. Rapeseed oil was sold under the name Colza oil, but rapeseed cultivation never caught on, and Colza oil was nearly as expensive as sperm whale oil.[49]

Another alternative was lard oil. Cincinnati was the hog capital of America, where, by midcentury, forty manufacturers were "consuming" more than 50,000 hogs and producing in excess of 1 million gallons of lard oil annually.[50] Lard oil was improved by heating it with alkali, a process that also produced a waxy substance called stearin that could be molded into less expensive candles than those made from whale spermaceti. Lard oil was safe, but it burned dimly in common lamps. It was not until the introduction of the solar lamp in 1841 that lard oil successfully challenged sperm whale oil as an illuminant, and then only in the Midwest and South. In the larger cities along the eastern seaboard, lard oil's price almost equaled that of the highest-quality sperm whale oil because of high transportation costs.[51]

The South was the source for another competitor to whale oils, turpentine. The vast pine forests supplied sap, which when heated produced turpentine (along with a thick residue called pine tar or resin). Turpentine was marketed as Sylvic or Rosin oil. While inexpensive, it tended to smoke and clog wick meshes and thus provided poor light. By "rectifying" or redistilling turpentine, a lighter, more-volatile fraction, called Camphene, was made. It too was inexpensive and widely used for illumination, but like turpentine, camphene was liable to smoke. To improve the burning qualities of camphene and turpentine, alcohol was frequently mixed in. The result was "burning fluid."

Burning fluid was the only major innovation in illumination of American origin during the first half of the nineteenth century. First marketed in 1830 following the patent of Isaiah Jennings of New York City, burning fluid became a popular lamp oil in the 1840s.[52] It burned brighter than other illuminants, except sperm whale oil, and it was affordable. Its major drawback was its volatility. Many deaths, severe burns, and extensive destruction, including entire apartment buildings, were caused by horrible explosions.[53] Adulterations were often publicly decried as the cause, but shoddiness was nothing new to American consumers. Government and business leaders called for laws to regulate lamp oils, but the melee of trade names made it nearly impossible to set safety standards. Despite the alarms and well-advertised dangers, the public persisted in buying cheap burning fluid, a testimony to the rapidly expanding oil market of the 1850s.

Kerosene thus "took off" because it offered three advantages over its competitors: safety, brightness, and cost. Kerosene's high boiling point (Gesner estimated 350° F) made it very unlikely to catch fire and explode. Improvements in refining and in the design of lamps further enhanced Kerosene's reputation for safety. In 1859 Rufus Merrill, Joshua's brother, introduced a new Kerosene lamp, which became the standard model.[54] After safety, "photometric value" was Kerosene's most marketable asset as well as a subject of scientific study. The most thorough research was conducted by Charles Wetherill, the chemist who had first identified the peculiar oil that could be distilled from the Albert min-

Table 5.1. Comparative Costs of Lamp Oils for July and August 1859 (per gallon)

Rosin	30¢–40¢
Camphene	44¢–47¢
Whale	54¢–57¢
Linseed	60¢–62¢
Lard	85¢–94¢
Coal	$1.12–$1.25 (July)
Coal	$1.12–$1.50 (August)
Sperm	$1.20–$1.40

Sources: "New York Markets," *Scientific American*, 1 (30 July 1859):75; (6 August 1859):91; and (20 August 1859):123. The weekly column on prices was discontinued after 20 August.

eral. In 1860, in a study for the Lafayette Gas Company of Indiana, Wetherill showed conclusively that coal oils supplied the brightest light for the buck, gas included.[55] While it is difficult to generalize about prices across the entire nation, data from New York City and Boston suggest that coal oil cost more than burning fluid but less than sperm whale oil. The price, however, fluctuated widely during the boom. In early 1858 it was relatively low, but it began to rise quickly as demand increased, such that by the end of the year, when Merrill was reporting a supply shortage, coal oil cost as much or perhaps more than the best sperm whale oil.[56] During the fall of 1859, the price plummeted. In places like Lafayette, Indiana, it was only seventy cents a gallon in December 1859 (table 5.1).[57]

Contemporaries gave two reasons for the sudden drop in price. The first was shoddiness. "In consequence of the urgent demand for [coal oil]," *Scientific American* reported, "great quantities in a half refined and adulterated condition, are being thrown upon the market."[58] Adulterations with alcohol produced the most flammable and explosive products.[59] One resident of Fall River, Massachusetts, complained about a hazardous coal oil called Helion Oil and another called Excelsior Oil. *Scientific American* responded that it had not seen the "quack" coal oils; "we cannot tell what they are, but we suppose they are coal oils with flashy names to astonish the marines. All such fluids should receive their true names in connection with that of the manufacturer, such as 'Breckenridge Coal Oil,' 'Newark (Ohio) Coal Oil,' & c."[60] The mention of particular manufacturers points to the second reason, production increase.

As late as 1857 there were only three main manufacturers, Gesner's Kerosene Oil Company, the Breckenridge Company, and the Downer Company. Each distilled crude coal oil and refined it into lamp oil and lubricants. A number of specialized refiners existed, but they concentrated on whale oil, lard oil, and other vegetable oils, as Downer had done before 1854. By late 1859 or early 1860 there were between 60 and 75 coal oil companies.[61] Some just distilled crude oil, others only refined lamp oils and lubricants, and a few big firms did both. For comparison, the coal gas industry (which was forty years old) numbered about 200 companies.[62]

The coal oil industry became concentrated in two regions: in the West along the Ohio River and its tributaries (roughly from Pittsburgh to western Kentucky), and in the East along the seaboard from Portland, Maine, to Philadelphia. Most western manufacturers used highly bituminous coals, such as cannel coal, containing roughly 50 percent bitumen. The largest East Coast manufacturers, however, relied on coals supplied by British companies from collieries in New Brunswick or Scotland. The two most popular were the Albert mineral and the Torbanehill mineral. This dependency was decried by industrial nationalists who regarded it as unwanted foreign interference in American affairs. "We would especially direct the attention of our American cannel coal companies to this subject," *Scientific American* editorialized, "because it would be a vast benefit to our citizens if this coal were obtained at cheaper rates."[63] British cannel cost $11.40 per ton on the New York wharf.[64] "Surely some of our western coal companies can institute measures to send their coal to the eastern seaboard, and sell it for $6 or $7 per ton," *Scientific American* rationalized.[65] But they could not. There were few, if any, railroads or canals capable of carrying western cannel coal to the East Coast.[66]

On the other hand, the manufacturing situation in the West gave industrial nationalists much to praise. *Scientific American* waxed poetic about "[v]ast beds of the rich coal from which this oil can be obtained [thereby] affording sources of supply for thousands of years to come."[67] The *Kanawha Star* newspaper calculated that an amazing 1,672,704,000 gallons of oil could be produced from western Virginian coal. At the going rate of $.60 per gallon, this meant an astounding $1,003,622,400,000.[68]

Not all prognostications were so precise, but to many observers the great Appalachian coal field offered innumerable opportunities, and several coal oil manufactories were established within its bounds. Most were small companies producing for local consumption, although there were a few large, vertically integrated firms such as the Lucesco Company and the North American Company, both of Kiskiminitas, Pennsylvania (near Freeport along the Allegheny River); the New York Coal Oil Company of New Galilee, Pennsylvania; the Newark Coal Oil Company of Newark, Ohio; and the Great Western Company of Kanawha, Virginia. These firms could manufacture up to 3,000 gallons of crude coal oil per day and refine it on site, some of which they distributed via railroads to New York City and Philadelphia.[69]

The main centers of coal oil manufacturing and refining, however, were Boston and New York City. Boston had as many as seven companies, and New York City, with the largest market, had at least fourteen.[70] The Kerosene Oil Company was the largest manufacturer in the country, but in late 1858 it underwent a major reorganization. As part of a new licensing agreement with Downer, Luther Atwood became the company's chemist, and Gesner left to take up consulting. The company then replaced Gesner's old revolving retorts with Atwood's patented "meerschaums," which could hold more coal, up to 100 tons, and distill it faster. In November 1858 Downer became a licensee of Young, and the Kerosene Oil Company began using the Torbanehill mineral, which it consumed at the rate of about 30,000 tons annually.[71] In 1859 the company produced about 5,000 gallons of Kerosene daily, a total in excess of 1.5 million gallons, nearly double its largest competitor.[72]

By early 1860 a booming coal oil industry was producing between 20,000 and 30,000 gallons of lamp oil per day, roughly 7 million to 9 million gallons annually. At least $8 million had been invested, about $1 million by the Kerosene and Downer companies alone, and approximately $750,000 in the Kanawha valley of Virginia. The number of employees engaged in oil works was estimated at 2,000, while an additional 700 worked in western coal mines. An unknown number were employed in specialized refining, cooperage, and lamp making, of which there were sixteen manufacturers. Optimistic observers were certain coal oil would soon rank among the leaders of American industry.[73]

Coal Oil Consultants

Kerosene fever generated a great number of questions. Prospective consumers and investors alike wanted to know more about "oil-from-coal": Which coal? What oil? How was it made? For general answers, the public could turn to the technological-scientific press, such as *Scientific American* or the *American Gas-Light Journal*, or Gesner's and Antisell's books. Antisell in fact had put his ideas in book form precisely because he feared the public might miss them if placed in a journal. Entrepreneurs and investors, on the other hand, could consult men of science.

Chemists and geologists were very active in coal oil during the late 1850s. "[D]uring a period of several years," Gesner recalled in 1860, "[I] was engaged as a consulting chemist."[74] His first engagement might have come as early as July 1856, when he was consulted by Dumas Grinand, the "practical chemist" of the Coal River and Kanawha Mining and Manufacturing Company of Virginia. Grinand had traveled to New York City with samples of Kanawha coal and "in company with Mr. Gessner [*sic*] . . . made several experiments." Grinand dutifully reported that "the scientific and technological question [was] fully solved."

Kanawha coal would produce a "brilliant and luxuriant light."[75] Gesner's other engagements came after he left the Kerosene Oil Company. One of these was the Columbia Oil Company of Brooklyn, literally adjacent to the Kerosene oil works at the mouth of the Flushing River. The company consulted Gesner when it began its operations in 1858 using Trinidad pitch. The next year Gesner made several consulting trips out west, especially to Ohio.[76] "[C]oal lands of every description," he explained, were explored "[because of] the introduction of *kerosene* in the market."[77]

In Pennsylvania, J. Peter Lesley did some of that exploring. In the spring of 1856, he consulted for several companies planning to exploit the "Darlington vein," a rich cannel coal outcropping in the western part of state. Lesley was excited by the vein because of its distinctive geology and commercial potential. It would be extensively mined by the New York Coal Oil Company, a large manufacturer of crude coal oil and one of the suppliers of the Columbia Oil Company.[78] Some years later, Lesley wrote to Leo Lesquereux that cannel coal was one of "two special studies" he wished to make: "to see exactly its places and properties." "If I were only rich," Lesley confided, "how you and I would work up these thesis questions together, & I should spend $10,000 on publishing them . . . with thousands of colored illustrations. A dream! a dream!"[79]

Cannel coals were also of great scientific interest to John Strong Newberry. A graduate of Western Reserve College in 1846, Newberry began studying fossil fish he found in the Ohio coal fields. His paleontology attracted the attention of James Hall, who became Newberry's mentor and encouraged him to publish. At the 1856 meeting of the American Association for the Advancement of Science in Albany, Newberry presented a new interpretation of the origin of cannel coal.[80] Rejecting the idea propounded by Lesley and Henry Rogers that highly bituminous coals came from peculiarly resinous vegetable matter, Newberry argued that cannel coals resulted from the bitumenization of fish. He imagined a large ancient swamp in Ohio, where the water was calm and low in oxygen; thus, large amounts of the carbon in the animal matter were preserved. In his paper, Newberry did not mention Charles Jackson's *Paleoniscus* discoveries at the Albert mine, but he did argue that his special process of bitumenization accounted for the coals rich in hydrocarbons and best suited to making oil.[81] Not surprisingly, Newberry was a highly sought-after cannel coal consultant. He boasted to Hall that he was very active in "prospecting" in Ohio on a "great number of pieces of coal land."[82]

Cannel coals, according to *Scientific American,* were "a wide field open for critical experiments by chemists who have the time, means and a good apparatus to conduct them."[83] James Dana thought the most "incredible" research had been done by Robert Peter (1805–1894), the chemist of the Geological Survey of Kentucky (1854–1860).[84] Peter published organic analyses of more than thirty

Table 5.2. Robert Peter's Comparison of Chemical Composition (1857)

	Breckenridge Coal	Boghead Coal
Carbon	82.355	80.487
Hydrogen	7.844	11.235
Nitrogen	2.749	0.874
Oxygen	7.051	6.726

cannel coals, including the Breckenridge coal, "Haddock's cannel," and the "Union [Oil] company's coal." Like Silliman before him, Peter found that the Breckenridge coal and the Torbanehill mineral (or Boghead coal) were the ones best-suited to oil production (table 5.2).[85] "The manufacture of oils, paraffine, &c., from cannel coal," Peter concluded, "has, since this survey commenced [1854], taken a wonderful expansion in this country, and is destined still more to increase as experience in the preparation and use of these valuable products of our cannel coal is acquired."[86]

Peter's prodigious research caught the attention of Henry How (1828–1879), professor of chemistry and natural history at King's College, Windsor, Nova Scotia. How had been consulted about a peculiar "oil-coal" discovered near Pictou. He analyzed the coal and found it yielded a profitable amount of oil, 77 gallons of crude per ton. (Breckenridge coal and the Albert and Torbanehill minerals yielded about 100 gallons of crude per ton.) After reviewing all the work on coal chemistry, How concluded that "oil-coals" presented a problem to men of science. "[T]he question as to what is and is not a coal, must be held to be an open one in those sciences in whose province the matter lies; and it will probably long remain so."[87] The coal oil boom had reinvigorated the classificatory controversy.

The "Coal-Oil Man"

During the fever, a "large number of 'experts' in coal-oil appeared," according to Gesner's son George.[88] And some were charlatans.[89] The most colorful of these confidence characters was the so-called Coal-Oil Man. Ready to advise on raw materials, the fastest distillation processes, and the best refining methods, the Coal-Oil Man was "the embodiment of all the learned and useful professions." "Formerly the Coal-oil man was merely a chemist and a man of science," observed the *American Gas-Light Journal* with a good deal of sarcasm, "but lately he has undergone a change, he is now a lawyer, doctor, druggist, wholesale and retail, sea captain, steamship owner, preacher, broker, pugilist, actor, sexton, and undertaker generally."[90] The *Journal* continued its satire.

> The very sight of [cannel coal] will cause him to exhibit signs of high excitement; his eyes grow bright and his tongue grows eloquent, and he pours forth a torrent of words concerning its *yield;* how much *crude* oil or *refined* from the ton, and cost of making it. . . . His age is between sixteen and a hundred. . . . He dresses in a variety of costumes—sometimes like a man about town, but generally in the garb of a Methodist minister, whose congregation is small, and not inclined to pay in cash. He talks well upon any subject connected with his business, and is a great favorite of the ladies, who declare that the smell of coal oil is delightful when toned down by age.[91]

Despite the detail, the public was not always able to spot the Coal-Oil Man or to distinguish him from the honest consultant.

On one trip to Ohio, Gesner took along his son, George Weltden Gesner (1829–1904), and they ran into some trouble. They had come to oversee the construction of a manufactory in Muskingum County and arrived from Cincinnati with stills, worms, and other equipment. Local residents saw the new project quite differently and began planting corn in the belief that it would fetch a high price at the works. Gesner's "coal oil," they suspected, was a cover to prevent the town ministers from discovering a whiskey distillery. But the ministers were not fooled. They denounced Gesner as "Satan's messenger" and called upon their flock to reject him and his works.[92]

The reactions of the Muskingum ministers and farmers to the Gesners may not have been different in other parts of rural America. Kerosene was a new commodity, and the consulting chemist was an unfamiliar character. The anecdote also provides insight into the way consultants saw themselves. The story was recounted by both Gesner and his son. Gesner spoke disdainfully of such "laughable farces" and patronized the locals for their "extraordinary experiments" made out of "ignorance of chemical manipulations."[93] Gesner, the man of science, wanted deference, not just attention. George, on the other hand, spoke approvingly of the locals' ability to learn the art of Kerosene manufacturing. It seemed to him that many "practical and intelligent men" became "very skillful in their business" without the benefit of science.[94] George described himself as a "consulting chemist and engineer." And while he was proud to advertise his association with his father during the geological survey of New Brunswick, George also emphasized his employment "in a regular Chemical Manufacturing business."[95] For George, the proliferation of practical men was a sign of a healthy and prosperous industry, not a challenge to scientific authority.

The case of the Coal-Oil Man thus reveals some of the ambiguities inherent in the relations between science and commerce. Amid popular distrust of "experts," there was much trust in science. Gullibility and respect produced the same effect, namely, the acceptance of authority, regardless of its authenticity. Confidence men merely exploited the public's predisposition to defer to those who professed a knowledge of technological science. Quacks and chemists can have much in common.

It is very evident that the earth was prepared with the special end in view of being man's abode, and the Great Architect of it has laid up stores in the bowels of the earth, from which man is to be supplied with light and heat, when our forests shall fail, and the whale cease to be chased by the daring mariners of Nantucket.
—*Scientific American*, 7 June 1856

WHETHER PROVIDENCE predestined Kerosene to replace wood and whales is unknown, but it is clear that Kerosene was a commercial success. It had effectively displaced whale oils from the lighting and lubrication markets. The *Whaleman's Shipping List* for 1860 dolefully noted that the number of vessels employed in the business was "considerably diminished" and the prospects for the whaling fleet were "very discouraging."[96] The *Boston Commercial Bulletin* also had bad news for whalers.

For many years New Bedford, Mass., has been known, not only as the greatest whaling port in the United States, but the whole world; it is now [1861], however, falling fast from its former oily greatness. [The] reduction [in vessels] has not been caused by losses of ships at sea, but by their withdrawal from the trade, as the business has been very unprofitable for the past four years. The price of whale oil has been greatly affected by substitutes, especially coal oil.[97]

The prospects for other animal and plant oils were equally dim. The production of lard oil was curtailed in 1860; in Louisville and Cincinnati, the slaughter of hogs was down 40,000 and 111,000, respectively, from the previous year.[98]

If a good time was coming for whales and hogs, it was not so for oil workers and investors. The Kerosene fever broke sooner and more suddenly than most expected. In March 1860 the biggest firm of all, the Kerosene Oil Company, went bankrupt. It was nearly $90,000 in debt. At the auction block in June, "[e]verything went off at very low figures."[99] The complete works for the distillation of crude coal oil and the refining of Kerosene sold for a mere $96,000.[100] According to the creditor's report, the company had spent $396,970.58 on buildings and equipment. In five years of operation, the company had manufactured oils to the amount of $568,951.05 and made a total profit of just $112,420.45.[101] Over-expansion and large loan payments had brought the downfall. "Thus ends the history of the great Kerosene Oil Company," observed the *American Gas-Light Journal*.[102]

The causes of the Kerosene Oil Company's collapse were symptomatic of the larger problems facing the industry. A coal oil glut drove down the price of lamp oil, which, combined with the rising costs of cannel coals, contributed to many other bankruptcies. Western manufacturers seemed to weather the crisis somewhat better, as did the Downer Company, which emerged as the largest Kerosene manufacturer in the country. As part of the bankruptcy agreement, Downer acquired the sole rights to the name Kerosene.[103] A commercial flier from early 1860, just prior to the bankruptcy, made clear the market power of the brand as well as the increasing competition.

> The NEW-YORK KEROSENE OIL COMPANY [are] the EXCLUSIVE Proprietors of the PATENT RIGHTS for the manufacture of Kerosene; also, owners of the TRADE MARK "KEROSENE," and the public are hereby notified, that any infringement of either will be vigorously prosecuted.[104]

Many coal oil companies ignored the legal threats and flagrantly sold their products as Kerosene, the name associated with the best quality. In the boom-and-bust frenzy, Kerosene was thus transformed from a patented trademark into a household product. In America, kerosene (little k) became the generic description for all mineral-based lamp oils, including those manufactured from a recently discovered raw material called petroleum.

CHAPTER 6

The Kerosene Cases

The discoveries of the chemist have astonished the world, and
from his laboratories have originated the most beautiful as well as
useful inventions of the day. Probably no more useful branch of
discoveries have been made than those tending to the production
of materials for artificial light. . . . [B]ut it was not until within a
very few years past that the idea of obtaining oil from coal was
conceived. The chemist, in his experiments for other materials, at
last hit upon this, and to his astonishment produced an oily substance from coal. Experimenting still further, a result was obtained
highly gratifying, and which has been and will be, productive of
immense profits.

—"Kerosene Oil," *New York Commercial Advertiser,* 24 August 1859

URING THE WINTER OF 1858–1859, as Kerosene was fast becoming the
best-selling lamp oil in America, a legal dispute erupted in the United States and
Great Britain. On one side was Abraham Gesner, the inventor of "Kerosene,"
and on the other stood James Young, a Scot, who held a United States patent
for "Paraffine." The high-stakes cases were about which patent, Kerosene or
Paraffine, covered coal oil.

But more than legal rights were in the balance. The dispute revealed great
tensions in the relations between inventors and men of science. Coal oil was
both a scientific discovery and a new commercial product. How "scientific" and
how "new" and "commercial" became matters of debate. In their patents, Gesner and Young had laid claim to inventions *and* to discoveries. What troubled
many observers was the *nature* of those discoveries. Were they scientific, practical, or economical? Were there any meaningful distinctions among the three?
Obviously, patents covered intellectual property. But what kind of knowledge
was being owned? And who would "own" the knowledge—the discoverer or the
patentee? Nor was it agreed on who should do the patenting. As the participants

dug into the history of coal oil, they exposed the foundations of what could and often did divide inventors and men of science. To continue the metaphor, the controversy became a sort of social archaeology uncovering the assumptions, ascriptions, and prejudices about the roles that men of science (and science in general) played, and could play, in invention and business.

So, at one level, the Kerosene cases were about profits and priority; at another, they revealed a struggle over the control of invention and innovation; and, at yet another, they delved into the meaning of science. Such wide-ranging significance was not missed by contemporaries. The cases made newspaper and journal headlines in America and Britain. Kerosene became a cause célèbre and brought into sharp focus the manufacturers, men of science, and inventors, who laid claim to coal oil. The experts once again (or, perhaps, as usual) disagreed. Men of science had been criticized in the past for their mercenary behavior, but the Kerosene cases cut deeper into their much-vaunted disinterestedness and heightened anxieties about ethics, propriety, and objectivity.

The Patent

James Young described himself as a chemical manufacturer and practical chemist.[1] He had been introduced to chemistry by Thomas Graham (1805–1869), professor of chemistry at Anderson's Institute (now the University of Strathclyde) in Glasgow in the early 1830s. Young became a favorite student of Graham's and, in time, his laboratory and lecture assistant. In 1837 Young moved to London when Graham became professor of chemistry at University College. An accomplished experimentalist with a sound theoretical foundation, Young was nonetheless dissatisfied with his subordinate position and future prospects. In the fall of 1838 he decided to leave the university for a career in the Lancashire chemical industry.[2]

Young began as a manager (1839–1844) for James Muspratt's new soda-making plant in Newton and then joined Tennants, Clow and Company as its consulting chemist (1844–1851). In addition to overseeing its plant at Ardwick Bridge, Young traveled to all the Tennants's subsidiaries to advise on dyestuffs, bleaching agents, and other heavy chemicals. In much the same role that Gesner played at the fledgling Kerosene Oil Company, Young took over responsibilities for designing improvements to production processes and introducing new products, one of which he patented.[3] After more than a decade, Young had not only acquired the managerial, financial, and technical experience necessary to work in the biggest and most advanced companies in the chemical industry but also gained valuable knowledge of how to formulate, register, and work patents.

Young's interest in oil dates to December 1847 when the chemist Lyon Playfair (1818–1898), a former Anderson schoolmate, invited him to examine a petroleum seepage in a coal mine at Riddings in Derbyshire. Playfair had been consulted by the mine owner about the oil's commercial possibilities.[4] After several experiments, Playfair found the petroleum could be distilled into a lubricant and lamp oil and encouraged Young to develop these. The following year Young built a pilot plant at Riddings, and in all likelihood his move into the oil business would have remained a small venture had not the problem of supply placed any future plans in jeopardy. The flow of petroleum at the mine had gradually diminished such that by December 1850 it had virtually ceased. For the business to continue, Young needed to find a new supply.

Young's search for a raw material for making oil provides a clear example of research and development in mid-nineteenth-century chemical manufacturing. As Young later recalled during the patent trials, he decided to experiment with various types of coal on the theory that "the petroleum might be produced by the action of heat on the coal."[5] An "artificial" petroleum, Young reasoned, might be distilled from coal. He tried many varieties, placing each one in a common gas retort and subjecting it to different rates and degrees of heat. He found that all coals, more or less, produced a crude oil when brought, slowly, to a low red heat (rather than rapidly to high heat, the process used in manufacturing gas). It was this discovery that Young intended to patent.

Writing the patent was a prolonged and deliberate affair, during which Young had the invaluable help of Edward Meldrum, his plant manager, and Edward William Binney. Binney was a well-respected coal geologist. His work on fossil plants, especially upright trees, had contributed much to the debate on coal's origin. Binney was also a shrewd attorney with an extensive knowledge of patent law. So while Meldrum ran experiments on coal, Binney studied patents, British and foreign. Young, for his part, made a search through the scientific literature on chemistry and chemical manufacturing. The result of these coordinated efforts was a streamlined patent specification filed on 17 October 1850.

> Be it known that I, James Young, of Manchester, England, have invented improvements in the treatment of certain bituminous substances, and in obtaining products therefrom, and do hereby declare the following to be a full, clear and exact description of the same. My said invention consists in treating bituminous coals in such manner as to obtain therefrom an oil containing paraffine (which I call paraffine oil), and from which I obtain paraffine.[6]

Purposely written to be both general and specific, it covered the process, distilling coal, *and* the product, the crude coal oil that Young called "paraffine oil" because it contained the mineral wax paraffine. Young made no claim to apparatus, plant design, refining technique, or materials (although he did note that "cannel coal" was best fitted for the purpose). He made no claim to any lubricant, lamp oil, solvent, or other commercial product, only the crude coal oil.

To exploit the patent, Young and his two partners established three companies: E. W. Binney & Company of Bathgate, Scotland; E. Meldrum & Company of Glasgow, Scotland; and James Young & Company of Manchester, England. The actual paraffine plant was located at Bathgate for the simple reason of its proximity to that peculiarly bituminous substance called the Torbanehill mineral or Boghead coal. In theory, and according to the specification, Young's patented process worked with any bituminous coal, but in experiments Young found that cannel coals—the type used in the gas industry—produced the most crude oil. In actual operations, Young relied exclusively on the Torbanehill mineral.[7]

Young's business prospered through sales of lubricating oil to Lancashire textile firms, but it was not long before he began to meet resistance, primarily from gas companies. Many already used cannel coals, and with a careful change in distilling procedure (lowering the temperature) they thought they too could make oil. This is precisely what the Hydro-Carbon Gas Company of Manchester and Salford embarked on, and in 1853 it started to sell a lubricating oil to nearby textile manufactories. This was the kind of competition Young needed to suppress, and so he sued the Hydro-Carbon Company for patent infringement.[8]

The trial lasted only two days, with few expert witnesses or lengthy discussions of previous patents and chemical publications. The gas company's star witness was Edward Frankland (1825–1899), its consulting chemist, who probably connived at infringing Young's patent in the first place. At trial, Frankland readily admitted that the gas company followed Young's process, but he denied that Young's patent was valid because the process was previously well known to scientific chemists like himself.[9] Young's expert witnesses, Thomas Graham, Lyon Playfair, August W. Hofmann (1818–1892), John Stenhouse (1809–1880), and Andrew Fyfe (1792–1861), flatly denied this assertion. Lord Chief Justice Campbell agreed with Young's experts and instructed the jury to recognize Young's specification as the exact, and only, description of the process for producing oil from coal. The jury complied, and in less than an hour's deliberation they found for Young.[10]

The verdict obviously did not satisfy the Hydro-Carbon Company or other gas companies wishing to diversify into oils. The defendants planned to appeal the decision, which undoubtedly would have resulted in a more determined and lengthy trial, but Young and his partners decided to settle out of court. For the sum of £3,000, reportedly for a license to the hydrocarbon process (a patented method for making oil from resin), the Hydro-Carbon Company agreed to abandon its oil venture. To many it looked like a bribe, and indeed Young had initially opposed the deal on just these grounds, but Binney and Meldrum persuaded him. They were not at all confident Young's patent could withstand an extended trial, and the settlement, they hoped, would forestall further competition from the gas industry and perhaps other oil refiners.

It did neither. Infringements continued, and the number grew rapidly during the late 1850s. The reason was money. During the trial, Young and his partners announced that the lamp oil, which after 1854 became their principal product, returned a profit of 100 percent. As demand for lamp oil increased, so too did the likelihood that gas companies or oil refiners would cash in on Young's coal oil. Young and his partners, meanwhile, did not yet feel financially secure or legally steeled to take on the infringers. Their decision to do so would come from an unlikely place—the United States.

The Infringers

The roots of American infringement go directly to Young, who, with canny commercial foresight, acquired a U.S. patent for his Paraffine in March 1852, two years prior to Gesner's Kerosene patents.[11] The remarkable expansion of the American coal oil industry coincided with, but began to outpace, the British boom. Young watched these developments with some anxiety, and in November 1856 he engaged the New York City law firm of Benedict and Boardman to secure settlements with companies infringing his American patent rights. From the beginning, Young and his lawyers realized they faced an uphill battle, made worse because Young was a foreigner whom many American manufacturers were patriotically predisposed to oppose. They treated Young's solicitations for royalty payments as unlawful and greedy attempts to tax them. For two years they managed to avoid his attorneys, during which time leading manufacturers such as the Downer Company and Gesner's Kerosene Oil Company experimented with a variety of raw materials and processes in a determined effort to evade Young's patent.

Gesner's and Downer's elaborate chemical researches proved unsuccessful or unprofitable, and in November 1858, with Kerosene sales booming, Young's attorneys came to terms with Downer and the Kerosene Oil Company. (It is very likely that Gesner left the Kerosene Oil Company at this time and became a consultant for other coal oil manufacturers.) The Americans contracted to pay Young a royalty of four cents per gallon on refined Kerosene in return for a license to use Young's patented process.[12] The agreement certainly appeared to be a coup for Young. He managed to bring under his control the largest and most influential manufacturers on the East Coast.[13] But the Americans also stood to gain. It was more cost effective, in the long run, to take a license than to go to court.[14] They also had protection from the best patent, a fact demonstrated by Downer's desire to purchase it outright from Young for $20,000. Young refused, but he did agree to press suit against infringers, at least two cases per year.[15] Young and his American licensees thus tried to constrain the competition, but they also opened the way to an extended legal, commercial, and scientific controversy.

Five days after the Downer-Young agreement, Benedict and Boardman placed announcements in the New York City newspapers of Young's invention and his intention to sue infringers. By December his lawyers began drawing up a case against the Columbia Oil Company of New York, the firm located near the Kerosene oil works. By January 1859 Young's lawyers had served notices to prosecute on five other manufacturers, including the Glendon Oil Company of Boston. These legal actions brought a flood of public inquiry, incrimination, and accusation. Benedict and Boardman informed Young that some of their correspondents "assume a tone of indignation alleging that Oil has been made from coal *in this country* for *twenty years* as they can prove!"[16] The lawyers, quite understandably, preferred to settle the cases rather than to litigate; this would save money, time, and trouble, and it would prevent the infringers from unifying. The gathering storm, however, could not be contained.

In March 1859 *Scientific American* published the "correct" account of the manufacture of coal oil. In response to a "great number of inquiries regarding the history of this invention," it reprinted the text of what it regarded as the first patent, Young's.[17] The article caused "considerable commotion" among American interests.[18] A Baltimore refiner, A. L. Fleury, asked peevishly if there were not "better and much quicker processes of manufacturing coal oils and paraffine, all invented in this country, and which can be secured at less cost than those of English importation."[19] Indeed there was, responded R.M. from Williamsburg, New York, and he pointed to Gesner's Kerosene patents. In contrast to Young's "tedious and troublesome" process, Gesner's was the "most excellent" method.[20]

Besides adopting a different method, a patent lawyer from Washington, DC, suggested another foolproof way to "breaking down" Young's patent based on the very fact that Young was not American. Under the U.S. Patent Law of 1836, aliens were required to put on sale their discovery within eighteen months of the patent's issue. Because Young had reportedly not built a plant or sold any of his Paraffine in the United States, his patent was void.[21] *Scientific American* regarded this legal loophole as simply "another illustration of the severity of our miserable system discriminating between foreign inventors and our own citizens."[22] Young's lawyers feared it might undermine the patent. Young himself considered it inconsequential. He knew he had advertised his Paraffine through a New York agent, James Lee & Co., but it took Benedict and Boardman three days of searching through newspapers, *after* interviewing Lee, to find the elusive advertisement. When they finally did, it was enough "to save" Young's case.[23]

What bothered Americans more than Young's nationality were the broad claims to discovery made in his patent. Many were unclear as to what was covered and what was not. The correspondent R.M., for example, believed that Young had provided only an "outline" of the process for distilling coal. Others, namely Gesner, had supplied far more detailed and useful patents. Young had not measured the temperature at which coals distilled oil. His patent contained the vague wording, "a low red heat." Gesner, by contrast, had specified the temperature at 800° F. "So far as we know," responded *Scientific American,* "the patentee [Young] is the discover of the method of obtaining oil from coal and bituminous shales, [and] [t]his seems to cover the manufacture of all coal oils by distillation in essence and principle."[24] An Ohio manufacturer, James Campbell, thought it was "sublime impudence for a man to claim the discovery of coal oil in essence and principle." Campbell found in the chemical researches of J. J. Berzelius and Justus Liebig the very same process "thoroughly investigated and published to the world."[25]

Campbell's argument was not well developed, and *Scientific American* dismissed it by changing its assertion: Young's claim to novelty rested on the "regular distillation of the coal," by which "a large amount of oil" was secured. The novelty lay not in the principle but in the "economical result."[26] This swift and clever maneuver disguised a crucial distinction. In shifting from principle to production, *Scientific American* provided a legal strategy to Young's attorneys as well as a fundamental question for all the participants in the controversy: how does the law (British and American) distinguish between the essence and principle of an invention and of a scientific discovery?

The legal terms *essence* and *principle* were ambiguous and ill-defined in both Great Britain and the United States.[27] In spite of this, or perhaps because of it, patent specifications were usually phrased so as to encapsulate the broadest claims to invention, just as Young had done. Case law in both countries recognized something called the principle of an invention,[28] which by precedent referred to a mechanical principle, not to a scientific law, which by definition was not patentable.[29] Patenting a mechanical principle, such as one used in cotton textile machinery, was relatively straightforward; there was no natural process for cloth manufacturing. With regard to chemical processes, the distinction between a scientific law and the principle of an invention was not nearly so clearcut. In the coal oil controversy, the critical point was whether the process for producing oil by heating coal was a chemical one, meaning scientific and natural, or a practical one, meaning mechanical and artificial. The legal waters were muddied still further by the imprecise criteria used to establish priority. Prior use—meaning, for example, the construction and operation of a cotton textile machine before the issuance of a patent—was solid ground for nullifying a patent claim. But did scientific publication of a chemical process constitute grounds for priority?

By the end of 1859, these questions were becoming increasingly worrisome to Young, his lawyers, his licensees, and even his infringers. Legal battles were going to be complex, costly, and exhausting. "[I]t appears the lawyers there [the United States] are as fond of money as those at home," remarked Young.[30] The decision to proceed thus rested with him. Young conferred with Binney, who advised him to pursue the Americans first, after which he could take on British infringers. Binney's reasoning was sound; it was more likely that an adverse judgment in the British courts would prejudice an American jury than an American loss would influence a British court.

In choosing to go after the Americans first, Young faced a possible liability not present in the British cases—the issue of mineral classification. American companies, like their British counterparts, relied on cannel coals in general, but East Coast firms were using the Albert mineral or the Torbanehill mineral exclusively.[31] Thus, even though these two minerals had been judged to be "coals" in New Brunswick, Nova Scotia, and Scotland, it remained to be seen whether U.S. courts would agree. Young knew that mineralogical classification was a possible weakness in his case; he therefore addressed the matter explicitly in all his contracts with American licensees. "[I]t is a fundamental element of this agreement and of the license to be granted in virtue hereof that all products of the bog-head and Albert mines are and shall be deemed and taken to be bituminous [coal]."[32] In spite of the impending taxonomic trouble, Young felt confident he could win against American infringers. On the advice of Benedict and Boardman, he decided to pursue two cases simultaneously, one against the Columbia Oil Company, the other against the Glendon Oil Company. Each was using the Albert mineral, but each represented a competitor in the two major markets, New York City and Boston, where Young's licensees were strongest and "very desirous . . . of establishing their monopoly."[33]

The first step in preparing for any court case was to secure "the opinions of scientific men of high reputation."[34] As Young knew from the trial with the Hydro-Carbon Gas Company, the quantity and quality of the expert witnesses were crucial to success. Young had several on retainer—Graham, Hofmann, and Playfair—all of whom would have to be deposed. "[B]ut it would be well to know what any of them would say," cautioned Young, "before giving their names to the enemy as we do not know what influence has been at work since we talked with them."[35]

Young would also need American experts. Early on, Benedict and Boardman consulted Daniel Treadwell (1791–1872), formerly Rumford Professor of the Application of Science to the Useful Arts at Harvard (1834–1845), and James Renwick (1792–1863), formerly professor of natural philosophy and experimental chemistry at Columbia (1820–1853).[36] Neither Treadwell nor Renwick had done research in the geology or chemistry of coal (although Renwick had taught chemistry), but both were experts on the relations of science and the arts and favored the position that progress in the arts involved more or something other than the application of scientific principles.[37] Benedict and Boardman recruited other unnamed expert witnesses, the total cost of whose services would come to nearly $2,000, "a large sum," according to the lawyers.[38]

For its counsel, the Columbia Oil Company called upon the New York City law firm of Birdseye, Sommers, and Johnson. Judge Lucien Birdseye was an experienced and knowledgeable patent lawyer, who coordinated a comprehensive and costly defense. In a brief filed with the Circuit Court of the United States for the Southern District of New York, Birdseye collected twenty-three scientific publications along with forty-nine British and seven French patents. Presumably he included American patents as well, but that list has been lost. Birdseye also submitted the names of several scientific and practical experts, considerably more formidable than Young had anticipated. The star witness was to be none other than Gesner.[39]

For the Glendon Oil Company, the defense was managed by the legal firm of Smith and Rollins. Although its records along with much of the correspondence concerning the case have disappeared, it is not unlikely that the Glendon's strategy would have followed the same lines as the Columbia's. In all likelihood, the two were coordinated. Benedict and Boardman suspected the American infringers had pooled their resources to finance the defense. And once again, Gesner would have testified.[40]

Faced with such a "skillful and determined" challenge, Benedict and Boardman urged Young to visit the United States to supervise and expedite the legal proceedings. Reluctantly, Young sailed for New York City in January 1860. He found his stay in the United States to be enjoyable overall but not especially profitable, for it was in the spring of 1860 that the American coal oil bubble burst and the Kerosene Oil Company went under. The company suspended all royalty payments to Young, which put Downer, the co-signer of the contract with Young, in financial difficulty. Unable to meet his obligation, but still willing to do so, Downer agreed to new terms. He and Young negotiated a lower royalty payment—one cent per gallon.

The kerosene bust played to Young's advantage in the end. Both the Columbia and Glendon companies found themselves in tight straights, and Young, through pressure and bluff, compromised with them. They settled for a license at two cents per gallon. In addition, Benedict and Boardman managed to license a further twenty-two American companies, most of which were probably small firms scattered along or near the Atlantic seaboard. The large and profitable manufactories located in Pennsylvania and places further west did not take licenses, nor did they heed Benedict and Boardman's threats of legal action.

The Trial by Press

With skillful negotiation, Young had forestalled a full trial of his American patent. In turn, the American controversy lost some of its focus, but not its heat. Instead of the confines of a courtroom, the controversy went public in a vicious exchange in print among three chemists: Gesner, Antisell, and Francis (Frank) H. Storer (1832–1914). This trial by press, in effect, expanded the controversy. These experts were now at liberty to craft their arguments for a broader audience, including other chemists and geologists, without threat of jeopardizing the legal cases. They took full advantage of the opportunity.

In 1859 Antisell published *The Manufacture of Photogenic or Hydro-Carbon Oils*. Although his exact relationship to Young is unknown, Antisell made a compelling case for Young and clearly attempted to sway public opinion in his favor. At the same time, Antisell presented a solid defense of the U.S. Patent Office—the agency that granted Young his monopoly and Antisell's employer. Since 1856, Antisell had been an examiner specializing in the growing field of chemical manufacturing. U.S. law required that each application be reviewed for novelty and originality before a patent could be granted. Because Antisell and the other examiners, as well as the commissioner, considered themselves men of science, American patents were effectively sealed with scientific approval. In his book, Antisell paid close attention to patent specifications because the Kerosene cases, in effect, challenged the Patent Office's scientific authority.

From the outset Antisell stated that he thought hostility toward Young was unfounded. "An impression had taken hold of the American manufacturing public," he observed, "that the patent of James Young has no force, as it was not a new invention at the date of the patent; and from the unfavorable effect of that patent upon the actual manufacture of coal oils in this country an ill-feeling has been produced against it."[41] Antisell allowed that Young had acted unwisely in trying to bully American licensees, but Young's shortsightedness had not affected his patent rights, of which there was "no shadow of a doubt." To prove his point, Antisell presented a "History of the Art," in which he reviewed more than a century of discoveries, both scientific and practical, in Britain, France, and Germany.[42]

The writing of histories, much as the writing of patents themselves, is a singularly constructivist moment—the fabrication of "truth," in this instance "the steps of the discovery of the production of photogenic oils from different materials."[43] According to Antisell, scientific experiments on the distillation of bituminous substances had been conducted sporadically throughout the seventeenth and eighteenth centuries. Not until the nineteenth century were the chemical products systematically investigated. The Swiss geologist and chemist Nicholas Théodore de Saussure (1767–1845) was among the first to do so. He distilled the "asphaltic limestones" of Travers in 1819 and 1820 and obtained several oils resembling petroleum found in Amiano, Italy. In the 1830s the French chemist Auguste Laurent (1807–1853) and the German chemist and *Naturphilosoph* Baron Karl von Reichenbach (1788–1869) conducted a series of important experiments. Laurent distilled samples of bituminous shale found near Neuchâtel and described their chemical products, while Reichenbach did the same for Moravian coal. Unlike previous investigators, Reichenbach heated his coal samples gradually, which produced a quantity of oil. After distilling this oil, Reichenbach isolated a white crystalline substance similar in appearance to wax, which he called "paraffine." The oil in which this wax was suspended Reichenbach named "paraffine" oil. These results were published in respected scientific journals in France and Germany, but that research, Antisell maintained, was not continued by others and subsequently forgotten during the 1840s.[44]

Antisell next turned to the work of several inventors, foremost among whom were Archibald Cochrane (1748–1831), ninth Earl of Dundonald and the French engineer Alexandre François Selligue (1784–1845). Dundonald held several patents for chemical manufacturing using Trinidad pitch. (The possible commercial exploitation of this asphaltum had spurred his son, Thomas Cochrane, tenth Earl of Dundonald, to consult Gesner.) In France, Selligue had distilled oils from bituminous shales and then purified them in order to make lamp oil, which apparently sold very well in the 1840s and 1850s.[45] In fact, Antisell regarded Selligue as the founder of the mineral oil industry in France.

To Antisell's mind, these early researchers and inventors only laid the groundwork for Young.

> [T]he practical manufacture of oils from coal is due to James Young, [and] really two great results were first demonstrated practically by the operation of Young's process, namely—1st. That coal was a material from which liquids could be manufactured economically, as tar, bitumens, and schists, had been hitherto employed; and 2d. That the liquids so formed were paraffin-containing compounds.[46]

The relevant part of Antisell's justification was use of the words "practically" and "economically," the same ones emphasized by *Scientific American*. The essence or principle of Young's invention was its commercial result. Antisell credited Reichenbach with the scientific discovery of paraffine and paraffine oils, but Young's was a practical, and therefore, different discovery. In a word, Young's patent was valid.

Gesner replied to Antisell with his own book, *Practical Treatise on Coal, Petroleum, and Other Distilled Oils* (1860). Gesner did not directly challenge Antisell's account of the history of mineral oils; it served his purposes perfectly well. Instead, he dealt with the disputed process for manufacturing oil from coal and attacked Young's "alleged improvements." The argument, in broad strokes, rehearsed the familiar question of mineral classification. Gesner insisted, again, that the Albert mineral was asphaltum, not coal. In his book, Antisell, not surprisingly, had agreed; it was the same position he took at the Albert mineral trial. In fact, Antisell defended Gesner's Kerosene patents on the grounds that Gesner had specified bituminous substances other than coal.[47]

But Gesner wanted to have it both ways. If the high courts of Nova Scotia and New Brunswick had ruled that the Albert mineral was coal, then Gesner believed he should be entitled to use this judgment to his own advantage. He asserted that he had distilled oils from the Albert mineral as early as 1846 on Prince Edward Island. Four years later, while working to perfect his patented Kerosene gas, Gesner claimed to have produced oil by slowly heating the Albert mineral, although he did not patent this process until 1854. Gesner gave no explanation for this crucial delay, but nonetheless, he argued that the experiments had been done and the process for making oil had been in use by him since 1850, the year he began working with Dundonald in Halifax. In strict legal terms, Gesner's Kerosene oil was the same as Young's Paraffine oil because the Albert mineral was coal. Like minerals produced like oils. Gesner could thus lay claim to the patent rights, which he did rather bluntly: "The first successful attempt to manufacture oils from coals in America was made by the author of this work."[48]

A more substantial and sophisticated rebuttal to Antisell was made by Frank Storer. Storer was among that generation of American men of science who, after their early education in the United States, had sought advanced training in Europe. He had developed his interests in chemistry while at Harvard as a student of Josiah Cooke and eventually had become Cooke's laboratory assistant. After an appointment as chemist with the United States North Pacific expedition in 1852, Storer spent two years (1855–1857) studying in Germany and France with such noted chemists as Robert Bunsen at Heidelberg, Theodor Richter at Freiberg, and Emile Kopp in Paris. Upon his return, Storer established his own laboratory in Boston and became a consulting chemist for the Boston Gas-Light Company, much as Jackson had done twenty years earlier. Between 1857 and 1865, the year he became professor of general and industrial chemistry at the newly founded Massachusetts Institute of Technology, Storer was a professional chemist earning his livelihood by practicing his science in commercial engagements.[49] In many respects, Storer was the perfect choice to defend American companies against Young's infringement suits. Well versed in chemical experiment and theory, he also had extensive practical experience in industry. As such, he was almost the American counterpart to Young himself.

Storer entered the controversy by way of a venomous review of Antisell's book published in the *American Journal of Science*. Storer's hostility might be taken as a measure of the importance of the subject, and it might also have reflected the views of other men of science. Many readers, including Antisell, believed this to be so. For Benjamin Silliman Jr. and James Dana had endorsed Storer's article with a short introduction admonishing readers to make a "careful perusal" for it would "amply repay" their effort.[50]

Storer opened with a personal attack on Antisell accusing him of a "most lamentable lack of familiarity with the chemistry of the subject" and ridiculing his book as "simply a jumble of badly selected abstracts, huddled together in a manner which must be anything but edifying to the student."[51] He leveled his harshest criticisms at Antisell's "History of the Art" and proceeded to give his own account of coal oil. Following the best-practiced approach to nullifying a patent, Storer recounted all the scientific and practical works published before the date of Young's patent in order to prove prior knowledge and practice.

According to Storer, the first coal oil manufacturer was Selligue. To support this assertion, Storer relied on a complicated argument about mineral classification. The term coal, Storer thought, was too broad and ill-defined to be the basis for a patent. "The amplitude of variation [of] this species," he noted, "is indeed so great that it would be a matter of no small difficulty to choose any single member of the medley as a central point, or even to conceive of an ideal coal to which all other varieties should be referred." Because the scientific community could not agree on how to define coal, Storer thought all bituminous substances should be treated as "a continuous series," whose end points were "hydrogenous matter" (lignites) and "earthy matter" (bituminous shales). Various mixtures of the two matters would produce any kind of bituminous substance, including, obviously, coal. Using this two-constituent linear system, Storer declared, "we have . . . an infinite number of substances, shading into each other by scarcely perceptible degrees."[52] Because both Young's Torbanehill mineral and Selligue's bituminous shales fell along the spectrum, the two were, theoretically, manufacturing the same oils. Storer concluded with a rhetorical flourish: "how far removed in anything but productiveness is the 'coal' (Boghead) upon which Mr. Young has operated from the 'shale' distilled by M. Selligue?"[53]

The issue of productiveness raised a second problem—specification. Storer attacked the discrepancy between what Young wrote and what he did. Although Young claimed to use coal, he relied exclusively on the Torbanehill mineral. Young's actual "economical" or "practical" discovery was that ordinary coal was *not* commercially successful. The Torbanehill mineral was the *only* profitable source of Paraffine. Young's entire business had therefore been mineralogically determined. Storer regarded Young's specification of coal as a fraud.

But more important to Storer was the whole idea of economical or practical novelty. How could the amount of oil produced by Young make the process itself patentable, if in fact the process was already known? Antisell had credited Reichenbach with the discovery of paraffine and paraffine oils, but refused to allow this scientific discovery to trump Young's practical discovery. Storer saw no difference between the two types of discovery. He listed "some twenty or more" articles that Reichenbach had published on the distillation of coal and other bituminous substances in the leading scientific journals of the day. Storer concluded:

> Looking at the question for a moment, solely in its scientific bearings, we cannot refrain from an expression of astonishment, that the details of Reichenbach's researches are so little known to the generality of chemists; while on the other hand, we are forced to confess, that it is indeed rare that scientific researches, conducted by a chemist in his laboratory, have so fully described a future art—have so accurately pointed out the methods to be followed and precautions to be observed by the practical manufacturer.[54]

Storer accepted scientific research as proof of Reichenbach's production of coal oil. Discovery was discovery, and there were no two ways to produce coal oil. Young's patent was therefore void.

Publication was key to Storer's argument about priority and property. Storer admitted that Reichenbach had not patented his research, but he did publish it and therefore made it public and his own. In this regard, Storer was supported by legal theory that argued that one reason for the existence of patents in the first place was to encourage individuals to make public their discoveries, thereby enriching the nation while at the same time giving due credit and protection to the discoverer. The question of whether scientific publication could be considered the equivalent of prior knowledge and practice was to be decided in British courtrooms.[55]

The Courtroom

Young left the United States in June 1860 and took with him some of the material from the American controversy. In preparation for a British trial he asked the Scottish chemist John Stenhouse to review the case. After "having carefully perused Dr Antisell's work on photogenic oils, and the review of that work by Frank H. Storer," Stenhouse advised Young to sue the infringers. If Young brought the case in Scottish court, where the Torbanehill mineral had been "judiciously declared to be a coal," he could avoid entirely the "quibbling" issue of classification, which made all prior patents using other bituminous substances void. The only issue to be decided then would be Reichenbach's research. With scientific expertise, Stenhouse concluded, Young should be assured of winning.[56]

In the first week of November 1860, Young and his partners brought suit against the Clydesdale Chemical Company for patent infringement. The "eminent chemists" Frederick Penny, William Odling, Lyon Playfair, Robert Kane, August Hofmann, Henry Letheby, John Stenhouse, and Thomas Anderson appeared as expert witnesses for Young. Equally "eminent chemists," Dugald Campbell, Alfred Swaine Taylor, and William Thomas Brande, defended the Clydesdale Company. Both sides tried to recruit Edward Frankland, the cunning chemist who had testified against Young in the Hydro-Carbon Company case. Young outmaneuvered Clydesdale. He paid Frankland and his research assistant, Robert Warrington (1838–1907), fifty guineas each as a retaining fee. In return, Frankland and Warrington refused "to listen to the oft repeated overtures" made by the Clydesdale Company. In the end, neither Frankland nor Warrington appeared in court.[57]

The trial lasted six days, and most of it was devoted to the chemistry of paraffine oils. Young's experts allowed that Reichenbach had discovered these oils and the process for producing them, but Reichenbach had not manufactured paraffine oils in "economic quantities." The Clydesdale's witnesses swore that coal oil distillation was well known to chemists and had been published in numerous works. It was thus public knowledge to which Young could not claim proprietary rights. In the opinion of one observer, "the whole of chemical science was ransacked by the defendants."[58] In fact, most of the arguments urged in defense of the Clydesdale Company were the same as those used by Frankland at the Hydro-Carbon trial and recently published by Storer. According to one observer,

> the defendants set up that the process described in the patent was well known to the scientific world before the alleged invention by Young. To support this, all the old fossil apothecary books, chemical works, encyclopedias, and French and English patents, of late so fatiguingly inflicted upon the public by Frank Storer and others, were dragged into Court, where a number of chemical experts (?) stood ready to swear, that out of the old, dry bones they could construct Mr. Young's living invention.[59]

In presenting his summation before the jury, the Lord President of the Court of Session, Duncan McNeill, asked it to consider

> whether there was anything, either in chemistry or in any other science, which might not be placed under some general law. It did not follow, however, al-

though the general law were known, that its application in any particular instance was also known. The patentee admitted that it was known that [paraffine] products could be obtained from coal; but it was not known that paraffine in merchantable quantity, could be obtained from it—and that was his discovery.[60]

In short, the "application" of science meant "merchantable quantity." As Antisell and *Scientific American* had argued, quantity was the practical principle of what Young had patented. On 7 November the jury returned a verdict for Young, and the judge awarded him £7,500 in costs and damages.

As Binney had predicted, the case was "of great interest to American manufacturers," who believed its outcome would set a precedent for the prosecution of similar cases.[61] The verdict pleased Young's band of American supporters, to say the least. "The present trial has proved," *Scientific American* boasted, "that the views which we first expressed, and with which Dr. Antisell coincided, have been confirmed."[62] The *American Gas-Light Journal* concluded that "the verdict may be regarded as finally and triumphantly establishing the validity of the [Young's] patent, both in Great Britain and in this country."[63]

Such confidence was misplaced. Unknown to Young and his partners, a confederacy of infringers was at work. Its leader was Ebenezer W. Fernie, a speculator who had made his money in railways and mining, especially coal, iron, and manganese. Fernie first heard of the idea of making lamp oil from coal after a visit to New York City in 1859–1860, at the height of the Kerosene fever. It was there that he met Gesner, the probable source of Fernie's information on American manufacturing and raw materials, in particular the Breckenridge coal and the Torbanehill and Albert minerals. Fernie might also have received some financial support. Young suspected that Americans had backed some of his British rivals. And Binney expressed similar anxieties: "I think it likely the Yankees will purchase all our opponents and thus set them on their legs."[64]

With scientific and technical knowledge in hand, Fernie returned to Britain to start manufacturing Kerosene. He established a plant at Leeswood, Flintshire, in which he used local cannel coal and a relatively new import—American petroleum. Fernie then bought controlling interests in several other English and Scottish oil manufacturers. The extent of Fernie's operations became clear to Young only when, in 1862, he decided to sue Fernie. As Young and his partners soon discovered, Fernie had made alliances with most of the disgruntled manufacturers in Britain, all of whom had been constrained by Young's patent monopoly, including the ever-litigious William Gillespie, the owner of lands on which the Torbanehill mineral was mined.

According to contemporary observers, the case of *James Young versus Ebenezer Fernie* was "a classic in Victorian patent litigation."[65] From February to May 1864, the trial was held in Chancery before Vice-Chancellor Sir John Stuart without a jury. There was a lengthy parade of expert witnesses: thirty-one for the plaintiff, forty-two for the defendant. Most of Young's experts were chemists, the very same ones who had appeared at the trial against the Clydesdale Company. Fernie resorted to the best scientific and practical experts from the gas and chemical industries. He bolstered these expert witnesses with references to forty-six scientific works and the specifications of six patents, including Dundonald's and Gesner's.

As with all previous trials, the case of *Young v. Fernie* focused on the question of scientific priority. The principal witness for Young proved to be August Hofmann, the acknowledged doyen of coal chemistry and an experienced courtroom chemist. Hofmann was under examination for three days, during which he spent a great deal of time discussing Reichenbach's research. Hofmann conceded that the German chemist had discovered paraffine and even kept "a commercial article in view." But Reichenbach's work was "scientific," and "his experiments on the distillation of coal were laboratory experiments." It was the "practical chemist" Young, in Hofmann's opinion, who first succeeded in producing paraffine and paraffine oils in commercial quantities.[66]

The Fernie trial, however, did have a twist to it. Because the case was brought in English courts, it presented the opportunity to debate the issue of mineral classification one more time. "[T]he men most distinguished in their several departments," observed the *Chemical News,* have been called upon to settle "a simple matter of fact, 'Is the mineral dug on the lands of Torbanehill coal or not?'"[67] Young turned to his distinguished chemists for an answer. Hofmann considered the mineral "a peculiar kind of coal"; however, he admitted that he knew of no coal "which will produce paraffine oil commercially."[68] Fernie's experts agreed on the chemistry, but not the classification. The Torbanehill mineral was a separate mineral species. Because it produced oil, it could not be coal.[69]

Young then turned to two distinguished geologists: Andrew Ramsay (1814–1891) and Archibald Geikie (1835–1924). Ramsay, the senior director for England and Wales on the Geological Survey of Great Britain, made the more persuasive witness. He asserted that the Torbanehill mineral was definitely coal. Geikie, the younger of the two, agreed that the mineral was coal, but he was disinclined to draw fixed boundaries between it and other bituminous substances, particularly the shales of the Lothian district in Scotland, where he had recently been surveying. As in the Albert mineral trial, the geologists, Ramsay and Geikie, relied on stratigraphy, rather than chemical composition, to classify coal.

In the end, neither side could give a compelling answer to the question of mineral classification. "We find one-half of the chemists answering in the affirmative," remarked the *Chemical News,* "and the other half in the negative."[70] In fact, the scientific debate over mineral classification would continue for several years. Appropriately enough, it would be Young's initial idea that artificial coal oil was the same as the naturally occurring petroleum that would fuel the new investigations into the chemistry and geology of coal and oil.

The Fernie trial came to an end after thirty-four days of testimony. In deciding the case, Vice-Chancellor Stuart summarized the chemical evidence, particularly that of Hofmann, and then cited the work of the "eminent American chemist, Dr. Antisell, who holds an important position in the Patent Office of the United States of America." In Antisell's book, Stuart found the most elegant exposition of the reasoning upon which to judge the case.

> "So remained paraffine until this hour [1850] in the collection of chemical preparations, but it has never escaped from the rooms of the scientific man."[71] Something therefore remained to be ascertained in order to [make] the useful application of this article for economical and commercial purposes. This illustrates the important distinction between the discoveries of the merely scientific chemist and of the practical manufacturer who invents the means of producing in abundance, suitable for economical and commercial purposes, that which previously existed as a beautiful item in the cabinets of men of science.[72]

The point was economic. Young's process represented new knowledge by virtue of its abundance of oil. Young's patent was therefore valid. Stuart ordered Fernie to pay costs amounting to £10,000 and damages of £11,422 14s. 6d., the equivalent of a royalty of 3d. per gallon on 913,818 gallons of paraffine oil. It was the largest sum Young was ever awarded, and it marked closure of the legal controversy in Britain.[73]

> Whenever we attempt to draw a line dividing between the *sciences,* usually so called, and the *arts,* it results in distinctions, which are comparative, rather than absolute. In many branches of knowledge, the two are so blended together, that it is impossible to make their separation complete. . . . Discovery is the process; invention is the work of art. So common, however, is the connexion of the two with each other, that we find both a science and an art involved in the same branch of study.
> —Jacob Bigelow, *Elements of Technology* (1829)

IN THE BRITISH COURTROOMS and in the American press, the Kerosene cases concerned the difference between chemical science and chemical manufacturing, or, as Bigelow put it, the line of distinction between science and art. Time after time, Young's expert witnesses championed the theory that his patent was original and novel by virtue of its results—the commercial quantities of oil. British judges and juries unanimously endorsed this interpretation: scientific discovery and the essence of invention were not the same.

But the legal trials and the public debate revealed another distinction, one based on a principal of a different sort, purpose. What was the purpose of discovery or invention? In other words, what motivated an individual to do it in the first place? The difference between inventors and men of science lay not in what they did—they both distilled oil from coal—but rather *why* they did it.

This point came out most clearly in discussions about Young and Reichenbach. Through each of the trials, Young maintained that he was a "practical" or "manufacturing" chemist. He consistently denied that he made any "scientific" contribution. In spite of his experimental method and his reading of the latest research and his university training, Young regarded his work as different from that of scientific chemists. Reichenbach, however, was not as consistent in his self-appraisal. In 1854, in preparation for Young's case against the Hydro-Carbon Gas Company, he had published a brief article in the *Philosophical Magazine* in which he defended Young.

> It is true that the discovery of paraffine is my own, and I have announced it. To Mr. Young, however, belongs the merit of a secondary discovery, the merit of having elaborated a method which furnishes a comparatively *large* quantity of the substance, and which is sufficiently remunerative to the manufacturer. . . . I hope Mr. Young will succeed in convincing the legal authorities of the priority of his practical discovery, which was not part of any purely scientific investigation, and which I cannot claim in any way.[74]

Six years later, in the Clydesdale case, Reichenbach stated under oath that his earlier statement was contrived. "The article was written to oblige Mr Young," he admitted, "*so* that it might be advantageous to Young." Since then, Reichenbach had reconsidered. The difference between his process and Young's lay in the raw material. Reichenbach produced oil "out of black coal and Mr Young produced it out of Boghead." When asked under cross-examination if there was "any other difference?" Reichenbach responded: "Paraffine is paraffine. . . . I know of no other difference." The about-face might have been very detrimental to Young's case had not Binney anticipated that Reichenbach would not give "so good answers as his printed article." Because Binney had been unable to "prepare" Reichenbach for examination, he thus tried hard to get Reichenbach to talk about commercial results and quantities of oil.[75]

A more effective counterbalance to Reichenbach's retraction was the testimony of Young's other expert witnesses. They unequivocally cast the German baron as a man of science. Hofmann stated this repeatedly at the Fernie trial. Frankland, who for monetary reasons did not appear at the Fernie trial, discussed Reichenbach's motivation at length in a public lecture before the Royal Institution in 1863. In Frankland's opinion, Reichenbach was a man of science because his motivation was noncommercial. "It is no part of his duty, it is not his function, to apply those [scientific] truths to the utilities of life. Success in this direction demands quite different powers of mind."[76] For Frankland, temperament determined motivation, and motivation determined social function and professional identity. Inventors sought profits; hence, they made patents through the application of chemistry. Men of science sought truth; therefore, they made scientific discoveries through research.

Such a classificatory scheme seemed plausible enough. Since the seventeenth century, natural philosophers had been appropriating to themselves the role of disinterested pursuers of truth. They were most often gentlemen who possessed the sort of breeding, education, and financial independence that comes with such social status.[77] They should not want or need not seek patents for profit.

Frankland's argument echoed Jackson, who ten years earlier, at the conclusion of the Albert mineral trials, had blasted Gesner for taking out a patent on Kerosene gas. In a speech before the American Institute in New York City, Jackson charged that patents degraded the honorable place that men of science sought for themselves. "No true man of science will ever disgrace himself by asking for a patent; and if he should, he might not know what to do with it any more than the man did who drew an elephant at a raffle. He cannot and will not leave his scientific pursuits to turn showman, mechanic, or merchant."[78] Gesner had compromised his scientific standing by confounding the disinterested search for knowledge with the self-interested pursuit of profit. The proof was in the patent.

"Is it a disgrace to be a mechanic and merchant, and exhibit a patent?" asked Alfred Ely Beach (1826–1896), editor of *Scientific American*. Beach did not think so. He wondered, "What kind of a view has [Jackson] of what he calls a scientific man?" Beach was an advocate of "democratic science" as opposed to Jackson's elite variety, and he knew "quite a number" of scientific men, "certainly superior to the learned Doctor," who had taken out patents and profited by them. The Philadelphia chemist Robert Hare, Union College president Eliphalet Nott, and of course Gesner were good examples. In short, Beach thought Jackson had the "wrong idea" of the man of science.[79]

And so too did Storer. He did not subscribe to the portrayal of scientific men as aloof from patents, profits, and production. From his perspective, this distorted the reality of midcentury relations between science and the arts, an intimate connection that was plain to see in Reichenbach's career. The German chemist was a consultant, much like Storer himself (and Jackson and Frankland, for that matter), who combined science with an active career in industry. From 1821 to 1836, Reichenbach in partnership with Count Hugo zu Salm of Vienna had established a profitable ironworks and steelworks at Blansko, Moravia, which operated a number of blast furnaces and coal ovens. There Reichenbach had pursued his "scientific research upon the commercial products of the distillation of bituminous [substances]."[80] Storer thought Young's side had grossly misrepresented this practical research by trying to make Reichenbach into a laboratory recluse. Antisell, in particular, had purposely and unnecessarily driven a wedge between invention and discovery, between research and application. Although the gap might have been greater in the past, Storer conceded, such separations were only matters of degree, not of kind—hence, the use of such terms as technological science.

American inventors took a very different view of the matter. To many of them, the most serious threat posed by Storer and his breed of professional chemist was their insistence on a place for themselves and their science at the table of invention. Practical men regarded Storer's artful (or artless) positioning as an encroachment upon their rights and privileges. One writer using the name "Paraffine" made the following, typical argument.

> It is a well known fact that the science of chemistry is rapidly becoming contemptible in the community, or at least among such portion of the people as are aware of the undeniable fact, that no new and useful invention has ever been made and patented that involved the discovery or new and useful application of anything appertaining to chemistry, that has not been attacked by professional chemists of apparently high standing among their brethren, and attempted to be overthrown on the score of want of novelty.[81]

Science applied was science reified. To equate publication with application, or research with invention, would allow men of science to steal the credit and cash from their rightful owners. It meant scientific interference in the traditional domain of inventors, artisans, and practical men.

Jackson saw the situation completely opposite. "Men of science devote their lives to increasing the sum of human knowledge," he asserted. They gave their discoveries freely to the world; however, they were the victims of "crimes brought on by selfish ambition." Jackson thought the discoveries of unsuspecting men of science were being stolen by "mere speculator[s] in inventions." These "cunning and artful" impostors appropriated scientific discoveries by patenting them. This was not application; it was thievery. In anger, Jackson demanded, "How can we protect the true discoverer?"[82]

Patent discoveries. That was Antisell's answer, and he was in complete accord with the inventors. In a scathing article published in the *American Gas-Light Journal*, Antisell denounced Storer and his ilk for their "neoteric intolerance," a not uncommon fault among European-trained chemists, who tended to evaluate all past inventions by the standards of current knowledge.[83] By doing so, they made it look easy to elicit the scientific principles on which invention was based. Such arguments, however, were as familiar as they were fallacious. To Antisell's mind, practical discoveries stood apart, and often preceded, scientific ones.

What was worse, in Antisell's opinion, was the disgraceful behavior of Storer and his publishers, the *American Journal of Science*. They were the ones who had brought disgrace upon men of science and American science itself. "[I]t is a matter of grave consideration . . . to find the earliest and the only professedly pure scientific journal departing from its legitimate functions and prostituting its columns, for an object which, whether worthy or not, does not come within the true sphere of its duties."[84] The duties of "a high-toned journal" did not include taking sides in patent disputes. This was neither pure nor disinterested. The *American Journal of Science* had sold its pages to clever lawyers.

On the surface, this hostile reaction by Antisell appears curious; after all, neither he nor Storer (and especially not Gesner) were very worried about the fact that men of science took out patents, consulted for companies, published those researches, or served as expert witnesses in patent cases. In these activities, many Americans and Britons actively participated. For his part, Antisell, the Patent Office employee, actively *encouraged* men of science to patent their discoveries for that would avoid just this kind of controversy. What so bothered Antisell, as well as many American inventors, was the notion that a scientific publication could be construed as the equivalent of a patent. If men of science wished to own the legal rights to their discovery, they had better patent them. Antisell had no time for sanctimonious special pleading of the sort Storer (or Jackson) had indulged in.

And neither had the Patent Office. In an address to the American Association for the Advancement of Science, Leonard D. Gale (1800–1883), the commissioner of patents (1852–1857) made it absolutely clear: "Mere discovery does not entitle the discoverer to letters patent." The discoverer needed to take "a step further" and point out how "the discovery subserves a useful purpose." Gale agreed with Jackson that men of science "held themselves aloof" from patents, but this situation was to be regretted, not rejoiced. Moreover, Gale thought patents contained a "store" of information that might prove useful to science. As with Antisell, Gale thought several of the arts, notably "illuminating materials," were in advance of science. Thus, he admonished men of science to work with patents, not against them.[85]

And so did the American press. *Scientific American*, a journal generally sympathetic to science, ridiculed Storer for his disingenuousness. "Under the appearance of much candor, smartness and chemical erudition," it discovered a "very shallow" article, whose author did not understand "the true nature of the patent."[86] Similarly, the *American Gas-Light Journal*, another press usually sympathetic to science, derided Storer; the "very disinterested" and "learned" chemist had produced "no other effect than to disgust" the public.[87]

Disgust was a good word to describe the public's opinion of the Kerosene cases. Observers of the Clydesdale and Fernie trials, especially, were wholly unconvinced of the impartial character of the expert witnesses. If they were supposed to be seekers of truth, these hired men of science certainly did not come off that way. Or, at least, the press found nothing so high-minded in their behavior. The *Chemical News* remarked: "Assuredly Science in the lecture-theatre and Science in the witness-box are two distinct beings. Of all sciences, perhaps Chemistry has, hitherto, cut the poorest figure in the courts of justice."[88] Witness-box quarreling seemed to reduce scientific authority to mere opinion, expert perhaps, but neither accurate nor impartial. The *Chemical News* summed up the controversy this way:

> The evidence of "Experts" is just now the object of general derision. Smart newspaper writers, wishing to indite a telling article, select the discrepancies in scientific evidence for a theme; noble Lords, anxious to enliven the dull debates of our hereditary legislators, find nothing so provocative of laughter as a story about the differences of "mad doctors;" and barristers, ready to advocate any opinion, and anxious, perhaps, for a monopoly of the "any-sidedness," when addressing a jury, dilate with well-simulated indignation on the fact that eminent scientific men are to be in the witness-box on opposite sides.[89]

The Kerosene cases had bloodied the public face of science. Not only did it seem to undermine ethical and scientific standards, but it also seemed to produce nothing more than indignation, frustration, and disillusionment on the part of the public.

Perhaps this was nothing unusual. Controversies always run the risk of exposing the ungentlemanly side of science when they go public. And patent disputes, especially those involving so much money and so many high-profile men of science, were always going to be big events. Thus the Kerosene cases revealed what many already knew; in spite of the popular rhetoric that science was the handmaiden of industry, the daily relations between science and industry were contentious, problematic, and unstable.

To the specific question at the heart of the cases, whether scientific publication could nullify a patent, the British courts, at least, had decided against it. For a man of science to own, as opposed to receive credit for, a discovery, he would have to take out a patent. To the question of whether a man of science *should* patent a discovery, the answer obviously depended on the individual. If a person patented a discovery, that person as well as the discovery itself could be cast as commercial. If a person did not, the discovery and the person might remain disinterested and pure, but the cash profits and the legal priority might belong to someone else. Patents could protect intellectual property but only at the risk of social demarcation.

The one thing patents did not cover was a distinction in types of knowledge. The Kerosene cases showed that different discoveries (scientific versus practical) and different activities (research versus application) were as arbitrary as the social distinctions. Some men of science, and some inventors, might struggle to demarcate and defend the boundaries that separated the two, but others tried equally hard to erase them. Patents, as practical discoveries, could be upheld, legally, only by recourse to a social fiction about the motivation of the knowledge producer. As Bigelow explained, the line separating pure from profitable research was a preciously thin one.

PART 3

PETROLEUM

CHAPTER 7

The Rock Oil Report

Gentlemen: I herewith offer you the results of my somewhat
extended researches upon the Rock Oil, or Petroleum, from
Venango County, Pennsylvania, which you have requested me to
examine with reference to its value for economical purposes.
—Benjamin Silliman Jr., *Report on the Rock Oil, or Petroleum,*
from Venango Co., Pennsylvania (1855)

I

F THIS MODEST INTRODUCTION to a routine consulting report hinted at
anything unusual, it was Silliman's mention of "somewhat extended researches."
The impact of those researches would prove to be anything but typical or timid.
Within a decade, commentators were pointing to Silliman's report as the cata-
lyst to developments in American petroleum.[1] By the turn of the twentieth century, historians were identifying Silliman with the start of the modern oil industry.[2] The *Report on the Rock Oil* is now the most famous consulting report
ever written.

For that reason alone, Silliman's report merits a close look. Oddly enough, it
has not received much attention. It is usually treated as a promotional tract, and
Silliman's role is reduced to mere endorsement. This perspective misses the
specific reasons *why* the report was commissioned in the first place and, equally
important, *how* it was written. Silliman, the consulting chemist, played the key
role in deciding what to make of petroleum. On the one hand, that meant outlining the possible products to be manufactured. On the other, it meant evaluating and extending petroleum analyses. Silliman's report thus conformed to
the commercial culture of midcentury America and, at the same time, widened
the stream of ongoing research on the chemistry of coal and oil.

Silliman's report has another somewhat famous feature, a rich correspondence surrounding it. Many letters among the "gentlemen," and between them
and Silliman, have survived; thus, historians have the rare opportunity to study
an engagement from the perspective of both the capitalists and the consultant.[3]

In that light, the business partners displayed all the anxieties and aspirations of adventurers staking their energy, cash, and reputations on the professional opinion of a distinguished man of science. Silliman, for his part, can be seen to be using an otherwise ordinary engagement to further his consulting practice and his own scientific research, right down to the purchase of new chemical equipment. That Silliman's report was published allows the public and private elements of a consulting engagement to be compared and contrasted. It also makes clear the historical significance of this genre of scientific-commercial literature. For Silliman's *Report on the Rock Oil* helped to make petroleum a science-based industry.

The Pennsylvania Rock Oil Company

In 1852[4] the physician Francis Beattie Brewer moved from Barnet, Vermont, to Titusville, a small town in northwestern Pennsylvania, to join the firm of Brewer, Watson and Company, "an extensive lumbering business on Oil Creek," of which his father, Ebenezer Brewer, was a partner.[5] Several oil springs were known to exist on the company's property, and one had been worked for several years to provide a lubricant for sawmill machinery and possibly an illuminant for torches used in open parts of the lumber mill. Brewer had been introduced to petroleum by his father, who had sent him several gallons to examine while he was still practicing medicine in Vermont. Upon arriving in Titusville, Brewer learned that it was common for families in the area to have a little bottle of it "to administer externally in Rheumatism, continuous eruptions of all kinds, burns, scalds, cuts, bruises, and superficial inflammation; and internally for [numerous] diseases."[6] Petroleum's "great efficacy in several diseases" spurred Brewer to explore the possibility of marketing it more widely as a "domestic remedy."[7]

In the fall of 1853, on a return visit to New England, Brewer took a bottle of "Creek Oil" to show his friends and relatives.[8] He gave a sample to his uncle, Dixi Crosby (1800–1873), professor of surgery and obstetrics at Dartmouth Medical School, where Brewer had studied medicine, and to Oliver P. Hubbard (1809–1900), professor of chemistry at Dartmouth College. Both pronounced the petroleum useful. Some weeks later, another former student, George H. Bissell (1821–1884), a lawyer from New York City, visited Dartmouth.[9] Professor Crosby showed him the petroleum and "expatiated with great enthusiasm" upon its "wonderful properties."[10] Bissell, however, was not immediately impressed. He waited until the summer of 1854 before commissioning Albert H. Crosby, Dixi's son, to go to Titusville to meet with Brewer. If Crosby returned with a favorable report on the condition of the oil springs, Bissell and his business partner, Jonathan G. Eveleth, were willing to organize a petroleum company and to try to "launch the enterprise on the New York stock market."[11]

Within a month, Crosby had journeyed to Titusville to inspect the springs, and, in the company of Brewer, traveled down Oil Creek roughly nine miles to its confluence with the Allegheny River (the site of the future Oil City). "[We] stood on the circle of rough logs surrounding [one] spring," Brewer recalled, "and saw the oil bubbling up, and spreading its bright and golden colors over the surface, it seemed like a golden vision."[12] Crosby hurried back to New York City with an attractive proposal for Eveleth and Bissell. Brewer was willing to "assign or deed to the Co. [Eveleth and Bissell's prospective petroleum company] . . . a certain 100 acres of land known here [Titusville] as the Hibbard Farm & embracing most of the oil territory as yet discovered & further other oil springs on our lands."[13] The lease cost $5,000. Eveleth and Bissell thought the terms reasonable, and Brewer shipped several barrels of petroleum "for exhibition."[14] When Brewer arrived in New York City several weeks later, he was dismayed to find that Eveleth and Bissell had decided to hold off on plans to form a petroleum company.

Perhaps the whole scheme was too visionary.[15] Or, more likely, the scheme was delayed for lack of investors. Eveleth and Bissell could not come up with the $5,000 for the lease. And even though they were "possessed of considerable means," as capitalists, they did not rank, according to one observer, among "the 'heavy' of New York."[16] Eveleth and Bissell did not think there was any need to rush "the thing through at 40 horse power."[17] In promoting a new joint-stock company, it would take "time to make all the necessary investigations of the premises."[18] Besides, raising money for a new company was especially difficult "in one of the most stringent seasons that has ever marked the financial history of our country."[19] The markets were in recession.[20]

In early October 1854 Eveleth and Bissell arranged to visit Titusville themselves, but instead of going directly to Oil Creek, they stopped in New Haven, Connecticut, to see if they could interest "certain capitalists" in investing in the petroleum scheme.[21] They contacted their old friend, Anson Sheldon, a retired minister, who introduced them to James M. Townsend (1825–1901), president of the City Savings Bank, and to other leading citizens, including Silliman, who was Hubbard's brother-in-law and might already have heard of the wonders of petroleum.[22] Townsend and others were interested; but before investing any money they made two demands: a scientific analysis of the petroleum was to be undertaken to determine its commercial value, and a committee was to be sent to Titusville to study the oil springs.

Eveleth and Bissell now pushed forward with the plans to form a joint-stock petroleum company. On 10 November 1854 Brewer deeded the oil tract to them.

The value of the Hibbard farm was raised to $25,000 because the two New York promoters thought that "if the amount proposed [$5,000] be given [publicly], we shall find it impossible to get stock taken at any price—the capital stock [$250,000] being so much larger than the amount paid for the land."[23] According to a contemporary writer, J. T. Henry, "the land was put at a figure far above its cost," which was "usually done in the formation of joint stock companies."[24] Henry's observations would seem to be consistent with the opinion of Josiah Whitney, who had griped that in any "swindling speculation" the land was "fixed at an enormous sum."[25] But Eveleth and Bissell were doing just the opposite. Cheap land and expensive stock smelled of a swindle. In contrast, land valued at $25,000 with a company capitalized at $250,000 would appear to be reasonable for a new mining venture. Eveleth and Bissell were shrewd businessmen behaving conservatively. Paradoxically, they had to practice deception in order to be perceived as being honest.

In doing the deal, no money changed hands. Bissell and Eveleth gave notes, and Brewer was given stock for his part of the purchase price of the lease. Albert Crosby, who could not produce a note to meet his share, dropped out of the scheme. On 30 December 1854 the certificate of incorporation for the Pennsylvania Rock Oil Company was officially filed with the state of New York; Brewer, Eveleth, and Bissell had formed the first petroleum company in the United States.[26]

Not Selling Stock in Oil

The Pennsylvania Rock Oil Company of New York was a midsized, joint-stock enterprise for mid-nineteenth-century America, whose capital was divided into 10,000 shares of $25 each. (Typically such companies were capitalized at somewhere between $100,000 and $1 million.) As with all stock ventures, this capitalization meant *potential* capital. If all the stock were sold at par value (the price on the certificate), the sum would be equal to the company's capitalization. This, however, never happened. First, there was an amount (sometimes large) of stock that was given to or held by trustees, directors, or even the company's consultants.[27] Second, the stock that was sold almost always had to be discounted. The entire art of stock promotion lay in selling as much stock as close to par value as possible. The larger the discount, obviously, the more worthless the stock. In the case of the Pennsylvania Rock Oil Company, Brewer, Sheldon, Bissell, and Eveleth sold what they could at prices ranging from $2.50 to $0.50, a 90 to 98 percent discount.

The trouble with such deep discounts was that they too smacked of swindling or, at least, haste. And it was not the intention of Eveleth and Bissell to rush into the market with their stocks and "operate on 'change,'" as Charles Jackson had put it.[28] They had from the beginning planned to prospect for and produce petroleum. "I do not believe," Crosby confided to Brewer, "[in] making it [Pennsylvania Rock Oil Company] pay all that it can in three months."[29] The reason for selling discounted stock was simply to build up some kind of a "Treasury" from which funds could be drawn "to raise, procure, manufacture and sell Rock Oil."[30]

The stock price and investors' willingness to purchase it depended on several interrelated factors: who was on the board of directors, and what was known (or unknown) about the product and its market? The answer to the first question could be found in the company's articles of incorporation; but there was nothing in these about rock oil or its uses. Some oil historians have assumed that the Pennsylvania Rock Oil Company intended to market petroleum principally as a lamp oil and, perhaps, only secondarily as a lubricant or medicinal agent.[31] The reason seems to be linked to accounts dating from the mid-1860s, about a decade after the establishment of the company. Bissell, for example, recalled his initial meeting with Dixi Crosby as follows: "I became greatly interested in the product . . . not . . . for medicinal purposes, but . . . [as] a good illuminator, and we [Eveleth and Bissell] sought it as an article of commerce. Illuminating oil from coal was just beginning to be talked of, but very little was made then."[32] The oil historian Henry, writing in 1873, made the same point. "Coal oil was just being introduced in the eastern states for illuminating and lubricating, and the *similarity* of products *naturally* suggested the question why Petroleum might not be used for the same purposes." Henry added: "Coal oil was selling for a dollar a gallon."[33]

The natural similarity between coal oil and petroleum, so self-evident to Bissell in 1866 or to Henry in 1873, might not have been so obvious to Bissell, Eveleth, or Brewer in 1854. Petroleum was thick, green-to-black in color, with a strong, foul bituminous odor. It burned very poorly and produced a great deal of smoke. If petroleum had been tried in Brewer and Watson's sawmill, they must have been aware of its problems, which probably explains why they burned it in the open parts of the mill. To be used as lamp oil, petroleum needed to be refined, but the Pennsylvania Rock Oil Company had been organized, initially, to market petroleum in its crude or unrefined state. Furthermore, coal oil was not an especially popular lamp oil at this time; Gesner's Kerosene Oil Company began production only in the summer of 1854. Most likely, then, petroleum would have been marketed, primarily, as a lubricant. "[T]he Co attach the greatest value to it as a lubricator," Albert Crosby remarked in the fall of 1854, "and think that [is] the trumpet to blow the loudest as there is a great desideratum of such an article."[34]

To be sure, the business partners would not have prevented the sale of petroleum as a lamp oil or medicinal agent, which after all had been the spur to

Brewer's involvement in the first place. In stock promotion, as Eveleth and Bissell knew from experience, it was better to advertise multiple commercial uses rather than just one. Gesner had done precisely this with his Kerosene oil. In November 1854, before the Pennsylvania Rock Oil Company was chartered, Eveleth spelled out the model he had in mind for a successful petroleum promotion.

> In fine we shall put everything in perfect shape, so as to bring it out in the papers, in the manner they did the Breckenridge Coal.
>
> We shall get some of your [Brewer's] best western men to examine and testify as to facts *there.* Treasure up all valuable facts respecting the springs and get a paper from that most important individual, the oldest inhabitant.
>
> We shall prepare a series of Questions to be answered by you and others there, all of which have been or will be asked by us.[35]

The Breckenridge Cannel Coal Company had been launched in February 1854 with a lot of publicity, a special charter from the legislature, and a large capitalization, just the kind of backing Eveleth and Bissell would have loved to have had. The Breckenridge Company had been set up, initially, to mine cannel coal, the variety highly sought for manufacturing gas. Making oil from Breckenridge cannel had not been part of the original plan, but by the spring of 1856, oil had become the principal product, and the company was renamed the Breckenridge Coal and Oil Company. One person who knew a lot about coal gas, coal oil, and how to launch a company "in the manner" of the Breckenridge was Silliman. In all likelihood, the reason Eveleth and Bissell had traveled to New Haven was to meet him.

Eveleth was optimistic about the next step in the promotion plans: if we can "convince the public that there is a large *quantity* of oil, then we can make the stock sell well."[36] But it did not. During the first months of 1855, Eveleth and Bissell tried by "Herculean efforts" to sell stock. "Every effort was made," observed Henry, "but the great stringency in the market, not less than the unusual character of the enterprise, placed the stock in the ever dangerous category of 'fancies,' and prevented its being taken to any great extent in the city of New York."[37] "Indeed, from the day the oil lands were sold to eastern capitalists," commented a resident of Oil Creek, Thomas A. Gale in 1860, "all circles in Titusville" had dubbed the enterprise "The Fancy Stock Company."[38] Even Brewer's father had "no confidence" in Eveleth and Bissell and cautioned his son against involvement in such a "doubtful transaction." Ebenezer Brewer warned: "You are associated with a set of Sharpers, and if they have not already ruined you, they will do so if you are foolish enough to let them do it."[39]

So despite all their careful plans and market maneuvering, Eveleth and Bissell could not move stock in the Pennsylvania Rock Oil Company. "The enterprise continued to hang fire."[40] Petroleum was an unknown risk, and investors were unwilling to buy stock, whether fancy or not. Brewer concluded that "times were hard and capital was timid."[41]

Eveleth and Bissell once again called upon Sheldon, their Connecticut contact and ardent petroleum enthusiast, "to fan the little flame of interest, manifested by a circle of gentlemen in New Haven."[42] But the New Haven capitalists were waiting, and so were Eveleth and Bissell, as it turned out, for a recently commissioned report on the commercial value of petroleum by none other than Silliman.

Consulting Silliman

After the banker James Townsend and other New Haven capitalists made clear the conditions for their investment, Eveleth and Bissell arranged for samples of petroleum to be sent to two of "the best chemists in the country": Silliman and Luther Atwood. "We shall have it analyzed," Eveleth optimistically informed Brewer, "and shall make use of their reports."[43]

Atwood, a practical chemist working at the Downer Company outside Boston, was trying to find ways to improve his patented lubricant, Coup oil. Petroleum was one of several bituminous substances he was experimenting with in an effort to remove Coup oil's dreadful odor. Atwood began his investigation of Pennsylvania rock oil in late October, and by the first week of November he had sent off his report to Eveleth and Bissell. No copy has survived, nor is there any record of what he did, or how much he charged. Atwood's report was probably no more than a short letter outlining the chemical tests he had done, no doubt various fractional distillations, along with a list of the quantities and qualities of the distillates. The partners were pleased with Atwood's analysis. "[H]is report of the qualities of the oil," Bissell reported to Brewer, "and the uses to which it may be applied are very favorable."[44] The principal use was as a lubricant, but Atwood's favorable report did not help Eveleth and Bissell sell stock.

Silliman's report was a different matter, in large part because Silliman was a very different kind of consultant, and he was particularly influential in New Haven, where he had recently returned to resume teaching at Yale.[45] It was most likely that the New Haven capitalists suggested his services, although Silliman himself might have wanted to study the relatively unknown rock oil. Silliman's services would be expensive and extensive, but in no way could Eveleth and Bissell have anticipated how costly and how thorough an analysis he would do. Nor would Silliman. Typically, such an in-house, laboratory analysis would have taken only a few days or maybe a week or two (the time required to test the Albert mineral, for instance, or for Atwood's analysis). Silliman, however, took six months. From November 1854 through April 1855, he worked, off and on, with rock oil.

Silliman's twenty-page *Report on the Rock Oil, or Petroleum, from Venango Co., Pennsylvania, with Special Reference to its Use for Illumination and Other Purposes* can be read as a chronology of his engagement with petroleum. Divided into five sections, each describing a different experiment, it guided the reader through the transformation of petroleum from a lubricant into a lamp oil. Silliman's report was not simply a confirmation of an earlier marketing decision made by Eveleth and Bissell; on the contrary, it contained a persuasive argument for remaking petroleum into a new product and thus changed the type of company that was formed.

The first section of the report was probably completed by the end of October, and in it Silliman reviewed the extant literature on petroleum, most of which could have been culled from articles in the *American Journal of Science*. On the geology of petroleum, Silliman noted that the "position of the rocks furnishing this natural product, is in the coal measures—but it is by no means confined to this group of rocks, since it has been found in deposits much more recent, and also in those that are older."[46] Such stratigraphical uncertainty would lead to a good deal of debate among geologists within the next decade. But it also tied in with a bigger question, the relations among petroleum, asphaltum, and coal. Silliman asserted that petroleum was "uniformly regarded as a product of vegetable decomposition," but he acknowledged that there was no consensus about "[w]hether this decomposition has been effected by fermentation only, or by the aid of an elevated temperature, and distilled by heated vapor."[47] Silliman was being cautious and tactful. He did not think consulting reports were the proper place to engage in a scientific controversy (although others, such as Jackson, did use reports in this way). For the time being, Silliman was prepared to sidestep geological questions about origin and occurrence in order to concentrate on chemistry.

Under "General Characteristics of the Crude Product," the title of the first section, Silliman recorded the results of some basic tests for inflammability. On the one hand, such tests were redundant, for in terms of mineralogical classification all bituminous substances were by definition inflammable. On the other hand, as the Albert mineral trial had shown, such tests could be used to determine where this particular kind of petroleum fit relative to other bituminous substances. Pennsylvania rock oil "takes fire with some difficulty, and burns with an abundant smoky flame."[48] Silliman, at least initially, did not regard its use as lamp oil as very likely.

Nor was he especially impressed with its lubricating qualities. Petroleum, he noted, was "frequently used in its crude state to lubricate coarse machinery."[49] It was not suitable for fine or high-speed machinery, for which sperm whale oil was the best and most expensive lubricant (which was one reason Gesner's Kerosene, Breckenridge's coal oil, and Downer's Coup oil hoped to target it). As a lubricant for coarse machinery, petroleum would have to compete against other less costly animal and plant oils already well established in the market. Again, Silliman did not think petroleum's prospects were good.

The second section, "Examination of the Oil," reported the tests on fractional distillation, a procedure very familiar to chemists, but because this report was for nontechnical readers, Silliman elaborated.

> Fractional distillation is a process intended to separate various products in mixture . . . having unlike boiling points, by keeping the mixture contained in an alembic at regulated successive stages of temperature as long as there is any distillation at a given point, and then raising the heat to another degree, &c.[50]

Silliman discovered that Pennsylvania rock oil contained two distillates that might be valuable as solvents. This bit of good news was immediately passed along by Bissell to Brewer.

> Prof. Silliman of Yale College is giving it a thorough analysis, and he informs us that so far as he has yet tested it, he is of the opinion that it contains a large portion of benzole and naphthalin and that it will prove more valuable for the purposes of application to the arts, than as a medicinal, burning, or lubricating fluid.[51]

The bad news was that Silliman had nixed the entire promotional plan for petroleum, the medicinal, burning, and lubricating oils. Other small worries about Silliman began to surface among the business partners. "Our expenses of a thorough analysis will be very heavy," Bissell admitted to Brewer, "but we think the money will be well spent . . . believe us."[52]

Silliman had been experimenting for nearly two months, and he was still in the midst of the second section of his report, when, a few days before Christmas 1854, he wrote to Eveleth and Bissell. "I am very much interested in this research, & think I can promise you that the result will meet your expectations of the value of this material for many useful purposes." He explained that the fractional distillation experiment had consumed "2 to 3 weeks, & is still in process." Best of all, he had extracted "six different oils."[53] In response, Sheldon assured Silliman that "[a]ll parties concerned are looking forward with no small degree of interest to the result of your experiments."[54]

Silliman, however, seemed far more interested in his scientific pursuits than in making any report to the company. The delay did not dismay Eveleth and Bissell so much as the fact that they were paying for it. In early February 1855, Eveleth sent Brewer a succinct note tinged with exasperation: "Hope to have Silliman's report soon."[55] A week later Eveleth expressed a similar urgency when noting that Silliman's "report is looked for with much interest. We shall have it printed the moment it is ready."[56] But that moment was to be delayed still further.

On 17 February Eveleth reported that Silliman had suffered "an explosion" in his laboratory. The partners were now being asked for money to purchase "a new still, and then a retort."[57] Although the cost of materials and supplies was usually passed on to the commercial patron, in most engagements explosions were not part of the analysis. The explosion might have served, unwittingly, as means to another end. Several months earlier, Silliman had written to the president and fellows of Yale College describing the "very dilapidated condition" of the laboratory and requesting $600 for "our most urgent wants." An explosion would seem urgent enough, but it might also have required Silliman to overspend (by $500) his $600 appropriation, for which he was duly reprimanded by the Prudential Committee of the College.[58]

The accident and the time required for securing new equipment delayed the "Distillation at higher temperature," the third section of Silliman's report. These experiments involved separating the lightest, most volatile fractions of the petroleum. Silliman then tested these distillates for use in lubricating and illuminating. One fraction caught his attention, and he immediately informed Eveleth and Bissell. "Having met unexpected success in the use of the distilled product of *Rock Oil* as an illuminator, I am solicitous to test its power in this respect in various lamps and also in comparison with *various oils*."[59] Unexpected was the word for it. The partners had been ready to abandon the idea of lubricants and lamp oils for solvents when suddenly they were put back on to lamp oil. Moreover, Silliman was now talking about refining petroleum. The partners would have to contemplate building a manufactory, or shipping the petroleum to one, in addition to getting it out of the ground.

Silliman also wanted *more* time and equipment to complete the photometric tests.

> If therefore you wish an accurate comparison by photometric valuation of the Rock Oil, with Sperm, Colza & Sperm Candles you will see that I am supplied with the following means and apparatus Viz—
>
> 1st A few gallons say 2 or 3 each of Colza & Sperm Oils . . .
>
> 2d *Two Carcel's Mechanical Lamps* . . . in every way alike & preferably of the standard size of 1 inch circular wicks. These may be *hired* to be returned, & should be accompanied by a supply of wicks and duplicate chimneys
>
> 3d Two of the best approved form of Camphene Circular wick double draft Lamps as nearly alike as possible, in size of wick &c &c with supply of wicks and chimneys.
>
> 4th A few pounds of *Judd's patent* Sixes such as they furnish to the Gas Companies for comparison of lights.[60]

It was very unusual for consultants to make such extensive and detailed demands, and the tone suggests that Silliman understood very well that the partners were anxious and willing to do whatever it took to get their hands on his report.

The photometric tests were part of an ulterior plan. Silliman was a member of the Lighthouse Board, a conclave of scientific men who, under the auspices of the Smithsonian Institution and headed by its director Joseph Henry, were commissioned by Congress to conduct a series of studies to determine the best lamp oil for use in U.S. lighthouses. Silliman mentioned the Lighthouse Board to Eveleth and Bissell, but he gave them little choice in the matter of whether to pay for the tests.

Silliman wanted the petroleum partners to buy "all new apparatus"—lamps, fixtures, and a special photometer.[61] This last piece had to be special-built by "scientific artists," as Silliman explained in his report; "I proceeded therefore to have constructed a *photometer,* or apparatus for the measurement of light, upon an improved plan." J. & W. Grunow, opticians of New Haven, designed the new instrument to Silliman's "entire satisfaction."[62]

Silliman's photometer would in fact become the subject of an article in the *American Journal of Science.*[63] With it, he conducted several experiments for the New Haven Gas-Light Company. Gas was manufactured using various coals, including "Hillsboro (New Brunswick)" and "Ohio Diamond Coal," samples of which he had gathered from previous consulting engagements. Silliman made no mention of using petroleum to manufacture gas. In his *Report on the Rock Oil,* he noted that petroleum would work, but "other products, now known and in use, for gas making, might be employed at less expense for this purpose, than your oil."[64]

Silliman was using Eveleth and Bissell to subsidize his research on gas lighting. By withholding his report, he bought extra time (and materials) for that research. "I spoke to Mr. Sheldon of rendering a preliminary Report of our Results so far obtained but on reflection I consider it inexpedient to do so—until these light comparisons are made. . . . Be so good as to deposit to my credit in Merchants Bank N.Y. $100 on a/c of having paid my assistants for services."[65] His assistants undoubtedly were his students from the Yale School of Applied Chemistry. He and John Addison Porter (1822–1866), the other professor in the laboratory, did analyses for commercial patrons as part of the program in industrial chemistry. Thus, Eveleth and Bissell were also subsidizing education.

It is very doubtful if any of these photometric experiments would have made Eveleth, Bissell, Brewer, and Sheldon feel better about the delay in receiving Silliman's report. The stock of the Pennsylvania Rock Oil Company was not selling, and the only capitalists possibly interested in buying any, the New Haven gentlemen, would not commit until Silliman had finished. To the business partners, Silliman's report seemed the sole hope for the floundering petroleum company.

Silliman finished the rock oil report on 16 April 1855, just "before leaving for his Southern tour." He and his father planned to go to the Blue Ridge Mountains of Virginia to examine copper mines. At age seventy-five, this was to be the last engagement for Silliman Sr. Presumably the southern tour also included a stop in Washington, DC, to discuss with Joseph Henry the progress of the Lighthouse Board.[66] Silliman, however, did not give his report to Eveleth and Bissell. He placed it "in the hands of a friend in N.Y." with explicit instructions not to deliver it to the company "until satisfactory arrangements were made for the payment of his Bill, which amount[ed] . . . to the round sum of $526.08."[67] This sum, in addition to what Eveleth and Bissell had already paid, roughly $100 for the apparatus destroyed in the explosion, amounted to "the whole expense of analysis exceed[ing] $600."[68]

Sheldon called the price "exorbitant," and it was. Compared to other engagements, Silliman was easily charging the petroleum partners triple his usual rate. Indeed, at precisely the same moment he was finishing the rock oil report he was completing negotiations with Headley, the organizer of the Kentucky Mining and Manufacturing Company, to write up a report for only $500. For the coal company, Silliman had visited Kentucky (with Hall) *and* had done a chemical analysis of the coal. He offered no such geological expertise to Eveleth and Bissell. In short, his petroleum price seemed far out of line.

Yet Silliman knew that he had the petroleum partners over a barrel: no money, no report. Sheldon recognized the squeeze for what it was, but he also recognized the importance of Silliman's report. It was favorable, "more favorable, even, than I had dared to hope."[69]

Silliman was indeed very optimistic about the prospects for Pennsylvania rock oil. In his professional opinion, petroleum was an excellent investment. "In conclusion, gentlemen, it appears to me that there is much ground for encouragement in the belief that your Company have in their possession a raw material from which, by simple and not expensive process [fractional distillation], they may manufacture very valuable products."[70] The business partners found the necessary $526.08. The report proved to be the turning point for the Pennsylvania Rock Oil Company. As Eveleth aptly summarized the situation for Brewer: "This report of Silliman's is doing the *right thing,* but it has cost a good deal of money."[71]

The right thing meant interesting capitalists. When Eveleth and Bissell got their hands on the report, they immediately sent it to a New Haven printer, with Silliman's approval. Within two days it was ready for distribution and "in the hands of the monied men in the city."[72] "The report will make a stir. Stock will then sell. Things look well," Eveleth exclaimed in telegraph-style to Brewer;

"Will send you 50 copies and more if you want. It is a good report."[73] The response of prospective investors was good, too. Sheldon received "a number of applications for Reports & also for sample specimens of the oil."[74] Not everyone, however, was favorably impressed by the report. Brewer's curmudgeonly father did not "consider the stock worth a straw" and flatly declared "the whole transaction a perfect failure."[75]

The stock was in fact the problem, but not for the reasons implied by Ebenezer Brewer. "Joint stock companies located in New York are in bad odour," Sheldon explained, "[because] in New York the individual property of the stock holder is liable for the payment of the debts of the company."[76] This meant that an investor buying a share at $2 whose par value was $25 might be called by the courts to pay the full $25 to the company's creditors should the company go bankrupt. The investor was responsible for the stated price of all stock owned.[77] While the New York laws were designed to prevent the sale of fancy stock, they could also be misused. In the early 1850s several apparently reputable New York joint-stock companies, in collusion with their creditors, had perpetrated such "enormous frauds" by declaring bankruptcy in order to collect the face value of discounted stock. Many "prominent men" in New Haven had "experienced losses" in swindles committed by New York companies, in particular the New York and New Haven Rail Road Company and the Western Empire Company. As a result, New Haven "monied men" were very wary of New York stock.[78]

Sheldon thought stock would be taken if the Pennsylvania Rock Oil Company was organized under the Laws of Connecticut and if New Haven were made "the place for its business operations," thus avoiding New York's financial liability laws.[79] He also proposed another reason for moving the company: "We can organize under Conn. Laws without giving notice in the Public Prints. Some of the companies now in operation in this city [New Haven] were organized in that way."[80] Secrecy might seem an odd thing for a public joint-stock company, especially if the promoters only wanted to boost the price of their fancy stock, but not if the New Haven capitalists wanted to monopolize what might be a very lucrative product. Whatever the reason, the desire to move the company to Connecticut trapped Eveleth and Bissell. "Under these adverse circumstances," Sheldon concluded, "there must be a change in organization . . . or the enterprise would prove a failure."[81]

By the end of May 1855, Eveleth realized that he and Bissell would have to form a new company. He decided to go to New Haven "& shall remain here till the thing sinks or swims."[82] He told Sheldon, whom he thought was being too cautious, "to talk it up." Eveleth knew that they had to get "first rate men interested" in the petroleum scheme.[83] By the beginning of July, Sheldon had managed to secure "the right kind of men[,] men of means—of good business habits & that will work harmoniously together." Upstanding capitalists, Sheldon boasted to Brewer, were "the class of men that are coming in."[84] Petroleum stock finally began to sell. Eveleth wrote to Brewer in a joyous mood: "The Stock for the New Co. is nearly all taken, & *all*—yes, *all*—*all* I say. . . . We will have our congratulations meeting before the first day of August."[85]

He rejoiced too soon. Eveleth had only verbal assurances from the New Haven gentlemen. They still wanted a committee to visit the oil springs and report on the quality and quantity of the petroleum. This was the second part of the proposal that had been made almost a year earlier, the first part being fulfilled by Silliman's report.[86] So Sheldon and Asahel Pierpont, a mechanic and businessman, formed a committee-of-two and journeyed to Titusville in the last weeks of July 1855. Everything rested on this committee's conclusions, so Eveleth and Bissell dashed off a letter to Brewer telling him to "[d]o all in your power to make them appreciate the true value of the springs. Work that *one*, if possible, won't you?"[87] Sheldon and Pierpont returned after a fortnight "spent at Titusville & vicinity in surveying the Oil Lands" and reported their findings.

> We took a general survey of the land, . . . & made some 25 or 30 excavations at different points over some 15 or 20 acres[.] In every instance we found undoubted marks of oil. After digging down into the hardpan, the oil mixed with water seemed to rise up as though there was a pressure beneath. The deeper the excavations, the more abundant the oil appeared. So far as we could judge, the excavations would each daily yield from 2 qts to 4 gallons & they might be increased to any number. From our investigations, it is evident that the oil can be found any where you are disposed to dig over a large tract of land in Oil Creek Valley.[88]

The committee was "well pleased," and so, too, were the other capitalists.[89] "There cannot be any doubt," Sheldon predicted, "that there will be a yield of oil sufficient to justify our company."[90] Pierpont declared that "*the oil was there & that he was satisfied*."[91] Eveleth concluded: "That fixes the thing certainly."[92]

By August 1855 the New Haven capitalists had their survey and their chemical report. They vacillated for a month longer while some members debated whether to travel to Titusville to see for themselves. After deciding to let the report of Sheldon and Pierpont stand, the New Haven capitalists formed a new joint-stock company. On 18 September 1855, with an increased capitalization of $300,000, divided into 12,000 shares at $25 each, the Pennsylvania Rock Oil Company was reincorporated under the laws of Connecticut.

"[A] first rate Co.," Eveleth thought, but the arrangement, "the best we could do," he grumbled to Brewer, "sacrifices our own [company] dreadfully."[93] Sheldon got 1,000 shares, as did Pierpont; Townsend took 500. Brewer, Bissell, and Eveleth each held 1,200 shares, and thus still retained a controlling interest.[94] But the bylaws and articles of association, which were made public, contrary to

Sheldon's speculations about secrecy, announced that the Pennsylvania Rock Oil Company was a New Haven enterprise. A majority of the directors were to be chosen from among New Haven stockholders, and the company's headquarters was to be in New Haven.

Pennsylvania laws, however, prevented out-of-state corporations from owning land within the state. So Eveleth and Bissell deeded the Hibbard farm to Pierpont and William A. Ives, who gave a bond for its value and then leased it back to the company for ninety-nine years. Thus the oil tract was, legally, not owned by the Pennsylvania Rock Oil Company of Connecticut.[95] These property laws were changed in the late 1850s and would greatly facilitate the organization of companies during the oil boom of the early 1860s. As for the Pennsylvania Rock Oil Company of Connecticut, the deal was finally done.

Silliman was elected president of the company and received 200 shares of stock. His role was nominal; the company needed only "the prestige of a name renowned in science."[96] But as Eveleth told Brewer, "we must have him."[97] Silliman brought credibility to the new petroleum company and, simultaneously, gave a boost of confidence to his consulting report. But, unlike his other business ventures, Silliman did not remain a member of the Pennsylvania Rock Oil Company for long. The reason, according to J. T. Henry, had to do with production. "[Silliman] never expected to see it obtained in any great quantity."[98] This explanation seems somewhat odd because Silliman joined the company, as did other New Haven gentlemen, after Sheldon and Pierpont made their report. Nonetheless, finding commercial quantities of petroleum *did* prove to be very time-consuming and difficult, and Silliman, the restless scientific entrepreneur, might easily have lost interest. In the event, a more determined Townsend replaced Silliman as president.

Boring for Oil

In 1856 Eveleth and Bissell contracted with David H. Lyman and Rensselaer Havens, partners in a prominent Wall Street real estate firm, to begin operations in Titusville. In return for a fifteen-year lease on the Hibbard farm, Lyman and Havens agreed to pay a royalty of twelve cents a gallon on all petroleum raised. But before digging could get under way, the firm foundered and then crashed during the financial Panic of 1857. Townsend had to find a new contractor.

Boarding at the same New Haven hotel as Townsend was Edwin L. Drake (1819–1880), a recently retired conductor from the now bankrupt New York and New Haven Railroad. Townsend liked and trusted Drake, so he leased the Hibbard farm to him for fifteen years for one-eighth of the oil. When Eveleth and Bissell learned of Townsend's actions, at the annual meeting of Pennsylvania Rock Oil Company stockholders on 8 January 1858, they were furious. "[A] stormy time we had from Gotham," Townsend recounted to Brewer, "our friends E & B on learning that we had leased the property Broke out in the most immoderate rage."[99] Townsend quelled the New York rebellion by renegotiating the terms of Drake's lease (a royalty of twelve cents a gallon). Townsend then organized a new company, the Seneca Oil Company of Connecticut, on 23 March 1858. Capitalized at $300,000 and divided into 12,000 shares, the Seneca Oil Company took charge of the oil tract and the production of petroleum. Townsend controlled the majority of Seneca stock.[100] In effect, he replaced Eveleth and Bissell as the driving force in the petroleum venture.

In the spring of 1858, Townsend sent Drake to Titusville to begin "active, practical work on the ground."[101] As the new superintendent of Seneca Oil (with a salary of $1,000 per year), Drake had no experience with petroleum, mining, or Pennsylvania. But he did have a railroad pass that allowed him to travel free, although he could not reach Titusville by train. He could go only as far as Erie, Pennsylvania, where he caught the mail stage for the rough forty-mile trip to the tiny lumbering town. Drake also had a title. Before leaving for Titusville, Townsend addressed several letters to "Colonel" E. L. Drake care of Brewer, Watson, and Company.

Despite the bankruptcy of Lyman and Havens, "Colonel" Drake and Townsend seemed confident of producing petroleum. After all, numerous examinations of the oil springs had confirmed its abundance and ease of recovery. Drake hired workers and started digging a well on the site of the principal spring. After several weeks, they reached a depth of about 150 feet, when they hit ground water. It flooded the excavations and forced Drake to take a new approach. "I shall not try to dig by hand any more," he informed Townsend in August 1858, "as I am satisfied that boring is the cheapest."[102]

Boring for petroleum was Drake's singular contribution to the fledgling business. But it was probably not his decision alone. In his personal account, Townsend took credit for directing Drake to bore for oil.[103] And so, too, did Bissell. He claimed to have gotten the idea back in 1856, when he saw an advertisement for medical petroleum by the Pittsburgh druggist Samuel Kier (1813–1874). Bissell then told Lyman and Havens and Townsend, who "formed a scheme to monopolize [boring]."[104]

The one person who apparently had nothing to do with the decision was Silliman. According to J. T. Henry,

> the first idea . . . *should* have been suggested to a mind cognizant of [the discoveries of Petroleum in the brine wells along the Muskingum and Kanawha rivers]; and yet, though himself editor-in-chief of the periodical in which the circumstances were described [i.e., *American Journal of Science*], he very candidly confessed, that throughout the five months he was prosecuting the analysis, the thought of artesian boring, never once occurred to him.[105]

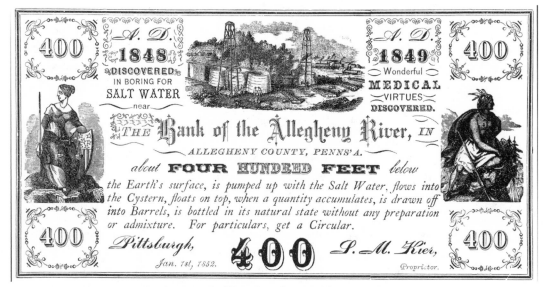

Copy of Samuel M. Kier's label for a bottle of "Petroleum" showing a derrick, which gave George Bissell the idea of boring for petroleum. *Source:* J. T. Henry, *The Early and Later History of Petroleum* (Philadelphia: James B. Rodgers, 1873), facing page 83.

Henry may have interviewed Silliman, but the confession seems forced. In his *Report on Rock Oil,* Silliman had noted that "wells are sunk for the purpose of accumulating [petroleum]."[106] And the *American Journal of Science* had published articles on well boring. But by the summer of 1858, Silliman was engaged in other business, in particular oil-from-coal not oil-from-rocks.

Regardless of who thought of boring, it was the persistent and hardworking Drake who made the idea work. The techniques and tools for boring through rock, though well known and widely available, were also well employed by salt companies. Drillers were busy boring for brine in western Pennsylvania and along the Kanawha River in western Virginia. Drake spent nearly ten months trying to recruit an experienced well borer from the neighboring town of Tarentum before he managed to secure the services of William A. Smith, better known to most Oil Creek residents as "Uncle Billy."

In May 1859 Drake and Uncle Billy made preparations for boring near an oil spring on an island in the middle of Oil Creek just below Titusville. By early August they had finally managed to set up a derrick and a newly acquired steam engine. Averaging less than three feet a day, a relatively slow rate, they had gone down only seventy feet or so by the end of the month. On Sunday, 28 August 1859, a day when there would be no work, Uncle Billy went to check the well, and there, in the pipe, was petroleum.[107]

> Thus link by link, was forged slowly, the chain of events . . . effect-
> ing what is known as the "discovery" of Petroleum. [B]ut inas-
> much as they did *not* suggest anything of practical importance to
> the very intelligent gentlemen who beheld them, it would seem to
> increase our obligations to the person who finally did grasp the
> simple idea of the philosophy of Petroleum.
> —J. T. Henry, *The Early and Later History of Petroleum* (1873)

THAT PERSON was Silliman. To the Yale chemist belonged the discovery of the true value of petroleum. Drake had merely bored a well and collected the raw material. And it was raw, for "the article had no certain market," according to Henry.[108] The world had to be made ready for petroleum, and the way to do that was through chemistry. "It is mainly to this science," crowed Henry, "that we owe those elaborate experimental researches which demonstrated the practical utility of Petroleum to the domestic comforts of refined civilization."[109]

To Henry, petroleum was the source of civilization, primarily because it produced the comfort of artificial light. This was not so much a product (i.e., lamp oil) as it was a testament to human progress. A good cheap illuminator was a necessity—economically, politically, and from a humanitarian perspective. It brought light to the darkness that once engulfed the "habitations of the poor."[110]

What turned petroleum into light was chemistry; or, as Henry put it, "[t]he refining influence—we might say [is] the civilizing influence."[111] Science was key to the age of petroleum. Silliman had determined the value of petroleum through elaborate experimental researches. For Henry, the chemistry of petroleum was of a piece with the chemistry of coal oil. In their timing and intent, Silliman's analyses of rock oil were little different from James Young's work on Paraffine or Abraham Gesner's on Kerosene. The application of chemical knowledge in all these cases was manifest in the manufacturing. Silliman himself made that clear at the end of his report: "There are suggestions of a practical nature, as to the economy of your manufacture, when you are ready to begin operations, which I shall be happy to make, should the company require it."[112]

Yet there was another aspect to petroleum that made it different from coal oil, and this was technological. As Drake had shown, the best method for extracting petroleum was boring, not mining. This meant derricks and drill bits, not pickaxes and shovels. Petroleum extraction required experienced well borers not an army of unskilled miners. According to Henry, "[t]he great difficulty [with coal oil] was the cost of labor."[113] Petroleum thus seemed to hold out the promise of production without toil and sweat and blood. The new oil business would be another example of American ingenuity; derricks and drills would serve as labor-saving machines.

In the historiography of American oil, Drake and Silliman have come to represent two very different sides of the business—production and refining. Drake is usually depicted as a man battling the odds, both natural and social. Boring for oil is set up as a fool's errand, so finding it is a triumph of determination and luck. Drake's brand of discovery thus becomes the seedbed for the heroic wildcatter and the plucky oil prospector. In contrast, Silliman is depicted as the methodical man of science who proves what everyone anticipated, namely, that petroleum will make a superior lamp oil. Silliman's discovery serves as a reassuring research result, which inevitably will lead to John D. Rockefeller and his Standard Oil monopoly. In short, Drake embodies the individualism of oil, while Silliman represents its corporate dimension.

Such a dichotomy is too simple, despite its appeal and influence. Drake might have been determined, but he was not in the dark about where or how to find petroleum. That is not to say his discovery was not exciting or surprising, only that his task was well defined and his methods well known. Silliman's discovery was no less exciting and surprising, and he, too, had to wrestle with material unknowns. Moreover, Silliman, like Drake, was slow and costly. Overall, the "discovery" of oil was a long process, beginning with Silliman in 1854 and culminating with Drake in 1859. That five-year period marked the start of the age of petroleum.

CHAPTER 8

The Elusive Nature of Oil and Its Markets

> The economic importance which petroleum has lately assumed gives a new interest to the chemical and geological history of this and various related substances. It is proposed in the following pages to bring together some facts and theoretical considerations bearing upon the nature, origin, and distribution of bitumens.
> —T. Sterry Hunt, "Contributions to the Chemical and Geological History of Bitumens," *American Journal of Science* (1863)

T HERE WAS SOMETHING OLD and something new about petroleum. Many regions where it occurred naturally in pools or springs had been mapped by geologists, and samples from around the world had been analyzed by chemists for their mineralogical compendia. Still, no one expected relatively large amounts to exist in the shallow subsurface until Edwin L. Drake bored his famous well. As the geologist T. Sterry Hunt observed, Drake's discovery spurred men of science to reconsider their theories concerning bitumens, the substances they had been wrestling with for decades. Petroleum had become a hot research topic.

That would not be the way most historians or modern scientists would describe geologists' response to rock oil. On the contrary, geologists have been depicted as indifferent or oblivious to petroleum developments.[1] Consider, for example, the astonishing opinion of Marius R. Campbell, president of the Geological Society of America in 1911.

> It is extremely difficult, if not impossible, to determine what were the opinions of scientific men of more than fifty years ago regarding the geologic relations of bitumens. . . . Even after the drilling of the first well and the resultant wave of commercial excitement, little was written on the subject for twenty or twenty-five years.[2]

Campbell's soupçon of science is belied by the volumes that nineteenth-century geologists wrote about petroleum. Still, his restricted retrospect is not unusual. Oil historians often talk about twitching sticks and divining rods as the major (if not the *only*) methods for finding oil.[3] For storytellers with a sensationalist streak, petroleum prospecting is often dramatized as a wild and woozy hunt, a drunkard's chase with drillers lurching from one spectacular strike to another and thirsting for gushers as unpredictable as the muddy streets of the many boomtowns. Petroleum, unlike chemical and electrical manufacturing, the other new industries of the second half of the nineteenth century, seems to be a throwback to less enlightened times, an irrational and irresponsible industry, not a science-based one.

Without denying that oil strikes were astonishing and frequently dangerous, it is nonetheless an exaggeration to characterize petroleum explorations as clueless stumbling in the dark. Boring for oil required money and equipment, hard labor and much time, and for those very material reasons, Americans wanted as much information as they could get before plunking down a derrick. Men of science might not have been consulted to the degree they were in the coal or kerosene industries (for reasons to be explained), but they published many useful studies of the geology and chemistry of petroleum, and within a year of Drake's discovery, theories of petroleum's origin and occurrence were being adopted and adapted in explorations for oil. Science was not so incompatible or incomprehensible as to be impractical. Often theories were developed with the cooperation of oil operators and well borers. "Science is busy giving us rules for gathering the oil," one observer remarked, "and labor and capital are busy showing Science how she is partly right and partly wrong."[4] It was an ongoing dialogue, one that provided stimulus to science and to petroleum developments—and to technology.

Drake's discovery raised the inevitable question of what to do with the stuff. Petroleum was known in commercial circles as a possible substitute raw material for lamp oils and lubricants, but in August 1859 these markets were dominated by Kerosene and other coal oils. Marketing petroleum-based products would mean direct competition with this large industrial infrastructure. It would not be a simple case of substitution—petroleum for coal—despite the optimistic predictions of its promoters. Petroleum manufacturing would require its own techniques and tools.

To stake such claims for the importance of geology and technological science is thus to take aim at prevailing historical interpretations. But the claims must not be overstated. Tens of refineries, hundreds of companies, and thousands of wells were started in the first five years (roughly 1859 through 1864); it is impossible to say what role science played in each. Nonetheless, the patterns in petroleum explorations can be examined for the scientific guidelines used by well borers. Likewise, petroleum manufacturers adhered to identifiable technological practices. The production of knowledge went hand in hand with the exploration, extraction, manufacturing, and marketing of petroleum.

Introducing Petroleum

After finding oil in his well, Drake began calmly to collect it. He ran a pipe down the bore hole (called tubing), attached a pump, and began filling all the barrels he could get his hands on.[5] The well yielded about 400 gallons a day, so Drake had to build two 1,000-gallon wooden vats. Soon, he had more than 10,000 gallons on site and the obvious problem of what to do with it. The Seneca Oil Company, surprisingly, had not made any plans in the event that Drake discovered petroleum.

In early September, the ever-resourceful Drake managed to sell some of his growing stock to the oil dealer Samuel Kier. Kier had devised a distillation process for turning petroleum into a lamp oil called Carbon Oil and had built a small plant (a five-gallon still) on the outskirts of Pittsburgh. In November 1859 Kier contracted with Drake to take *no more* than 1,000 gallons a week delivered to him for sixty cents per gallon. Kier also agreed "to use his endeavors to sell the said Oil in preference to any other."[6] By early 1860 Kier was taking delivery of nearly 2,000 gallons a week (at a lower price of forty cents per gallon), but still Drake needed more outlets.[7]

In February 1860 Drake headed to the growing industrial cities of Cincinnati and Chicago. According to the trade card he handed out, "Petrolium, or Penn. Rock Oil is a Superior Lubricator, more lasting than other Oils; is entirely free from Acids, is not affected by extreme cold, and will not dry or gum."[8] Despite such worthy qualities, Drake encountered resistance. "There is a great prejudice against it," his Chicago agent reported, "[Mechanics] object on account of the odor. Another says don't use it—you will spoil your mashinery [*sic*]." The agent suggested running off petroleum's most volatile parts for use as lamp oil and purifying the remainder as a lubricant, advice remarkably similar to Benjamin Silliman Jr.'s consulting report. The agent also recommended lowering the price. "They are making coal oil in this state," he informed Drake, "and I am told sell the crude oil for 25 cts a gallon. If this is so it will be hard to find any margin for yours at 60 cts."[9]

During his western sales trip, Drake had the good fortune to meet George Mowbray, the consulting chemist for Schieffelin Brothers, a New York City wholesale drug firm. Drake took Mowbray to Titusville to show him the new oil well. Mowbray was impressed and agreed to draw up a memorandum whereby Schieffelin Brothers would take Drake's excess supply. Schieffelin would either sell the petroleum (at a 7.5% commission) or refine it "for illuminating and lubricating purposes" (at a cost of ten cents per gallon).[10]

That same month, March 1860, Drake received other good news. The Cleveland & Erie Railroad placed an order for some petroleum, and the master mechanic of the Sunbury & Erie Railroad pronounced petroleum "a first rate Lubricator." In an optimistic mood, Drake reported to James Townsend, the president of Seneca Oil: "It takes time and work to introduce [petroleum], but I shall succeed."[11]

The big hurdle to introducing petroleum, as Drake knew from his Chicago agent, was the already plentiful supply of cheap, safe, high-grade coal oils. In early 1860 these oils were doing a booming business. To gain a market share, petroleum would not only have to come down in price; it would also have to be improved in its consistency, odor, and color. To do this required manufacturing, but there were very few companies besides Kier and Schieffelin equipped to handle petroleum. The only ones with any experience were the coal oil companies, and they had invested considerable capital in designing plants to use coal, not petroleum. According to one 1860 estimate, the big manufacturers consumed 60,000 bushels of coal per day, from which they distilled 75,000 gallons of crude coal oil.[12] Big firms such as Lucesco and North American, both located near Pittsburgh, had established bona fide reputations for their *coal oil* products. "[N]o better nor cheaper light than good coal-oil can be produced," opined the *Pittsburgh Post,* "and the fact that this is the centre of the trade, is an important one for Pittsburgers. It was thought that the [petroleum] discoveries would hinder the business of its production, but, as yet, it has had no perceptible effect, the sales [of coal oil] being constantly on the increase."[13] Drake's daily pumpings were paltry by comparison, but the production of petroleum was increasing.

Pumping Wells

News of Drake's discovery spread quickly. According to Thomas Gale, an Oil Creek resident, Titusville soon became "the rendezvous of strangers eager for speculation."

> The capitalist, as well as that large class of men not so rich as ready to venture, are streaming in from all quarters. Here, too, are men who *have toughened their constitutions* in the coal-beds of Ohio, the lead mines of Galena [Missouri] and the gold placers of California. . . . Never was a hive of bees in time of swarming more astir, or making a greater buzz.[14]

The buzz was all about oil. In November 1859 the Barnsdall well struck oil. Located about 1,000 feet up the creek from Drake's, it yielded about five barrels a

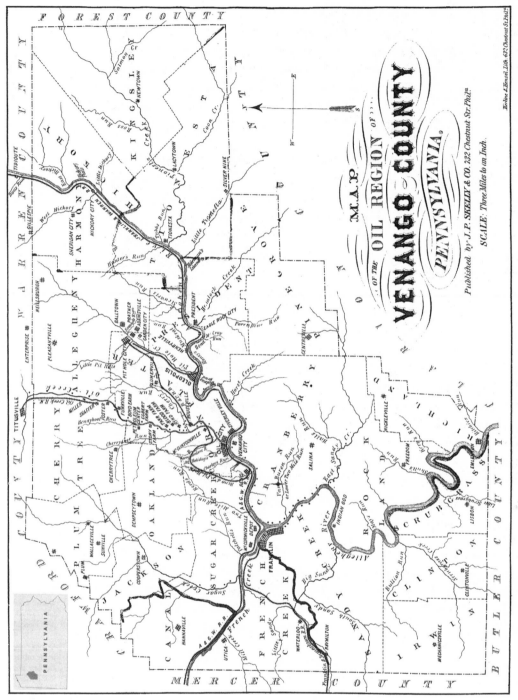

The oil region of Pennsylvania (1866). *Source:* S. J. M. Eaton, *Petroleum: A History of the Oil Regions of Venango County, Pennsylvania* (Philadelphia: J. P. Skelly, 1866), frontispiece.

day. In February, the Barnsdall well was deepened to 112 feet, which doubled its yield. "Barnsdall's was the lion of the valley," reported Gale; "Drake's having become an old story."[15] In March, a third strike, the Crossley well (located about half a mile down the creek from Drake's) came in with seventy-five barrels a day; Gale called it "a *whopper*." Together with Drake's and Barnsdall's, it formed the "trinity of contiguous wells."[16] By the summer of 1860, Gale estimated "several hundreds of wells" had been started along Oil Creek and the Allegheny River.[17]

But Oil Creek was not the only region to be hit "with a *furor* in excitement, or a *mania* in speculation."[18] In neighboring Ohio, western Virginia, and Canada West (present-day Ontario), wells were going down (as the practice of boring was described). In the summer of 1860, the geologist John Strong Newberry made a reconnaissance of oil operations around Mecca in northeastern Ohio. He counted 200 wells completed or in progress.[19] By November, there were 600 or 700.[20]

One reason for these rapid developments was the comparatively low cost of entry in the oil business, starting with the land. Gale reckoned only a quarter acre was required for a well, much less than the twenty or thirty acres needed in coal mining, although he suspected that all one really had to have was enough room to squeeze in a derrick.[21] It was not uncommon for wells to be bored within fifty feet of one another, a practice driven as much by land speculation as by geology. Wherever a well struck oil, nearby property jumped in value, thus smaller and smaller parcels were sold off. To many observers, derricks seemed to be stacked on top of each other.

The basic tools and techniques for oil boring were likewise relatively inexpensive and readily transferrable from the artesian and brine well businesses. "The drilling tools and other mining apparatus used by salt miners, in a more modified and simple form, furnished all that was requisite," reported two eyewitnesses.[22] One of the most popular early methods was the spring pole. This contraption consisted of a forty-foot length of green timber secured at one end to the ground and propped on a post ten or fifteen feet along, which acted as a fulcrum. Attached to the pole's free end was a tool string (iron bits and bars tied to a length of rope). These tools were dropped on the ground by pulling down on the pole; the pole would then spring up with the tools. This arduous process of pounding out a hole was known as "jigging down" or "kicking down" a well. Drake was one of the few operators during the first year to employ steam power because he had planned to bore down 500 feet, a not uncommon depth for brine wells, and an engine was required to raise long, heavy tool strings. (Spring poles used tools weighing about seventy pounds.)

Besides the tools, there was the labor. The cost of boring a well depended on depth and rock type. "In drillings for oil," Gale explained, "*a change in rock* is

"Kicking Down a Well." The spring pole was an inexpensive method of boring for petroleum in the shallow subsurface (less than 100 feet). *Source:* Ida Tarbell, "An Unholy Alliance," *McClure's Magazine, 20* (February 1903):390–403, 398.

frequent."[23] "Soft" rock could be chiseled away at a rate of five to six feet a day for as little as $1.50 a foot. "Very hard rock" received "a thousand blows" and a full day's work to advance two or three inches, which cost a bit over $2.00 per foot.[24] This rock-type accounting made borers very knowledgeable about oil strata, its hardness, thickness, color, and composition. Within the first year, borers were correlating subsurface conditions from well to well.

To complete a well might cost as little as $657.50 or as much as $2,000, "especially such as are *unlucky ones* and of great depth."[25] Most Oil Creek wells went down no further than 200 feet before they were abandoned. In parts of Ohio and Canada, wells were usually less than 100 feet. A well was completed when the borer got a "good show," enough oily bits in the bailer (an ingenious device sent down the hole to fish out rock shards) to warrant hooking up a pump. The borer's skill largely determined what was considered a "good show."

The first oil wells required pumping, and pumps needed steam engines, but these were hard to get in Oil Creek and hence very expensive, between $300 and $500. Pumping wells also needed coal, pipes, and something to hold the oil, all of which immediately increased the cost of a producing well. But once the equipment was installed, operating expenses were trivial. Gale estimated between $5 and $30 daily; Newberry pegged them at $27.[26] Profits thus seemed assured. "A 4-barrel well is called a paying institution," Gale remarked. With careful economy, he reckoned even a two-barrel well might pump out "a fine little income."[27] Newberry declared that "a good well soon pays for itself."[28] And he knew from personal experience. As an early visitor to Drake's well, Newberry had interviewed Brewer and Eveleth about oil springs along Oil Creek. What he learned convinced him to become an oil operator, a person like Drake who managed the boring of a well.[29]

"Catch the Hare"

All the cheery predictions of profit were predicated on striking oil. As S. J. M. Eaton, a Presbyterian pastor from nearby Franklin, Pennsylvania, put it: "In a popular cook-book there is this sage direction prefixed to a receipt for cooking a hare:—'First catch the hare.' So in regard to the preliminaries to boring an oil well, the first thing is to obtain a site that will be at least promising."[30] Locating a well meant a "knowledge of where and how rock oil occurs in nature," and Newberry had much to say about this.[31] So, too, did less scientific writers, including one from the *New York Tribune,* the first reporter of Drake's well.

> [Mr. E. L. Drake] came out here [Titusville] in May, last year . . . to find the source of the oil, which is so common along the banks of Oil Creek. Last week, at the depth of 71 feet, he struck a fissure in the rock through which he was boring, when, to his surprise and the joy of everyone concerned, he found he had tapped a vein of water and oil.[32]

Petroleum was framed in the familiar terms of salt geology. Oil was like brine. Both flowed in underground veins that were punctured much as surgeons opened blood vessels.[33]

In common articulations of this theory, borers described oil veins running underground parallel to surface rivers or streams. "It was generally supposed," reported Eaton, who along with Gale was an eyewitness historian of Oil Creek's first years, "that the oil was found running in slender channels and leaders through the rock, like water veins near the surface."[34] In this subsurface picture, oil veins were oriented *horizontally.* The supposed correspondence between oil veins and water courses translated into boring along and sometimes in rivers and streams. Derricks soon dotted Oil Creek's banks.

A more popular theory focused on rock fissures. The *Tribune* account mentioned a fissure, and the *New Haven Register* of 24 October 1859 elaborated: "at last [Drake] succeeded in sinking a shaft about seventy feet, when the drill suddenly sunk about four inches."[35] Fissures were commonly encountered in salt and water wells, but unlike veins, they were thought to run *vertically.* By the fissure theory, petroleum occurred in cracks, much like gold, silver, and other minerals. Oil fissures were filled and fixed, not flowing; they were tapped like beer kegs.[36]

Where to find oil fissures was debated. According to Gale, borers thought the best place was back from the water's edge, under the high, bold bluffs of Oil Creek. Where the hillsides met the flatlands was "indicative of broken rock beneath," especially if there existed a ravine.[37] "This is certain," Gale declared, "the rock must be *open* or it cannot hold oil: the more strata have been broken to pieces by upheaving and subsidence the more seams there are, the more probability of tapping one."[38]

"There is no doubt some geology in this, if not philosophy," mused Eaton. He too favored the fissure theory and dismissed the notion that "oil-courses would correspond with the water-courses." Oil Creek, he thought, was not "in the slightest degree" influenced by the underlying rocks but was the result of erosion; the creek had carved its own valley out of the surrounding hills.[39] For this geological reason, Eaton rejected Gale's recommendation; "ravines are generally not due to the force of circumstances beneath, but above." Surface depressions were merely superficial accidents, products of running water, not indicators of underground oil fissures.[40]

In propounding these geological explanations, Eaton claimed little originality and much diffidence. His views were "most plausible and consistent" with those in common use. He had consulted certain "gentlemen," unnamed local experts, as well as Henry Rogers's *Geology of Pennsylvania* (1858).[41] Rogers was often the cited authority on petroleum geology, and he appeared by name or was excerpted at length in newspaper articles, books, and oil company prospectuses. Paradoxically, Rogers's survey contained only two pages (out of 1,200) on petroleum. For many oil historians, this is the (missing) evidence of geology's uselessness to exploration. But readers like Eaton were not looking for well sites. Rogers provided oil interests with the broad geological framework for Pennsylvania rocks, on which they could superimpose their more detailed local knowledge.

By the summer of 1860, Oil Creek borers had identified two "sand rocks" at relatively consistent depths and thicknesses with good shows of oil. The first was shallow (less than 100 feet deep) and corresponded with Drake's strike. The second, a much more productive one, lay about 200 feet down. According to Gale, borers worked on the theory that oil fissures and/or oil veins were found *within* the first and/or second sand rocks. Using borers' accounts, Rogers's *Geology,* and James Hall's New York State geological survey, Gale determined the stratigraphical position of the second sand rock. "[T]he oil yielding rock," he concluded, was the Vergent and Cadent "stratum." The Vergent and Cadent formations were Rogers's idiosyncratic terms (which, Gale thought, "appear to be quite poetical") for the Lower Devonian or, according to the New York System, the Chemung, Portage, and Hamilton groups.[42] In designating them as *the* oil rocks, Gale judged that petroleum was older than coal.

Rogers came to a different conclusion. From his Regius Chair at the University of Glasgow, he offered an explanation of Drake's discovery.[43] Oil springs, he thought, occurred along the entire northwest margin of the Appalachian coal field, from western Pennsylvania through western Virginia to Tennessee. The springs coincided with "anticlinal flexures," places where uplift had fractured the rocks and thus permitted communication between the surface and subsurface. Petroleum reservoirs, on the other hand, occurred in places *without* fissures. Reservoirs required two conditions: a porous sandstone and an overlying shale bed. Petroleum was distributed within the pores of the sandstone, and the impervious shale served to "hold down" the oil. Rogers placed these sandstone reservoirs "in the more superficial strata of the coal-measures." In other words, petroleum occurred in the Upper Carboniferous.[44]

But the petroleum itself was *younger* than the rocks enclosing it. In other words, sandstone reservoirs were not the places petroleum originated but the places it collected. According to Rogers, petroleum had been distilled from underlying coal beds during the paroxysmal uplift of the Appalachians. The rising gases were then "arrested" and cooled in the pores of the overlying sandstones. In essence, petroleum was crude coal oil.

Rogers's theory of petroleum's origin was of a piece with his theories of mountain building and debitumenization. In his new geography of bituminous substances, anthracite was located in eastern Pennsylvania, bituminous coal in the state's western parts, and petroleum "in the regions north-west of the inland frontier of the Appalachian coal-field," where steaming and uplift had been less intense. Rogers predicted that more petroleum would be found on the western edges of the great coal field. It was "simply one phase in this gradation."[45]

Rogers's theory held considerable appeal among well borers and the general public. The *American Gas-Light Journal* told its readers in March 1861: "The most natural supposition would be that [petroleum] is a distillation of coals conducted in nature's laboratory, under modified conditions of heat and pressure."[46] A year later, *Scientific American* was reporting that "[m]any practical men in the Allegheny and Ohio valleys believe that petroleum has its origin in coal beds."[47] And as late as 1873, the oil historian J. T. Henry reported that the "coal theory" was the one "that has mostly obtained, [and] it is in many respects exceedingly plausible."[48] The plausibility or naturalness of the theory also indicated the familiarity Americans had with coal oils.

A slightly different distillation theory, one that challenged Rogers's idea and eventually replaced it within scientific circles, was proposed by Newberry. Newberry's geology was based on his Pennsylvania experience and his Ohio explorations. From his calculations of well depths and rock thicknesses, he estimated that in Ohio borers struck oil in the Waverly Group, a series of sandstones of the Upper Devonian or the Chemung and Portage groups. The Waverly sandstones were "*the* oil rocks of Ohio," meaning "the geological level at which we must look for new discoveries of petroleum" (table 8.1).[49]

The Waverly sandstones lay on top of the Black Slate, a layer of bituminous

Table 8.1. Correlations of Geological Formations

Ohio Series	New York System	Pennsylvania System
Waverly Group	Chemung & Portage Groups	No. IX (Sandstones)
Black Slate	Hamilton Group	No. VIII (Shales)

shale, and Newberry reasoned that petroleum was distilled from this shale, not from overlying coal. Further, he thought "the theory of the formation of petroleum should not include the element of extreme heat," as Rogers had described. Newberry postulated a distillation process that occurred at low temperatures, over a long time, and under pressure. In this way, "*liquid* hydro-carbons," not gases, were distilled. Nonetheless, Newberry's analogy, like Rogers's, was coal oil. "What we thus do by art rapidly," he concluded, "nature does as effectually when left to herself, but very slowly."[50]

The artificial reproduced the natural. Such thinking inspired numerous petroleum theorists. "[T]he oils and gases now flowing from the earth," Abraham Gesner explained in 1860, "are only the results of [processes] going forward in Nature's laboratory."[51] Likewise, Alexander Winchell (1824–1891), state geologist of Michigan, described "the slow spontaneous distillation" of bituminous shales as the source of the petroleum he found during his fieldwork in 1860.[52] The geologist E. B. Andrews (1820–1880) used Rogers's version of the distillation theory to explain the springs found along Duck Creek and the Muskingum and Little Muskingum rivers in southern Ohio. According to Andrews, petroleum formed when rising gases got trapped in fissures, not porous sandstones.[53]

Well borers in Ohio and Pennsylvania regularly reported sudden drops of tool strings. More importantly, they noticed that wells bored in close proximity to a producing one often did not strike oil, and if an adjacent well did make a strike, rarely was the depth the same or the oil identical in consistency and color. Andrews took these phenomena as evidence of self-contained oil fissures. These fissures were all vertically oriented and large compared to the thickness of the rocks. In other words, oil fissures cut across several stratigraphical layers. "[T]here is no such thing as an 'oil rock,'" Andrews asserted. "The oil is found in any kind of stratum."[54]

Any stratum, he would add, containing lots of fissures. Andrews considered this something of a geological law: "the quantity of oil is in direct ratio to the amount of fissures."[55] The place to look for fissures was on hills. Uplift created cracks, and Andrews had discovered a "line of uplift," or long anticline, running nearly north-south from Washington County, Ohio, to the main producing territory in western Virginia along Burning Spring Creek, a tributary of the

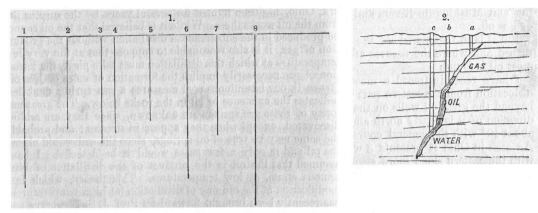

Left, Schematic of the different depths of oil wells along the Burning Spring Creek in West Virginia. *Right*, Vertical oil fissure. Three different types of wells (a: gas, b: oil, c: water) were possible depending on which part of the vertical fissure was struck by the drill bit. In Andrews's depiction the oil fissure cuts across many layers of rock; it is not contained within a particular stratum. *Source:* E. B. Andrews, "Rock Oil, Its Geological Relations and Distribution," *American Journal of Science, 32* (1861):85–93, 86, 87.

Kanawha River. Andrews thought the chances of finding oil were "very small" in the horizontal strata of Wirt or Wood (later renamed Ritchie) counties of western Virginia. But in the broken rocks along the "great uplift," he predicted, borers would find the "largest quantity of oil."[56]

Andrews's fissure theory was very popular among borers in Ohio and western Virginia, no doubt because it confirmed their experiences. "The petroleum is stored in fissures formed by the upheaving of the earth's crust by volcanic action," reported *Scientific American;* "and these fissures are perpendicular rather than horizontal in tendency, as it is proved by the fact that at wells, but a few rods apart, the oil is 'struck' at very different depths."[57] Geologists agreed. "It seems very certain," Gesner explained in an article to the Geological Society of London, "that the reservoirs of oil are fissures penetrating certain oil-bearing strata and the intervening deposits."[58] T. Sterry Hunt (1826–1892), the chemist and mineralogist of the Geological Survey of Canada, reinforced Andrews's theory that petroleum "concentrated along certain lines of elevation." In the winter of 1860, Hunt made a reconnaissance of the oil regions around Enniskillen in Canada West. He counted nearly 100 wells with a total production between 300,000 or 400,000 gallons.[59] (For comparison, Oil Creek produced 200,000 *barrels* in 1860.)[60] Hunt found that producing wells were located on the crest of an anticline running roughly east-west through Enniskillen. He also decided that the wells struck "separate fissures," which ran vertically downward through the topmost formation (shale) and connected with "a deep-seated source."[61] Discovering this source became Hunt's main concern.

Using stratigraphical maps of the Canadian survey, Hunt calculated that the Enniskillen oil came from the Corniferous Limestone, a porous Devonian-age formation immediately *beneath* the Hamilton shales and their Ohio equivalent, the Black Slate. The source of Canadian petroleum was different from that of Ohio, Virginia, or Pennsylvania petroleum. Hunt therefore rejected all the distillation theories and proposed a new theory based on the latest chemical analyses of coal done by the German chemical geologist Carl Gustav Bischoff. Hunt explained that coal's composition (carbon, hydrogen, and oxygen) was distinct from petroleum's (carbon and hydrogen). "It is the more necessary to insist upon the distinction," Hunt explained, "inasmuch as some have been disposed to regard the former [coal] as the source of the petroleum found in nature, which they conceive to have originated from a slow distillation of these matters."[62] Coal and petroleum formed under completely different conditions. The former was composed of *freshwater* or *terrestrial* plants, whereas petroleum, because it originated in limestones (at least in Enniskillen), was produced from the decomposition of *marine* plants or animals. And because petroleum contained no oxygen, Hunt thought the chemical conversion must have taken place in deep water. In short, petroleum was not a displaced product of distillation, but rather it formed in situ.

Deciding between Hunt's in situ theory and the more popular distillation theories meant coming to grips with petroleum's fundamental characteristic: it was liquid. Unlike such stratigraphically stationary minerals as coal, petroleum moved through the subsurface, after it was formed. Hence the place where it was found by well borers was not necessarily, or not likely, the location of its origin. Hunt noted that petroleum could move *vertically* via fissures as well as *horizontally* via the pores within a rock layer. In the latter case, petroleum migrated to a stratum's highest point, for example, the crest of an anticline. By this logic, Hunt became the first geologist to explain the anticlinal theory of petroleum accumulation.[63] Hunt, however, did not emphasize petroleum's horizontal migration (later he would), because in 1860 the more pressing and practical problem concerned vertical migration.

Oil springs were generally considered indications of petroleum at depth. Drake, for instance, had located his well near an oil spring. But surface indications were not an "infallible indication of success," observed Eaton.[64] "Only a short period of time was required," commented two oil historians in 1870, "to prove the fallibility of the first surface indications."[65] The reason lay *not* in any disconnection between surface and subsurface but in the very fact that there was communication. Oil springs meant leaky fissures.

Finding sealed fissures was the challenge, and in 1860 borers and geologists alike began to realize this. Both groups were able to identify and predict the depth of "oil horizons"—limestones or sandstones associated with petroleum—but not the precise location of fissures. Andrews had a theory that applied to the "great uplift" in southern Ohio and western Virginia, and Hunt had one for the "break" in Enniskillen. But along Oil Creek, wells struck oil along the "bottom lands" of the valley, not on the hills. Pennsylvania petroleum seemed to accumulate in places where rocks had been thrown down, not up.

In all the oil regions, wells tended to cluster wherever and whenever a strike was made. Despite their oft-remarked autonomy, adjacent wells had a powerful commercial appeal (as opposed to a scientific or practical rationale). As a result, they drove land speculation. And land was a surer way to make money than oil. The rate of successful boring along Oil Creek was low. "As the season advances," Gale observed in the summer of 1860, "instances of failure multiply." He reckoned that not more than one in five wells struck oil, and he could only name twenty that were pumping.[66] Around Mecca, Newberry counted about dozen wells (out of 200) that struck oil.

The low rate of success emboldened some independent-minded explorers. According to two well-respected journalists, not a few borers "scouted at theory." But "this same class of skeptics," the journalists were tickled to relate, was usually among the first to procure leases adjacent to a paying well.[67] Nonetheless, skeptics were common and audacious enough to warrant watching. "With precipitation they dig almost anywhere as though the chance was about the same in one place as in another; and off, a dozen miles from the place where oil is found, men will be heard to say, the prospect is as good in their town as anywhere else, and so they begin to drill."[68] Gale referred to these unorthodox prospects as *country wells*. They functioned as a challenge to predominant theories and practices as well as an antidote to clustering.

Precipitous diggings were also a sure sign that oil fever was spreading. "Excitement is king now," Gale proclaimed.[69] Even Rogers in far off Scotland could bear witness to the "petroleum fever," which had reached "as high a pitch as ever did the 'gold fever.'"[70] "A tide of speculators and operators began to set in," observed Eaton, "which would have overpowered that of California."[71]

Flowing Wells

In late 1860 drillers began extending their wells to depths of 400 or 500 feet. Perhaps it was a practical move in response to a "dry hole" or to a sudden drop in a pumping well's production, but in any event, digging deeper was less costly than starting anew. A slightly different rationale was provided by two journalists: "About this time some reflective operator expressed the opinion that, as the supply of oil seemed to come from great depths below the earth's surface, deeper wells would reach the main reservoir, or source of supply and greater quantities obtained. The theory so opportunely stated, was soon put to a practical test."[72]

The first flowing well was struck near the Kanawha River in western Virginia in October 1860. Several other large wells, some yielding 300 or 400 barrels per day, followed in January, February, and March 1861.[73] But then the Civil War completely disrupted production.[74] Bands of Confederate guerrillas burned derricks and terrorized operators, and as late as the spring of 1865, when the war was all but over, West Virginia prospects were still regarded as very risky.[75]

Deep drilling came to Oil Creek in April 1861, when Henry Rouse decided to extend his 150-foot pumping well.[76] He had gone down more than 300 feet when suddenly gas rushed up the hole. A nearby steam engine ignited it, and fire began to spew "with terrible fury" over the derrick. Then "the well exploded with a shock like that of an earthquake."

> It seemed as though the earth was vomiting flame threatening to fill the whole valley as with a sea of fire. The fiery column reached above the derrick that was soon consumed, accompanied with dense volumes of smoke, roaring like a hurricane, turning and bending in every direction as the wind veered from one point to another.[77]

The Burning well claimed the lives of at least eighteen persons, including Rouse, and took five days to extinguish.[78]

Of the lessons learned from the tragedy, the most obvious concerned the gas. It was highly inflammable, and therefore any kind of fire, including a lit cigar (Rouse had been smoking one), was banned near oil wells. It was also forceful. Never before had gas come rushing out of a well, or, for that matter, had oil. Wells had gurgled and hissed and sputtered for a while, but the Burning well was notorious for its fury. Gas threw oil sixty feet high, and the well gushed thousands of barrels for days on end. "Here was a new feature in oil production," Eaton flatly announced.[79]

Dry holes soon became the test for deep-seated sources, and drillers with nothing to lose but their time and effort began going down 500 feet or more. In June 1861 the Funk well struck oil and flowed 250 barrels a day, "to the astonishment of all oil-borers" along the creek.[80] Two months later the Phillips well came in with 2,000 barrels a day, and in September the Empire well flowed "at the modest rate" of 3,000 barrels daily. By October 1861 the huge Phillips well no. 2 was gushing 3,000–4,000 barrels a day.[81] These "leviathans" were soon followed by others—Noble (2,500), Caldwell (800–1,000), Maple Shade (1,000–1,500), Jersey (500), Coquette (1,500).[82] According to J. Peter Lesley, the quantities were "astonishing." "The earth literally spouts oil as a whale spouts brine."[83]

By the end of 1861, Oil Creek was famous for its flowing wells (hence their personal names), but more famous still were the farms on which they clustered. The Egbert farm, for example, embraced the Maple Shade, Jersey, and Coquette

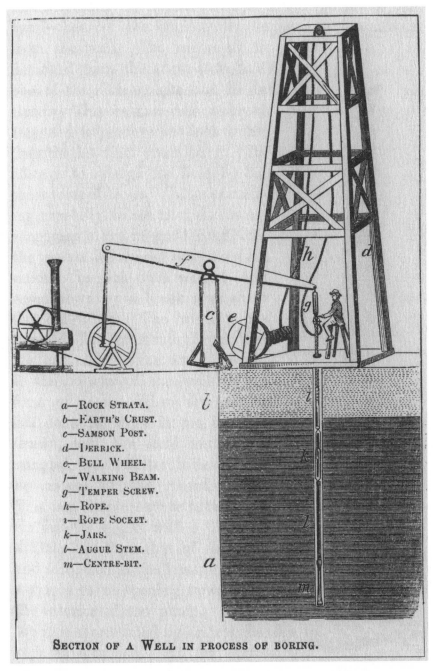

a—Rock Strata.
b—Earth's Crust.
c—Samson Post.
d—Derrick.
e—Bull Wheel.
f—Walking Beam.
g—Temper Screw.
h—Rope.
i—Rope Socket.
k—Jars.
l—Augur Stem.
m—Centre-bit.

SECTION OF A WELL IN PROCESS OF BORING.

An oil well setup, with steam engine and derrick. The well borer rotated the rope a one-quarter turn (or less) with each lift of the tool string to assure a round bore hole. *Source:* S. J. M. Eaton, *Petroleum: A History of the Oil Region of Venango County, Pennsylvania* (Philadelphia: J. P. Skelly, 1866), facing page 112.

wells. The twelve-acre Blood farm had thirteen flowing wells; by 1864, it had thirty.[84] As one observer wryly put it, "these wells were unequally distributed along the Creek."[85]

Flowing wells made a few people very rich, and a great many others very poor. During the winter of 1861–1862, Oil Creek produced somewhere between 8,000 and 20,000 barrels a day.[86] According to one observer, "[i]t was truly too much of a good thing."[87] Oil prices plummeted. By July 1862 a barrel cost only $2.00; the wooden barrels were worth more than the oil inside. "Thousands of barrels [of oil]" literally flowed into the creek for it was worthless to gather and store it all. "A panic seized the smaller institutions," Eaton lamented.[88] Pumps and derricks were abandoned and left to rust and rot. Petroleum had gone bust by the summer of 1862.

What little drilling continued was confined to sites adjacent to flowing wells. In this regard, flowing wells did not change the pattern of petroleum development. Wells continued to be located on bottom lands of the creek, but the clusters of derricks became tighter and more wasteful.[89] According to an 1862 estimate, thirty-one wells produced 20,000 barrels a day, of which three-quarters flowed back onto the ground or into the creek.[90] Samuel Downer, the Kerosene manufacturer, was dismayed.

"See here," said Mr. Downer, "don't you know you are wasting a hundred barrels an hour here?"

"Yes," said the interested party addressed, "but what am I to do with it? You won't give five cents a barrel for it; and I can stand a loss of five dollars an hour rather than let you have it at that price!"[91]

Flowing wells made the degradation even worse because they were deep. To lift the heavy tool strings, they needed coal-fired steam engines—noisy, smelly, and dirty. Coal was shipped from Pittsburgh along muddied roads and waterways. Any trees on an oil property were cut down and used for building derricks and engine houses. Lesley recoiled at the transformation of Oil Creek.

The once quiet, beautiful valley became a noisy den, a hideous desert. Derricks, scaffolds, and pumping gear took the places occupied by the tall forest trees or blooming orchards. . . . Not a blade of grass was to be seen, and nothing to be heard but the clanking of the pumps, the blowing of some new well in its first energy, the shouting of drivers urging miserable mules and horses through the nauseous mud, dragging empty barrels to the wells, or full ones down to the stream[,] a stinking bog of mud and salt mingled with oil.[92]

Big Cavities with Gas

If flowing wells inspired such poignant and painful views of the new industrial landscape, they also brought revisions in the subsurface. No longer was petroleum conceived in terms of vertical fissures. "The discovery of flowing wells," Eaton remarked, "destroyed this theory."[93] To hold such vast quantities, there needed to be very large subterranean spaces, "cavities," or "caverns." Eaton envisioned big caverns containing gas, oil, and water separated into three layers. He then pictured thin fissures leading away and upward from the caverns. Interestingly, Eaton did not think oil wells could reach the deep-seated caverns.

Most drillers and geologists did. The dynamics of flowing wells were explained best by E. W. Evans, professor of mathematics at Marietta College and a colleague of Andrews. Evans posited large completely sealed cavities containing a volume of high-pressure gas sitting atop layers of oil and water. Depending on where the drill struck the cavity, different types of flowing wells would be produced.[94]

By adopting this new theory of high pressure gas, drillers rejected one of the laws governing artesian water wells. In those wells, water flowed because of the pressure of the water itself. In other words, there existed a source (the "head") for artesian water on higher ground. If flowing oil wells had heads, it meant a petroleum source somewhere above Oil Creek valley on the surrounding hills. But in 1861 no one was willing to bore on the highlands.[95]

In southern Ohio, by contrast, hills *were* the places to look for flowing wells. According to Evans, big cavities, like little fissures, occurred on the crest of the great uplift. Likewise, in Enniskillen, the low hill continued to be the favorite location for wells. "This anticlinal structure," Hunt remarked after visiting in December 1862, "appears to be a necessary condition of the occurrence of abundant oil wells." But few Enniskillen oil wells were flowing. Hunt agreed that "[l]arge quantities of light carburetted hydrogen gas" powered flowing wells, but he dismissed any suggestion that the gas had "any necessary connection with the oil," for this smacked of the distillation theory of the origin of petroleum.[96] Hunt argued that the gas *and* the petroleum formed separately and in situ. Further, he contended that *all* petroleum and gas came from the conversion of shells, corals, and other marine animals.[97] Accordingly, all limestones, whatever their age, might contain petroleum. As with some of the early theories of coal's origin, Hunt's in situ theory treated petroleum as a chemical process, not a geological product.[98]

This idea found an unlikely supporter in Rogers. In an 1863 *Harper's Magazine* article, Rogers presented a new theory of petroleum. In typical Rogers style, he began by praising the immense scale of "the great petroleum tract" of North America. It encompassed 50,000 square miles, from Canada to Kentucky and

---

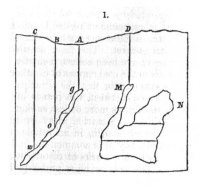

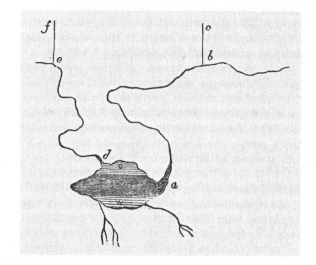

Left, E. W. Evan's 1864 scientific schematic of large oil caverns. The four wells (A, B, C, D) struck the caverns directly and depending on which part of the cavern different types of wells were produced. Wells A and D were gas wells; Well C was water, and Well B was a flowing oil well because the pressured gas in the cavern pushed out the oil below with great force. *Source:* E. W. Evans, "On the Action of Oil-Wells," *American Journal of Science,* *38* (1864):159–166, 162. *Right,* The Reverend S. J. M. Eaton's 1866 schematic of a large oil cavern containing gas (G), oil (O), and water (W). In Eaton's popular rendering, the wells (f-e) and (c-b) did not reach the deep cavern but rather the fissures leading from the cavern. The kind of well was determined by the kind of fissure. Well (f-e) was a gas well; Well (c-b) was a flowing oil well. *Source:* S. J. M. Eaton, *Petroleum: A History of the Oil Region of Venango County, Pennsylvania* (Philadelphia: J. P. Skelly, 1866), 145.

from western Pennsylvania to eastern Ohio, "by far the most abounding district" in the world.[99] He then described a distillation process incorporating the recent findings of Hunt, Andrews, and Newberry. "[C]hemists inform us," Rogers noted, that "there are some differences" between natural petroleum and manufactured coal oils. Rogers abandoned coal as the ultimate source of petroleum and decided instead on "the hydrocarbon elements" resident in rocks *underlying* the Carboniferous, "that is to say, impregnating the Silurian black slates [and] the Devonian black shales."[100] To help his readers understand his new distillation process, Rogers asked them to imagine the Earth as a "hypothetical oven." During the Appalachian uplift, the rocks of eastern North America had been heated like "puddings in a cook's unequally-heated oven." The slates and shales nearest the fire (in the anthracite region of eastern Pennsylvania) were "overcooked and rendered juiceless," while those farthest away (in western Pennsylvania) were "less and less baked, retaining larger and larger proportions of their primitive juices." The "juices" were the gaseous ingredients expelled from the slates and shales that had risen, condensed into petroleum, and now saturated the overlying sandstones. Perhaps the most startling part of Rogers's cookery theory was his claim that distillation "is *even now in progress*." (Hunt had not gone so far with his petroleum-as-process theory.) Deep down in "the great bitumen brewery," Rogers ruminated, "the fermentation is in full activity." Because petroleum was continuously being brewed, it could never be exhausted.[101]

Finding such inexhaustible supplies was another matter. Although it was widely occurring, Rogers knew that petroleum was not uniformly plentiful. He proposed a law to explain its "zones of comparative abundance and scarcity." Basically he reiterated Andrews: "most oil is found where the strata have been most disturbed," which meant that under Oil Creek there had to exist "an anticlinal arching."[102]

Lesley scoffed. The strata of western Pennsylvania exhibited an "almost unchanged horizontal posture."[103] Likewise, he rejected Rogers's distillation theory. "The oil is never found," he asserted, "[in] any connexion with coal beds, nor even with coal slates or bituminous shales." The juxtaposition of the oil regions and the great Appalachian coal field was "a geographical deception."[104] And as for Andrews's and Evans's ideas about flowing wells, "[i]t is impossible to postulate the gas first and the oil afterwards," Lesley carped, "for that order would require the generation of pressure sufficient *afterwards*." Condensation reduced gas pressure; it did not create it. The gas must therefore have been distilled from the oil.[105]

"Whence, then, comes the oil[?]" Lesley considered Hunt's in situ theory the best. "[G]elatinous sea organisms, both animal and vegetable, seem to have constituted the principal, if not sole, apparatus for generating petroleum."[106] The only problem was the reservoir rocks. Along Oil Creek, "the amazing discoveries of subterranean reservoirs of oil" had not been made in limestones but in sandstones belonging to Formation VIII and Formation IX (the Chemung and Portage groups). This lumpen fact suggested that decomposing vegetable matter (freshwater and terrestrial) might have formed petroleum in situ. Lesley concluded: "The theory of the genesis of [petroleum and] the structural difficulties attending the solution of the problem remain."[107]

From Coal Oil to Petroleum

If the science was contested, the economics was not, at least not by Lesley. He confidently declared that coal oil (as of November 1862) had been rendered "needless."[108] Flowing wells replaced the "refining machinery." This interpretation was appealing, and many writers since Lesley have relied on this sort of raw material determinism: cheap and plentiful petroleum substituted for costly coal.

From the perspective of coal oil companies, petroleum was not such an inexpensive or easy substitute. These companies were certainly aware of the surplus production, but some obstacles blocked the path to petroleum. First, it needed to be transported from wells to manufactories, most of which were located near western coal mines or in East Coast cities. At best, transportation was seasonal. In winter, Oil Creek was frozen; in summer, it had too little water; in spring, it flooded. When navigation was possible, crashes and fires were common. Overland, roads were execrable, and in the rain, they were impassable. As Joshua Merrill, superintendent of Downer's Kerosene oil works outside of Boston, explained, although petroleum might be "very cheap" at a flowing well, getting some to the plant was "very costly."[109]

The second obstacle was technological—turning petroleum into marketable products. In theory, the process was straightforward. Petroleum, as all scientific and practical chemists knew, contained several volatile fractions. By heating petroleum in a wrought-iron still, various distillates could be run off. Lighter fractions were suitable for burning in common lamps; heavier ones served as lubricants. Both types of oil could be improved by refining. For the price of a postage stamp, Schieffelin Brothers, the firm that contracted with Drake, would send "directions for distilling and refining the Oil pumped from the recently discovered Oil Springs of Pennsylvania."[110] Gale estimated that a small manufactory running about five barrels a day would cost only $200; "works which will clarify 10 barrels a day," $300.[111]

In practice, running a petroleum plant was more complicated. As every well borer could attest, there were various kinds of petroleum. Those from Franklin on the Allegheny River, for instance, were much thicker and blacker than the thin, green oils of Oil Creek. "It may be said, in general," concluded one observer, "that there are different species of oil, as there are different species of coal. Some, as those found in Canada and some parts of Kentucky, contain sulphur and other offensive ingredients, which makes them more difficult to purify and less valuable."[112]

Different petroleums required different degrees of heat and different amounts of time (and care) to run off the distillates. "It is therefore extremely difficult to obtain any one specific oil, of which the aggregate is compounded," Gesner explained; "the exact rate at which the boiling point does increase, according to the proportions of carbon and hydrogen present in the oils, has not been accurately discovered."[113] Moreover, the distillates required treatment with acids, alkalies, and other washes to purify and deodorize them. Sometimes, they required another round of distillation and refining to get an appealing color, odor, and burning quality. This complexity, Gesner concluded, made it "a difficult task to prescribe a mode of purification to meet the requirements of the oil refiners."[114] Petroleum manufacturing, in short, required experience, skill, and some chemical know-how, all of which were noticeably lacking among the fifteen or so small plants established along Oil Creek by the end of 1860.

On the other hand, large coal oil manufacturers, of which there were about sixty at that time, had perfected distilling and refining techniques, but they had no intention of giving away their trade secrets, a situation that obviously irked the petroleum booster Gale.

> [Kerosene] companies are proverbial for keeping shady as to the cost of purifying rock oil. . . . none but themselves can understand the business. Why envelop yourselves, gentlemen, in this ocean of fog, if not that you may monopolize the profits which belong to those who produce the raw material? You think them too poor to go into the expense of refining, which you mystify and magnify, that you . . . may take the lion's share of the profits.[115]

Coal oil companies had their own technological system beginning with coal. Most companies had long-term, yearly contracts with mines. Downer, for example, had an exclusive licensing agreement with the Albert Coal Company of New Brunswick, and according to Merrill, the Boston plant used the mineral "with marked success" through 1862.[116] Downer's Portland plant, "one of the best and most perfect Coal Oil works in the United States," according to its superintendent William Atwood, distilled 10,000 tons of the Albert mineral annually.[117] Moreover, all parts of Downer's business—from raw material to refined kerosene—were covered by patents, licenses, and other legal constraints. Converting from coal to petroleum would require capital investment.

Nonetheless, Downer thought it wise to keep an eye on petroleum developments. In the summer of 1860, he made a tour of the oil regions of Pennsylvania, Ohio, and western Virginia, prompted perhaps by the lengthy royalty negotiations he had just completed with James Young.[118] By October, Merrill had run some experiments in the Boston plant using Virginia petroleum. (Rail connections were closer to the wells along the Kanawha and its tributaries, and therefore transportation was more reliable and reasonably priced than that from Oil Creek.) In running these trials, Downer regarded himself as taking a technological lead. None of the large coal oil manufacturers around Pittsburgh (Alladin, Lucesco, or North American) were testing petroleum, and the New York City companies, Downer remarked, had not yet begun to "think of petroleum as a competitor to Albert [coal]."[119]

Downer overestimated his own foresight. Gesner was interested in petroleum, and in his *Practical Treatise* (1860), he predicted that "if the [petroleum] springs maintain their present supply, they will materially affect the distillation of oils from coals, bitumens, and other kindred substances."[120] He also offered chemical expertise. "Improvements are constantly advancing," he assured would-be petroleum manufacturers. "[T]he distilled hydro-carbon oils [will] attain the commercial and economic value they are destined to reach."[121] By 1862 a refining industry had begun to establish itself around Oil Creek; about thirty-five companies were in operation.[122] The largest was the Humboldt refinery in Plumer, at the upper end of Cherry Run, an Oil Creek tributary. The Ludovici Brothers of New York City erected the plant, and it had a daily capacity of 1,000 barrels. A. N. Leet, an accomplished practical chemist, supervised the operations.[123]

Downer opened his own Oil Creek refinery in 1863. The plans had been laid back in October 1861, after Confederate raids in western Virginia had cut off his Kanawha petroleum supplies. Downer's move to Oil Creek also came after he was able to modify his contract with the Albert Coal Company and to sell his remaining stock to the Boston Gas-Light Company.[124] Downer chose Corry, a town at the junction of the Philadelphia & Erie and the Atlantic & Great Western railroads, where he put up a state-of-the-art works (using steam for distillation and a second distillation for purifying). The Downer petroleum refinery could handle 1,800 barrels a day.[125]

With Downer leading the way, other large coal oil companies began to convert to petroleum. How they did so can be gleaned from the second edition of Gesner's *Practical Treatise* (1865), which was revised by his son George.[126] The treatise described the equipment and processes used to refine petroleum at

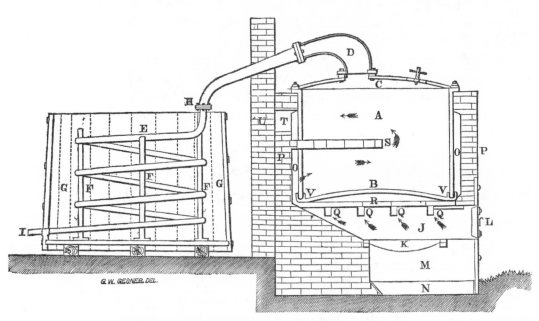

Petroleum still and condenser. *Source:* Abraham Gesner, *A Practical Treatise on Coal, Petroleum, and Other Distilled Oils,* 2nd ed., revised and enlarged by George Weltden Gesner, (New York: Baillière Brothers, 1865), 78.

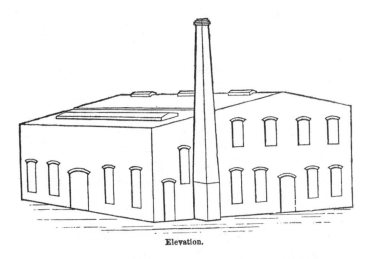

Elevation.

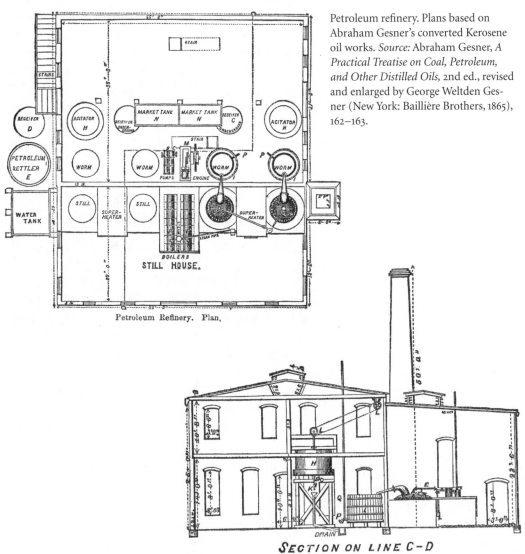

Petroleum refinery. Plans based on Abraham Gesner's converted Kerosene oil works. *Source:* Abraham Gesner, *A Practical Treatise on Coal, Petroleum, and Other Distilled Oils,* 2nd ed., revised and enlarged by George Weltden Gesner (New York: Baillière Brothers, 1865), 162–163.

Petroleum Refinery. Plan.

SECTION ON LINE C-D

Downer's "very extensive and perfect" plants. According to George Gesner, petroleum had one big advantage: it needed no retorts, the costly equipment used to cook coal.

By 1864 the conversion of coal oil manufacturers to petroleum was well underway. This astonishing development was made clear to James Young, who in that year made a second visit to the United States in order to obtain a seven-year extension on his Paraffine patent. Young decided to visit Downer's petroleum refinery in Corry and to tour the Pennsylvania oil regions. According to Merrill, Young was "somewhat alarmed" by what he saw: American oil works ran on petroleum. Not surprisingly, Young received his patent extension without any resistance.

The replacement of coal oil by petroleum was an important technological change, one that contemporaries and later historians and economists have often noted. In their analyses, both groups tend to focus on the techniques and tools used in refining. As a result, the conversion seems inevitable: it was only a matter of time and supply before coal oil manufacturers realized the savings to be gained by abandoning retorts and embracing stills. "The cheap rate at which refined Petroleum could be furnished, the seemingly inexhaustible supply of the crude product, soon overshadowed coal oil," wrote two journalists. "In a very short time its manufacture became unprofitable, and Petroleum reigned supreme as an illuminator."[127] According to the oil historian J. T. Henry, writing in 1873, "the way for [petroleum's] reception at home and abroad [had] been opened by the previous extensive introduction of coal oil; [but] the discovery of petroleum prostrated the whole [coal oil] business, and threatened its projectors with overwhelming loss, from which they were happily rescued by converting their oil factories into refineries, which was done with very little trouble."[128] Coal oil and petroleum were in competition, a point that warrants highlighting because it might explain why it took three years *after* the first flowing well to establish the predominance of petroleum. Coal oil companies were as much a hindrance as a help to petroleum. Technological change was not sudden or smooth—and perhaps not inevitable.

Consider the coal gas industry. Petroleum could have been substituted for coal in gas manufacturing, or perhaps the natural gas that shot out of wells could have been used. But neither happened. Explaining the cause for something that did not occur is always historically suspect, but it seems fair to conclude that the costs of using petroleum or natural gas were too high. The gas industry had a more firmly entrenched technological system—from mining and manufacturing through distribution and marketing to politics; many cities had municipal monopolies. With regard to the legal system, coal oil had a distinct disadvantage, Young's Paraffine patent, a taxing constraint that might have encouraged petroleum substitution by way of avoiding royalty payments or law-suits for infringement. There was no single patent covering gas manufacturing or petroleum refining.[129]

Science might also have had a role to play in conversion. Every geological explanation that associated natural petroleum with artificial coal oil reinforced the idea that the former could replace the latter. Lesley and Hunt complained of nothing so much as the misguided theory that petroleum was distilled from coal. Ironically, when technological conversion occurred, it served to strengthen the popular scientific notion that coal and petroleum were similar. And finally, an explanation for the disappearance of coal oil might rest on the commercial success of Kerosene. In 1859 mineral oils had already been introduced to consumers as substitutes for animal and plant oils. The coal oil boom made kerosene (small *k*) a household name for safe and inexpensive lamp oils. Downer could not pursue legal actions against petroleum refiners who appropriated his brand name; it would have been financially futile to prosecute small infringers. As he told his superintendent of the Corry plant, "No jury would ever convict in Pittsburgh."[130] The result was a marketing free-for-all. Kerosene became the generic name for *all* mineral oils regardless of which mineral was used. Petroleum thus replaced coal, and kerosene consumers might never have noticed the switch.[131]

Oil Boom

If flowing wells motivated manufacturing interests, producers found them depressing. Throughout 1863 the oil regions of Pennsylvania and elsewhere were relatively gloomy places. A small number of big wells glutted the market. Oil prices remained low and stable.[132] Incentives to bore new wells vanished. The attention of most Oil Creek inhabitants, like much of the nation, focused on the Civil War.

The petroleum market began to improve in 1864. At the start of the year, a barrel fetched $3 to $4 at the well. In July the price had rocketed to $13. By year's end, following a string of Union victories, the reelection of President Abraham Lincoln, and continued inflation due to the war, prices settled in the range of $10 to $12 a barrel. At the same time production steadily declined from a peak of 3 million barrels in 1862 to fewer than 2 million in 1864.[133] The combination of high prices and low supply produced a burst of renewed interest in oil production. Operators began cleaning out old wells and started looking for new locations. By the summer of 1864, an oil boom was underway.

The oil boom was far larger in terms of people, places, prices, and press coverage than the rush after Drake's strike. Historians have often depicted it as a frenzy of land and stock speculation, and, accordingly, the pattern of development has been described as random, widely dispersed, and irrational. Here the drunkard's chase seem to fit very well.

Another perspective on the boom can be taken through the lens of practical geology. In the boom years of 1864 through 1866, production spread beyond the confines of Oil Creek valley to the surrounding hills. The expansion of oil territory came as a series of relatively small, cautious steps, all of which were made within familiar theoretical frameworks. Thus, amid the rush to buy and sell lands, stocks, and companies, there was almost an orderliness to the way drillers went about searching for oil. Science, as many observers noted, moved slower than commerce.

In dating the start of the boom, contemporaries often pointed to the strikes along Cherry Run, a small Oil Creek tributary. In July 1864 William Reed bored a well along the flat lands of the run; the Reed well, as it was known, flowed 300 barrels daily.[134] The location of the well suggests Reed's adherence to prevailing ideas about the correspondence of oil and water courses and/or the association of fissures and cavities with bottom lands. Reed had already sunk several other wells (all dry or small pumpers) along the bottom lands of French Creek and the Allegheny River. Reed's try on Cherry Run had been postponed until that summer only because of petroleum's low price.

News of Reed's strike spread quickly and brought other drillers to Cherry Run. Subsequent developments took an unusual turn because of two hitherto unprecedented features. First, Cherry Run was such a little stream that there was hardly any flat land along it. To get near Reed's well, drillers were forced to locate on the hillsides. Boring on hillsides had probably been done before, but Cherry Run was the first area where it was widely practiced. Eaton reported that wells "have been bored some distance up the face of the hill." He called it "Experimental boring."[135] The second feature of Cherry Run was the "uniform success" of the hillside wells. According to Eaton, not a single well failed.[136]

Hillside drilling, although an important change in well location, was only one lesson drawn from Cherry Run. In the fall of 1864, explorations began along other tributaries of Oil Creek and the Allegheny. As one observer noted, "These tributaries were at first unnoticed by the oil-seekers, but as the property on the creek became inaccessible to new adventurers, they turned their attention to the smaller streams. . . . The branching ravines have been found little, if any, inferior in productiveness, in proportion to extent, to the creek itself."[137] According to an 1870 survey, most of these tributaries had producing wells along both the flats and the hills, and some still had operating wells.

The most famous or infamous tributary was Pithole Creek. Pithole has become synonymous with the excesses of the oil boom. In terms of production, Pithole was home to half a dozen flowing wells, which for a few months in 1865

Hillside drilling along Pioneer Run (1865). *Source:* John A. Mather Photograph Collection, Drake Well Museum, Titusville, Pennsylvania.

poured out 6,000 barrels a day, roughly two-thirds of the total production of the Oil Creek region.[138]

The first flowing well along Pithole Creek, the United States well (sometimes called the Frazier well), struck oil in early January 1865. It was located a few yards from the creek on a spot reportedly chosen using a witch hazel stick.[139] According to a correspondent for the *Pittsburgh Chronicle* who visited Pithole that month, "It is the prevalent opinion that wells should be sunk at the base of some lofty and precipitous hill, on the lines of violent upheavals and rocky distortions. Nothing of that kind here."[140] Pithole Creek, like Cherry Run, was a brawling trickle, although it had gently sloping sides. The United States well had been bored to a depth of more than 600 feet. According to the same *Chronicle* correspondent:

> This well was sunk through four distinct strata of sandstone rock, instead of three, as customary in other localities. The first sand-stone was reached at one

Pithole (1865). *Source:* John A. Mather Photograph Collection, Drake Well Museum, Titusville, Pennsylvania.

hundred and fifteen feet; the second, at three hundred and forty-five feet; the third, at four hundred and eighty feet; the fourth, at six hundred feet; and the oil itself at six hundred and fifteen feet.[141]

The numbering of sandstones went back to the earliest days of well boring, and the identification and correlation of rocks from one well to the next was an ongoing process. By 1864 drillers were routinely referring to three. The first was a thin sandstone associated with Drake's and other shallow wells. The second lay 200–300 feet deep. Both these sandstones contained small amounts of oil that had to be pumped out. The third sandstone, lying about 400–500 feet down, was thought to be the source of the flowing wells. Pithole had apparently "pierced the fourth sand rock," a depth that had "never been reached on Oil Creek."[142]

Thus, in addition to the floods of oil (and money and people), Pithole helped to create a new vision of the underground. For most Oil Creek drillers and

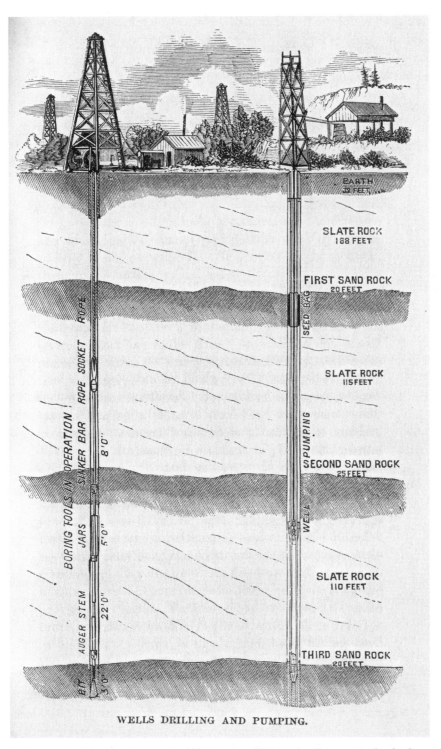

WELLS DRILLING AND PUMPING.

Ideal cross section of a well going down (*left*) and a completed well (pump in the third sand rock) along Oil Creek (circa 1865). The three sand rocks (or sandstones) corresponded to the three productive oil rocks or oil horizons. Andrew Cone and Walter R. Johns, *Petrolia: A Brief History of the Pennsylvania Petroleum Region, Its Development, Growth, Resources, etc., from 1859 to 1869* (New York: Appleton, 1870), facing 138.

operators, petroleum reservoirs were no longer being pictured as big cavities or vertical fissures or flowing veins but rather as thick, porous, horizontal sandstones. William Wright, a correspondent for the *New York Times,* explained the new thinking rather cleverly:

> Some experienced men regard [the loss of tools] as a happy omen of the early discovery of petroleum; since it is usually found in such caverns or pockets, which are believed to have originated in geological dislocations. . . . Experience does not fully bear out such expectations; and hence the loss of tools is apt to occasion bitter regrets rather than congratulations.[143]

A drop in the tool string now meant a loss of money and time, not an indication of oil.

Pithole's other remarkable feature was its height above the Allegheny River. "We are informed," reported the *Franklin Citizen,* "that this is the most elevated part of the country between Cherry Run and the Allegheny river. If so, the theory that oil existed only on low and flat lands is exploded."[144] Wright was more definitive: river bottoms, "*as such,* have no connection whatever with the deposition of oil. . . . [T]he hills will yield as freely as the low-lands."[145]

In the spring of 1865, petroleum explorations began to spread across the hilltops and table lands. Sandstones (third or fourth, depending on who was doing the counting) were as accessible from those elevated perches as on the valley flats, the only difference being a greater depth. In some instances, reported Wright, operators began "at elevations of two hundred or three hundred feet."[146] Drillers' willingness to go uphill reflected a general belief in the consistency of stratigraphy. Curiously, a firm foundation in geology encouraged, rather than restricted, *wildcatting*—the boring of wells in areas untested. For, as a matter of practice, there was much more untested territory now that drillers were willing to look outside Oil Creek valley.

> There only remains to be noticed that anomalous deposit of the Albert coal of New Brunswick, made famous by long litigation and the discussion of geologists.
> —J. Peter Lesley, "Coal Oil," *Report of the Commissioner of Agriculture for the Year 1862* (1863)

THE DEVELOPMENT of petroleum had a profound effect on the Albert mineral, not the mining of it, but the science. For despite the disappearance of the American coal oil business, the Albert mineral continued to be raised and shipped to Boston and New York City in relatively constant amounts throughout the 1860s. In the three years between 1863 and 1865, Albert Mines produced 56,289 tons, a little less than 19,000 tons annually.[147] In the five years between 1865 and 1869, the average annual yield was more than 17,000 tons.[148] Albert Mines flourished, and so presumably did the three mining companies (East Albert, Prince of Wales, and Princess Alexandra) subsequently established within its immediate vicinity.[149]

The commercial success of these ventures was an enticement to James Townsend, who, in late 1862, was thinking of selling his Oil Creek properties (Drake's pumping wells were now unprofitable) and so sent a business partner to Albert Mines. "[A]s for the coal business," the agent reported, "I consider our chances good for a valuable mine. This region abounds in mineral wealth and [make] no mistake—I consider such a mine as the Albert Mine, worth *more money* than a Cal. Gold Mine in the *long run.*"[150] With the glut of flowing wells, the Albert mineral seemed a safer investment than petroleum.

This was also the thinking of a congeries of Boston capitalists, among whom was the marine zoologist and mining entrepreneur Alexander Agassiz (1835–1910), son of the Harvard geologist Louis Agassiz. The capitalists wanted to expand their operations at the Princess Alexandra mine, and Agassiz consulted Lesley.[151] (Agassiz would soon become more interested in Michigan copper and, in 1871, would become the president of the very rich Calumet and Hecla Mining Company.) In the spring of 1865, though, Agassiz (like Townsend) was looking to capitalize on U.S. demand for the Albert mineral as a raw material for manufacturing or enriching *gas*. There was no demand for the mineral for manufacturing *oil*.

This point was brought home to New Brunswickers by the failure of the Caledonia Coal Oil Company. Located just three miles west of Albert Mines, the oil works had been constructed "at very considerable outlay of capital, [and] fitted with all the necessary apparatus for the manufacture of oil." By 1862 the Caledonia Company was failing, and within two years the oil works were "to all appearances, profitless, and on the rapid road to ruin."[152] Pennsylvania petroleum had killed coal oil manufacturing.

It had the opposite effect on oil explorations. Since the summer of 1850, geologists had been discussing the oil springs near the Albert mine. James Robb, for example, thought the springs were intimately connected with the asphaltum, as he classified the mineral. The connection was straightforward; oxidation or exposure to air of petroleum resulted in the formation of asphaltum.[153] Benjamin Silliman and Augustus A. Hayes, both witnesses for the defense, agreed with Robb's chemistry, but they did not think the Albert "coal" had any connection to nearby springs. Coal and petroleum, just as coal and asphaltum, were separated geographically, geologically, mineralogically, and chemically.[154]

And so the matter stood until the early 1860s when the oil springs became possible sites for wells (another reason for Agassiz's interest and Lesley's engagement).[155] Such was the case with the East Albert Company. In late 1864 the company consulted Charles H. Hitchcock (1836–1919), son of Edward and formerly the state geologist of Maine (1861–1862). Hitchcock examined the company's property, wrote a report, and then published a scientific version of his survey.[156] Hitchcock realized that the origin and occurrence of the Albert mineral were "vexed questions in geology," but he was ready to declare, contrary to "seemingly well-established theories," that the mineral was a vein. "To disclaim a bedded character casts no reflections upon the observations of the distinguished geologists who have decided otherwise," Hitchcock was quick to add.[157] His distinguished, but erroneous, predecessors had failed to observe the anticlinal fissure in which the mineral occurred.[158] From this fact Hitchcock concluded that "Albertite," as he called it, was "originally in a liquid state, was injected into a vertical fissure, and subsequently hardened." Hitchcock's new theory sounded much like the old one Taylor and Robb gave at the Albert trial, and Hitchcock did cite their joint report. Hitchcock, however, made a crucial addition: "carbonaceous veins are analogous to veins of petroleum." He then explained the analogy. "The borings for petroleum in Ohio and Western Virginia are most successful along lines of fracture, particularly an anticlinal axis. . . . similar to [that] respecting the Albert vein."[159]

Hitchcock modeled his new theory of Albertite explicitly on Andrews's theory of petroleum. But unlike Andrews, Hitchcock did not put in print any recommendations for boring wells (although he might have given such advice to the East Albert Company), nor did he propose any theories about the ultimate source of the Albert mineral. But he did note an important parallel case, the Ritchie mineral.[160]

Located in Ritchie County, West Virginia, about eight miles south of Cairo (the nearest station on the Baltimore and Ohio Railroad), this mineral had first come to public attention during the coal oil boom. In his *Practical Treatise* (1860), Gesner had described it as the "Cairo asphalt," which "will no doubt be found valuable for the manufacture of oils."[161] By March 1863 the Cairo property (400 acres) was under the control of some Philadelphia capitalists, who consulted Lesley. Lesley used the engagement to study some "geological facts of more than ordinary interest," which he presented at a meeting of the American Philosophical Society.[162] First, he noted that the Ritchie mineral was oriented vertically while the surrounding stratification was horizontal; and, second, it was "a solid bitumen vein rather than a coal bed."[163] The vein filled a fault, "which was the most interesting part of the phenomenon for structural geologists," and in fact, three-quarters of Lesley's presentation was devoted to structural geology, not to the Ritchie mineral. Nevertheless, based on his general knowledge of Appalachian coal, Lesley made a bold assertion: "There seems to be no escape from the conclusion that the substance filling this vertical vein is a product of the gradual oxidation of [liquid] oil once filling the open fissure. It is not impossible therefore that the lower regions of the fissure are still filled with liquid oil."[164] Such optimism, even though guardedly expressed, contributed to the formation of a very large oil company. The Grahame Crystallized Rock Oil Company of New York was capitalized at $6 million and purchased 5,000 acres in Ritchie County. In the company's prospectus, Lesley featured prominently. The Ritchie mineral, he thought, was similar to the Albert mineral.[165]

By 1863 Lesley had decided that the Albert mineral, or "Albertine," as he called it, was "not properly a coal bed . . . but a mass of hardened [petroleum]." Unlike the Ritchie mineral, however, Lesley did not think Albertine formed in a fissure. "It was originally a horizontal bed or *lake of petroleum*, hardened and covered up by sand and clay deposits of carboniferous age, and afterwards upturned."[166] (By April 1865, when Agassiz consulted him, Lesley had apparently changed his mind. Privately, if not publicly, he was willing to state that Albertine occurred in a fissure.)[167]

Henry Wurtz (1828–1910), a self-described "general scientific expert," and Benjamin Silliman Jr.'s former assistant in the chemical laboratory at Yale, thought Lesley's Albertine theory was preposterous. "[T]he great subterranean 'Sea of Petroleum,' of which some have fondly imagined," Wurtz declared, "is an idea which I believe will receive but meagre acceptation among those chemists whose minds are free from the trammels of pet hypotheses."[168] What was remarkable about Wurtz's attack on Lesley was not the disciplinary division (chemists and geologists had been split over the Albert mineral since the trial), but that it appeared in a consulting report to a rival company, the Ritchie Mineral, Resin, and Oil Company of Baltimore. The company had consulted Wurtz in late 1864, and he had examined the vein and done a thorough analysis of it. "[A] consideration of the facts," Wurtz explained, led to "the total subversion of the prevalent hypothesis implied in the current designation of 'crystallized petroleum.'"[169] According to Wurtz, neither the Ritchie mineral nor the Albert mineral was ever in the condition of fluid petroleum. "[N]o such substance [was] ever known to be formed, or to have formed, by the oxidation and inspissation [hardening] of petroleum."[170]

Wurtz, like many consultants, did not think a report was "the place for a scientific treatise." But he could not resist the opportunity (nor could other consultants) to explain his own theory. Wurtz described the Ritchie mineral as "a resinoid substance derived from, or formed by, some metamorphosis of unknown fossil matter," which had been forced by a very moderate pressure into the fissure.[171] Wurtz called the substance "Grahamite," a new mineral species. Grahamite was "neither coal, nor asphalt, nor albertite, but [is] chemically and mineralogically distinct from all and either of these."[172] Grahamite could be used for manufacturing lamp oil or gas, but its presence on the surface was *not* indicative of petroleum underground.

This conclusion was probably not what the Ritchie Mineral, Resin, and Oil Company was hoping for, which might explain why the company prospectus included extracts of Lesley's American Philosophical Society presentation[173] and a paragraph from Gesner's revised *Practical Treatise* (1865): "The Cairo asphalt . . . is evidently Petroleum, which has at some remote period issued from the earth and been hardened by evaporation, and exposure to the oxygen of the atmosphere. The oil springs frequently occur in the immediate vicinity."[174] Thus, within one company's prospectus was encapsulated a scientific debate over petroleum and its natural products.

In an attempt to resolve the debate over the connections (if any) among petroleum, asphaltum, and coal, the New Brunswick government sponsored *two* geological reconnaissances in the summer of 1864.[175] The House of Assembly granted $500 to Henry Youle Hind (1823–1908), a former professor of chemistry and geology at Trinity College in Toronto and an experienced explorer of the Canadian northwest, to study the remote, northern part of the province.[176] Meanwhile, the lieutenant governor commissioned L. W. Bailey, professor of chemistry and natural history at the University of New Brunswick in Fredericton (the position previously held by James Robb, who had died in 1861), to examine the more settled southern counties.[177] Both Hind and Bailey were angling for appointment on a new geological survey; both, naturally, emphasized mineral resources; and both, necessarily, confronted the conflicting interpretations of New Brunswick's most famous mineral. Hind followed the researches of Lesley, Hunt, and Andrews in reaching his conclusion that "Albertite [was] an inspissated or altered petroleum injected from below (from the Devonian) into fissures situated along an anticlinal axis."[178] Bailey was more cautious and deferential. In his preliminary report of 1864, he reiterated the decision reached during the trial, which he considered "undoubtedly the correct view": "Albertite occurred in true strata of the coal measures, and was therefore really a highly bituminous coal."[179] A year later, in his final report, Bailey took a different stand: "Albertite is neither *coal* nor *jet*, but an *oxydized oil*, derived from the decomposition of fish remains, and subsequently changed by chemical action." Furthermore, Bailey reported that "bucketsfull" of petroleum had recently been discovered by two oil companies.[180]

Despite such optimistic predictions of mineral wealth (reminiscent of Gesner), neither Bailey nor Hind received a government commission. Indeed, Bailey and Hind became locked in a bitter dispute. Amid charges and countercharges of incompetence and plagiarism (reminiscent of Gesner and Jackson), the legislature could not agree on funding a new survey. New Brunswick would have to wait until Confederation in 1867, when the Geological Survey of Canada would cover the ground and all the costs.

In the meanwhile, *the* authority on New Brunswick geology, John William Dawson, would have his say. In 1868 Dawson published a revised and enlarged edition of his *Acadian Geology.* In the extensive discussion of the Albert mineral, he began by reprinting his 1855 interpretation, which "[n]ow," he confessed, "is somewhat modified." Dawson considered Hitchcock's, Hind's, and Bailey's opinions "as established." The deposit had "unmistakenly the aspect of a vein or fissure." Dawson concluded that "the substance [was] a variety of asphalt or a solid hydrocarbon, originally fluid, like petroleum, and derived from the decomposition of vegetable or animal products." As to the likelihood of petroleum reservoirs, Dawson, unfortunately, had to report that in borings in Albert and Westmorland counties, on either side of the Petitcodiac River, "the amount of oil so far obtained has not proved sufficient to be remunerative."[181] Bailey's buckets did not amount to barrels.

Still, there was always the profitable Albert mineral or Albertite, the now-accepted name for a new variety of asphalt. Gesner had apparently been vindicated; it is unfortunate that he was not alive to see it. In April 1864, soon after receiving an appointment as professor of natural history at Dalhousie College in Halifax, Nova Scotia, Gesner had died. Nonetheless, the reclassification of Albertite speaks to Gesner's scientific perseverance and his commercial foresight. It also speaks to the powerful scientific, technological, and cultural impact of the developing petroleum business. According to Lesley, "[i]t is probable that *all* instances of solid bitumen found on or beneath the surface of the earth have resulted from the hardening of [petroleum] reservoirs."[182] Petroleum was the new paradigm.

CHAPTER 9

The Search for Oil and Oil-Finding Experts

I would most earnestly invoke men of science . . . to come
to the oil regions, and remain there for weeks and months,
collecting pebbles, fossils, fragments, and all other materials
obtainable for the nether world. Let them spend their time and
labor as enthusiastic explorers of truth, not with a view to lend
their names to this or that Mammoth Gas Bubble Company,
for a consideration in dollars or dollars' worth [of stock].
—William Wright, *The Oil Regions of Pennsylvania* (1865)

IN FEBRUARY 1865 James Hall was in a quandary. The oil boom was in full
swing, and he was "[p]ressed with outside business."[1] But he had strained the
tendons in his foot and could not walk. "To get my foot in a condition to travel,"
he told J. Peter Lesley, he would have to refuse many engagements.[2] It was all
"very disheartening," Hall lamented, "to be unable to take advantage of this state
of things."[3]

So Hall took hold of something else. A friend offered to take leases on properties in upstate New York if he thought they were good petroleum prospects.
The gentleman, whom Hall had "every reason to believe honorable," would find
the capitalists, organize the company, and bore the wells. All Hall had to do was
name the places. The scheme was enticing; "there is an abundance of Cash, hydrogen gas & oily indications."[4] He got to rest his lame foot *and* collect half the
"proceeds" from any discoveries. Frankly, he did not see anything "very objec-
tionable in the matter." Still, Hall felt compelled to justify his involvement to
Lesley (and possibly to himself) as a simple necessity: "In these times a small
salary must be eked out or no [one] can live."[5] He even offered Lesley the chance
to join in. On second thought, "you already know plenty of capitalists who will
do better by you than this proposition."[6]

In deciding to stay home, Hall was banking on the oil boom. "I think in the next year or two," he reassured Lesley, "or so long as the currency continues, we are to have more of this kind of work than ever before."[7] What Hall had in mind was consulting, and he knew that geologists were doing a red-hot business in oil. Hall too wished "very much to go into the Oil region next summer."[8] In the meantime, he was happy to do armchair explorations. But if Hall inadvertently ended up endorsing the prospects of some Mammoth Gas Bubble Company, as the *New York Times* correspondent William Wright feared such lending of names might lead to, it would be bad for business and worse for science.

Men of science, of course, were not the only ones doing explorations in oil. Others less scientific and sometimes less honest were hired by companies, capitalists, or whoever was willing to pay them. For historians, the oil boom reveals a marketplace, in which men of science jostled against other oil-finding experts. This marketplace was on display in the numerous company prospectuses and other pamphlets printed during the oil boom. A study of these publications challenges hackneyed depictions of oil explorations as lawless and random. In fact, the numerous consulting engagements helped to create a systematic body of knowledge called petroleum geology.

Fever and Fraud

Midcentury Americans liked to compare the oil boom to the California gold rush. Both dramas drew on stock accounts, both enthusiastic and excoriating, of colorful characters—rags-to-riches prospectors, unlucky (but honest and hardworking) miners, gullible investors, and audacious swindlers. According to one sanguine oil boomer: "People will come, people will buy, people will make money, people will lose it, people will get crazy—and the consequences are, discomfort, ill-temper, bad beds, worse meats, and worser whiskey."[9]

The most vivid image of the oil boom was the fabled boomtown, one minute bursting with activity, the next, busted, dry, and deserted. The boomtown captured the flow and ebb of oil and the tides of people. The most infamous was Pithole. Denounced as the dirtiest, greediest, and meanest place in America, Pithole, as its name luridly suggested, epitomized oil fever.[10] But visitors need not go to Pithole to see oil's ruinous effects. Oil City, at the confluence of Oil Creek and the Allegheny River, would do just as well:

> If you wish to live in mud, to walk in mud, to ride in mud, to see nothing but mud, to have the color of your clothing obscured by mud, to inhale nothing but air burdened with gas and petroleum, and to see what a livid, hungry, anxious genus of animals men are when they are bitten by this money-getting tarantula, by all means come to Oil City. . . . Here King Petroleum reigns, seated upon his muddy throne.[11]

During the boom, Oil Creek got its sobriquet: Petrolia. The name reflected the obvious source of the region's renown as well as the mania. Petrolia was a place unto itself, "five hundred miles from the centres of civilization."[12] Americans went there to escape. "The war, with its mighty events and momentous issues," noted one Petrolian, "fails to attract more than passing notice from the eager adventurers in pursuit of oleaginous wealth."[13] Even so, the Civil War's end generated much excitement. Thousands of decommissioned and paid-off soldiers swelled into a flood of single men looking for work and wealth.

To many outside Petrolia, the boom was nothing more than "the efflorescence of humbuggery," a gigantic bubble. "The prevailing *passion* for speculation in oil is scarcely excelled by any mercantile epidemic known in the history of enterprise," remarked one critic. "The tendency to discovery and speculation is not confined to the original centre of petroleum in Pennsylvania, but has radiated throughout the geology of the Western states."[14] The critic mentioned California, where Benjamin Silliman Jr. was causing quite a stir. But other oil regions—Canada, Ohio, West Virginia, and Kentucky—were engulfed by the boom. From the fall of 1864 through the summer of 1866, speculation swept the nation.

Part of the passion was fueled by inflation. During the war, the U.S. Treasury printed more than $400 million in paper money, the "greenbacks." Prices for oil lands rose to dizzying heights; speculators often transacted business *in cash* to the tune of $50,000 or $100,000.[15] Properties changed hands for more and more money without any digging or any intention of digging for oil. One buyer remembered:

> A gentlemen of New York was describing to me a piece of property he had that day bought . . . for $10,000, when my right-hand neighbor . . . who appeared as if dozing, suddenly brightened up and inquired the exact locality of the property. Plans were produced, title-deeds examined, and in less than half an hour the property was re-sold for $14,000, the seller appearing very doubtful about the wisdom of his step.[16]

The other enticement to speculation was stock. What made the oil boom an unparalleled, commercial epidemic was the profusion of joint-stock companies. The Reverend S. J. M. Eaton of Franklin counted more than 600; the oil reporters Andrew Cone and Walter R. Johns of Oil City estimated more than 1,000.[17] The promiscuous generation of oil companies reflected recent innovations in corporate law. Before the Civil War, most states required special charters from their legislatures to incorporate; in addition, Pennsylvania prohibited "foreign" or out-of-state corporations from owning property. By the early 1860s, new laws in Pennsylvania and New York allowed for a simplified process of incorporation, and foreign companies could operate merely by filing papers and paying fees.[18] "As the oil-region supplied the oil," wrote one observer, "so distant cities were relied on to supply the capital with which to develop it, as well as to stimulate and feed the speculative fever."[19]

The public's passion for stock had more to do with the oily hubbub than with changes in corporate law. Striking a flowing well was a newsworthy event. "The trouble now," complained William Wright, "is that . . . the successes are blazed abroad by telegraph, newspaper, and private epistle; while the failures are glozed over, or at best only touched upon, as if they were matters of which the public must be kept in profound ignorance."[20] The ceaseless sensationalism beguiled the public into believing oil was easy to find.

Oil companies also made it easy to invest. Low-budget stock sold for as little as twenty-five cents a share, "which had the effect of attracting an entirely new class of speculators," noted one reporter. "The cook and chambermaid who had only ten dollars to invest, had now the opportunity of becoming rich. The stock was rapidly taken, and, in most cases, doubled in value within a week. Yet few or none of these had a single well in operation."[21] Philadelphia, the largest oil stock exchange, was struck by the fever most violently. On average 600,000 shares traded monthly during the first part of 1865.[22]

"[M]aking stock companies," remarked a jaded jobber, was a "very common practice." "A number of gentlemen get together, and agree to form a company. Somebody is sent to Oil creek to buy some land; it may have oil or not; that is immaterial. It is sufficient that they own so many acres on Oil creek. The company is then formed, and the acres which cost them one hundred thousand dollars, are turned in at a nominal value of five hundred thousand, or a million dollars, and stock issued representing a capital of that amount."[23] Wright drastically discounted most petroleum ventures. "Most persons have come to believe that a certain amount of training, of experience, as well as natural capacity, is requisite to employment in any situation requiring knowledge, skill, and judgment. Not so with many of the oil companies."[24] The typical company was capitalized at $250,000 to $500,000, although many were valued at $1 million.[25] These sums reflected inflated property prices, which constituted most of the assets, no matter how small the acreage. Individuals could not afford to buy land in Petrolia; it was held by companies or original owners, who refused to part with their farms.[26] Some companies could not afford land. Their property consisted solely of fractions of an interest, from one-third to one-thirty-second in an oil well, which might be flowing, pumping, going down, getting started, or none of the above.

Regardless of capital or property, all companies operated at the level of the individual well, the "unit" of trade in the early oil industry. A company usually succeeded or failed depending on whether its well struck oil.[27] And even if a well did flow, it did not do so for long, perhaps a year or two, and all the while the flow diminished. Hence the existence of any oil company was correspondingly short, a transient affair that heightened the sense of a rush.

Like all mining booms, fast money was the modus operandi in the oil business. That is not to say that striking oil was unimportant, but bags of cash were made speculating in land and stock. Critics constantly admonished readers. "The unwary," sighed one, were "fleeced to an unmerciful extent."[28] Eaton offered more pastoral reflections: "[I]n stock companies, as in other things, there is the good and the evil—the true and the false."[29] Beware a plague of frauds.

"Famous Oil Firms."

Words by E. Pluribus Oilem   Music by Petroliana

With great solemnity

There's "Ketchum & Cheatum," and "Lure 'em & Beatum,"
And "Swindle um" all in a row;
Then "Coax 'em & Lead 'em" and "Leech 'em & Bleed 'em,"
And "Guzzle 'em Sink 'em & Co."

There's "Gull 'em & Skinner," and "Gammon & Sinner;"
"R. Askal & Oily & Son,"
With "Sponge um & Fleece um," and "Strip 'em & Grease 'em,"
And "Take 'em in Brothers & Run."

Chorus: Oh! oh! Oily firms pay, in Pennsylvania Jest so.

Slow, with expression

There's "Watch 'em & Nab 'em," and "Knock 'em & Grab 'em,"
And "Lather & Shave 'em well," too;
There's "Force 'em & Tie 'em," and "Pump 'em & Dry 'em,"
And "Wheedle & Soap 'em in view."

There's "Pare 'em & Core 'em," and "Grind 'em & Bore 'em,"
And "Pinchum good, Scrape um & Friend,"
With "Done um & Brown um," and "Finish & Drown um,"
And thus I might go to the end.

Chorus: repeat.[30]

To those who lost their money, the famous firms were hardly humorous. Bogus companies and fancy stock were the evil twins of the boom and the targets of the self-righteous. "The insane desire for oil *is* demoralizing," Wright conceded. "It leads to every imaginable kind of misrepresentation and cheating. In every transaction involving profit and loss, falsehood is expected, is looked upon as the rule, truth the exception." Wright reckoned petroleum was one gigantic swindling operation, "a system which has been reduced to both a science and an art. It is exquisite, magnificent, stupendous, brilliantly successful."[31]

The "machinery of deception" had several parts, but the principal one was information.[32] High-speed presses cranked out heaps of "windy" prospectuses, "lying" reports, "worthless" guidebooks, newspaper "puffs," and "long-winded" histories.[33] "Nearly every day," wrote the *Meadville Republican,*

> we find on our table some new weekly or monthly, and with few exceptions they are unreliable, sensational, stupid gas and twaddle. In many instances these publications are manifestly catch-penny concerns, issued by adventurous speculators for the purpose of puffing worthless oil-territory which they are anxious to sell. It is amusing to read their learned disquisitions on oil, their sa-gas-eous advice to persons seeking investments, and the chunks of wisdom displayed by them must have severely tested the elasticity of the remarkable pericraniums of the profound editors thereof.[34]

An oily cornucopia inundated anonymous investors and ingenious getters-up of stock companies alike, neither of whom had firsthand knowledge of oil wells or an earthly idea of what made a good prospect. They *needed* advice, which made them vulnerable. Information was the screw to turn the machinery of "moral swindlers."[35]

To escape the machinations, one needed a trustworthy expert, and in Petrolia there was no shortage of handsome individuals styling themselves as experts. For a price, anyone could buy one. Newly formed companies in Philadelphia, New York City, or Boston were especially eager to do so. They usually sent a representative to the oil regions to buy lands or take leases, to investigate properties and wells, and to consult an expert. Whether that person was competent and honest was another matter.

Peter O'leum, Swindler

Swindlers, of course, never advertised themselves as such. And no one, it seemed, knew one personally, which was why swindlers, as a group, were so easy to denounce. They came in any number of guises, including "sharpers," "predatory guerillas," and "barefaced liars." In the dread jargon of academe, they were always coded as the Other. Swindlers displayed all the bad features of the boom—greed, dishonesty, and coldheartedness. They represented "the bankruptcy of moral character, [and] the shipwreck of all conscientiousness, honor, and honesty."[36] Worst of all, their habits were infectious. Otherwise honest Petrolians often got caught in their deceptions and dishonesty.

Or sometimes Petrolians willingly colluded in the game. When they were not being branded as devils, swindlers could be cast as charming rogues. Although not entirely harmless, they were admired for their skill in humbling the arrogant and outwitting the know-it-alls. In this depiction, swindlers represented the wily natives of Oil Creek contriving to foil predatory interlopers.

Of all the commentators on the boom, Wright alone made an effort to analyze the swindler through a greasy fictional character known as Peter O'Leum. Peter was "the most disinterested fellow in the world, and clever withal." Peter was not a charlatan. Unlike the "Coal Oil Man," Peter, "a dull-looking, phlegmatic Pennsylvanian," pretended to know *less* than he really did. "The fact is," Wright declared, "in twenty-four hours the stranger has become almost a Petrolian; and Peter is careful to flatter his vanity, by congratulating him on the rapidity of his naturalization."[37]

Peter was also technologically adept. The telegraph was key to fast-flowing information and, not surprisingly, a useful swindling tool. Whenever a strike was made, the lines "caught up the joyful intelligence, and whispered it to all the outskirts of Petrolia, [and] also to interested parties in Wall, Chestnut, and State streets."[38] Lines would soon be run to the well, which meant information flowed in both directions. Drillers could be forewarned of inquisitive visitors, an advantage to someone like Peter who might wish to "enhance" a well's production. Pipes were sometimes run from oil tanks back to wells so that a sort of "perpetual motion" was achieved. Barrels of oil could be poured down holes only to be pumped back up in front of unsuspecting investors. Finally, Peter often spilled a barrel or two to create a "good" surface indication.

Peter liked to practice his art on Yankees, a group that a *London Morning Post* correspondent described as "[w]eary men . . . all wearing long-legged boots, and all plastered with mud from head to foot."[39] They were "[s]harp-eyed, trim-dressed, and eager," thus easy prey to Peter O'Leum.

> Hundreds of sharp adventurers from the great eastern cities, who fancy themselves familiar with all the business wisdom of the times, will wake up some bright morning to the chilling fact that they have been "taken in and done for" by the unsophisticated denizens of oildom, who know nothing of the mysteries and tricks of the moneyed men of Gotham.[40]

Of all the slick tricks, none produced more laughs and more denunciations than the sale of worthless lands. "[T]here are wild-cat lands in out-of-the-way places . . . in vicinities where oil has never been heard of, which have obtained a fictitious value, and forced through the representations and efforts of unscrupulous and irresponsible parties, [who] effected sales of properties so utterly worthless, except as the basis of a swindle."[41] The pejorative sense of wild-cat is a reminder that boring on untested land was as foolhardy as investing in fancy stock or fraudulent wells.[42]

Smellers and Diviners

"A new profession of men," Wright announced, "claiming to be gifted with extraordinary powers, has arisen in Petrolia, namely, 'oil-smellers' or 'diviners.'"[43] To call the practice a "new profession" was journalistic hyperbole. Persons claiming to possess special powers to locate underground waters or minerals had been around for millennia. "It is no new thing under the sun," Eaton remarked drily.[44] Nor was it a profession; smellers and diviners did not master or monopolize any systematic body of knowledge; their theories and practices were as idiosyncratic as themselves. At worst, they were not "legitimate science" and frequently described their powers as supernatural. At best, they were strict empiricists; what worked, worked.

Contemporaries ridiculed them as "wizard characters," "prophets," or "professional humbugs," yet they were often accorded a dollop of indulgence. "[S]o vast and wonderful has been the Petroleum development," the journalists Cone and Johns noted, "that it has afforded an excellent field for the exercise of the exploded superstitions of less enlightened times."[45] Smellers and diviners played on the gullibility of humankind, just as the coal oil men had done. They were popular and colorful, but they were not identified beyond such vague names as "Messrs. P&D" or "professional locator"; thus it is difficult for historians to figure their number or to judge their impact on developments. Oil company prospectuses did not mention them or their "magical power," for that would have been regarded as unorthodox, unscientific, and thus unhelpful in selling stock. Given this low opinion why would anyone consult one? "If there be any thing in oil-smelling, we may as well avail ourselves of it as not; for the diviner charges only from twenty-five to one hundred dollars for his services in examining a tract; and this is an inconsiderable item in the general expense, seeing we mean to bore any how."[46] They served as reassurance. "The experiment costs but a trifle," Cone and Johns observed, "and operators, consequently were willing to test its efficacy."[47] Wright, however, thought smellers and diviners had "more real power among the operators than the latter are willing to openly concede."[48] Nor was $100 such a trifling investment.

Smellers sniffed the earth or waved a "magic stone." Diviners used a twitching stick made of hazel or peach tree. "In passing over the vein or basin of oil, it is the creed of the believers in the faith of the witch-hazel that the twig suddenly and irresistibly reverses its position, and points toward the north."[49] Diviners applied "the hazel tree theory" or practiced "hydro-geology."[50] "The explana-

---

tion given by its experts is, that there is a kind of magnetism by which the rod is disturbed," Cone and Johns wrote, "and the possession of the same is limited to certain persons. We give the explanation for what it is worth, and are free to confess our inability to see any philosophical reasoning in it."[51] A more witty commentator noted that "[a]s the same method has been adopted for years in discovering hidden springs of water, the borer not infrequently obtains a copious yield of water instead of oil; but such accidents will happen in the best-regulated systems of science, and they only strengthen the faith of the true believers."[52]

Another group often derided as charlatans were the spiritualists. These oil finders were guided by misty revelations or oily sensations, and they became popular after the astounding strike of the Harmonial well in Pleasantville in the fall of 1867. An ardent spiritualist was on his way to Titusville from Pithole when his horse suddenly halted and he was grabbed by a "spirit-guide," who lifted him out of his buggy and dropped him in a nearby field. On that spot, the spiritualist recommended boring a well, which produced 100 barrels daily and subsequently opened the Pleasantville field.[53]

Such dramatic and unique discoveries aside, the effectiveness of smellers, diviners, and spiritualists is difficult to judge. Petrolians agreed they found oil. "Numerous wells located by [hydrogeology] professors, proved successful," Cone and Johns admitted, "enabling those who made it a regular business, to realize handsomely."[54] Success justified their theories, practices, and employment. "[O]ne or two lucky guesses having been made by the professors of the art, [and] there is an increasing demand for their services."[55] "Still, if we had any advice to offer," wrote the always reserved Eaton, "it would be not to trust [them] too implicitly."[56]

Practical Oil Men

Practical oil men were the heroes of Petrolia. They bored the wells, identified and numbered the oil rocks, and developed and expanded the productive territory. They too were charismatic.

> About the box-stove, the crowd that triangulates itself . . . hears a philosophical dissertation on the rise, progress, and development of petroleum, and the geological eccentricities necessary to the presence of that precious fluid, by a fat man in a greasy collar, who spits tobacco-juice between his sentences against the opposite wall.[57]

With experience as their guide, they could be counted on as "reliable": "Those who have begun with nothing, and by hard labor, amid grease and dirt, learned all about putting down wells so as to make them work, or by diligent selection of likely lands have made them known to capitalists."[58]

"Practical oil men," as a type of expert, appeared in a large number of company prospectuses. The Hard Pan Oil Company, for example, declared that its land had been selected "with great care by experienced oil men,"[59] and similarly the Shreve Farm Oil Company emphasized that its properties had been examined by "gentlemen thoroughly acquainted with the entire oil region."[60] In citing them, en masse, companies traded on common knowledge or standard practice. Investing was therefore a safe bet *because* it was practically unexceptional.

Locating a well adjacent to a producing one is an example of the expertise practical oil men offered. The one-page prospectus of the Rockbottom Oil Company put it succinctly: its property was close by "paying Oil Lands," a "sufficient guarantee of ultimate success."[61] The Boston Petroleum Company was a bit more specific; its land "immediately adjoins the celebrated Egbert Farm, on which are located the *Maple Shade* and *Jersey Wells*, too well known to need comment."[62]

Another practical theory was the in-line or belt theory, by which wells were located between two producing ones.[63] The First National Petroleum Company,

Prospecting for oil. The gentleman in the middle is the capitalist or prospective investor and on the left is his keen-eyed secretary or scout with a traveling pouch containing maps, deeds, and other important papers. On the right, pointing in the distance, is the local expert or practical oil man. All three are wearing high boots because of the ubiquitous mud. *Source:* "Petroleum: What it is and where we get it; with a glance at the oil regions of Pennsylvania," *Frank Leslie's Popular Monthly,* 2 (1876):513−527, 514.

for example, explained to prospective investors that its property was "in a direct line between the 'Noble Well,' which flowed 2,000 barrels daily, and the new 300 barrel well on Pithole Creek."[64] Practical oil men seemed to specialize in drawing lines connecting flowing wells.

Outside Petrolia, practical oil men were not often cited in company prospectuses. One rare example was the Boston Petroleum Oil Company whose property along Duck Creek in southern Ohio was selected by "practical experts."[65] The absence reflected a lack of development and dearth of experienced practitioners. Companies, like the Harmon Petroleum Company, had to combine the expertise of practical oil men with other knowledge. "This land [in western New York] affords positive evidence, throughout its whole extent, of its great value as Oil-producing territory; and this view has been confirmed by the Reports of Geologists, as well as by the most practical Petroleum men of Oil Creek."[66] Combining experts, by juxtaposing extracts, was a familiar selling technique. When the Harmon Petroleum Company referred to the "Reports of Geologists," it meant the published state survey. In New York, Ohio, West Virginia, and Kentucky, regions with large stretches of undeveloped though promising oil territory, engineers and men of science were usually consulted.

Engineers

Unlike swindlers, smellers, diviners, or spiritualists, self-described civil and mining engineers offered their professional services in print. Through advertisements and endorsements in newspapers, journals, and pamphlets, they spelled out their technical training, job experience, and competence in surveying and map making. At the back of F. W. Beers's *Atlas of the Oil Region of Pennsylvania* (1865), the best-known and most highly regarded among the many oil atlases, Beers, a civil engineer, printed the trade cards of many fellow engineers. D. Larned, for example, a mining engineer from New York City, advertised his expertise and experience in making "accurate" topographical maps anywhere in the United States "with Promptness and Despatch."[67] For a fee (not stated in the advertisements), engineers were willing and able to examine oil properties, write reports, and oversee the actual boring and working of oil wells. Such onsite and ongoing expertise was characteristic of engineers. Other oil-finding experts for reasons of personal temperament or professional identity did not want to become company employees. Engineers were the only ones selling technical advice and day-to-day management skills.

Engineers usually worked for relatively large and well-financed companies owning hundreds or perhaps thousands of acres. But their reports were slightly different from those of men of science. Typically, engineers' reports included extensive descriptions of topography and lengthy histories of an area. The reports

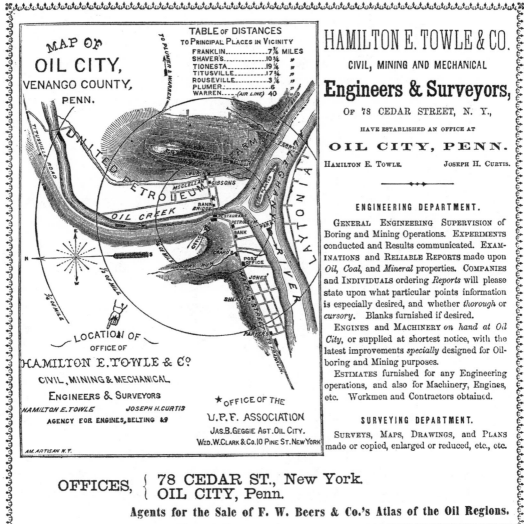

Advertisement for oil-finding engineers. *Source:* F. W. Beers, *Atlas of the Oil Region of Pennsylvania* (New York: F. W. Beers, A. D. Ellis, & G. G. Soule, 1865), 1.

were thorough but often did not display a firm grasp of stratigraphy or structural geology, despite the inclusion of many well-drawn diagrams and maps. Formations were described in general lithological terms, such as sandstones, shales, or limestones, with little or no reference to the surrounding geology or scientific nomenclature. Engineers focused on the particular property, not its relation to other places or phenomena, and they frequently became experts in certain regions.

A good example is Charles Richardson, a civil and mining engineer from Charleston, West Virginia. Richardson made a detailed study and map of the area around the Little Kanawha and Elk rivers in West Virginia. His report was published in the prospectuses of three local oil companies, and a shorter version appeared in a fourth prospectus.[68] Similarly, William Toshach, a civil engineer originally from New York, surveyed Elk County, Pennsylvania. His report and map appeared in the prospectuses of at least seven oil companies.[69]

To the ever-critical reporter Wright, engineers' tendency to draw diagrams and maps, "marked in blue and gold" for petroleum locations, was suspicious.

> In describing the swindling operations or sharp practice resorted to in the oil regions, it would be a glaring oversight to omit mention of those perpetrated by engineers (civil or uncivil) who have visited that country, bought lands, leased lands, or secured the "refusal" of lands for a time, and then gone East or West to organize companies for "developing" their territory.[70]

To Wright, "C.E. Esq." meant a confidence engineer, whose project "takes" because the public was duped by the authoritative title.[71] In Wright's moral universe, the problem was one of professional practice. Engineers not only consulted for big companies; they worked for them as well.

Consulting Men of Science

Judging from Petrolia prospectuses, men of science played no role in the oil boom. Stocks and properties sold easily; local oil finders supplied the expertise, and, if science were needed, extracts could be pulled from published sources. Consulting geologists were apparently too slow and too costly. In the time it took to arrange an engagement, an oil company could organize and start a well. In 1865 boring cost $4,000 to $6,000;[72] geologists charged $500 (plus expenses) for a survey and written report. On paper and on the ground, geologists were missing.

Beyond Petrolia, geologists were in the thick of an "extraordinary excitement in the speculating world," according to Lesley.[73] From his experience, solicitations were never-ceasing, which made it difficult to do anything else. "Business comes before pleasure," he complained to Josiah Whitney.[74] Most of the business came from well-capitalized companies based in Philadelphia, New York, or Boston, which owned lots of land close by, but not within, developed oil regions. In these untested areas, science served as a sort of investment insurance as well as a sound basis for boring wells.

In the fall of 1864, the Tennessee Petroleum and Mining Company engaged James M. Safford (1822–1907) to survey 5,000 acres along the Cumberland River northeast of Nashville. Safford had been the state geologist from 1854 to 1860 and was the authority on Tennessee coal.[75] In his report, Safford reviewed the region's topography, described the major geological formations, drew a vertical section, and added a "Special Description" of the oil tracts. Tennessee, although an "undeveloped region," was "full of promise" because it bordered "the great oil field" of southern Kentucky and embraced a "disturbed region," a line of great uplift or anticline, which was of "peculiar importance with reference to the occurrence of petroleum."[76] Safford concluded on an astonishingly upbeat note: "[Y]our lands are as promising as . . . the oil lands of Pennsylvania."[77]

The Kentucky oil fields received closer attention from Charles Hitchcock, the geologist who had consulted on Albertite in the fall of 1864. In the summer of 1865, Hitchcock surveyed 100,000 acres for the New York and Pennsylvania Petroleum, Mining and Manufacturing Company. (He also undertook two more engagements covering another 100,000 in West Virginia, Pennsylvania, and New York.)[78] Likewise, Albert D. Hager (1817–1888), a former assistant on the Vermont geological survey (1856–1860), visited eastern Kentucky that summer and surveyed 15,000 acres for the Kentucky National Petroleum and Mining Company. He found Carboniferous and Devonian strata, which he thought "promising" because petroleum was found in these rocks in Pennsylvania and Canada. Hager reported to the company's directors that "your chances of success in boring for oil [are] fully equal, if not superior, to those of the pioneers in petroleum, who have astonished the world with their unprecedented success in Venango County, Pennsylvania."[79]

The Kentucky National Petroleum and Mining Company also owned 5,000 acres in southern Illinois. To examine this property, the directors engaged George G. Shumard (1825–1867), former U.S. geologist on the exploring expedition to the sources of the Red River. Shumard made a quick survey noting deposits of coal, limestone, and lead. As to oil, he regarded the presence of anticlinal axes as "favorable" indications, and he identified several sites for boring wells.[80]

Another big firm that engaged several men of science during the oil boom was the Neff Petroleum Company. Peter Neff wanted a detailed report on several thousand acres in central Ohio, so he consulted Alexander Winchell, John Newberry, and Hamilton L. Smith (1819–1903).[81] Smith was professor of astronomy, natural philosophy, and chemistry at Kenyon College, in Gambier, and the local authority. He wrote a history of petroleum developments in the region and advised Neff to continue to sink wells, even though Neff's first four wells had struck gas. Not to worry, Smith advised, only one in twenty wells was "fully remunerative."[82]

Likewise, Winchell tried to frame his evaluation of Neff's prospects against the relative success of regional petroleum developments. Since 1855 Winchell had been professor of geology, zoology, and botany at the University of Michigan, and between 1859 and 1861 he had directed the state survey, during which he had surveyed the oil springs in St. Clair County, Michigan.[83] As a consulting

IDEAL SECTION of the Strata from East to West through the Lands of the NEFF PETROLEUM COMPANY.

W.   KNOX CO., O.   COSHOCTON CO., O.   E.

Vertical Scale: 544 Feet to the Inch.

Horizontal Scale: about 14,600 Feet to the Inch.

I. Black (Genesee) Shale
II. Chocolate Shales and Flags
III. Third Sandstone
IV. Second Shales
V. Second Sandstone
VI. First Shales
VII. First Sandstone
VIII. False Coal Measures
IX. Conglomerate
X. Coal Measures

Ehrgott, Forbriger & Co. Lith, Cin.

Alexander Winchell's ideal geological section of the Neff Petroleum Company's property in Ohio. The five wells were all of different depths (and to different oil horizons); each ended in a layer of shale, not sandstone. *Source: Prospectus of the Neff Petroleum Company* (Gambier, OH: Western Episcopalian, 1866), 10.

geologist, Winchell had been engaged by oil companies owning land in West Virginia, Ohio, Michigan, and Tennessee.[84] In the summer of 1865 he visited Enniskillen and then traveled to Neff's property. Using records from Neff's four wells (plus those from another company), Winchell drew a vertical section indicating the oil rocks and other formations. Knox County, Ohio, Winchell decided, was stratigraphically similar to Venango County, Pennsylvania; therefore, "a large reservoir of petroleum is liable to be struck." But it was his "duty," Winchell reminded Neff's directors, to caution against "an undue amount of confidence."[85]

Newberry displayed a similar cautious optimism. Recently appointed professor of geology and paleontology in the newly established School of Mines of Columbia College in New York City, Newberry had surveyed oil lands in southern Kentucky around the town of Burkesville in the fall of 1865, and the following spring he was engaged by Neff.[86] Newberry regarded Neff's property as suitable to the accumulation of petroleum. The company had "fair ground for encouragement," although he added, "there are no infallible signs which prophesy success in oil enterprises."[87]

But there were "geologically right" locations, according to E. B. Andrews.

During the oil boom, Andrews was an active consultant in the region around Marietta, Ohio.[88] One of his engagements was for the West Virginia Oil and Oil Land Company, which held property along the famous "break" or anticline that ran from Marietta to Parkersburg, West Virginia. Andrews was keen on the company's prospects in large part because he was keen on his own ideas. His consulting report contained a version of his fissure theory and even referred readers to his 1861 *American Journal of Science* article. Accordingly, Andrews concluded that "the [company's] chances of striking oil fissures [were] greater than on Oil Creek." He then took the very unusual step (insofar as announcing the fact) of agreeing to take three leases on which wells were to be bored "as proof of my individual confidence."[89]

Conversely, when Andrews thought a company's prospects were geologically *not* right, his pronouncement "was almost as effective as a dry hole in condemning a property."[90] Andrews often made such pronouncements in articles for the *Marietta Register*. During the oil boom, newspapers and the popular press became Andrews's favored media for his petroleum prognostications. On one occasion, when he wrote that land along Duck Creek was favorable, he started a small oil rush.[91]

Other consulting geologists were equally comfortable with the press. Winchell and Newberry, for example, wrote favorable articles for local newspapers on the oil prospects of central Ohio, which, not surprisingly, boosted the Neff Company's stock.[92] Companies, however, were not always pleased to see reports on their lands in the press. A lot of competition in the immediate neighborhood might not be good for production, although it would undoubtedly give a big boost to property prices. Likewise, some geologists were uncomfortable with having their opinions spread across the broadsheets. Anybody could pick up a newspaper article and edit it for republication or perhaps even fabricate one, which is what happened to Lesley in the summer of 1864. The *Philadelphia Mercury* began running a series of petroleum articles under his name. Lesley, however, never wrote them, nor did he read such a "vile sheet" as the *Mercury*. When he "accidently" discovered the articles, he fired off a furious letter to the editor demanding an immediate withdrawal, which he got.[93]

The *Mercury* incident reflected a growing problem for Lesley; as the authority on Pennsylvania geology, he was *expected* to have an opinion. Six months later Lesley was "startled" to find his name attached, once again, to certain petroleum "facts." Lesley's name had appeared in a prospectus for a mining company in Ridgeway, Pennsylvania. The unauthorized source this time was Lesley's good friend Peter W. Sheafer. Lesley reminded Sheafer that the information had been given in a private letter. As for the public: "My mouth is shut about oil in Oil Creek."[94] That Sheafer and the *Mercury* were willing to trade on Lesley's name, illegally in one case and embarrassingly in the other, reflected a well-grounded belief that Pennsylvanians wanted to hear from their best-known geologist.

A Cautious Consultant

Lesley had not published on petroleum since 1862, despite numerous offers.[95] Charles Eliot Norton, the liberal and reform-minded editor of the *North American Review*, had asked him repeatedly for an article,[96] and his good friend Leo Lesquereux pushed him to publish.[97] Lesley's reasons for withholding involved a mixture of scientific and professional prudence. He thought Andrews, Newberry, and Hunt had rushed into print theories that were "*unrefined*—a little pitchy."[98]

Lesley chose to enter the scientific debate through consulting. Beginning in the summer of 1864 and continuing for the next year, he undertook no fewer than twenty major engagements, roughly two per month, a rate that placed him far ahead of any other consultant, with the possible exception of Silliman in California.[99] By Christmas 1864 Lesley could look back at the "extraordinary number" of surveys that had yielded him "a harvest of fees," which he pegged at $16,121.[100] He bragged to his nephew Benjamin Smith Lyman that it would be "strange" if he did not collect at least $10,000 in professional fees the following year. He had already booked $3,000 worth of engagements for the spring of 1865.[101]

Lesley worked mostly for coal companies in 1864, although petroleum figured prominently in his surveys. It was simply too important and potentially profitable to omit. As Lesley told Lesquereux, "[W]ho thinks about anything but oil now?"[102] In the opening line of his Boston and Pictou Coal Company report, for example, Lesley reiterated the wide-ranging scope of the engagement. "You desire me to give you my professional opinion upon [your] property and also whether there is any possibility of its being *oil* country." Lesley did not have "the least doubt" that the company had good coal on its land, but as to oil, he "pronounce[d] against it."[103] In contrast, Lesley devoted nearly half of his report on the Warner tract in Beaver County, Pennsylvania (which had some of the best coal for iron manufacturing), to petroleum. He concluded with cautious, yet optimistic, advice. "I do not assert that the Oil Creek Sand Rocks *are* underneath the Beaver River. But I assert that *similar* sandrocks, charged with oil . . . do exist beneath the Warner Tract, and that a well sunk at the river bank, from seven to eleven hundred feet should produce oil."[104]

In early 1865 Lesley decided to concentrate on petroleum, and with Lesquereux's help, he fashioned a sort of research program. Lesquereux suggested a serious "consideration" when choosing among possible engagements, namely, petroleum's connection "with every formation of coal of any epoch whatever."[105] Lesley objected to the idea that petroleum was distilled from coal; however, he was intrigued by petroleum's frequent presence in rocks associated with coal. Lesley thought oil might be found "*everywhere* inside the coal limits," meaning the great Appalachian coal field, and in at least fifty different rock layers.[106] Another "question of importance," Lesquereux posed, "is the formation and possibility of formation of oil reservoirs by *disturbances*." Both Lesley and Lesquereux rejected the theory of anticlines, mostly because the oil rocks of Oil Creek lay horizontally. Further, they doubted whether an anticline existed in southeastern Ohio or western Virginia, the places where Andrews had identified his well-publicized "break." But the presence of an anticline in eastern Kentucky, Lesley thought, was "possible." Moreover, Lesley was unsure how far south the Oil Creek formations extended along the western margin of the Appalachian Mountains.[107] To resolve these questions, Lesley accepted an engagement from a petroleum company on the Paint Lick Fork of the Sandy River in eastern Kentucky.

The Paint Lick survey came to Lesley on the recommendation of Louis Agassiz.[108] Following his standard procedure, Lesley drafted a three-page preliminary report, in which he outlined the general geology of eastern Kentucky. Without having seen the company's property, Lesley thought it likely that there existed Devonian shales and limestones, which "ought to yield oil."[109] The company liked this prediction and so engaged Lesley to examine the property. In late March, Lesley surveyed three tracts totaling nearly 55,000 acres. Petroleum was present, he believed, but its location was not associated with an anticline. Lesley considered the eastern Kentucky reservoirs to be horizontal, and he described "three distinct horizons or layers of petroleum": Lower Carboniferous conglomerates, Upper Devonian sandstones (analogous to those beneath Oil Creek), and Lower Devonian limestones (analogous to those beneath Enniskillen). (Newberry, by contrast, examined lands about 130 miles to the west of Lesley and thought petroleum would be found along an anticline in the Lower Silurian Blue Limestone.)[110] Lesley concluded his report on a reserved note. "I can not promise that any well shall be a five thousand or even a five hundred barrel well; and yet I can assign no certain reason against one being bored. . . . I must here leave the subject in the hands of practical experiment."[111]

The difficulty for Lesley was a lack of information about the subsurface. As mentioned previously, geologists were consulted by companies located in areas outside the producing oil regions where the exact number, identity, and thicknesses of the strata below the surface were largely unknown. Geologists thus had to make reasoned estimates as to the depths of likely oil rocks. "This is the only guide we have, as yet," Lesley explained to one company. "According to this guide, the *plane* of the Top sand of Venango, would be, theoretically, eight or nine hundred feet beneath [your property]."[112] This plane or oil horizon, as it was called, was the target when boring a well.

Stratigraphy was more reliable as a guide to places where oil would *not* be found. "There is not a shadow of a chance for oil wells in all southeastern Pennsylvania," Lesley told one capitalist.[113] Or as Wright put it more colorfully, "there is quite as good a chance of striking a rich deposit of . . . milk, or even butter and cheese, at any point east of the Blue-Ridge in Virginia, Maryland, Pennsylvania, New Jersey, and New York, as in opening a ten-barrel vein of petroleum."[114]

Stratigraphy, however, was an imprecise tool, especially when fissures might affect oil accumulation. "[T]here is always the crevice," Lesley reminded one company. "The augur may strike a crevice, bearing oil, at any depth. But no reliance can be placed upon crevice oil, for any great length of time; except in some few cases where crevices have been struck of such unusual magnitude, that they contain hundreds of millions of barrels of petroleum, and are practically inexhaustible."[115] Lesley did not doubt the existence of oil fissures and cavities, but their size and location were unpredictable. No general theory could be adduced from such random occurrences. "Andrews's diagram of vertical caverns," Lesley complained to Lesquereux, "is all bosh."[116]

Lesley thought most oil would be found in "undisturbed districts [with] horisontal horisons [*sic*]."[117] To test this hypothesis he accepted an engagement from the Brady's Bend Iron Company, whose property was located along the Allegheny River about forty miles south of Oil Creek. Although the area was regarded as favorable territory because of its topographical similarities with Petrolia, oil had not been discovered south of Venango County. Lesley wanted to resolve the practical question of "the extension of the Oil Creek sandrocks Nos. 1, 2, and 3, as far south as Brady's Bend." Were the oil rocks there? After an extensive survey of the property, Lesley decided that they were. "The three established horisons of oil-yield on Oil Creek are about 200, 400 and 600 feet below water level there. They must necessarily be 750, 950 and 1150 [feet] below water level at Brady's Bend."[118]

Lesley's conclusions echoed those of his Paint Lick survey. He again found no association between anticlines and oil and reasserted his theory of the horizontal structure of petroleum reservoirs. With regard to fissures, he thought they were small, numerous, and distributed unevenly throughout the immediate area. "It is this last feature," he conceded, "that makes the local success of any given well, even in a productive oil region such a lottery, and obliges every well regulated oil boring organization to arrange for sinking a number of wells over a considerable area."[119] Lesley suggested five locations and advised the Brady's

Bend Iron Company to keep accurate records of the rocks through which they bored.

By April 1865 Lesley had finished his first batch of engagements, but at a high cost. He described the "pressure of professional business" as "severe," and near the end of the month, his nerves suddenly "gave way."[120] This created a problem. He had agreed to survey lands along Slippery Rock Creek in western Pennsylvania. A company had contacted him in late February, and he had written a preliminary report for which he had received $100. The region looked promising to Lesley; the valley of Slippery Rock Creek corresponded, geologically, to the valley of the Allegheny River just above Brady's Bend.[121] Here then was Lesley's chance to extend westward his knowledge of the Oil Creek formations. But he felt too weak to do fieldwork. To help him fulfill his professional obligations, he called upon Lesquereux.

Lesquereux was eager to help because he was unemployed. He had fully "expected to have . . . something to do in explorations for oil land,"[122] but companies were unwilling to pay his $50 per week fee, which was very low for a man of science of his caliber. Lesquereux, who had arrived in America in 1848, attributed his unemployment to his deafness and his poor command of English.[123] Lesley asked Lesquereux to do the survey, and together they would write the report.

In early May, Lesquereux surveyed the Slippery Rock property and sent a report to Lesley containing his calculations of the depth of the Oil Creek sandstones below Slippery Rock Creek.[124] From his memory of the area and his recent work on Brady's Bend, Lesley figured something different and so asked Lesquereux to reexamine the property.[125] "This was a bad source of delay," Lesley explained to the Slippery Rock Petroleum Company. Lesley forwarded copies of his Paint Lick and Brady's Bend reports to the company, while he and Lesquereux sorted out their discrepancies.[126]

The different depths reflected a big stratigraphical problem: the separation of the Carboniferous and the Subcarboniferous formations of western Pennsylvania. Lesley and Lesquereux agreed that the identification of the Tionesta Sandstone—the first thick sandstone of Formation XIII (the Great Coal formation) above Formation XII (the Great Conglomerate)—was key to the number, thicknesses, and depths of the Slippery Rock sandstones. But Lesley and Lesquereux used different methods to identify the Tionesta Sandstone. As a structural geologist, Lesley traced the sandstone from Oil Creek westward using lithology and succession, whereas Lesquereux worked from fossil plants. It soon became clear to both of them that a proper scientific resolution of the stratigraphy was beyond the scope of the engagement. In early July, a temporary solution was reached by using Lesley's estimates;[127] that, at least, allowed Lesley to send the Slippery Rock report, in which Lesley explained the source of the delay. "There is a question of much geological interest which we have not yet certain data for settling, & which the palæontologists in the end must help us settle. But it is of no practical importance in this part of Slippery Rock Creek. I mean, the recognition of the Tionesta Sandstone."[128] Lesley's self-restraint probably had more to do with the fact that the report was *not* going to be published than for reasons of practicality.

Lesley decided not to print the Slippery Rock survey because he wished to avoid another disagreement with Lesquereux. He had wanted to cite Lesquereux's contributions in a scientific article. But Lesquereux did not consider it "right." Lesley should not "let anyone suppose that you [Lesley] accept assistance from any other geologist for any cause whatever." "Your position at Philadelphia is at the head of the Geologists," Lesquereux insisted, "and must be carefully kept as such."[129] Lesley, though, was very particular about assigning credit where it was due because of his painful experience with Henry Rogers. In the end, Lesley mentioned Lesquereux's assistance at a meeting of the American Philosophical Society, and he made sure to split the $1,000 fee.[130]

The collaboration did have a fortunate spin-off for Lesquereux. Soon afterward he was engaged by the Columbus National Petroleum Company to survey lands in the central part of Ohio.[131] In his published report, Lesquereux discussed the horizontal structure of the reservoir sandstones and the source of the petroleum, which he attributed to the decomposition of freshwater and saltwater plants.[132] Lesquereux hoped to present this research at the newly founded National Academy of Sciences meeting in Northampton, but he fell ill. "I have been unwell the whole summer," he lamented, "especially suffering of chills and over fatigue of field work."[133] Lesquereux, like Lesley, seemed to suffer from the pressure of business during the oil boom.

Petroleum Geology

Lesley's and Lesquereux's debate over the Tionesta Sandstone was just one example of how petroleum engagements furthered other research agendas. During his work for Neff Petroleum, Winchell found in the Waverly Group a specimen of *Chonetes mesoloba*, a fossil characteristically associated with coal. He subsequently proposed to reclassify the Waverly Group with the Subcarboniferous rather than the Devonian.[134] Similarly, Lesquereux, during the Slippery Rock survey, discovered among the coal beds a specimen of *Fucoides Cauda-galli*, a Devonian seaweed, which suggested a possible role for marine vegetation in coal genesis or, alternatively, an older age for the coal beds, which was the crux of the disagreement with Lesley.[135] Lesley, for his part, incorporated his Paint Lick and

Brady's Bend surveys into a larger study of the uplift and erosion of the Appalachian Mountains.[136]

Engagements thus served as the means to do fieldwork on petroleum geology. And the scientific results of geologists' participation in the oil boom were manifold and long-lasting. Winchell's survey of the Michigan oil regions appeared in the *American Journal of Science,* as did Safford's report on Tennessee's petroleum formations and Newberry's on the Cumberland oil region of Kentucky.[137] Likewise, Andrews distilled his work on West Virginia and southern Ohio into "Petroleum and Its Geological Relations," an influential article in the *American Journal of Science.*[138] And Hitchcock summarized his private surveys in the grandly titled "Petroleum in North America" for the *Geological Magazine.*[139] By 1866 geologists had reached some tentative but important theoretical and practical conclusions—in other words, an American science of petroleum geology.[140]

With regard to stratigraphical correlations, geologists decided that petroleum was not confined to a single or even a group of rocks of a particular age. "A little oil is to be found almost everywhere in our country," Andrews explained, "and in almost every geological formation, from the Lower Silurian upwards."[141] Hitchcock counted no less than fourteen oil horizons, although eleven of them occurred between the Silurian and the Carboniferous.[142] Newberry simply stated that "oil may occur at any geological level."[143] Such indeterminate stratigraphical distribution was one reason oil was so hard to find. It was also a fundamental characteristic of petroleum; hence Newberry made it the first point in his "physical theory of oil wells."[144]

To explain the stratigraphical distribution of petroleum, geologists focused on the fact that it migrated. According to Newberry, "[f]rom its point of origin, petroleum always tends to rise to the surface," which was the second point of his physical theory.[145] Petroleum was not stratigraphically bound like coal; there would be no "Petroleum Measures of the Hydrocarboniferous Age."[146] Petroleum could theoretically occur in any stratum in which the liquid was stopped in its vertical rise.

The term geologists used to describe the stratum where oil stopped was *horizon*. Its exact origin was obscure, and its definition was imprecise. Newberry called the Waverly Group, a formation several hundred feet thick in Ohio, "the oil horizon." Whereas Lesley identified three oil horizons in northwest Pennsylvania, by which he meant the first, second, and third sandstones, each was on the order of ten to twenty feet thick. Lesley's was the more common use of the term, but regardless of differences in thicknesses, the depth to an oil horizon and its geographical extent might be calculated. Horizons were large-scale structures.

Various geological positions of rock oil (circa 1865). The diagram is an ideal section of the numerous formations or horizons in which oil had been found in North America circa 1865. Most of the petroleum is contained in large caverns, although fissures are shown leading upward from these. *Source:* Samuel Harris Daddow and Benjamin Bannan, *Coal, Iron, and Oil; or, The Practical American Miner* (Pottsville, PA: Benjamin Bannan, 1866), 662.

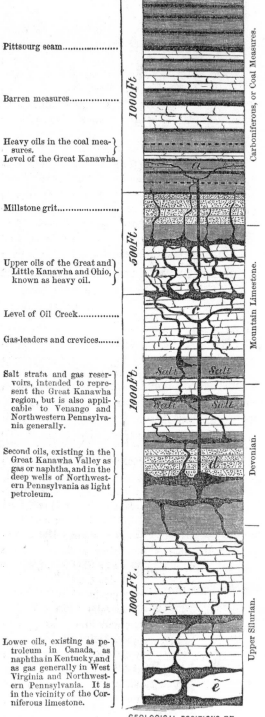

Pittsburg seam....................

Barren measures..................

Heavy oils in the coal measures.
Level of the Great Kanawha.

Millstone grit.......................

Upper oils of the Great and Little Kanawha and Ohio, known as heavy oil.

Level of Oil Creek................

Gas-leaders and crevices........

Salt strata and gas reservoirs, intended to represent the Great Kanawha region, but is also applicable to Venango and Northwestern Pennsylvania generally.

Second oils, existing in the Great Kanawha Valley as gas or naphtha, and in the deep wells of Northwestern Pennsylvania as light petroleum.

Lower oils, existing as petroleum in Canada, as naphtha in Kentucky, and as gas generally in West Virginia and Northwestern Pennsylvania. It is in the vicinity of the Corniferous limestone.

1000 Ft.
500 Ft.
1000 Ft.
1000 Ft.

Carboniferous, or Coal Measures.
Mountain Limestone.
Devonian.
Upper Silurian.

GEOLOGICAL POSITIONS OF ROCK-OIL.

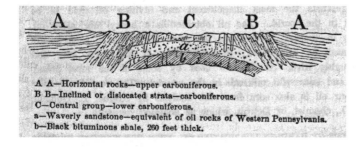

A A—Horizontal rocks—upper carboniferous.
B B—Inclined or dislocated strata—carboniferous.
C—Central group—lower carboniferous.
a—Waverly sandstone—equivalent of oil rocks of Western Pennsylvania.
b—Black bituminous shale, 260 feet thick.

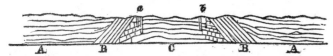

E. B. Andrews's illustrations of the anticline or "break" along the Ohio River. The top illustration was published in a consulting report; the bottom is a generalized version for scientific audiences. Both show that the place to bore for petroleum (*a and b on the bottom figure*) was not at the top of the anticline but along its flanks where the rocks were fissured. *Sources:* E. B. Andrews, *Report . . . to the West Virginia Oil & Oil Land Co.* (Detroit: Advertiser and Tribune Company, 1866), and E. B. Andrews, "Petroleum in its Geological Relations," *American Journal of Science, 42* (1866):33–43, 34.

Striking oil, however, required an understanding of small-scale structures. What geologists looked for were "natural and closed reservoirs," as Newberry phrased the third point of his theory.[147] Andrews, Hitchcock, Hunt, and Newberry favored the anticlinal theory, by which oil accumulated on subterranean hills. But there were two different interpretations for why this occurred. Hunt maintained that *within* a particular stratum petroleum moved to the highest point because of hydrostatic pressure. Andrews posited the existence of numerous oil fissures formed by the bending and breaking of the strata.[148]

Andrews claimed to have found anticlines in western Pennsylvania. Rogers also thought they existed in the Oil Creek region.[149] Hitchcock, who had done "a careful study of the distribution of the producing wells upon Oil Creek," decided there existed a *synclinal* basin with fissures.[150] Lesley, however, was vociferously opposed to the anticlinal theory.

> Stress has been laid by some geologists of note upon a supposed genetic connection between the accumulation of petroleum and anticlinal axes. But there are no anticlinal axes in the Pennsylvania oil region of the French and Oil Creek wells, nor in the Pennsylvania and Ohio oil regions of the Beaver River, nor in the E. Kentucky oil region of the Sandy and Licking waters.[151]

Lesley conceded the possibility of an anticline in the oil regions around Marietta, but still he did not think that the oil was confined to Andrews's "break."

Lesquereux and Winchell agreed with Lesley. They argued that porosity, not fissures, was the crucial feature of oil reservoirs. Porosity referred to the spaces between the grains of sand. Lesley explained:

> [A]ll sand and gravel beds are mere sponges, perpetually saturated with oil and water, the mingled fluid being slowly driven towards every available outlet by the gas which is generated with and from the oil. Such spongy rocks must be enormous reservoirs of petroleum, which it is in fact almost impossible to exhaust.[152]

Whether in porous, horizontal sandstones or numerous, anticlinal fissures, petroleum required "an impervious cover by which these reservoirs are hermetically closed," the fourth point of Newberry's physical theory.[153] Basically, if oil migrated upward, something had to stop its escape to the surface. Impervious beds of shale or some other fine-grained rock were needed to form a cap or seal on a reservoir. And if oil were trapped in a "hermetically closed" reservoir, there would be no surface indications. "The absence of strong surface indications," Winchell explained to the Neff Company, argued for "a retention of the product within the original receptacles."[154] By the end of the oil boom, geologists (with the possible exception of Hall) recommended *against* boring in places with surface oil.

Wherever a well went down, geologists asked drillers to keep samples of rocks and measurements of thicknesses. In this regard, consulting contributed to a technical innovation: the collection of well records. Winchell used them in constructing his ideal section of the strata for the Neff Petroleum Company. Lesley thought they were necessary "for any knowledge of the condition of things in the Devonian underground," because in western Pennsylvania the Devonian did *not* outcrop on the surface. Lesley criticized drillers for not keeping well records. "It is impossible to estimate the loss which geology has suffered during the last six years from this reckless ignorance."[155]

Lesley wanted a state or federal agency to enforce the keeping of well records in order to map the subsurface geology of the oil regions. This desideratum was fulfilled only when the Second Geological Survey of Pennsylvania was organized under Lesley's direction in 1874. Until then, Lesley made it a condition to his acceptance of any engagement that the company keep thorough records.

If well records were important to understanding the structure of the subsurface, they were even more so to theorizing about the origin of petroleum. For unlike coal or Albertite, petroleum could not be studied in situ. Geologists could examine only the materials—water, oil, gas, rock shards—that came up the bore hole. The process of boring necessarily destroyed any fossils in the enclosing rocks, and because petroleum was a liquid, it contained no fossils. Even chemistry was "slow" to help "in such difficult and important cases." Analyses,

Lesquereux complained, proved "nothing" because petroleum's composition varied as widely as its geographical and stratigraphical locations.[156]

Nonetheless, geologists agreed that petroleum was ultimately composed of organic matter. It was still open to debate whether petroleum was derived from distillation or decay. The two principal theories were practically the same as those first articulated by Newberry and Hunt in 1860. Among the geologists who supported some type of distillation theory, Andrews was the most forceful advocate.[157] He argued that the most persuasive evidence lay in the fact that oil occurred in so many different rocks. Either one had to assume that special conditions existed in each and every one of those rocks or that oil originated in one place and then migrated to the others. Andrews preferred the latter. He maintained that petroleum was the natural analogue to artificially distilled coal oil. Petroleum's ultimate source had to be bituminous strata such as the Black Slate of Ohio or similar carbonaceous rocks beneath other oil regions.[158]

Hunt remained the major proponent of the decay theory. He firmly believed that petroleum formed in situ from marine animals, and its only source bed was limestone. If it were found in other formations, it must have migrated from some deep-seated limestone.[159] Hunt received some support from Lesley, Lesquereux, and Hitchcock, who also envisioned an in situ chemical process, but they included marine plants along with animals.[160]

One thing all American geologists could agree upon was that petroleum had been produced in the past (except Rogers, who still believed in the bitumen brewery) and that it now existed somewhere in the subsurface. And there were lots of these possible places. "The immense territory in North America," Hitchcock crowed, "assures the world that the petroleum of the New World, like the coal, is probably practically inexhaustible."[161] It was a simple fact of scale. America had big oil.

> I happened to be employed by Brady's Bend Company to examine their property, and to give them among other items, an opinion upon the probable existence and depth of oil beneath it. To do this I merely did what any geologist who had thoroughly studied that country would have done. . . . Yet, when after a few months oil was actually struck at Brady's bend within feet of the depth which I had actually assigned to it, the astonishment of all classes of oil men was ludicrously extravagant; a score or two of copies were made from the manuscript report, and these copies passed from hand to hand as precious things, and their author was looked upon as a prodigy of mental penetration, and was offered large sums of money to locate wells in different districts; none of which

offers, of course, were accepted. . . . Let "practical men" believe in
and respect the slowly, carefully reached conclusions of "theoretical men" enough to take them into consideration, so far as to
comprehend them, and to govern themselves by them in their own
collection and collation of facts relating to their own pecuniary
interests.

—J. Peter Lesley, *The Geology of the Oil Regions* (1880)

THE MOST VEXING historiographic question about the early years of the oil industry is the practical result of all the science. Historians have repeatedly asserted that geologists were unable to find oil.[162] Geologists did predict strikes
in certain locations, a point to be discussed shortly; however, they did not make
any large discoveries in northwest Pennsylvania. The explanation for this is not
hard to find: geologists did not consult for Oil Creek companies. Those firms
relied on other oil-finding experts. And without commercial engagements, men
of science missed the chance to work in the most productive region in the world.
As a result, the history of Petrolia praises the practical oil men or spiritualists or
diviners and criticizes men of science as useless by virtue of their absence.

By contrast, nineteenth-century Americans seem far more savvy. They realized that "striking ile" could justify any theory, from oil-smelling to geology. And
it was the oil not the explanation that people were interested in. "Who cares,"
wondered Wright, "about the mystery of the genesis of petroleum? Of what
consequence to the purchaser whether the drill has struck the aorta of a half-
petrified whale, which is making its last and greatest spout . . . or whether the
greasy liquid has been distilled from coal, or is a new chemical combination going on in the basement story?"[163] Oddly enough, Wright cared. He hoped that
the "best talent, joined to the utmost disinterestedness of purpose [would se-
cure] a body of scientific truths."[164] And so did men of science. Consultants and
their theories were judged by other geologists and by scientific criteria—disinterestedness, evidence, rigor, and so on.

In general, men of science did not fault each other for a lack of practical results, and oil companies did not complain either. On the contrary, many expressed their appreciation. The superintendent of the Brady's Bend Company
thanked Lesley for pointing out the possible locations of petroleum. The company bored four wells on his recommendation, and each produced about one
barrel a day—hardly a result to make history, but one that apparently inspired
confidence. The company continued boring and on its seventh or eighth try
struck two-hundred-barrel-a-day wells (table 9.1). By the late 1860s many large
wells had been struck in the Brady's Bend area, known as the Lower Region (embracing Clarion and Butler counties), in contrast to the Upper Region around

Table 9.1. The Oil Wells at Brady's Bend—Lower Region, Pennsylvania

Height of well mouth above engineers' number datum (feet)	Depth of well (feet)	Depth below river, highest water mark (feet)	First yield in barrels per day (bbls.)	Present yield per day
1. 96	1,400			1 bbl.
2. 232	1,111	1,268	5+	No sand rock
3. 97.62	1,262	1,113		1 bbl.
4. 97.69	1,105	1,264	7	Abandoned
5. 100.31	1,290	1,105	5½	2 bbls.
6. 300.48	1,414	1,090	9	4 bbls.
7. 437.41	1,345	1,077	840	8 bbls.
8. 379.18	1,065	1,066	4½	150 to 200 bbls.
9. 101.38	1,300	1,066	1	3+ bbls.
10. 330.27	1,200	1,070		Abandoned
11. 111.13	1,212	1,189		Powerful gas blow
12. 216.50	1,402	1,095½	12	13 bbls.
13. 426.38		1,076	3	2 bbls.

Source: J. T. Henry, *The Early and Later History of Petroleum* (Philadelphia: James B. Rodgers, 1873), 255. The engineers' datum was an assumed level one hundred feet lower than a mark on the Brady's Bend Iron Company's warehouse, showing the extreme height reached by the Allegheny River during the flood of March 17, 1865.

Oil Creek.[165] Lesley took some credit for his predictions about Brady's Bend, and years later he turned his engagement into the aforementioned parable.

The practical results of other engagements were not quite as astonishing as Lesley's self-proclaimed triumph. Winchell did not find oil in Michigan, and he and Newberry were proved incorrect in Ohio, where the Neff Petroleum Company struck gas, not oil. (In the 1870s Neff became a profitable gas company.)[166] Andrews's record along the "break" was spotty. Oil was found, but not a lot. Years later, Edward Orton (1829–1899), the state geologist of the third Ohio survey (1882–1899), reported that Andrews had failed to map, in detail, the anticline. Had he done so, he would have been more successful. Nonetheless, Orton praised Andrews by calling him "the father of petroleum geology" for introducing the anticlinal theory, which Orton and others later used to much success.[167]

Lack of success did not undermine a general faith in the efficacy of geology or in the trustworthiness of geologists. Success was not the measure of science. Yet one of the unstated assumptions with regard to science-based industry is its commercial success. How can science be unprofitable or unproductive? In the oil explorations of the 1860s (and throughout the nineteenth and early twentieth centuries), there was no conveyer belt from fieldwork to flowing well. The involvement of geologists in oil did not translate directly into discoveries. But men of science did prosper, financially and scientifically, as consultants during the boom. Moreover, they managed to maintain their professional integrity and to create a field called petroleum geology. These were important developments in the relations between science and industry.

CHAPTER 10

California Crude

The *real* enemy to the Survey comes almost exclusively
from the fact that I cannot be used as a tool to forward
nefarious speculations & it has its basis in this "petroleum
swindle" as I call it.
—Josiah Whitney to J. Peter Lesley (1870)

WHEN JOSIAH WHITNEY penned those lines, he had been director of the California Geological Survey for a decade, and for nearly half the time he had been predicting its termination because of the wicked influence of petroleum interests. Whitney abhorred the oil boom and rejected any predictions of oil being produced in southern California. In stark contrast, his former friend, Benjamin Silliman Jr., reveled in the boom and in 1864 went out to California as a consultant and found a fabulous wealth of oil.

On the surface, then, the disagreement between Whitney and Silliman was scientific; was there oil in southern California? Deeper down, their differences were personal and professional. What role were men of science supposed to play in the discovery and development of mineral resources? It was a familiar question, but this time the stakes were very high. The government geologist was pitted against the consulting chemist, public versus private science.

The answers to those questions were debated in the most appropriate of places, the National Academy of Sciences where in 1874, after the California survey had ended, Whitney brought charges against Silliman of having abetted a great petroleum swindle. Whitney wanted the academy to censure Silliman for his commercial speculations and to expel him. Silliman asked the members to condone consulting and to exonerate him. The Silliman-Whitney controversy scandalized men of science and forced them to confront the most significant moral question of all: what was the proper place of science in America?

Surveying

California rushed into statehood in 1850 with the discovery of gold and all those seeking its glitter. Within five years, gold production faltered and with it the state's economy. In the Mother Lode region along the Sierra Nevada's western flank, prospectors' pans, rockers, and sluices had cleaned out the nuggets from the easy-to-sift gravel beds. Bigger, more mechanized operations took over the streams and rivers, while other mining companies began to chisel out quartz veins and drive shafts deep into the hard rock. These large-scale activities required capital, and so the governor called on the California legislature to establish a survey to lend "practical" aid to the mining interests. The governor also made sure to include farmers in his plea for a survey by assuring them of the usefulness of soil examinations. The legislature responded in April 1860 with a bill authorizing "an accurate and complete geological survey" and directing the state geologist to furnish a report with "proper maps and diagrams" along with "a full and scientific description of [the state's] rocks, fossils, soils, and minerals, and of its botanical and zoological productions, together with specimens of the same."[1]

The California survey was a big event in American science and has been treated as such by historians, yet not for its scientific achievements but rather for its struggles. From the moment of his appointment in April 1860, Whitney began complaining about the dirty necessities of public accountability and the conniving of commercial interests. Sympathetic scholars have responded accordingly. They have pitched the survey's story as a contest between "pure" and "applied" science, an unavoidable conflict between elite knowledge and democratic values. Petroleum fueled the drama, which, inevitably, reads like a tragedy for Whitney, the pristine man of science whose higher ideals were corrupted by the baseness of gritty politics. The survey's end punctuates the argument about nineteenth-century Americans' so-called indifference to basic research.[2]

But no state survey was ever meant to be pure or permanent, and Whitney, like most geologists, was well aware of the clamorous politics of nineteenth-century science. Compared to contemporary surveys, Whitney's struggles seem unremarkable. Surveys were designed to be useful. Whitney's practical shortcomings thus cannot be redeemed by his theoretical high-mindedness. In one regard, Whitney's survey might be seen as a relative success: it lasted fourteen years, from 1860 to 1874, longer than any previous one.[3] Such longevity, ironically, suggests some lobbying skills and political adroitness on Whitney's part. It certainly warrants an examination of how he kept the survey going against so many supposed enemies.

The first feature of the California survey worth noting is the ordinary language of its organic act, which Whitney himself claimed to have drafted, a not uncommon practice.[4] It was a sweeping scientific mandate, in which practical benefits were not stated explicitly on the assumption that they would flow unimpeded from an accurate and complete investigation. This was the bedrock of all surveys, and while altercations occasionally arose between legislators and surveyors, more often the silences reflected agreement on scientific means toward practical, and hence political, ends.

In two respects, the survey was unusual. First, there was no date for its completion. Most states placed limitations, in terms of either money or years, on surveys. The lack of a deadline probably reflected a provision in the California Constitution restricting government offices, like the state geologist, to four years.[5] Conversely, the open-endedness left Whitney and legislators with ambiguities as to how and when the science would be deemed complete. Whitney certainly hoped the survey would continue beyond 1864. Second, the size of the appropriation was unprecedented. Whitney's annual salary was $6,000, a figure equal to the governor's and twice that of any other state geologist. Survey assistants were paid additional, unspecified amounts, and an extra $20,000 was allotted for the first year's expenses.[6] "No similar enterprise in the United States has ever been set on foot on a more liberal and enlightened basis," announced the *American Journal of Science*, "or fraught with more interesting scientific and practical problems."[7] Therein lay the rub or, more precisely, the expectations that would rub the wrong way. Each year Whitney had to appeal directly to the governor and legislators for further appropriations. In his annual reports, he had to make a persuasive case that the survey was solving scientific and practical problems. Whitney understood these challenges when he accepted the position and, in many respects, was the ideal man to meet them.

At forty-two, Whitney was in the prime of his scientific career. An accomplished chemist and geologist, he had worked as a consultant and written the definitive study of American mines, *The Metallic Wealth of the United States* (1854). He had extensive experience on government surveys—the Lake Superior Copper District under Charles Jackson and later with John W. Foster, the Iowa Survey under James Hall, and most recently the Wisconsin Survey with Hall again. (Whitney had only just finished his Wisconsin report in October 1860.) In letters of recommendation, prominent men of science, including Louis Agassiz, James Dana, and the Sillimans (father and son) attested to Whitney's upstanding character and scientific reputation. In California, Whitney had the support of Stephen J. Field, chief justice of the state supreme court and a family friend. But Whitney also had a histrionic prickliness. He was the sort of person who could not get into a personal disagreement without entering a major controversy. He disliked politicians and detested getters-up of stock companies, and his antagonism toward speculative enterprises did not win him many friends in the California mining community, although he was liked by well-bred Boston capitalists. Whitney deflected his commercial distrust with disinterestedness, of which he boasted in his inaugural address before the California legislature in March 1861.

> Although for twenty years constantly engaged in the examination of mines and mining properties, I have never been the owner of a share of mining stock or a foot of mining ground, or, either directly or indirectly, pecuniarily interested in any enterprise or undertaking in any way connected with the mining interest. If I were, I should consider myself unfitted for the position I now occupy.[8]

Whitney's trustworthiness would assure the value of California mines and the geological survey.

To assist him in surveying, Whitney initially chose two well-trained and promising men of science: William H. Brewer (1828–1920), a graduate of Yale's scientific school and an agricultural chemist; and William Ashburner (1831–1887), a graduate of Harvard's scientific school and a coal geologist. In 1861 Whitney enlarged his corps with Charles F. Hoffmann (1838–1913), a German-born topographer; James G. Cooper (1830–1902), a zoologist; and William M. Gabb (1839–1878), a paleontologist and protégé of Hall. In 1862 and 1863 three unpaid young volunteers joined up: a French geologist, Auguste Rémond, and two recent graduates of Yale's scientific school, James T. Gardner (1842–1912) and Clarence King (1842–1901). With the exception of Cooper, none of these assistants had any experience in surveying. Also absent from Whitney's corps were any Californians. Most surveys employed, in some fashion, local experts, so as to gain information and the goodwill of residents. But Whitney snubbed California's mining engineers, especially William P. Blake (1826–1910), a prominent consulting geologist well regarded by the mining community, but also a rival for Whitney's job.[9] The California survey was an East Coast scientific operation.

Whitney arrived in San Francisco in November 1860 accompanied by Brewer and Ashburner. They headed south to reconnoiter the Los Angeles area, the easiest part in geological and geographical terms. Whitney planned to run his survey along four north-south lines. The first followed the Coast Ranges (where elevation points for maps could be tied to the Coast surveys), the second skirted the Sierra Nevada foothills (the all-important gold mining districts), the third traced the high peaks (practically unmapped), and the last ran down the eastern border with Nevada (the Washoe silver mining region). These lines were meant to establish the general topography and geological structure of California.

For the most part Whitney stuck to his plan, and it proved a politically and scientifically effective one. During the first year, 1861, the survey moved up the coast to San Francisco and surrounding counties. Whitney reported on the famous quicksilver mines at New Idria and New Almaden and assured legislators about another very useful mineral—coal. Around Mount Diablo in Contra Costa County, coal deposits were of sufficient quality and extent that "the State will soon cease to be dependent upon other regions for her supplies of fuel."[10]

Whitney requested $30,000 for 1862, but he received only half that amount. Heavy rains and extensive flooding wrecked large parts of the state's taxable property.[11] The survey was thus forced to concentrate on the Bay area, especially the Mount Diablo coal. But Whitney knew it was important to make a start on the gold fields and the Sierra Nevada. "[T]he general geological structure of the State," he explained to the legislators in his annual report for 1862, was a necessary prerequisite to all detailed examinations of mines. "It is not, however," he warned them, in language remarkably reminiscent of Abraham Gesner's comments on coal, "the business of a geological surveying corps to act, to any considerable extent, as a prospecting party." To do so would mean confining the survey "to a very limited area," precisely the kind of unbalanced geographical and topical treatment that had undermined William W. Mather's Ohio survey back in 1839. Whitney's broad approach had distinct practical advantages in "preventing foolish expenditures of time and money in searching for what our general geological investigations have determined not to exist in sufficient quantities, in certain formations." As with other state surveys, geology of a negative variety—where *not* to mine—was what Whitney could show for his survey so far.[12]

The legislature appropriated $20,000 for 1863, "an amount," complained the *American Journal of Science*, "entirely inadequate to carry on the Survey in as complete a manner as contemplated in the Act of the Legislature."[13] Nevertheless, whatever Whitney completed in the third year would, most likely, determine whether his survey was extended beyond its constitutional deadline. And so Whitney hurried to finish an exploration of the entire state, but he covered only forty-six of forty-nine counties; three were too dangerous because of "Indian difficulties." He also rushed to map "the unknown regions" of the High Sierras.[14] By default rather than by design, Whitney's California survey became more topographical than geological, a situation reminiscent of Rogers's Pennsylvania survey. Whitney harped on California's immensity, nearly 200,000 square miles, twenty-four times the size of Massachusetts and more than the combined area of the three largest states east of the Mississippi. Scale, once again, formed the character of the geology.

In spite of these difficulties, "[t]he main object of the survey," he reassured legislators after the 1863 fieldwork, remained "the elucidation of the mineral resources of the State, including everything which bears on the working of the mines, and reducing their products to a marketable condition." In making his plea for four more years, "with a liberal appropriation of not less than $40,000 per year," Whitney moved back and forth between threats of suspending the survey without completion and promises of "sufficient benefit" and "value" should his appointment be renewed. He promised five large volumes embracing general geology, economical geology, paleontology, botany and zoology, and a collection of maps.[15]

In April 1864 the legislature granted a two-year extension and $25,600 for field operations with an additional $6,000 for printing costs. Whitney had won the political battle; now he had to produce something. In May he headed back East for an eighteen-month adjournment from fieldwork to write his *Geology*, vol. 1: *Report of Progress and Synopsis of the Field-Work from 1860 to 1864* (1865).

Engaging

Silliman arrived in San Francisco that same April, and most historians have assumed that he was there as the agent of Thomas A. Scott (1823–1881), vice-president and director of the Pennsylvania Railroad.[16] Scott was a shrewd speculator in mining ventures, and in the late fall of 1863 he had organized an expedition to explore western Arizona and parts of Nevada and New Mexico territories. The expedition's leader was John Wyeth, founder of the well-known Philadelphia chemical firm John Wyeth & Son and Scott's brother-in-law. "The object of the enterprise," according to Scott, was "to discover, locate and secure land, minerals, and mineral rights of every nature; to hold, develope [*sic*] or dispose of the same; to mine, reduce, separate and dispose of gold, silver, metals, or other valuable substances that may be discovered, secured, or acquired." The consultant on whom Scott and Wyeth called was not Silliman, but Benjamin Smith Lyman.[17] Well educated in mineralogy, mining, metallurgy, and assaying, Lyman had assisted his uncle, J. Peter Lesley on many coal engagements, but it was his other uncle, Joseph Lesley, who worked for Scott and the Pennsylvania Railroad, who probably got him on the expedition.[18] From Scott's perspective, Lyman came with the right skills, experience, and price—$250 per month plus a one-twenty-fourth part of the stock in any mining company organized. Lyman thus became the "Geologist and Topographer of the expedition."[19]

Silliman went to California for other reasons and by engagement to other parties. According to the geologist James Hodge, Silliman was engaged for $5,000 "in hand before starting,"[20] which means he was brought out to California by capitalists who needed a man of science for more than a mineral expedition. In two places, Silliman's name would be worth as much as his science, in a court of law and in a stock company prospectus, both of which pointed to Washoe.[21] The Comstock Lode had been discovered in 1859, but by 1864 the silver boom had busted. Stock prices plummeted, while lawsuits over mining claims skyrocketed. One observer reckoned that the dozen leading Comstock companies were party to more than 200 suits.[22]

The day after his arrival, Silliman hurried to Washoe to begin an investigation for the Potosi Company, which was embroiled in a series of long-running cases against the Chollar Company. Silliman then served as an expert witness for the Gould and Curry Company, known to be sympathetic to the Potosi. Gould and Curry had brought suit against a neighboring company, the North Potosi, for taking its silver. In the case *Gould and Curry v. North Potosi*, Silliman testified that the two companies were mining one vein, not two, as another expert witness, Blake, had testified. The court agreed with Silliman, an important legal decision that helped to solidify a consensus on how silver veins formed.[23]

In the scramble to control Comstock silver, San Francisco capitalists underwrote many of the lawsuits as well as the technological improvements, both risky ventures providing another clue to who hired Silliman. William C. Ralston (1826–1875), financier and soon-to-be president of the powerful Bank of California, was keenly interested in the Comstock. He has been called the "master of Machiavellian tactics" and certainly fit the mold of a capitalist who could use Silliman.[24] Years later, Ralston referred to Silliman as his "good friend" and congratulated him on his "forecast" with respect to his silver theory. "The Comstock is looking splendidly," Ralston bragged, "the most remarkable development ever known."[25]

Ralston was also a leading force behind the Quicksilver Mining Company. For several years, that company had been trying to take over the New Almaden mines. By 1864 a series of lawsuits against the British-Mexican owners had reached the U.S. Supreme Court. The Quicksilver Company hired Silliman to report on the mines in part to reassure its stockholders, one of whom happened to be Scott, that after the takeover they would continue their profitability.[26]

Silliman had reasons besides silver and mercury for going West. After his Comstock engagements, he spent several days examining the Bodie gold mining district, 100 miles south of Washoe. In his report to the Empire Gold and Silver Mining Company, Silliman declared Bodie to be "one of the most valuable localities for the precious metals hitherto discovered in the United States."[27] Capitalized at $1 million, the Empire Company was incorporated in July 1864 by several prominent New York and Boston bankers. Presumably it was this engagement to which Hodge referred when he told Hall that Silliman got "$2,500 in gold for a single report."[28]

Whether Silliman had gone West under contract with the Empire Company or Ralston or some other capitalists, it is worth noting that his consulting work, with the exception of his Quicksilver Company report, focused on mines *outside* of California. This suggests some regard on Silliman's part for Whitney's domain. Likewise, Whitney instructed Brewer to refrain from consulting for Comstock or other mining companies because of their official positions.[29] Consulting, however, was not expressly prohibited in the survey's enabling act.[30]

On the other hand, Silliman had prepared for his western venture by forwarding a copy of his recent consulting report on gold mines in Nova Scotia to the leading San Francisco–based trade journal, the *Mining and Scientific Press.* Clearly, he expected to book other engagements, and in this respect Whitney's absence might have encouraged local capitalists to consult him. By June 1864 Silliman had made two new contacts—Tom Scott and Edward Conway. Scott wanted him to examine the silver and gold prospects selected by Lyman months earlier. Conway asked him to report on prospects of a different sort—oil. It was Silliman's acceptance of Conway's engagement that set in motion the entire controversy over the nature of California crude, the value of Whitney's survey, and the ethics of consulting.

Consulting

California was known to have bitumen. Usually it was found oozing out of the rocks as semisolid tar or in very thick asphaltum pools, which were intermixed with sand and water. During the 1853 Pacific Railroad Survey, Blake had examined the coast between San Francisco and San Diego and reported that "there are numerous places in the Coast Mountains . . . where *bitumen* exudes from the ground and spreads in great quantities over the surface. These places are known as *Tar Springs,* and are most numerous in the vicinity of Los Angeles."[31] Thomas Antisell, the Albert mineral expert and later Patent Office examiner, had covered much the same ground as Blake as the geologist-chemist appointed to the 1855 Pacific Railroad survey. Antisell concluded that "bitumen was *par excellence* the mineral of southern California, being found in almost every county south of San Francisco."[32]

Such an exalted resource did not go unnoticed by the survey. During its first year, Whitney sent Brewer to investigate the bitumen. Not surprisingly, Brewer found the amounts impressive. But the question before the survey was not one of quantity, but quality: what to make of it? Whitney sent samples from La Brea and other springs to Frank Storer, the "highly skillful" Boston chemist and Antisell's nemesis. After running a series of experiments, Storer informed Whitney that "the answer to this question was in the negative. . . . The asphaltum cannot be profitably used for the manufacture of burning or lubricating oil."[33] There was too much water intermixed; the asphaltum frothed when heated.

Silliman had shown very little interest in petroleum since his renowned report on Pennsylvania rock oil. But, like other consultants, he could not resist the oil boom. And, more to the point, he was getting $2,500 from Conway.[34] As the chief clerk in the office of the U.S. surveyor general of California, Conway and his associates had acquired, through either purchase or patent, 100,000 acres to the east of San Buenaventura. Silliman arrived in Santa Barbara on 26 June and spent three days exploring the springs around San Buenaventura and then visited the Cañada Larga and Ojai ranchos, two properties just north of Conway's. On 2 July Silliman sent a letter to his brother-in-law John Church in New York City. "[There] are at least twenty natural oil-wells, some of the largest size. The oil is struggling to the surface at every available point, and is running away down the rivers for miles and miles. Artesian wells will be fruitful. . . . As a ranch, [the Ojai] is a splendid estate, *but its value is its almost fabulous wealth in the best of oil.*"[35] On 4 July Silliman went by stagecoach to Los Angeles where he spent several days investigating other oil springs, including La Brea. On 11 July he submitted his twelve-page report to Conway.

Silliman, obviously, had not had time to examine the property closely, but he gave it "all the attention in my power." He discovered six groups of oil springs, which, he decided, were "Nature's indications . . . favorable to the development of artesian wells." He also found numerous places of "hardened asphaltum," indications, he thought, of former oil springs. Here was "a remarkable and almost unrivaled source of supply," he enthused. The only limitation to the amount of oil was "the number of artesian wells."[36]

As for geology, Silliman described a "uniformity of structure," by which he meant, rather simplistically, alternating beds of sandstone and bituminous shale dipping to the north at a high angle, 70 to 80 degrees. The strata formed a 2,000-foot ridge running thirteen miles east-west. The oil, Silliman thought, was confined to reservoirs in the shale because it was impervious and thus acted like a dam to arrest the oil's vertical migration. The pervious interbedded sandstones contained water. Silliman did not comment on how the telltale springs were connected to the shale, but he did explain the chemical relation. Once California petroleum reached the surface, it thickened from evaporation; the hot sun literally baked it. Given this direct relationship, Silliman saw no reason why the asphaltum could not be distilled into profitable lamp oil. Nonetheless, he recommended sinking wells to draw the oil "in its thin or natural condition."[37]

According to Silliman, natural petroleum originated in situ in each layer of bituminous shale. He cited the work of his former student T. Sterry Hunt in Canada although he disagreed with Hunt on one point: petroleum was formed by the chemical transformation of woody fiber, not animal sources. More importantly, Silliman thought petroleum was "by no means necessarily connected with any particular geological period." Here was the familiar argument of process over product. Woody material of any age, under the right chemical conditions, might be transformed into petroleum. Silliman noted that in the eastern United States and Canada, petroleum was found in Silurian and Devonian

Map of the Ojai rancho in Santa Barbara County and surrounding ranchos in southern California. The gray lines were drawn by William Brewer to indicate the roads and possible (limited) routes taken by Benjamin Silliman during his explorations for oil. *Source: A Description of the Recently Discovered Petroleum Region in California, With a Report on the Same by Professor Silliman* (New York: Francis & Loutrel, 1865).

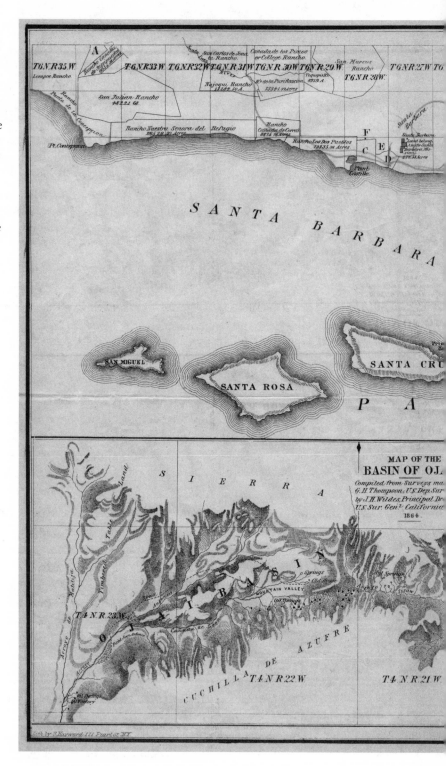

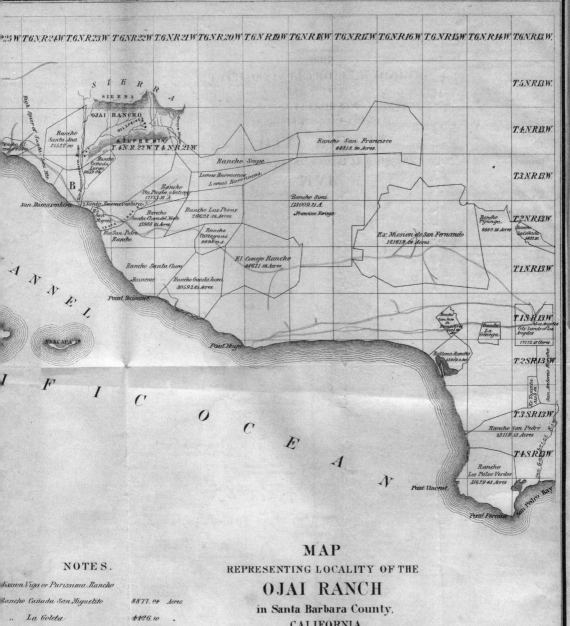

T.5 N R 13 W.

T.4 N R 13 W.

T.3 N R 13 W.

T.2 N R 13 W.

T.1 N R 13 W.

T.1 S R 13 W.

T.2 S R 13 W.

T.3 S R 13 W.

T.4 S R 13 W.

SIERRA

SIERRA

OJAI RANCHO

OIL SPRINGS

High Sierra of Santa Inez Mts.

Rancho Santa Ana 21522.00

A SUPRE MO
T.4 N R. 22 W T.4 N R 21 W

Rancho Cañada Larga 6659.04

B

San Buenaventura

Santa Buenaventura

Rancho San Miguel

Rio San Pedro Rancho

Rancho Sta. Paula y Saticoy 17773.35 A.

Rancho Santa Clara del Norte 13988.91 Acres

Rancho Sespe

Lomas Borrantes, Lomas Barrancas.

Rancho San Francisco 48813.90 Acres

Rancho Las Posas 26623.36 Acres

Rancho Calleguas 9998.29 A.

Rancho Simi 113009.21 A.
Francisco Noriega

Ex. Mission de San Fernando 121619.24 Acres

Rancho Tujunga 6660.21 Acres

Rancho La Cañada 4521.96.

Rancho Santa Clara

Saticoy

Rancho Guada Josa 305.93.83 Acres

El Conejo Rancho 48671.56 Acres

Point Hueneme

CHANNEL

ANACAPA I.

PACIFIC OCEAN

Point Mugu

Rancho San Jose de Buenosayres 4439.06

Rancho La Cienega

Ballona Rancho 13919.9 Acres

City Lands of Los Angeles 17172.81 Acres

San Antonio R.o

Rio Tijuanita 3869.96

Rancho San Pedro 43119.13 Acres

San Gabriel R.o

Point Vincent

Rancho Los Palos Verdes 31629.83 Acres

San Pedro Bay

Point Fermin

NOTES.

Mission Vega or Purissima Rancho

Rancho Cañada San Miguelito   8877.04   Acres
   „   La Goleta   4426.10
   „   Las Positas y las Cateras 3281.30
Senegital \Land belonging to St.a Barbara Miss.n\ 1.13
Vineyard of San Jose \Land belonging to St.a
                \Barbara Mission   7.47   „
Springs.

MAP

REPRESENTING LOCALITY OF THE

OJAI RANCH

in Santa Barbara County,

CALIFORNIA.

belonging to the

CALIFORNIA PETROLEUM COMPANY

1865.

rocks, but in southern California, petroleum was "certainly no more ancient than the cretaceous." According to Whitney's survey, California had no Silurian or Devonian rocks. Silliman concluded that California petroleum was the youngest in North America.[38]

After his petroleum engagement, Silliman headed off with Wyeth on a two-week adventure in Arizona.[39] Scott, however, had become more interested in oil than in gold. Upon Silliman's return to Los Angeles, Scott asked him to write up his views on the Ojai rancho, a property that Scott was trying to buy. For a fee of $1,500, Silliman prepared a "professional report,"[40] which he submitted on 1 September. It was practically a word-for-word rehash of the Conway report, with the notable addition of an "ideal section" showing the "oil series." Silliman ended his Scott report on a highly optimistic note: "Suffice it to say, that having made the first researches on the products of Oil Creek, long before any wells were bored there, I am of opinion that the promise of a remarkable development at Buenaventura is far better than it was in the Pennsylvania or Ohio regions—since so famous."[41]

This prediction was not remarkable enough for Scott. He wanted some calculations showing expected production. Silliman obliged with astonishingly precise numbers on asphaltum manufacturing and oil boring. Assuming the asphaltum covered a square mile with a minimal thickness of one yard, Silliman figured that the Ojai contained at least 2,890,000 tons, which, at 50 gallons per ton, meant 144,500,000 gallons of oil. "If an effort is made to estimate the money value of this product," Silliman boasted, "it will be found to reach a figure almost fabulous." Even more fabulous were Silliman's petroleum numbers. If the company bored ten wells, each *flowing* 100 barrels daily for a year, the income would be $1,365,000. "It is easy to see from these figures," Silliman gushed, "the successful exploration of an oil-producing district is far beyond all ordinary calculations."[42] What was even more staggering, if that were possible, was Silliman's conclusion. "Every mine of metals is a magazine of limited supply. . . . Not

IDEAL SECTION FROM BUENAVENTURA TO MUPU ARROYO.

Benjamin Silliman Jr.'s ideal cross section showing the oil rocks in a near-vertical orientation. Scales: horizontal, 8 miles to an inch; vertical, 8,000 feet to an inch. *Source: A Description of the Recently Discovered Petroleum Region in California, With a Report on the Same by Professor Silliman* (New York: Francis & Loutrel, 1865), 1.

so with petroleum. It flows on year after year, and still the source of supply seems unimpaired."[43] How could Scott not be enticed into oil?

During the fall of 1864, Scott, through his California agents, leased or purchased nearly half a million acres in the counties of San Luis Obispo, Santa Barbara, and Los Angeles. Silliman meanwhile accepted several more silver- and gold-mining engagements in Nevada and California. By mid-December, Scott wanted Silliman to examine his lands. Setting out from San Francisco by steamer, Silliman traveled along the coast to San Luis Obispo, but inclement weather prevented him from going ashore, so he made a quick study from the ship. He then spent three days exploring Santa Barbara and returned to San Francisco to draft a short report. On the morning of 4 January 1865, Silliman headed back East, his pockets stuffed with gold.

According to his federal tax statements, Silliman's 1864 income amounted to a whopping $54,288, roughly twenty times his academic salary of $2,765. This enormous figure, or rumors thereof, was and is often cited as proof of Silliman's haste, greed, and gung-ho commercialism.[44] But before ascribing motivations, it might be useful to consider logistics. Judging by the number of written reports, Silliman consulted for at least thirty well-capitalized companies. If he had $5,000 in hand before leaving Yale, and an additional $2,500 for the Bodie report, a fair estimate of his professional fee would be $1,500 per engagement. That was top-dollar for any consultant and no doubt reflected the monopoly held by Silliman, the only famous East Coast professor in California, as well as wartime inflation. Still, thirty reports is a heap of work for nine months, an average of one engagement per week, even *if* big time consulting was Silliman's sole objective in going to California.[45] But he might have gone West for the same reason Whitney had, to see and study it. Consulting engagements served, once again, as a scientific expense account. Likewise, another possible source for Silliman's large income might have been the sale of assets; he did own stock in other mining companies. Or, perhaps, he inherited property when his father died in November 1864, which was the reason for his hasty return home.[46]

Boosting

Shortly after arriving in New York City in January 1865, Silliman learned of two newly organized companies: the Philadelphia and California Petroleum Company of Philadelphia and the Pacific Coast Petroleum Company of New York City. Both were based on lands recently acquired by Scott, and both consulted Silliman. The president of the Philadelphia and California Company was John C. Cresson (1806–1877), former superintendent of the Philadelphia gas works, former president of the Franklin Institute, current vice-president of the American Philosophical Society, and trustee of the University of Pennsylvania.

Cresson asked Silliman for a report on the company's three large ranchos in Santa Barbara County: Las Posas (26,623 acres), Simi (113,009), and San Francisco (42,800).[47] Silliman had not explored these lands; he had seen them only during his travels by stage between San Buenaventura and Los Angeles. Nonetheless, he was willing to do it.

Silliman's report began with a geological overview of California oil, the scale of which was "unlike what is seen upon the eastern side of the continent." There were "natural wells of petroleum and tar" forty feet in diameter, and hillsides were covered, "often for hundreds of acres, with hardened asphaltum." Silliman mentioned sixteen "oil outcrops," from one of which, supposedly, a five-gallon sample had been taken.[48] Silliman did an analysis of this sample on very short order in the chemical laboratory at Yale. The oil sample had a strong naphtha odor, was dark brown, and "thin and mobile" as water, which was remarkable, Silliman noted, for in California's climate "the thin and more volatile portions of the Petroleum evaporate on exposure, leaving only the thicker and heavier oils, the further evaporation and oxydation of which leaves asphaltum." Moreover, this sample, "even in its natural condition . . . and after exposure to the air," burned with a bright flame and strong light.[49] Silliman distilled the sample into lighter fractions and described the burning, lubricating, and explosive qualities of each. He congratulated the Philadelphia and California Company on the "extremely satisfactory results" and predicted that "fresh oil" flowing from wells would contain more light fractions than the sample.[50] He then suggested where those wells should be located: asphaltum on the surface "emphatically" indicated oil at depth.[51]

Silliman was likewise impressed by the Pacific Coast Company's property, five ranchos embracing about 75,000 acres in San Luis Obispo County, which, once again, he had not explored, but neither did he have any samples to analyze. His report was more a general discourse on California petroleum geology. "[T]he probable origin of the Asphaltum of California," he began, "is in my judgment, of the greatest practical importance." He reiterated the idea that "the Asphaltum found along the coast range . . . is referable to the former flow of oil" and then explained how past and present oil flows were associated with "axial disturbances."[52] What he meant was hills, but Silliman did not mention any theory of anticlines. On the contrary, he thought the best place to bore wells was in valleys, at the base of hills, or along hillsides, *not* on top.[53] Silliman was convinced that this practical guide would meet with "every probability of success."[54] He was very optimistic. "It is difficult to give even a plain statement of the facts," he reflected, "without seeming to be carried away with enthusiasm or infected with a spirit of exaggeration."[55]

Success for the California oil companies was realized, initially, in selling stock. Each had huge capitalizations—the California, $10 million; the Philadelphia and California, $10 million; and the Pacific Coast, $5 million—which made these behemoths the biggest in a market already awash in oil stocks. But as with all mining companies, big money reflected big land, which meant big unknowns. The companies relied heavily on Silliman.

In February, Silliman began giving popular scientific lectures on the West in cities from Boston to Baltimore. He was enlightening and entertaining and well equipped with maps and diagrams. His presentations, like his visit itself, mixed natural history (mountains, deserts, animals, and plants) with commercial opportunities (silver, gold, and oil), and he always carried samples. According to Hodge, who attended Silliman's lecture at Cooper Union, he lit a lamp "filled with crude California oil and . . . displayed its illuminating properties." The oil "was of such extraordinary purity," observed Hodge, that "it burned in a kerosene lamp like Pennsylvania refined oil."[56]

Silliman was not a shill for the oil companies, but undoubtedly these displays aided stock sales as well as increased interest in all things Californian. To Whitney, at home in Northampton, Massachusetts, and still scribbling away on his *Geology*, Silliman's demonstrations seemed to be direct attacks on the practical failure of his survey. In a completely opposite situation to that experienced by Gesner thirty years before in New Brunswick, Whitney, the government geologist, had *failed* to find a rich mineral resource. "If Silliman's reports are correct," Whitney worried, "I am an idiot and should be hung when I get back to California."[57]

The chance to hang Silliman came in March 1865, when D. L. Harris, a Massachusetts legislator, wrote to Brewer, now a professor in the renamed Sheffield School of Science at Yale, about whether petroleum and asphaltum existed in southern California and, if so, were they related? Brewer's response appeared in the *Springfield Republican* and was soon republished in several eastern and western newspapers. Based on his observations, Brewer asserted that petroleum did *not* occur in southern California. Asphaltum did, but as far as he knew, the two substances were unrelated. Brewer defined petroleum as a mineral oil that did not harden on exposure. California asphaltum, moreover, could not be distilled into lamp oil or lubricants.[58]

Dismayed by this denunciation of his science and his character by someone he considered a friend, Silliman responded in a calm, confident, but not-too-polite manner with a long letter to the *Springfield Republican*. Brewer's definition, Silliman noted, was "not in accordance with the established opinions of chemists." The natural association of petroleum with asphaltum was "well-nigh universal," *except* in Pennsylvania, where, Silliman conceded, asphaltum was not found. Nonetheless, Silliman wanted to correct any misunderstandings about the source of California oil; it was not the asphaltum so plentiful on the surface, but the subterranean "zones of oil" to be reached by "Artesian borings." Silliman then turned to distillation. In his experiments on the California sample, he obtained "ninety-six percent of commercial products," a remarkable result subsequently confirmed by two distinguished chemists—Cyrus M. Warren (1824–1891) of Boston and John Maisch (1831–1894) of the U.S. Army Laboratory in Philadelphia. Warren had recovered 93.8 percent of commercial products and Maisch a full 100 percent from the sample of California crude. "[T]his it appears," Silliman concluded, "confirm[s] all I have said on the subject."[59]

In May 1865 Silliman published his and Warren's analyses in the *American Journal of Science.* The results were identical to those in the Philadelphia and California Company prospectus, with the notable addition that Silliman identified Peter Collier as the Sheffield School laboratory assistant who ran the tests. Silliman also noted that his "method of practical distillation," while sufficient for commercial purposes, was inferior to the new scientific process developed by Warren. Silliman's admission revealed a difference in standards but also the use of a scholarly journal for something other than science.[60] Faced with such a definitive defense, Brewer recanted. In a letter to the *Springfield Republican,* he admitted to missing the California localities visited by Silliman and to chemical ignorance. Silliman's reputation remained intact, but the damage done to the California oil companies proved irreparable. The Pacific Coast Company failed to attract investors and folded.[61] The California Company and the Philadelphia and California Company suffered severe slowdowns in stock sales. To regain momentum and "in vindication of Professor Silliman from the injurious statements which had been circulating respecting the truth of his report,"[62] Scott solicited an authoritative boost from two distinguished men of science, John Torrey and Charles Jackson.

Torrey and Jackson were visiting California in the summer of 1865, each for personal reasons, when they agreed to a joint examination of the Ojai rancho. Jackson had gone to see a friend in San Francisco, but, like Silliman, he let it be known to the *Mining and Scientific Press* that he would be willing to serve "the public . . . as a mineralogical and geological expert."[63] Torrey, by contrast, was looking for plants, not engagements, and was wary of involvement in the growing oil controversy and the speculation boom. He did not want "to run any risk of giving an opinion that will cause loss to Company, or individual—& I take warning by the case of Silliman, who has obtained . . . a very unenviable reputation."[64] Despite such scruples, Torrey signed a "certificate" with Jackson attesting "that the territory in question contains an immense quantity of Asphaltum, Maltha, and Petroleum," which were all the same substance, varying only in "degrees of thinness."[65]

Jackson, with Torrey's approval, then drafted a more thorough report, which explained, in the clearest terms yet, the scientific debate over California petroleum's origin and occurrence:

In Pennsylvania and New Brunswick, where the true coal formations exist, the Petroleum occurs in the . . . devonian to the silurian formations. In California, and indeed on the whole Pacific coast, the only coal formation that exists is in the tertiary and cretaceous rocks, the coals belonging to the species of brown coal or lignite.

From analogy, it seems probable that Petroleum in California should occupy the same relation to the coal formation of the tertiary age as those of the Eastern States bear to the true coal formation; and hence we ought to expect it to be found in the lower portions of the tertiary and in the cretaceous rocks.[66]

Jackson's inclusion of New Brunswick was an oblique reference to Albertite, which he still believed to be coal, despite the scientific consensus. His point about the relative ages of coal and petroleum, on the other hand, was generally accepted. California had lignite (not true coal); likewise, it had *young* petroleum (not the true Devonian or Silurian kind). In the East, petroleum resulted from the decomposition of marine plants. (The decomposition of land plants formed true coal.) In California, the same process of decomposition took place, but the vegetable matter was more recent, hence the genesis of lignite and of a different kind of petroleum. Regarding geological structure, Jackson agreed with Silliman that California's oil rocks were steeply inclined. Pennsylvania's were more or less horizontal. This meant that the gas, which drove Oil Creek's spectacular flowing wells, had probably escaped in California. Jackson, the shrewd consultant, did not foreclose entirely on the possibility of flowing wells in southern California; "there will undoubtedly be found some locations where the gases are really pent up," but he thought most wells would have to be pumped. Like Silliman, Jackson did not worry that surface springs might indicate that the oil had escaped. "I would advise borings to be made nearer to the outflows of the Petroleum springs than has thus far been made." In conclusion, Jackson opined, "the prospects here are favorable."[67]

Busting

In November 1865 Whitney returned to California having finished his *Geology,* a summary of four years of fieldwork and the official statement on oil. "Our line of exploration," Whitney explained, "did not happen to lead us to those ranches which have become celebrated by the formation of monster oil companies on them."[68] This, however, did not impede his pontificating. California petroleum originated in the "great bituminous slate formation, of Tertiary age," the same one identified by Silliman, Torrey, and Jackson. But Whitney noted what none of them dared say: "the strata of this formation . . . are turned up at a high angle. . . . For this reason, a large flow of oil on the surface cannot be considered as a favorable indication for boring wells and much less can heavy accumulations of asphaltum be so regarded." There were no sealed reservoirs. The oil "escapes and is lost."[69] Whitney, however, refrained from drawing a direct connection between escaped oil and asphaltum. "Whether the asphaltum of California is derived from the thickening, or oxidation, by exposure to the atmosphere, of exactly the same substance, chemically speaking, as that which is called petroleum in Pennsylvania is a matter for chemical investigation."[70] Whitney deferred to chemistry regarding asphaltum's origin, but he assumed moral authority on "foolish mining enterprises." "Within the last year the 'oil-excitement' of the Atlantic States has penetrated to the Pacific Coast, . . . by unprincipled speculators [and] by the creation of stock companies with an immense number of shares, which have been disposed of to a credulous public, [and] the luckless stockholders will never receive even a moderate return for the money invested."[71] Whitney's depiction of oil companies as stock swindles was misleading and self-serving. The California Petroleum Company and the Philadelphia and California Company, and many smaller firms, had purchased equipment and hired workers to bore wells. As Silliman explained, "powerful Companies [have] the expectation of doing a large and permanent business."[72]

By the end of 1865, however, none of the California oil companies had struck a flowing well, and those that did find oil pumped small amounts. Furthermore, the asphaltum had proved as difficult to refine as Storer had predicted. In May 1866 the Philadelphia and California Petroleum Company folded; within eighteen months the California Petroleum Company shut down. In California as everywhere else, the oil boom went bust. "The bursting of the oil bubble was imminent," one observer soberly recalled, "The oil companies fell, in the Fall and Winter of 1866 and 1867, one after another, like a row of bricks. Thousands who had purchased the stocks were overwhelmed in the ruin that followed. . . . All classes were affected by it."[73] The failure to find profitable petroleum seemed to confirm the value of Whitney's survey and helped to convince legislators that he deserved another two-year appropriation: $30,000 for fieldwork and $15,000 for publication. For Silliman, the bust marked the beginning of a series of unfortunate events.

Silliman's first setback involved a young Rhode Island chemist named Stephen F. Peckham (1839–1918). Peckham had studied chemistry at Brown University for two years before enlisting in the Union army in the summer of 1862. In January 1865 he got transferred to the U.S. Army Laboratory in Philadelphia, where he worked under Maisch and assisted in the analysis of California petroleum. Peckham also read Silliman's consulting report. "It was stated," he later recalled, "that Silliman had explored the country and had concluded it contained a fabulous wealth of oil."[74] After a chance meeting with a California Petroleum Company stockholder, Peckham was hired as the company's new oil refiner.

Peckham's residence in southern California lasted only a year (July 1865–June 1866), during which he moved from disappointment to skepticism to suspicion of Silliman's science. "Good Lord," Peckham quipped, "how much Silliman can see after dinner and a good bottle of Porter."[75] Peckham's job was supposed to be refining, but he spent most of his time looking for oil springs, which led Thomas R. Bard (1841–1915), the assistant superintendent of the California Company, to complain.

> Our chemist, a name by which our refiner is called, [is] "a scientific Cuss" as we say in Cala. [and] as much use here as "scientific Cusses" generally are where practical men are needed, & though he draws a salary of $250 in greenbacks per month, [he] has refined *five* gallons of oil, and has retained for his refinery a splendid engine & boiler, which . . . could have been . . . engaged in the work of developing [and] pumping oil in abundance.[76]

But there was no abundance of oil, and Peckham began to think that Silliman's report was "either stupidly or willfully false from beginning to end."[77] Not surprisingly, such ideas put him at odds with Bard and the California Company, which released him.[78]

Whitney was happy to hire the petulant Peckham and encouraged him to write up his findings, which he did in May 1867 for the *American Journal of Science*. Peckham's article was a detective story about how Silliman, Warren, and Maisch had been deceived. The tale began with Peckham's year-long hunt for "green oil." "I visited more of the bituminous outcrops in Santa Barbara and Los Angeles counties," he declared, "than had ever before been visited by any single person who had written on the subject."[79] What Peckham found was dark green and viscous. "I have seen no oils from natural outcrops that could properly be called "thin and mobile," a direct reference to Silliman's consulting report.[80] Peckham then undertook to prove that the Silliman-Warren-Maisch samples had been salted with kerosene. The first clue was density. His dark green oil had a specific gravity of .918. The other chemists had measured .863.[81] According to Peckham, kerosene sold in San Francisco with a specific gravity of .810. Peckham did the math; half dark green oil (.918) plus half commercial kerosene (.810) "equals .864 as the average density."[82] Another clue was chemical composition. Dark green oil did not contain paraffin; Pennsylvania petroleum did, and so did Silliman's sample. Peckham's final piece of evidence was the distillates. In their experiments, Silliman, Warren, and Maisch had reserved between 42 and 50 percent of the yield as lamp oil, which, more importantly, proved identical in its illuminating qualities to kerosene. By contrast, ordinary fractional distillation of dark green oil produced only 3.5 percent lamp oil. Most telling, Peckham's distillation of California crude produced no naphtha and very little lubricating oil. "The absence of either very light, or very dense oils," Peckham announced, "is a marked peculiarity of the distillate of Southern California petroleums." "These differences," he declared, "all point to the falsification of the oil examined by those gentlemen."[83]

Peckham did not know who salted the samples; he merely offered his findings to correct "an error" and to assist "in the dissemination of reliable information respecting California Petroleums."[84] Silliman was certain that no mistakes had been made, scientifically, commercially, or morally. As coeditor of the *American Journal of Science,* he had the great advantage of previewing Peckham's article before it went to press. This allowed him to insert a disclaimer at the end of his own article on California tar, which, conveniently, appeared in March 1867, the issue immediately *prior* to the one containing Peckham's findings.

> [N]o question of . . . authenticity, until a very recent period, ever reached me. It is now confidently asserted, by certain parties, that the samples in question were sophisticated by the addition of refined commercial petroleum. An inquiry instituted privately by me has elicited from the parties immediately concerned in its transmission only an emphatic denial of the charge of falsification, and if any such fraud has been perpetrated I am well persuaded that the responsibility falls elsewhere.[85]

If the public's trust had been compromised, Silliman swore to discover the truth.

Meanwhile, he had further results to convey about the distillation of California crude. Silliman chose some "surface oil," dark and viscid like Peckham's sample, and was able to distill 60 percent commercial products, mostly heavy oils for lubrication with less than 20 percent lamp oil and no naphtha. The absence of very light oil troubled him, as it did Peckham. The frothing, Silliman noted in an aside aimed at Storer, did not. Silliman recommended "cracking" the heavy oils either by heat or by a new method patented by the Scottish practical chemist James Young whereby pressure (ten to fifteen pounds per square inch) was added.[86]

Silliman was deploying the *American Journal of Science,* once again, to support his commercial science, but this time he seemed to be aware of the possible ethical improprieties. Twice he admitted his chemical examination had "chiefly a technical object." The scientific interest was "subordinate in a great degree," although he hoped the results would contribute to the knowledge of hydrocarbons. Peckham, too, had distinguished between "technical" and "scientific." The former referred to the distillation of marketable oils; the latter dealt with chemical compounds. Silliman's technical experiments had, again, been conducted by an assistant, A. J. Corning, formerly Warren's assistant. Corning worked in "research" at the "operative laboratory" of Downer's Kerosene oil works outside of Boston. In a note of disclosure, Silliman thanked Downer and his superintendent, Joshua Merrill, for allowing Corning to do the tests, especially on cracking, and the California Petroleum Company for supplying the samples. Basically, Silliman was the front man for industry-based chemistry.[87]

Silliman's second unfortunate event coincided with and perhaps was encouraged by Peckham's announcement of salting. In the spring of 1867 two lawsuits were filed against Scott's oil companies. In the first, a Philadelphia stockbroker, Lewis Ashurst, sought to recover his investment in the Philadelphia and California Petroleum Company.[88] In the second, another irate investor, Eugene Lynch, sued John Church, president of the California Petroleum Company.[89] Such lawsuits for fraudulent stock were not uncommon,[90] but in both suits Silliman was named as co-defendant, and his consulting reports were submitted in evidence. Ashurst et al. "do not know what interest said Silliman had with the other defendants, whether he was a party with them or some of them to the original conspiracy, or whether he received from them a pecuniary inducement to make the false and fraudulent statements that appeared in the said report."[91] The plaintiffs also targeted Silliman's "most brilliant lecture," delivered in Philadelphia in March 1865, where he exhibited a sample, "but that said oil must have been purchased for the purpose [and] pretended to have been received from California." Accordingly, Silliman's sole reason for lecturing was to "heighten the excitement and increase the demands for shares."[92]

Silliman's responses to these suits are unknown. The court records have also disappeared. But it is known that in both cases the plaintiff lost. Silliman and the companies were acquitted of any wrongdoing.[93] Nonetheless, the legal actions were unprecedented to the extent that no man of science of such stature had ever been sued for consulting. (The plaintiffs in fact acknowledged that Silliman had "a high scientific character" and "a reputation for personal integrity and respectability.")[94] Cresson, the president of the Philadelphia and California Company, for example, was not named in the suit. Furthermore, two of the plaintiffs were men of science, the wealthy entomologist John L. LeConte (1825–1883), who lost $2,000, and the Pennsylvania coal geologist Peter W. Sheafer. Silliman was being singled out for the excesses of his commercial behavior during the oil boom.

Upon his return from California, Brewer had been trying to recruit other Yale faculty to the anti-Silliman cause. His main ally was William Dwight Whitney (1827–1904), Josiah's younger brother. William was an outstanding linguist, professor of Sanskrit, and since 1861 head of Yale's modern languages department. He and Brewer had been unsuccessful in undermining Silliman's position until the lawsuits. Now they could point to a very public accusation of swindling, and this was a grave concern to Yale. In May 1870 the Academical Faculty voted to remove Silliman's chemistry class from the college, thus bringing down on Silliman and his family the most severe of misfortunes. When Silliman learned of the decision, he felt compelled, by honor, to resign.

> As the Faculty have, intentionally or thoughtlessly, done an act which it is now too late to recall, I feel justified in expressing to them my view of their conduct and my surprise that among so many officers of experience and wisdom there was no one to interfere to save a colleague who had grown up among them, and who had cherished an inherited zeal for the interests of the College—from injury so deep and, apparently, so heedless.[95]

That same month Silliman was taken off the governing board of the Sheffield Scientific School and his courses canceled. This last act was an especially cruel insult in light of the fact that he had founded the school and trained many of its current faculty, including Brewer. All that was left to Silliman was his title (professor of practical chemistry) and the position of professor of chemistry and pharmacy in the medical school. Yale College and the Scientific School, it seemed, had decided that commercial engagements were too embarrassing and improper for its faculty.

Classifying

The salted samples raised serious doubts about Silliman's integrity *and* posed important scientific questions about the differences between California crude and Pennsylvania petroleum. During the late 1860s and early 1870s, chemists focused on the processes by which liquid oil changed into solid asphaltum. These investigations led to new theories and, predictably enough, to a reexamination of Albertite.

In the spring of 1867, Silliman left Yale for another nine-month visit to California. The suddenness of his departure suggests some urgency to resolve the salting situation. Upon arriving in San Francisco, however, Silliman made no trips to Los Angeles or to Washoe, his other area of expertise. In fact, he had far fewer engagements than three years earlier. Most of his time was spent working on prospects in the Mother Lode region, including his own gold mine at Quail Hill. In March, Silliman examined the mine with Blake, recently appointed professor of mineralogy and geology in the College of California, and wrote a joint report predicting a rich yield in gold and silver. Within two years, however, the Quail Hill mine was "dead dead dead."[96] Silliman's fortunes fared better in the Grass Valley area, where he was engaged by a group of San Francisco capitalists, led by the merchant W. H. V. Cronise, to examine the Empire mine.[97] Silliman thought the prospect very favorable, and indeed, the Empire proved to be the second most profitable mine in Grass Valley. A few years later, an investor congratulated Silliman on keeping "the pot boiling in a professional way."[98]

As for his scientific pursuits, Silliman presented several papers before the California Academy of Sciences. All concerned gold and silver, except one. On 1 April he read his *American Journal of Science* article on California tar and highlighted its suitability as fuel. "This use of the material," he remarked, "is suggested [because of] the depression in the commercial value of the petroleum. It is notorious that the supply of oil from Pennsylvania has so far exceeded the demand of the markets." This was Silliman's explanation for the California bust. He then added a few remarks on the "chief controversy between the advocates of the supposed animal origin of petroleum against those who have suggested the theory of its derivation from vegetable matter." He ended with a discussion of the French chemist Marcelin Berthelot's (1827–1907) recent hypothesis that petroleum was *inorganic*. Berthelot had conducted experiments that showed how heat, water, and alkali metals could combine to form curious compounds akin to natural hydrocarbons. Silliman did not put much stock in Berthelot's theory, and his summary was more a show of erudition than of interest. After 1867 Sullivan seemed to lose his passion for petroleum.[99]

Peckham, by contrast, became increasingly obsessed by it. At the August 1868 meeting of the National Academy of Sciences, his grand theory of bitumens was announced, by Whitney, because Peckham was not a member of the academy.[100] "[T]he study of petroleums," Peckham declared, was *the* key to understanding all bitumens. "In all the more extended treatises upon chemistry and mineralogy," he explained, "naphtha, petroleum, maltha, and asphaltum [are given] as varieties of bitumen." Such a system was confusing, hence Peckham proposed a new classification based on the assumption that "petroleum [was] the normal form of bitumen." All other forms were derived from it, by either distillation, evaporation, or oxidation.[101]

According to Peckham, petroleum originated in situ from the decomposition of animal matter. He rejected any distillation theory and Berthelot's inorganic idea for want of "observed facts."[102] As for plants being the source, he regarded Leo Lesquereux's latest algae theory as mistaken; "[i]f petroleum were invariably derived from marine algae we should expect to find it identical in composition wherever found," which was not the case. California petroleum, Peckham explained, contained lots of nitrogen, while Pennsylvania petroleum had only a little. Nitrogen was crucial, for it indicated an animal source. On this point, Peckham agreed with Hunt, but he thought Hunt's in situ theory applied only to petroleum found in Silurian and Devonian limestones; the oil contained little nitrogen because the source animals were primitive. (The decomposition of Silurian polypi, for example, was indistinguishable from that of Devonian algae, which, Peckham thought, explained Lesquereux's error.) For Peckham, the clearest evidence of petroleum's animal origin was found in California. "[W]hen we pass up from those early formations," he explained, "to a deposit rich in remains of the higher marine animals, in which cetacean bones are frequently met, [we] find an oil comparatively rich in nitrogen."[103] In this sense, California crude *was* whale oil.

Peckham next tackled the scientific and practical questions surrounding asphaltum. "Whether the accumulations of asphalt were a residuum from evaporation or a product of oxidation," Peckham later recounted, "[was] of great importance to those engaged in prospecting for petroleum in the [southern California] region." If asphaltum were produced by evaporation, as Silliman argued, "it is an indication of vast quantities of petroleum a short distance below the surface." If it resulted from oxidation, "petroleum can only be obtained below the reach of oxidizing influences."[104] To determine which process applied, Peckham examined the two kinds of petroleum, Pennsylvanian and Californian. The former contained paraffin and little nitrogen; it was chemically stable and formed asphaltum very slowly by evaporation, which accounted for the absence of any asphaltum along Oil Creek. California petroleum contained no paraffin and lots of nitrogen; it was chemically unstable and formed asphaltum rapidly and easily by oxidation—hence, the ubiquitous deposits in southern California.[105] The surface deposits of asphaltum thus could not be connected to shallow pools of petroleum. Silliman was wrong. Here was the "conclusive reason," Peckham announced, "why borings for oil in the vicinity of asphalt beds were not likely to prove successful."[106]

The centrality of petroleum to the new understanding of bitumens was echoed by James Dana. In 1868 he published the fifth edition of his influential *A System of Mineralogy*. Fourteen years had passed since the last edition, and "[c]hemical researches had been carried forward in connections with almost every [mineral] species."[107] The most stunning changes involved coal and petroleum. In 1854 Dana had classified coal, diamond, and graphite as carbon minerals and shunted bitumens (petroleum and asphaltum) to an appendix. Now the Carbon class was gone and in its place Dana put petroleum, or, more precisely, Hydrocarbon Compounds (table 10.1). For the very first time in the history of American science, coal was cut out of the main classes of mineralogy.

Dana had ousted coal because it was not a "*native* hydrocarbon," by which he meant a pure chemical compound. Coal was a mixture of hydrocarbons (simple and oxygenated) in the way that granite was a mixture of quartz and feldspar. Mineralogy concerned quartz and hydrocarbons, not granite and coal. "Most of the kinds [of coals] hitherto recognized in mineralogy," Dana declared, "are more analogous to rocks than minerals."[108] Mineralogy meant hydrocarbon chemistry, and petroleum was the paradigmatic material. Derived from animal and vegetable sources as well as from the destructive distillation of "hydrocar-

Table 10.1. James Dwight Dana's General View of the Mineral Classes (1868)

I. Native Elements
II. Sulphids, Tellurids, Selenids, Arsenids, Antimonids, Bismuthids
III. Compounds of Chlorine, Bromine, Iodine
IV. Flourine Compounds
V. Oxygen Compounds
VI. Hydrocarbon Compounds

Appendix to Hydrocarbons
   Asphaltum
   Mineral Coal

---

bonaceous [shales]" (Dana was typically ecumenical), petroleum was a simple compound of hydrocarbons.[109] By fractional distillation, chemists had determined "with great exactness" the composition of those hydrocarbons, which Dana classified as separate mineral species (table 10.2).[110]

If coal was cast out, so too was asphaltum. Dana regarded it as a chaos of compounds. "[T]he true line of investigation [i.e., hydrocarbon chemistry] is so little appreciated," he declared, "that new resins or asphalts are from time to time brought forward as species in mineralogy upon characters that only prove them to be mixtures."[111] Chief among such specious species was Albertite. Dana blasted Lesley, Henry Wurtz, and Charles Hitchcock, as well as James Robb and Charles Wetherill (from the Albert trial) for their roles in forwarding Albertite as a mineral. Curiously, Dana singled out the Torbanehill mineral as a separate species, Torbanite, which, "although related to cannel coal," had "a very nearly uniform composition." But Albertite and Grahamite and all other asphaltums were petroleum derivatives.[112]

"I do not believe that the origin of albertite, grahamite, or any such substance," Peckham retorted, "has the remotest connection with the formation of any [asphaltum] or with petroleum of any description."[113] Albertite contained paraffin; therefore, if it was hardened petroleum, it had to be the Pennsylvania variety. But Pennsylvania petroleum thickened by slow evaporation, and Peckham thought it "impossible" that Albertite, in liquid form, had entered the fissure and then somehow been exposed to the atmosphere. Peckham also rejected the idea that Albertite had been distilled from petroleum. "This hypothesis presupposes a subterranean sea of petroleum," he gibed, in a direct reference to Lesley's 1863 idea. "If such is the case," Peckham continued, "why is not Hillsborough Co., New Brunswick, one of the richest petroleum regions in the world?" Albertite, Peckham concluded, was a special variety of solid bitumen, *not* a petroleum derivative.[114]

Dana had to agree. In the second edition of his *Manual of Geology* (1876),

Class VI: Hydrocarbon Compounds
    I. Simple Hydrocarbons
        1. Marsh-Gas Series, C_nH_{2n+2}
            1. Naphtha Group (liquids)
            2. Beta-Naphtha Group (liquids)
            3. Scheererite Group (solids)
        2. Ethylene Series or Olefines, C_nH_{2n}
            1. Pittolium Group (liquids)
            2. Paraffine Group (solids)
        3. Camphene Series, C_nH_{2n-4}
            1. Fichtelite Group (solids)
        4. Benzole Series, C_nH_{2n-6}
            1. Benzole Group (liquids)
            2. Könlite Group (solids)
        5. Naphthalin Series, C_nH_{2n-12}
    II. Oxygenated Hydrocarbons
        1. Geocerite Group
        2. Succinite Group
        3. Retinite Group
        4. Scleretinite Group
        5. Pyroretinite Group
        6. Unnamed Group
        7. Dysodile Group
    III. Acid Oxygenated Hydrocarbons
    IV. Salts of Organic Acids
    V. Nitrogenous Hydrocarbons

Dana acknowledged the consensus surrounding Peckham's interpretation. "In the lower part of [the] Subcarboniferous," Dana explained, "there are . . . 'false' Coal measures . . . contain[ing] numerous remains of fishes. [The Albert mine, New Brunswick] affords a peculiar coaly material, pitch-like in aspect, which has been named *Albertite;* it fills a fissure, instead of constituting a true-coal bed."[115] Dana cited Dawson as the person who coined the name and Wetherill as the source for the mineral's chemical composition. Albertite was (and ever has been) an organic mineral species.

The classification controversy had thus finally reached closure. It had taken nearly a quarter of century for men of science to decide that the Albert mineral was neither coal nor asphaltum nor hardened coal oil nor petroleum derivative. Each interpretation had mirrored a particular moment in the industrial development of nineteenth-century America. As always, classification was an acute commentary on context. Thus it is not so much ironic as indicative of the intimate relations of science and industry that it was only in the 1870s, when Albertite had practically no commercial value, that the mineralogical debate over its identity came to an end.

Retaliating

As hydrocarbon chemistry and petroleum geology moved from contested to consensual, the stage was being set for a heated argument about the use (or misuse) of those sciences. Whitney's attacks on Silliman would coincide and escalate with every setback faced by the California survey.

In March 1868 Whitney suffered his first defeat; the California legislature failed to make appropriations for his survey. "'Petroleum' is what has killed us," Whitney told his brother William at Yale. "By the word 'petroleum' understand the desire to sell worthless property for large sums [of money]."[116] Whitney's appraisal of California politics was highly improbable. The petroleum business had been dead for two years. Whitney's own production, or lack thereof, was more likely the cause. He had published only two volumes and one map. He had promised much more, but nothing was in press.

With the survey suspended, Whitney retreated back to a comfy chair as the Sturgiss-Hooper Professor of Geology in the School of Mines and Practical Geology at Harvard.[117] He continued working on the survey and, with his own funds, published two more scientific volumes (one on Pacific Coast land birds and another on paleontology), a revised map of the San Francisco Bay area, and a popular tract, *The Yosemite Guide-Book* (1868). In November 1869 Whitney, with his publications in hand, headed back to California on the newly completed transcontinental railroad, determined to win an appropriation. The legislature had not discontinued the office of state geologist and the governor was favorably disposed to geology. In January 1870, however, a bill was introduced to abolish the survey. Whitney suspected the machinations of malign petroleum interests and called on his scientific friends, including Lesley, to defend him. The campaign worked. In March the legislature authorized a generous $25,000 to reimburse Whitney and cover publication costs. In addition, Whitney got $2,000 per month for the next two years.[118]

Restarting the survey proved difficult, not least because his former assistants had dispersed. The exception was Charles Hoffmann, whose topographic mapping continued at the highest level, in terms of standards and elevation, the Sierra Nevada. Other work went slowly, and Whitney wasted valuable time overseeing field parties. Ironically, he had the money but not the experienced men of science to spend it on. When it came time to renew the survey in early 1872, Whitney, as usual, projected the publication of many volumes: conchology, ornithology, paleontology, and Brewer's long-awaited botany. All were based on previous fieldwork and written by the old corps. Noticeably absent was anything on geology, especially on mining, by Whitney himself. Nonetheless, in March 1872, Whitney got $48,000 and another two years, but it was not smooth sailing ahead. The newly elected governor, Newton Booth, a San Francisco businessman, thought too much public money had been spent on too few results. He informed Whitney that the tide was running against his ship of science.

In October 1872 Whitney retreated back East to reinforce the anti-Silliman campaign. Besides his two principal allies at Yale, his brother William and Brewer, he had the support of faculty at Harvard, the chemist Wolcott Gibbs (1822–1908) and Louis Agassiz. Whitney's plan was to remove Silliman from the National Academy of Sciences, a decision that reveals as much about the status of the academy as it does about American science.

The National Academy was born in secret at a time of war. In March 1863 in the last minute rush of legislation of a lame-duck Congress, Senator Henry Wilson of Massachusetts introduced a short and simple bill. Its first part named fifty incorporators of the new academy, its second gave those members the power to make rules and fill vacancies, and its last part gave Congress the power to consult the academy about "any subject of science or art." Members did not receive pensions or payments for such consulting work. The mastermind behind the academy was Alexander Dallas Bache, superintendent of the Coast Survey and foremost patron of scientific men in the federal government. The list of members was drawn up by Bache, Agassiz, and a few of their friends. The group called themselves the Lazzaroni, and they had set the goal of elevating American science by doing away with charlatans and amateurs. "We have a standard of scientific excellence," Agassiz told Bache, "Hereafter a man will not pass for a Mathematician or Geologist [lest he] be acknowledged as such by his peers."[119]

Most of those chosen by Bache and his peerless Lazzaroni were completely surprised by Wilson's letter telling them of their selection to the academy.[120] Many opposed the new organization, including William Rogers, Dana, Torrey, and both Sillimans. Joseph Henry (1797–1878), director of the Smithsonian Institution and the most famous scientific man in America, was "not well pleased with the list or the manner in which it was made. . . . I do not think that one or two individuals have a moral right to choose for the body of scientific men in this country who shall be members of a National Academy."[121] Joseph Leidy called it an "illiberal clique, based on Plymouth Rock," a reference to the fact that many of the Lazzaroni were from Harvard.[122] Even Whitney, in far off California at the time, complained. "No end of mathematicians & as(s)tronomers with small show of chemists geologists &c!"[123]

The membership was indeed a problem. Of the fifty incorporators, thirty-two did physical sciences (mathematics and physics) and only eighteen did natural history (including geology), a ratio reflecting neither the relative sizes of the communities nor their accomplishments. Given such manifold reasons for resentment, the academy would most likely have withered away were it not for Henry. In 1867, upon Bache's death, he took up the presidency and undertook some much needed reforms. Original research, regardless of field, became the criteria for election, and, most importantly, twenty-five new members, most in natural history, were elected in 1872. By that time, Whitney had changed his opinion of the academy.

The event that precipitated Whitney's attempt to revoke Silliman's membership was the failure of the Emma silver mine in Utah. Located in the Wasatch Mountains about thirty miles from Salt Lake City, the Emma mine had been the scene of a great silver strike in 1868. By 1871 the owners were looking to sell out, so they organized a stock company and hired three experts. In July, Blake reported very favorably on the mine and received a $2,000 fee. In August, Rossiter Raymond (1840–1918), U.S. commissioner of mining statistics, examined it and pronounced it promising; and in October 1871, for a $15,000 fee, Silliman journeyed west to examine the mine. He too was impressed.[124] In November, the Emma Silver Mine Company was incorporated with a capital of $5 million and a prospectus highlighting Silliman's optimistic opinion. For more than a year the Emma Company paid regular monthly dividends, but in the early spring of 1873, the directors announced a suspension of payments. The stock price dropped, and angry investors demanded an investigation. In June, Clarence King, Whitney's former assistant, visited the mine. He found it in ruins. "The great Emma 'bonanza,' the object of such wide celebrity, the basis of such extravagant promises," King testified, "is with insignificant exceptions worked out."[125] To Whitney, the Emma mine was a gigantic stock swindle, and the last piece of evidence in the case he was building against Silliman.

In the fall of 1873 Whitney returned to California in a desperate, last attempt to renew the survey. Governor Booth was opposed, and so too was the legislature. According to the *Mining and Scientific Press*,

> [Whitney] made up his mind to carry the survey on in the order which best suited him and paid no attention whatsoever to the suggestions of the people. . . . If he had deferred even a light degree to the wishes of the community, the more scientific part of his labors might have been accomplished at leisure, and with a liberal appropriation to assist them. As it is, the public has gradually . . . become prejudiced against him.[126]

Whitney's high-handed attitude, if not his high-minded science, had lost him the patronage of the people. The California Geological Survey was over.

Before leaving the state, a disagreeable Whitney sat down in his San Francisco office and drafted a bill of wrongs against the "scientific and moral character" of Silliman and sent the nine-page diatribe to the National Academy.

Arguing

Whitney's indictment arrived at the meeting of the council of the academy on 28 October 1873.[127] As the governing body, the council comprised five officers (president, vice-president, foreign secretary, home secretary, and treasurer) and six elected members. At the time, the council was well disposed to Whitney. Gibbs was vice-president; the Harvard mathematician and Lazzaroni Benjamin Pierce (1809–1880), was a newly elected member, and so too were LeConte, a plaintiff in the *Ashurst et al. v. Scott* case, and Whitney's brother William. Silliman's supporters numbered but one, the foreign secretary and Yale graduate F. A. P. Barnard (1809–1889). Both Lesley and Hall had just stepped down, and there were no other experienced consultants on the council. After some discussion, the council resolved *not* to read Whitney's letter. Instead, it sent a copy to Silliman. Upon receiving the "amazing document," Silliman began preparing his defense. The council gave him three months to collect evidence and write a rebuttal. At its next meeting, the council would read Whitney's charges and Silliman's defense together.[128]

"I charge Benjamin Silliman," Whitney began with typical brusqueness, "with having been for years the ready and efficient aid of operators and in operations which deserve no greater name than that of 'swindling.'"[129] "These operations," Whitney explained, "are connected with the sale of mines and supposed mineral lands . . . in the far West, especially in California[,] where I have been professionally engaged since 1860." For his first example, Whitney turned to the Emma mine, such a scandal that "it would be proper to put Professor Silliman on trial on ground of this transaction alone." But it was only one in a series, "of which at least one was of a magnitude and conspicuousness beyond even the Emma mine fraud. I refer to the notorious California petroleum swindle." This was "one of the most skillfully planned and extensive frauds ever brought before the public." It involved millions of dollars and innumerable investors across America. Even more money would have been lost, Whitney declared, had not his survey "persistently refused to sanction the swindle." Whitney presented his stand as a brave act. "[A]buse was poured on the heads of those who endeavored to enlighten the public in regard to the true nature of these swindling operations." He then described the operations of three other Silliman swindles: the Empire Gold and Silver Mining Company of Bodie, California; the Burdett mine, near Nevada City, California; and the New York and Nova Scotia Gold Mining Company. In each of the five swindles, "[t]he gist of the thing" was easy to summarize.

I) The enormous fee paid for a report on the property;

II) The extraordinarily favorable character of that report;

III) The immense nominal capital of the company;—that is to say, the high price at which the property was offered, in the form of stock, to the public;

IV) The total failure of the company to realize anything from their operations;

V) The utter worthlessness of the property and the fact that the report for which so large a sum had been paid by the promoters of the enterprise was nothing but a tissue of falsehoods, from beginning to end.

As Whitney expounded these steps, he made three assumptions, all "naturally," "clearly," or "undoubtedly" revealed by the five cases. The first and most important was the link between money and motivation. "I have had many years experience and acquaintance with the business of consulting," Whitney testified, "and have never known such exorbitant fees being paid or received except when there was a distinct understanding that the report was to be made a favorable one in proportion to the amount of fee received." The exact amounts of Silliman's fees were unknown to Whitney, but he was eager to relay what he had heard from others. For the Emma mine survey, Whitney said Silliman got as much as $5,000 cash down and $45,000 in cash or stock contingent on the mine's sale. For the Bodie report, Whitney learned that Silliman pocketed $10,000 in gold, "a sum which is much greater than was received by any mining engineer for an honest, legitimate report." Three times Whitney referred to Silliman's "highly flattering" Bodie report as "a tissue of intentional falsehoods" and pointed to various "grossly exaggerated" statements. The most flagrant case of "flattery," however, was California oil. "Silliman's statements of 'fabulous wealth,'" Whitney declaimed, were published in prospectuses, advertisements, and newspapers, "all easily to be obtained." It was these false reports that "induced" investors to purchase stock. Silliman was thus the reason the companies were able to get started, which was Whitney's second assumption. As he read off the enormous capitalizations printed on the prospectuses, he targeted Silliman as "the guilty party." Silliman "effected the sale of the properties." When the companies failed, the reason was obvious, or so Whitney assumed; the property was worthless. No gold was found in the mines, and no oil in southern California. In short, Silliman had been willing "to sacrifice truth to gain."

"Now it may be asked," Whitney concluded, "why has Professor Silliman this power for evil[?]" The answer was threefold: Silliman possessed the good name and reputation of his father; he had belonged to the Yale faculty; and he was a member of the National Academy. Whitney noted how Silliman deployed all three when signing his consulting reports, thereby vesting his opinion with authority, or "power for evil," another phrase Whitney liked to repeat. Each factor was thus implicated; Whitney himself felt he was *particeps criminis* by virtue of his academy membership. And each factor suffered severe repercussions. For the family, Silliman proved to be a "degenerate son"; Yale lost respect as a place of learning; and the academy was most damaged, for not only was the "high standing" of its members compromised, but the name American was discredited. "To maintain the purity of American science," Whitney cried, Silliman must be expelled.

In view of the palpable venom of Whitney's accusations, Silliman's response was a model of self-restraint and decency. He thanked the council for its "considerate" manner and praised its members for their just, sober, and sound judgment. Beyond such pleasantries, Silliman knew perfectly well that his standing as a scientific man, "a matter which may be dearer to me than life," depended on a strong and swift counterpunch. And he delivered it. He denied each and every charge and set out the "facts," in contrast to Whitney's "perversion of facts."[130]

Silliman had the evidence, and a lot of it, too, more than fifty items (maps, letters, pamphlets, depositions, newspaper articles) totaling more than five pounds of paper. He wielded it purposefully for its "moral effect." Much was technical and came from miners, managers, engineers, and geologists. Significantly, Silliman presented complete copies of all his consulting reports, in contrast to Whitney's selected "extracts and comments." He also included other consultants' reports, and these were particularly poignant for not only did they support Silliman's evaluations but they showed that other men of science, some of whom belonged to the academy, were actively involved in evaluating mines and mineral lands for fees. Thus Silliman gave to the council Raymond's report on the Emma mine, Torrey's and Jackson's on California oil, Blake's on Bodie and Burdett, and T. Sterry Hunt's and Henry Youle Hind's on Nova Scotia gold. The point was sharp. If Silliman had erred in his professional advice, then so too had others.[131]

But Silliman was not ready to concede that his advice or the properties were worthless. Companies might have failed, but as the other reports revealed, the failures might have nothing to do with the science. As to his personal involvement, Silliman welcomed an examination into his "real part" in the companies' establishment or their stock sales. Finally, he came to the heart of the matter, money, "which must interest all professional men." "I shall aid and not object to the consideration of the question: what are proper retainers or fees?" In Silliman's view, money for science was not evidence of corruption but of the necessity and value of "technical opinions." Here, then, was the true disinterested consultant. In a ringing conclusion, Silliman told the council he could not allow Whitney to accuse him of fraud or unjust gain without making it appear "that my whole life, for the last few years, has been a lie."[132]

After reading Whitney's charges and Silliman's defense, the council now faced an ugly and perhaps insoluble problem. "One member of the Academy cannot make such accusations against another," Silliman warned, "without grave responsibility upon the accuser, the accused, and their tribunal." Whitney had asked the council to pursue an investigation. Silliman likewise had asked it to conduct a fair and thorough trial. Neither the council nor the academy, however, had the legal right or power as specified in its constitution to conduct such an investigation. Silliman, in fact, had reminded the council of this glaring, legal obstacle. He was willing to proceed with an inquiry but pressed the council to conduct it with "proofs under oath, by depositions and oral examinations."[133]

"There is no need that the examination be conducted with legal forms or so as to lead to a legal conviction," William Whitney complained to the academy's home secretary Julius E. Hilgard (1825–1891). William, who had been unable to attend the meeting, castigated the council for being swayed by Silliman's "cloud of irrelevant verbiage." William was especially angered, or perhaps afraid, that the council would "throw the matter into the open Academy." That would invite public scandal; instead, William offered to examine Silliman's evidence. "I am in no sense or degree an enemy of Mr. S.," he declared with supreme disingenuousness. "You may rely upon my sense of justice."[134]

The council chose not to rely on William Whitney. It appointed Gibbs as the committee-of-one to "digest" Silliman's documents. Silliman was concerned. "Dr. Gibbs is a colleague and old personal friend of Mr. J.D.W." Still, Silliman was willing to carry on; "I believe him [Gibbs] incapable of an unfair action."[135] Gibbs, however, had no time, interest, or stomach for the investigation. In April 1874 he told the council that it was "impossible" to summarize Silliman's defense because of the "purely technical character" of the evidence. Gibbs recommended that someone familiar with "the subject of mining engineering and with practical geology" be appointed, and, more importantly, that Whitney be required to submit "positive evidence . . . in support of the charges."[136] Thus, Gibbs set in motion a long and increasingly irritating quest to extract some kind of proof from Whitney. For the next eight months Whitney delayed, denounced, and in the end defaulted on his obligation to comply with the council's request.

It was a painful time for Silliman. He had worked diligently and quickly in soliciting and collecting evidence from a network of far-flung correspondents, many of whom were in hard-to-reach places in the West. He had even prepared digests of his evidence for each of the five cases. Gibbs had never even bothered to look at these.[137] Worst of all was the secrecy. The council enjoined Silliman to keep the matter private. Henry promised it was a "secret transaction" to be kept within council confines.[138] Silliman agreed to observe great "circumspection" in consideration of his colleagues.[139]

The brothers Whitney did not. In December 1873 Whitney gave a copy of his indictment to E. L. Godkin, editor of the *Nation,* while William let Samuel

Bowles, editor of the *Springfield Republican*, read another copy. Both Godkin and Bowles were among the most ardent liberal reformers of the Gilded Age, and they ran editorials accusing Silliman of swindling and corruption. In an especially cruel cut, Bowles used Silliman's silence as evidence of his guilt. Upon reading the "offensive, false, and scandalous statements," Silliman was "very angry" and "with cause."[140] Trial by press, he knew, weakened his case with a "jury of [his] peers." Moreover, he feared the public would not wait patiently for the council's due process.[141] Silliman felt compelled to take action and gave notice to the editors "to withdraw the same or produce their proofs in court."[142] On threat of suit for libel, Bowles agreed to meet with Silliman and subsequently printed a retraction. Godkin refused.[143] Silliman consulted the council about making public his defense.[144] Hilgard, Gibbs, and presumably Henry thought Silliman had every right to defend himself.[145] William Whitney did not: "I object to, and strongly protest against, his printing any part of the *acta* of the case before the Council." In another expression of supreme disingenuousness, William upbraided Hilgard: "why should [Silliman] be in such a hurry?"[146]

William's cunning bought time for his brother to gather evidence while the press simmered about swindling. Whitney, however, had little interest in proof. The cases were too "notorious" to require it. In July 1874 Whitney decided to take an extended European tour, thereby deserting William and practically destroying their plans. Silliman thought this behavior wholly reprehensible. In a moving letter to the council, he denounced Whitney's "silent contempt" for the academy and spelled out the high costs of Whitney's stealth.

> Comment is needless. The plain facts tell their own story. It is time which kills us all and I have had enough of this kind of medicine.
>
> The Council have been the unconscious & innocent victim of a policy of procrastination instituted by my detractors to place me in Chancery and keep me there silent and powerless while public judgement crystallized and hardened into irreversable [*sic*] decrees of prejudice. Pardon my freedom of speech. . . . You see the fruits of a conspiracy of *10* years growth which I must crush or it will crush me.[147]

Judging

Joseph Henry, too, had had enough of Whitney's medicine, and as council president, he called for a special meeting to be held at the Smithsonian Institution to decide the issue. William Whitney balked; he had a sore throat and meetings of the Oriental Society to attend. He could not possibly make it "until the week after New Year's."[148] Gibbs was also busy and suggested April 1875. Whitney, just returned from Europe, blasted the council in his usually unrestrained manner:

"this investigation [was made] as difficult for me as possible." He would not come. Only Silliman responded cordially. "I should know the day selected," he requested, "that my counsel may so arrange his engagements to be with me."[149] The day was 30 December 1874.

William Whitney was able to make the meeting after all, and he brought a most remarkable document, an account of the great California oil swindle. Written by Brewer, it sounded like a collaborative effort, but Whitney had had no hand in it. The council, unaware of this deception, listened as William read the "positive evidence" it had requested eight months earlier.

Brewer's "petroleum draft" was long (more than 5,000 words), repetitive, and unprecedented.[150] The California oil companies, Brewer began, had failed, a "notorious" fact requiring neither argument nor proof. Silliman was "the scientific authority on which these companies forwarded their claims to public confidence." The purchase of property, organization of companies, and sale of stock were all "legitimate" results of Silliman's statements. Brewer, though, added some historical context. "The wild speculation in oil property in Pennsylvania had . . . unsettled and excite[d] the public mind and put it in a condition to be deceived more easily than before." That is why "the speculation was notoriously successful."

To understand how Silliman effected the speculation, Brewer took the council through a page-by-page analysis of Silliman's 1864 consulting report to the California Petroleum Company, the largest in terms of capitalization. He drew the council's attention to Silliman's extraordinary statements, particularly to the line about the "fabulous wealth in the best of oil" and the estimate of $1,365,000 annual profit. Brewer quoted Silliman at length about the limited supply of gold mines in contrast to the unlimited flow of petroleum wells. "The relation of this statement to actual facts needs no explanation to scientific men."

Next, Brewer retraced Silliman's lecture tour along the East Coast in the spring of 1865. He characterized Silliman as a showman intent on deceiving the public and added his own touch of melodrama to Silliman's performances. "[M]uch was invested . . . much distress followed. In one case, suicide was believed to result and a widow and large family of children were left impoverished." Silliman, apparently, was willing to say anything for oil.

"Now, was the worthlessness of these investments and the consequent loss and suffering of the losers, and the resulting damage to scientific authority and influence, owing to [Silliman's] statements being *untrue?* We think emphatically yes." To prove it, Brewer turned to pages 13 and 14 of Silliman's report. "Here then," he declared, "are at least *fifteen* actual misstatements of fact, or else gross exaggerations of facts, or else unwarranted assumptions and suppositions." To help the council see the errors, he numbered them. (Brewer, however, seems to have identified only thirteen on the pages he handed out to the council.) Then, he juxtaposed (literally set in parallel columns) Silliman's statements with "*facts now well known and not disputed.*" So, where Silliman wrote "full of tarry oil" (#2 & #2a), Brewer argued that "the *full* means 8 feet of water and 6 inches of maltha" and "the tarry *oil*, is maltha so tenacious that it could be rolled up in balls of a foot in diameter." In such excruciating detail, Brewer worked his way through Silliman's entire consulting report.

It is not necessary to replicate Brewer's exercise, but it is important to note that he made at least two interesting misstatements of his own. Brewer testified that Torrey and Jackson had denied, "when privately questioned," that their reports confirmed Silliman's. Because Torrey was dead and Jackson confined to an asylum, the council would have been hard pressed to verify this. Similarly, Brewer asserted that the California survey had shown Silliman's ideal geological section "to be entirely wrong . . . before he visited it." Where the survey made this declaration or published its section Brewer did not say, nor could he.

Brewer then turned to hydrocarbon chemistry. Relying on Peckham, he accused Silliman of misleading the public about the chemical differences between Pennsylvania petroleum and California crude and regaled the council with a blow-by-blow retelling of the salted sample saga. "*To this day he has never explained to the public, or to his scientific collaborators, his connection with this fraud.*"

"In conclusion," Brewer swelled to his peroration, "a great wrong (and alleged *swindle*) was perpetrated." Silliman had "willfully misrepresented" the facts or was "unworthily influenced" in making his statements. As with Whitney, Brewer did not elaborate on the important distinction between being a knowing accomplice and an unwitting tool. He simply moved to his bigger point: "the good name of science was injured, and particularly in California, great discredit was brought on science."

While William Whitney read Brewer's indictment, Silliman sat outside in the hallway. He had been promised the chance to address the council, but he never got to make his case. Instead, the council read his "personal statement" on California petroleum, a summary of his evidence that in no way could have responded to Brewer. The council then read letters from the absent members. Gibbs worried about "the ethics of the matter," of allowing Whitney to make "charges without any testimony on his side." He recommended a face-to-face hearing between Whitney and Silliman.[151] Barnard reiterated that neither the academy nor the council had the legal power "to inquire into the moral character of any of the acts of the members." And he doubted whether "the Academy as a body shall be in any manner compromised by any acts of its individual members." This was not a case of self-defense; the standing of American science was not at risk. Nevertheless, because the council had taken up Whitney's charges, Barnard could see only two possible conclusions. The charges were either "groundless [or] at most unsustained."[152] Likewise LeConte had "great doubt" about the council's jurisdiction. He feared a "public scandal," which would be "no good result either for science, for the community, or for the Academy."[153]

The council agreed and resolved that the National Academy of Sciences had "no jurisdiction and no power which would enable it to investigate such a matter and to do justice to the parties."[154] They admitted that they had "informally considered" Whitney's charges against Silliman, but it would be "improper" for them to take definitive action. The council, therefore, could not expel Silliman. But neither could it exonerate him.[155]

In deciding not to decide, the council had satisfied no one. Brewer and the brothers Whitney regarded it as an endorsement of Silliman and resigned from the National Academy. Silliman was "greatly disappointed" and thought he had been "badly treated."[156] He demanded a hearing. "The Cowardly attacks of the Messrs Whitney made under the cover of secrecy of your private deliberations," he protested, had done "all possible injury." His hands had been tied for more than a year; "it must now be understood," Silliman warned, "these proceedings may of necessity have to be made public."[157]

> A still more bold and successful mode of misrepresentation and swindling is by the employment of mercenary professors. . . . Not that every penny-a-liner is necessary "particeps criminis;" for some have told the truth to the best of their knowledge, and only lacked time, patience, and perhaps means to remain and make their inquiries more thorough. Still, there is little doubt that others, of the sensation class, have deliberately lent themselves to mystify the outside public.
> —William Wright, *The Oil Regions of Pennsylvania* (1865)

NINETEENTH-CENTURY American men of science never tired in their moralizing. They considered themselves individuals of the highest character embodying everything from honesty and patience to seriousness and disinterestedness. The very pursuit of science refined them. Thoroughness, accuracy, and hard work were among the popular bourgeois virtues reproduced by doing science. As moral beacons upon a hill, men of science expected the public to look up to them and to trust them.[158]

Silliman's conduct seemed to the Whitneys and Brewer to break this bond of trust. It shattered the moral economy of science. To maintain "the purity of American science," Whitney's oft-used phrase, it was necessary to expose the fraud and extirpate Silliman, like some cancerous tumor, from the body scientific. This proved more difficult than they imagined. Whitney and Brewer needed to show that Silliman's statements were false *and* that he was fully aware of this. Brewer tried to do this through his rigorous reading of Silliman's consulting report. But Brewer's comparisons were false ones. He juxtaposed Silliman's science of 1864 with the science of 1874. A decade of hydrocarbon chemistry made a great deal of difference. All men of science can be shown to be "wrong" by later research. This is not a mark of moral delinquency of the original investigator. If the council had decided to censure Silliman based on future "rightness" or "wrongness," they threatened to stifle any research.

Silliman's science, however, was not really the issue; it was Silliman, the man of science. Whitney and Brewer had to prove Silliman's intent. In his petroleum draft, Brewer never asked *why* Silliman did it; in fact, he never mentioned money. For Whitney, it was all about the money. The almighty dollar had corrupted Silliman; or, as Whitney put it, Silliman had sacrificed truth to gain. This was a more grievous charge. Silliman was not party to a swindle; he himself was a fraud.

None of the council members (with the exception of William Whitney) were willing to go that far. "I am not prepared," Henry concluded, "to adopt the opinion that [Silliman behaved] with the design to defraud." Henry did think Silliman "gave false impressions as to the value of the mines." But this was a commercial miscalculation, the result, perhaps, of an overly optimistic outlook, not a moral failing.[159] In refusing to judge Silliman's character, the council revealed the limits of the National Academy's purpose. It was an honor society, not a professional one. It did not have (and still does not have) a code of ethics for scientific behavior. The council could not pass judgment on how a man of science *should* act beyond determining whether a person's research was worthy of recognition. The council thus preserved the academy as a national institution.

By downgrading the Silliman-Whitney controversy from a debate over American science to a scientific and personal dispute over oil,[160] the council avoided the question of the commercial relations of science. Yet the most troubling episode remained the salted sample. According to Wright, the mercenary professor might not be fraudulent, but every penny-a-liner was vulnerable. Silliman had been manipulated by those more interested in money than science. Whitney and Brewer never addressed the difference between an accomplice and an unwitting tool. The former is a user; the latter is being used. Consultants could not always control how others less scrupulous would use the science or its authority. Silliman was not so much a rogue chemist as a tobacco scientist (to import a modern analogy). That was Peckham's interpretation.

Early in 1864 the region [between Santa Barbara and Los Angeles] was visited by an eminent eastern chemist, who was so far misled by false local representations and by gross deceptions practiced upon him as to induce him to make a report upon this as a petroleum-producing region of great richness. This report, and others of similar character, led to the formation of mining companies representing stock to the value of millions of dollars, all of which it is needless to add, was lost to the *bona fide* investors.[161]

Peckham reached that conclusion in his official report on the petroleum industry for the 1880 U.S. Census. "In compiling this report," he explained, "I wish to make it an authentic statement of *fact* and shall deem it a great favor if those who have primary knowledge will assist me so far as they are able."[162] He interviewed most of the major players in the early history of oil, including Silliman, whose *Report on the Rock Oil* along with "the Albertite vein in New Brunswick" Peckham highlighted for their roles in starting the industry, which, by 1880, was beginning to take hold in southern California. Several flowing wells had been struck, and total production amounted to almost 200,000 barrels annually. Peckham hoped for "a fair return and a permanently profitable investment" on California crude.[163]

Despite the apparent vindication, Silliman never again worked in oil or in southern California. But he did continue to consult. During the early 1880s, he went to investigate gold and silver mines in the Arizona and New Mexico territories and presented papers on these engagements at meetings of the National Academy, the American Association for the Advancement of Science, and the newly founded American Institute of Mining Engineers.[164] His renewed energy, however, was cut short in October 1884 when he suffered a severe heart attack. Silliman died at home in New Haven the following January.

To the end, Silliman was a scientific entrepreneur, and the successes and failures of his career highlight the difficulties in defining the proper role of science in industry or in a commercial society in general. Capitalists were always eager to cash in on the credibility of his science. Hiring Silliman or any consultant was a strategy for winning the public's trust, and its money, of course. Conversely, men of science could gain credibility if the companies for whom they worked proved profitable. Whenever he was questioned, Silliman instinctively pointed to his rock oil report as evidence of the mutual benefits of scientifically informed industry. Whitney and Brewer, on the other hand, noted again and again the failure of the California oil companies as definitive proof of Silliman's corruption and the damage done by commercially influenced research. Empiricism was powerful evidence.

But it was (and is) dangerous to science and men of science to judge the success or failure of their consulting practice by its commercial profitability. To do so undermined all notions of disinterestedness, independence, and objectivity. The bedrock of science—the truthfulness of its claims—was the moral behavior of the man of science. Even the content of one's science was inextricably linked to the character of one's conduct. Credible knowledge was a matter of trust. And nineteenth-century Americans understood very well the precious balance between trust and credibility. Whitney was right to suspect getters-up of mining schemes and all those companies and capitalists looking to make money first and putting everything else second. The profit motive was not the proper motive of science.

Epilogue

Americanization of Science

[I]n considering the history of science in relation to civilization,
[we call] attention to the growth of the utilitarian spirit, which
is gradually substituting immediate, practical, wealth-yielding
studies for the more elevated, disinterested, and ennobling intellectual pursuits which have been cherished in past times. . . . This
influence [is] strengthening in Europe, but [it is] so predominat-
ing in this country that it is now generally known by the term
Americanization.
—*Popular Science Monthly* (1878)

HOW TO MEASURE CIVILIZATION? There were many ways: fine arts, literature, religion, government, education, industry. Not surprisingly, *Popular
Science Monthly* decided that "the best criterion of the position which a nation
has gained in the scale of civilization is the contributions which its men [of sci-
ence] have made toward the understanding and conquest of Nature." By this
standard, the United States ranked well below its European counterparts and,
what was more troubling, was falling further behind. "The science that gives
promise of immediate results, that can be turned into money, is appreciated,"
Popular Science Monthly lamented, "that which aims only at the extension of scientific truth wins little support." At best, then, Americanization meant an ill-
advised inversion of the practical over the theoretical; at worst, it portended a
neglect for truth and ultimately "the perversion and degradation of civilization
itself."[1]

For many post–Civil War Americans, their civilization, rife with graft, fraud,
and disillusionment, did seem degraded. Corruption bedeviled the two admin-
istrations of President Ulysses S. Grant and Republican governments at all lev-

els, with particularly brutal results in the South. In the Credit Mobilier scandal of 1873, to take one spectacular example, half a dozen prominent members of the House of Representatives were caught enjoying the profits plundered from government railroad contracts. Later that year the collapse of Jay Cooke and Company, the most respected bankers in the nation, brought down the Northern Pacific Railroad and brought on the Panic of 1873 and an economic depression from which the United States would not emerge until 1879. The debacle of the 1876 presidential election only deepened the despair.[2]

One root of these evils, to be sure, was the pursuit of the almighty dollar, "the vulgar passion of Americans for money," carped *Popular Science Monthly*. The cries of greed, bribery, and swindling were widespread and alarming. Josiah Whitney's excoriations of Benjamin Silliman's "exorbitant" consulting fees were of a piece with journalists' exposés of lawmakers on the take. "We are greatly stirred," Andrew D. White, the president of Cornell University, soberly explained "as this fraud or that scoundrel is dragged to light; and there rise cries and moans over the corruption of the times." Corruption, White believed, festered on indifference, "indifference to truth as truth," and materialism, "that struggle for place and pelf, [which] is the very opposite of the spirit that gives energy to scientific achievement."[3]

Jeremiads about American scientific achievement were very popular around the nation's centennial.[4] Simon Newcomb (1835–1909), superintendent of the Naval Observatory in Washington, DC, bemoaned "our contributions to the exact sciences [are] nearly zero." In accounting for this stunning lack of progress, Newcomb conflated exact science with theoretical science, a not uncommon tactic among mathematicians, physicists, and astronomers. For Newcomb, America was theoretically bankrupt.[5]

Joseph Henry, the senior statesman of American science, was not so dismissive of American contributions, especially in geology, but he was no less concerned about recent trends.[6] For Henry, the danger facing American science was not intellectual poverty but money.

> The man imbued with the proper spirit of science does not seek for immediate pecuniary reward from the practical applications of his discoveries, but derives sufficient gratification from his pursuit and the consciousness of enlarging the bounds of human contemplation, and the magnitude of human power, and leaves others to gather the golden fruit he may strew along his pathway.[7]

Henry, the idealist, conjured up an implausible scenario: men of science doing science at their leisure rather than as their livelihood. Henry, the realist, knew it had never ever been thus.

As president of the National Academy of Sciences, Henry had had to steer the members around the crisis over scientific ethics raised by the Silliman-Whitney controversy. In addition, he had had to guide the academy to a new policy on membership. The original restriction limiting the size of the academy to fifty members was removed, and after 1870 younger men of science were elected with more current interests, both scientific and commercial. One of these was the chemist Charles F. Chandler (1836–1925). Chandler held a PhD from Göttingen, but he often practiced his science as a consultant. During the 1860s, Chandler had twice been denied election because, according to the Harvard chemist Wolcott Gibbs, "he uses his science only as a means to make money."[8] Chandler finally became a member in 1874, the year of the Silliman-Whitney controversy.

In his final address to the National Academy in 1878, Henry took the chance to reflect on the state of American science. He reminded the members that the basis of their selection was original research, "positive additions to the sum of human knowledge." But he admonished them to keep in mind another criterion, "the purity of [the academy's] character." New members, Henry advised, had also to possess an "unimpeachable moral character." "Indeed, I think immorality and great mental power in the discovery of scientific truths are incompatible with one another."[9]

The emphasis on purity and the true spirit of American science were themes taken up and propounded most energetically by a younger member of the academy, Henry A. Rowland (1848–1901), elected in 1881 and the first professor of physics in the newly created Johns Hopkins University. In an address before the physics section of the American Association for the Advancement of Science in 1883, Rowland made his now famous "Plea for Pure Science." "American science," he began, "is a thing of the future, and not of the present or past." By science, Rowland meant physical science. He admitted that he knew nothing of natural science but thought his remarks would apply nonetheless.[10] Rowland blasted Americans for their mediocrity, a self-satisfaction borne of successful applications of science. Telegraphs, electric lights, and other conveniences brought wealth; and "[e]verybody can comprehend a million of money," he sniffed. But that was the problem; Americans confounded applications with intellectual advancement. "[H]ow few can comprehend any advance in scientific theory?"[11]

For many historians of American science, Rowland's plea marks a point of departure.[12] For the first time, distinguished "scientists" (Rowland's term of choice and the title increasingly employed by the 1880s)[13] began to speak publicly of "pure science," science for the sake of science. No longer would scientists need to "sell" themselves to the American public or to various levels of government on the basis of their usefulness or practicality.[14] Scientists could now focus on theory and basic research and thus begin to strive for parity with Europeans, especially in the exact sciences.[15] Emblematic of this pure-science ideology was the founding of the Johns Hopkins University. When it opened its doors in 1876, it marked the beginning of a new form of higher education, the graduate school. Universities would assume the leading roles in scientific research and the training of PhD scientists.[16] American science was launched by 1880.

The laudable, long-term consequences of the pure-science ideology have played a central role in historians' characterizations of the virtues and values of the new American scientist, the "profession" of the future.[17] What is lost or overlooked in these forward-looking celebrations are the underlying causes, the connections between Rowland's plea and "the corruption of the times." As Rowland made clear, pure science was meant as a reform. And reform was as much a part of the Gilded Age as was corruption. But the corruption that Rowland sought to reform was not American scientists' supposed indifference to basic research.[18] It was their interest in money. In other words, the opposite of pure science was not applied but impure science.

Rowland was quite certain that commercialization had corrupted American science. And the most obvious evidence of this was consulting.[19] "There are also those who have every facility for the pursuit of science . . . yet who devote themselves to commercial work, to testifying in courts of law, and to any other work to increase their present large income. Such men would be respectable if they gave up the name professor, and took that of consulting chemists."[20] Rowland railed against the waste of time and talent in the pursuit of objects he did not deem fit for "those in the chairs of professors." If professors, like Silliman, chose to consult, the reason was obvious, at least to Rowland (and to Whitney); they were interested only in "increasing their fortunes." Consultants were greedy, dishonorable, and disloyal to the true spirit of science. Scientists did not "have wealth before them." Indeed, they had "no such incentive to work." Scientists were the antithesis of lawyers and doctors, the old professionals who took fees for service from their clients and patients.[21] Scientists possessed the highest "moral qualities." "Commerce," Rowland explained, was "a curse to those with high ideals." Scientists' "duty" was to advance science, not to "earn their living" doing science. But scientists did need "a career worthy of their efforts."[22] That career, according to Rowland, would be found in the university, where scientists would do science. And to keep such science professors in their seats, Rowland, interestingly, proposed to combat the lure of filthy lucre with cold cash. To forgo "commercial science," pure scientists had to be paid "a suitable and respectable salary to live upon." In addition to money, they needed time. Pure-science professors "must not be overburdened" with teaching or else they could not afford to concentrate on research.[23]

Rowland's ideal professor was to be the pinnacle of a pyramid—the beacon of truth so to speak—below which might be found assistants (who took on teaching duties), poorer science teachers (who "do commercial scientific work under some circumstances"), and, at the base, those whose "mind possessed the necessary element of vulgarity" to make the applications of science.[24] Here was a division of labor and a hierarchy of knowledge designed to prevent the corruption of the pure by the impure and to preserve the rigid ranking of moral authority.

This graduated separation between science and application was taken one step further by Rowland's colleague Ira Remsen (1846–1927), the first professor of chemistry at Johns Hopkins. Remsen was also a proponent of pure science and an Academician, elected in 1882; but, unlike Rowland's physics, Remsen's science was deeply involved in industry. Nearly 90 percent of American chemists were employed outside the walls of academe. Industry, Remsen realized, was not distracting chemists from their scientific pursuits; it was practically determining their research agenda and professional identity.[25]

> We cannot pick up a paper . . . without finding that the renowned chemist Professor So-and-So has analyzed our product and found it to be the purest article in the market. In the same paper we are likely to find an equally strong statement in regard to some rival product and this is shown by the analysis of Professor Thus-and-Thus to be purer than anything of the kind known to man.[26]

For Remsen, the problem was not consulting per se, but its promiscuous practice. Professors "So-and-So" and "Thus-and-Thus" were examples of "reckless recommending." There was so much demand for chemistry by industry that neither scientists nor industrialists could restrain themselves. As in the oil boom, the market for experts, any kind of chemical expert, was getting out of control.

> The country is overrun with . . . incompletely trained men, who are trying in every way possible to make a living out of chemistry. Now when a manufacturer wants a chemist, he will generally take as cheap a one as he can get, as generally he cannot distinguish between a chemist proper and a chemist improper, if I may be allowed the expression.[27]

Germany was Remsen's model, where "[t]he chief chemists in the principal chemical factories are university men." American universities needed to "manufacture" large numbers of good-quality, industrial chemists. Good quality meant proper training, which, of course, meant "a thorough grounding in pure chemistry."[28] Remsen held out a plan for science-based industry, where the high values of science would uplift the chemical factories. In effect, Remsen's plan would allow pure professors to concentrate on science by serving up their grade-A students to the maw of industry.[29]

In the rarefied world of academe, pure science was the highest ideal precisely because it was nonpecuniary. By definition, then, such scientists were incorruptible. Oddly enough, such an un-American value could survive only in a well-endowed university, like Johns Hopkins. Modeled on the German variety, which Rowland had studied firsthand,[30] the new research university was literally an ivory tower, untainted by commercial work and a fortress against the pressures of profit. Here, then, was an institutional solution to an individual moral problem. Where men of science are weak, the walls of universities had to be strong.

From a historical viewpoint, the appeal of Rowland's and Remsen's reforms lay in their target, the university, where, not surprisingly, most historians also work. In this rendering, pure science (or pure history) is preserved in a very special not-for-profit place. The university was still connected to the outside world in a very practical way through its students and in a powerfully theoretical way through the imagined conveyer belt that ran straight from basic to applied.[31] The pure-science ideology tapped into the old assumption that knowledge was useful, only now it need not be immediately so. In time, the purists argued, applications and inventions would trickle down from the ivory tower.

The rise of the pure-science ideology in the new research university would thus seem to be the antidote to Americanization. As a conclusion to this book, it would suffice, but it would be incomplete, for at least two reasons. First, it misses some of the larger concerns among American scientists about corruption and the wider reforms required to confront it. And, second, it disregards the positive aspects of Americanization and therefore cannot explain why or how "immediate, practical, wealth-yielding studies" persisted. In short, not all scientists were pure professors in research universities. This was especially true for geologists.

In nineteenth-century America, geology was the one science that measured up to European standards.[32] Theoretical and practical, basic and economic, geologists had made contributions both at home and abroad. Geology thus rarely came up in criticisms of Americanization or in visions of pure science. Moreover, its principal social location—the survey—made geology distinctive. Geologists, to be sure, were keenly aware of the "corruption of the times," but retreating to Rowland's "mountain"—the university of high ideals—was neither practical nor possible. In the Gilded Age, surveys had to take on new roles and new meanings.

The two big state surveys of the post–Civil War era, Ohio and Pennsylvania, were both products of the oil industry. In 1869 the Ohio legislature created a survey, and the governor appointed John Strong Newberry as chief geologist and E. B. Andrews, John H. Klippart, and Edward Orton as assistants. All had worked at one time or another as consultants for oil companies. In language common to all antebellum surveys, the Ohio geologists were instructed to pursue scientific examinations and, simultaneously, to evaluate useful materials, especially petroleum, for economic purposes. What was new about the second Ohio survey, though, was the reason for its establishment.

> During the 30 years that had now elapsed since the suspension of the first survey the resources of the State had slowly been developing. . . . Information gained from private experience was monopolized by those who paid for it. Instead of being used to inform the landowner as to the mineral wealth underlying his possessions, such knowledge was made subservient to the speculator.[33]

Surveys had always had a public function, but now the public needed to be protected against the private use of knowledge. Moreover, surveys, which also had always been organized to bring science to industry, now needed to be enacted to prevent industry from monopolizing science.

The idea of using a survey to counteract industry's creeping secrecy was also advanced in Pennsylvania. "In the . . . 20 years between 1854 and 1874," J. Peter Lesley wrote, "a [great] change took place." Petroleum was discovered, a Civil War was fought, the demand for coal intensified, and a "multitude of private surveys took place in all parts of the State." But many consulting reports were kept confidential. "Business refused to give away its valuable secrets." Such behavior was understandable for business but bad for science and worse for the public. Lesley lobbied for a new survey, "not so much for the discovery of the unknown as for making known to the public discoveries which in a multitude of private hands awaited publication."[34]

Significantly, Lesley's argument was rejected by "the three . . . great mineral industries"—anthracite, coal, and iron, which "had their own geological advisors," who "understood" the strata and "desired no interference."[35] Lesley's plea for public science did find favor among oil men, who in 1873 were in a muddle of confusion and "in a state of highest excitement" because of the recent discovery of the Bradford field along the border with New York. The oil interests wanted "the State legislature [to] provide for a scientific examination of the [oil] phenomenon."[36] Their clamor, along with the support of counties in "mineral poverty," resulted in the creation in 1874 of the second geological survey of Pennsylvania with Lesley as its director. The survey lasted sixteen years, employed nearly ninety scientists, and published sixty-seven volumes, all at a cost of roughly $700,000. It was the largest state survey of the nineteenth century.

But it was not the largest geological survey. That honor was reserved to the federal government. In 1879 Congress consolidated the four surveys working on the trans-Mississippi West (those headed by Clarence King, Ferdinand V. Hayden, John Wesley Powell, and Lieutenant George M. Wheeler) into the United States Geological Survey (USGS) under the direction of King. In its organic act, Congress required the USGS to examine the geological structure and mineral resources of the national domain, a directive consistent with all state surveys. In addition, Congress laid down the guidelines for ethical behavior of USGS employees. "[T]he Director and members of the Geological Survey shall have no personal or private interests in the lands or mineral wealth of the region under survey, and shall execute no surveys or examinations for private parties or corporations."[37] Here was a clear statement about conflict of interest designed to protect the scientists' research as much as the public's interests. Ironically, King would leave the USGS in 1881 to pursue a professional career as a consulting geologist.[38]

State surveys had on occasion incorporated similar ethical guidelines, but never so emphatically. The second Ohio and Pennsylvania surveys, for instance, had no such language in their enabling acts, although the Ohio legislature did stipulate the appointment of a chief geologist "of known integrity."[39] Lesley, in his capacity to employ survey assistants, made clear what was "expected" of them. They were "to have no private professional business within the limits of the State." Outside Pennsylvania, they could work "in some professionally profitable way." But Lesley's surveyors were enjoined from accepting fees "from any capitalist or company for taking up one line of survey in preference to another or out of its proper order."[40] Public and private science did not mix.

As bulwarks against the invasion of industry, the new geological surveys had much in common with Rowland's research university. As institutions for the promotion and preservation of ethical behavior, they resembled Henry's picture of the National Academy. Geology thus gained a new look as part of the reforms to American science in the Gilded Age; but surveys also lost something, their educational role. Antebellum surveys had been the main training ground for geologists and other men of science. The surveys supplied the experts to all those interested in capitalizing on the land and its resources. In the post–Civil War era, new institutions arose to fill that growing demand, in particular the Columbia School of Mines.

Founded in 1864 in New York City, the Columbia School of Mines provided a three-year education to aspiring mining engineers. Much of that education was provided by former consultants. In 1866 Newberry became professor of geology and paleontology at the School of Mines (a position he held concurrently with his Ohio survey position), and Chandler, the prominent consultant, became professor of chemistry. That same year, Brewer wrote to Whitney in California that the School of Mines was "creating a great sensation." By 1873 it enrolled more students than did Columbia College.[41]

The other leading school for the training of engineers was the Massachusetts Institute of Technology (MIT), founded by William Barton Rogers, former state geologist of Virginia and Henry Rogers's older brother; William also became president of the National Academy of Sciences upon Joseph Henry's death in 1878. At MIT, Frank Storer, the controversial oil analyst, became professor of chemistry. T. Sterry Hunt, the chief advocate of the anticlinal theory of petroleum accumulation, was appointed professor of geology. And Cyrus Warren, who had backed up Benjamin Silliman on the California oil sample, became the first professor of organic chemistry. Other members of the first generation of consultants also took academic positions. Lesley became professor of geology and mining at the University of Pennsylvania in 1872, and three years later he assumed the post of dean of the Towne Scientific School, which he held until 1883.

The exception that makes the point about the influence of consulting in academe was Whitney. In 1865 he accepted a prestigious chair in geology at Harvard University and organized its School of Mining and Practical Geology. The Harvard school proved to be a complete failure. In the era of increasing demand for mining engineers, Whitney could not attract students. It is tempting to put down Whitney's failure to his personal hostility toward capitalists and their companies. In Whitney's case, the pure science of geology and the business of mining were incompatible.[42]

In general, though, the flourishing of a new specialized occupation called mining engineering and of the new schools where its practitioners were taught showed that theory and practice could thrive together.[43] And so too could the scientists, engineers, and businessmen. In 1871 the American Institute for Mining Engineers (AIME) was founded in Wilkes-Barre, Pennsylvania, in the heart of the anthracite district. The AIME was the first fully functioning society for "professional" engineers. (The American Society of Civil Engineers [ASCE] was founded in 1867, but it did not begin to operate regularly until the early 1870s.)[44] AIME "members" were identified as "professional mining engineers, geologists, metallurgists or chemists." "Associates" were superintendents, managers, and businessmen, "desirous of being connected with the Institute." All were assumed to be interested in "the advancement of professional knowledge." At the opening meeting, one founder announced: "The time has come when scientific research is to assume its true position [in mining]." In the first volume of its *Transactions,* the AIME published articles by Hunt, Silliman, William Blake, and Benjamin Smith Lyman. In 1877 Hunt was elected president of the AIME.[45]

The AIME was one of the so-called founder societies along with the ASCE, the American Society of Mechanical Engineers (est. 1880) and the American Institute of Electrical Engineers (est. 1884). For many historians of American technology, the establishment of these societies and the many engineering schools marks their discipline's point of departure.[46] Civil, mining, mechanical, and electrical engineering moved from the shop culture of apprentices, journeymen, and masters to a school culture of technical education, industrial jobs, and managerial careers. Science and technology began to converge as engineers brought knowledge to bear on the practice and organization of business. Professional engineers thus occupied the central position between science and industry.

In many ways, the new engineer embodied the old consultant. Both were professionals who identified themselves by their expertise and experience and disinterestedness.[47] In the case of mining, both engineering and consulting had roots in the geological surveys (going all the way back to the advertisement by William W. Mather and James Hall) and grew out of the intersection of scientific research and industrial problems. Gilded Age engineering represented the formalization and institutionalization of the informal and individualized pattern of antebellum consulting. It is no small matter of semantic coincidence that the ideal of the professional engineer is occupational independence—to be a consultant.

Consulting *scientists* did not, by any stretch of the imagination, disappear in late nineteenth-century America. "It thus frequently happens that the man of science is consulted on all matters of a scientific nature," observed the physicist Thomas Mendenhall in 1891, "[e]xamples of this condition are by no means wanting, and they are not confined to, as might first be assumed, the lower ranks of science."[48] Industry still came looking for expertise from the most distinguished professors and best surveyors. Only now professional engineers were filling the rapidly multiplying roles created by the demands of industry. For Lesley, this was a welcome development. The quantity and quality of the new mining engineers, he explained in 1886, "exercised an important influence upon the sentiment of the [people] toward geology as an applied science."[49]

Lesley's synonym signals one final aspect in the reform of American science during the Gilded Age—the emergence of an applied-science ideology. Its chief advocate was the mechanical engineer and first president of the ASME and dean of Cornell University's Silbey College of Engineering, Robert H. Thurston (1839–1903). In an address before the mechanical science section of the American Association for the Advancement of Science in 1884, the year after Rowland's plea, Thurston propounded "The Mission of Science." Applied science, he explained, embraced "the cultivation of science, in its relations to every branch of the arts, and to every department of the industries of the world." Thurston envisioned a partnership between science and industry—equal, harmonious, and fruitful. The steam engine, the electric telegraph, and the railroad *were* the measures of "the real progress of science." As energetic and dogmatic as Rowland, Thurston preached a sort of sci-tech evangelicalism and anointed applied science "the distinguishing characteristic of the age."[50]

Late nineteenth-century America was the age of applied science *and* pure science. They were both reforms—determined efforts to manage the relations between science and industry. The pure-science ideology was an expression of pessimism, even fear. It proposed to keep the two in separate spheres. Science, for its part, was to be pursued in not-for-profit places like universities and government surveys. Otherwise, it would be transformed from the handmaiden to the prostitute of industry, and engineers would become corporate tools. Truth needed to be protected from capitalist greed.[51] Those were the lessons to learn from the Silliman-Whitney controversy.

Applied science bespoke of an optimism about the ability of scientists and engineers to work with industry without losing their identities or integrity or purpose. Geologists had always prided themselves on their ability to balance the scientific standards of excellence with industrial (and larger societal) demands for usefulness. Applied science captured the new sense of profitable industrial engagement. It, too, was a lesson drawn from the Silliman-Whitney controversy. In the decisive meeting of the council of the National Academy, William Brewer virtually shouted out why scientists needed to be engaged with industry: "*cool and cautious science should have exercised a conservative and protecting influence;* such are its claims, and this the public reasonably expected."[52] Silliman, as usual, described in more measured tones how science and industry should interact.

> [The "Scientific Expert"] is, if fit for his responsible duty, a neutral in the case, viewing it, as nearly as human judgment permits, in a perfectly impartial aspect. In this manner it is possible for the man of real science, and skilled in his art, to confer great benefit, not only upon the parties in interest, but upon the subject; while science itself often gains by the researches which are called out by these contests upon points which have before been overlooked or imperfectly investigated.[53]

Reciprocity was the ideal. Industry gained trustworthy advice; scientists got to work on interesting problems. The public benefited from material progress and the advancement of knowledge. For many (perhaps most) Gilded Age Americans, the moral and material efficacy of science could be united only by the active engagement of science in the industrial society. Engineering would become the single largest profession in the United States.[54] Such engagement, however, was not without potential as well as real ethical problems. Scientists and engineers were not blind to "the corruption of the times," whatever times they may be. The question for them was not *whether* science can engage with industrial capitalism, but *how*.[55]

The Americanization of science is thus Janus-faced. On one side, it speaks the warning of pure science, the sacrifice of truth to gain, and the degradation of civilization. On the flip side, it smiles on applied science and engineering, new knowledge and technologies, and the progress of an industrializing world. Either way, Americanization marked a new era in the relations of science, technology, and industry. The coin of consulting has thus been well circulated.

Notes

Abbreviations

AAAS	American Association for the Advancement of Science
AG-LJ	*American Gas-Light Journal*
AJS	*American Journal of Science*
APS	American Philosophical Society, Philadelphia
BJHS	*British Journal for the History of Science*
BSNH	Boston Society of Natural History
Giddens, *Documents*	Paul H. Giddens, ed., *Pennsylvania Petroleum, 1750–1872: A Documentary History* (Titusville: Pennsylvania Historical and Museum Commission, 1947)
Giddens, *Sources*	Paul H. Giddens, ed., *The Beginnings of the Petroleum Industry: Sources and Bibliography* (Harrisburg: Pennsylvania Historical Commission, 1941)
GSL	Geological Society of London
Merrill, "Reminiscences"	Joshua Merrill, "Joshua Merrill," in *Derrick's Hand-Book of Petroleum: A Complete Chronological and Statistical Review of Petroleum Developments from 1859 to 1899*, 2 vols. (Oil City, PA: Derrick Publishing Company, 1898)
NYSA	New York State Archives, Albany
NYSL	New York State Library, Albany
SLB, LR	Secretary's Letter Books, Letters Received
SUA	Strathclyde University Archives, Glasgow, Scotland
S-W NAS	"Silliman-Whitney Controversy" file, National Academy of Sciences, Washington, DC
T&C	*Technology & Culture*
YUL	Yale University Library, New Haven, Connecticut

Introduction

1. For all its importance to the pursuit of science, money has not been a central topic of concern to historians of science. As the inimitable Roy Porter noted, the exclusion of "such a vulgar subject" might reflect the "gentlemanly" breeding of historians of sci-

ence more than the absence of monetary interests. Porter, "Gentlemen and Geology: The Emergence of a Scientific Career, 1660–1920," *Historical Journal*, 21 (1978):809–836, 832 n. 66. Howard S. Miller, *Dollars for Research: Science and Its Patrons in Nineteenth-Century America* (Seattle: University of Washington Press, 1970); W. H. Brock, "The Spectrum of Scientific Patronage," in G. L. E. Turner, ed., *The Patronage of Science in the Nineteenth Century* (Leiden: Noordoff International, 1976), 173–206. By contrast, for students of modern science and science policy makers, money is of supreme importance. David S. Greenberg, *Science, Money, and Politics: Political Triumph and Ethical Erosion* (Chicago: University of Chicago Press, 2001).

2. Margaret W. Rossiter, *Women Scientists in America: Struggles and Strategies to 1940* (Baltimore: Johns Hopkins University Press, 1984).

3. George P. Merrill, *The First Hundred Years of American Geology* (New Haven: Yale University Press, 1924); A. Hunter Dupree, *Science in the Federal Government* (reprint, Baltimore: Johns Hopkins University Press, 1987); Michele L. Aldrich, *New York State Natural History Survey, 1836–1842: A Chapter in the History of American Science* (Ithaca, NY: Paleontological Research Institution, 2000); Patsy Gerstner, *Henry Darwin Rogers, 1808–1866: American Geologist* (Tuscaloosa: University of Alabama Press, 1994); Benjamin R. Cohen, "Surveying Nature: Environmental Dimensions of Virginia's First Scientific Survey, 1835–1842," *Environmental History*, 11 (2006):37–69; Walter B. Hendrickson, "Nineteenth-Century State Geological Surveys: Early Government Support of Science," *Isis*, 52 (1961):357–371; Stephen P. Turner, "The Survey in Nineteenth-Century American Geology: The Evolution of a Form of Patronage," *Minerva*, 25 (1987):282–330; Anne Marie Millbrooke, "State Geological Surveys in the Nineteenth Century" (PhD dissertation, University of Pennsylvania, 1981); Morris Zaslow, *Reading the Rocks: The Story of the Geological Survey of Canada, 1842–1972* (Toronto: Macmillan, 1975); Suzanne Zeller, *Inventing Canada: Early Victorian Science and the Idea of a Transcontinental Nation* (Toronto: University of Toronto Press, 1987); William E. Eagan, "The Canadian Geological Survey: Hinterland between Two Metropolises," *Earth Sciences History*, 12 (1993):99–106; Sally Gregory Kohlstedt, *The Formation of the American Scientific Community: The American Association for the Advancement of Science, 1848–1860* (Urbana: University of Illinois Press, 1976).

4. Professionalization can be characterized by learning, licensing/legitimation, and livelihood (usually employment in universities, government agencies, or corporate R&D labs). See Charles Coulston Gillispie, *Science and Polity in France at the End of the Old Regime* (Princeton: Princeton University Press, 1980), 84–89, 549–551; Samuel Haber, *The Quest for Authority and Honor in the American Professions, 1750–1900* (Chicago: University of Chicago Press, 1991); Nathan O. Hatch, ed., *The Professions in American History* (Notre Dame: University of Notre Dame Press, 1988); Thomas L. Haskell, ed., *The Authority of Experts: Studies in History and Theory* (Bloomington: Indiana University Press, 1984); Gerald Geison, ed., *Professions and Professional Ideology in America* (Philadelphia: University of Pennsylvania Press, 1983); Andrew Abbott, *The System of Professions: An Essay on the Division of Expert Labor* (Chicago: University of Chicago Press, 1988); Joseph Ben-David, *The Scientist's Role in Society: A Comparative Study* (Chicago: University of Chicago Press, 1984); Magali Sarfatti Larson, *The Rise of Professionalism: A Sociological Analysis* (Berkeley: University of California Press, 1977); Burton J. Bledstein, *The Culture of Professionalism* (New York: W. W. Norton, 1976); George H. Daniels, "The Process of Professionalization in American Science: The Emergent Period, 1820–1860," *Isis*, 58 (1967):151–166; Nathan Reingold, "Definitions and

Speculations: The Professionalization of Science in America in the Nineteenth Century," in Reingold, *Science, American Style* (New Brunswick: Rutgers, 1991), 24–53.

5. Such "commercial science" was not undertaken by scientific gentlemen. Aileen Fyfe, "Conscientious Workmen or Booksellers' Hacks? The Professional Identities of Science Writers in the Mid-Nineteenth Century," *Isis, 96* (2005):192–223; James A. Secord, *Victorian Sensation: The Extraordinary Publication, Reception, and Secret Authorship of Vestiges of the Natural History of Creation* (Chicago: University of Chicago Press, 2000), 437–439; Adrian Desmond, *Huxley: The Devil's Disciple* (London: Michael Joseph, 1994).

6. Cited in Hugh Richard Slotten, *Patronage, Practice, and the Culture of American Science: Alexander Dallas Bache and the U.S. Coast Survey* (Cambridge: Cambridge University Press, 1994), 33.

7. Jack Morrell and Arnold Thackray, *Gentlemen of Science: Early Years of the British Association for the Advancement of Science* (Oxford: Oxford University Press, 1982).

8. Robert V. Bruce, "A Statistical Profile of American Scientists, 1846–1876," in George H. Daniels, ed., *Nineteenth-Century American Science: A Reappraisal* (Evanston: Northwestern University Press, 1972), 63–94; Clark A. Elliott, "Models of the American Scientist: A Look at Collective Biography," *Isis, 73* (1982):77–93.

9. The flier was distributed both privately, as a handbill, and publicly, in the Albany newspapers and in the *American Journal of Science*. Hall Mss, box 39: Mining Reports, NYSL.

10. Men of science usually identified themselves as "consulting chemists/geologists," or "professional chemists/geologists," or sometimes "mining engineers." They did not call themselves consulting *scientists* and rarely "consultants"; nor was the term scientific consulting used. My use of the term consultant thus indicates more of an analytical and descriptive category than a strictly historical one.

11. On scientific entrepreneurship, see Charles E. Rosenberg, *No Other Gods: On Science and American Social Thought* (Baltimore: Johns Hopkins University Press, 1976), 135–152, 153–172; Crosbie Smith and M. Norton Wise, *Energy and Empire: A Biographical Study of Lord Kelvin* (Cambridge: Cambridge University Press, 1989); Carroll Pursell, "Science and Industry," in George H. Daniels, ed., *Nineteenth-Century American Science: A Reappraisal* (Evanston: Northwestern University Press, 1972), 231–248.

12. The practice has not gone unnoticed, but there is no systematic study of it. Paul Lucier, "Commercial Interests and Scientific Disinterestedness: Consulting Geologists in Antebellum America," *Isis, 86* (1995):245–267; Julie Renee Newell, "American Geologists and Their Geology: The Formation of the American Geological Community, 1780–1865" (PhD dissertation, University of Wisconsin at Madison, 1993); Robert V. Bruce, *The Launching of Modern American Science, 1846–1876* (Ithaca: Cornell University Press, 1988); Elliott, "Models"; Carroll Pursell, "Science and Industry," in Daniels, *Nineteenth-Century American Science*, 231–248. For Britain, see Jack Morrell, *John Phillips and the Business of Victorian Science* (Aldershot: Ashgate, 2005); Colin A. Russell, *Edward Frankland: Chemistry, Controversy, and Conspiracy in Victorian England* (Cambridge: Cambridge University Press, 1996); Katherine D. Watson, "The Chemist as Expert: The Consulting Career of Sir William Ramsay," *Ambix, 42* (1995):143–159; Geoffrey Tweedale, "Geology and Industrial Consultancy: Sir William Boyd Dawkins (1837–1929) and the Kent Coalfield," *BJHS, 24* (1991):435–451; Robert F. Bud and Gerrylynn K. Roberts, *Science versus Practice: Chemistry in Victorian Britain* (Manchester: Manchester University Press, 1984); Robert H. Kargon, *Science in Victorian Manchester:*

Enterprise and Experience (Baltimore: Johns Hopkins University Press, 1977); Hugh S. Torrens and William R. Brice, "James Buckman (1814–1884) English Consulting Geologist and His Visit to the Guyandotte Coal-Fields in 1854," *Southeastern Geology, 38* (1999):191–201.

13. Slotten, *Patronage, Practice, and the Culture of American Science;* Bruce, *Launching of Modern American Science;* Miller, *Dollars for Research.* Cf. Mario Biagioli, *Galileo, Courtier: The Practice of Science in the Culture of Absolutism* (Chicago: University of Chicago Press, 1993); Paula Findlen, *Possessing Nature: Museums and Collecting in Early Modern Italy* (New Haven: Yale University Press, 1994); Pamela H. Smith, *The Business of Alchemy: Science and Culture in the Holy Roman Empire* (Princeton: Princeton University Press, 1994); E. C. Spary, *Utopia's Garden: French Natural History from the Old Regime to Revolution* (Chicago: University of Chicago Press, 2000).

14. Hugh Torrens, "Patronage and Problems: Banks and the Earth Sciences," in R. E. R. Banks et al., eds., *Sir Joseph Banks: A Global Perspective* (Kew: Royal Botanic Garden, 1994), 49–75; Torrens, "Arthur Aiken's Mineralogical Survey of Shropshire, 1796–1816, and the Contemporary Audience for Geological Publications," *BJHS, 16* (1983):111–153; Jack Morrell, "Economic and Ornamental Geology: The Geological and Polytechnic Society of the West Riding of Yorkshire, 1837–53," in Ian Inkster and Jack Morrell, eds., *Metropolis and Province: Science in British Culture, 1780–1850* (Philadelphia: University of Pennsylvania Press, 1983), 231–256; Roy Porter, *The Making of Geology: Earth Science in Britain, 1660–1815* (Cambridge: Cambridge University Press, 1977); Porter, "Gentlemen and Geology." On France, see Gillispie, *Science and Polity;* on Germany, see Abraham Gottlob Werner, *Short Classification and Description of the Various Rocks,* trans. with an introduction and notes by Alexander M. Ospovat (New York: Hafner, 1971). Cf. Morrell, *John Phillips and the Business of Victorian Science.* Phillips did not object to consultancies, but he did not do many because he was very busy with other activities.

15. Bruce, *Launching of Modern American Science,* 98.

16. See *The Papers of Joseph Henry,* vols. 1–5, ed. Nathan Reingold (Washington, DC: Smithsonian Institution, 1972–1985), and vols. 6–11, ed. Marc Rothenberg (Washington, DC: Smithsonian Institution, 1992–2007); Daniel J. Kevles, *The Physicists: The History of a Scientific Community in Modern America* (New York: Vintage, 1978).

17. Chandos Michael Brown, *Benjamin Silliman: A Life in the Young Republic* (Princeton: Princeton University Press, 1989), esp. ch. 6.

18. John Lauritz Larson, *Internal Improvement: National Public Works and the Promise of Popular Government in the Early United States* (Chapel Hill: University of North Carolina Press, 2001). Larson, however, had nothing to say about men of science.

19. Tal Golan, *Laws of Men and Laws of Nature: The History of Scientific Expert Testimony in England and America* (Cambridge, MA: Harvard University Press, 2004); Golan, ed., "Law and Science," special issue, *Science in Context, 12* (1999); Carolyn C. Cooper, ed., "Patents and Invention," special issue, *T&C, 32* (1991).

20. Cf. Walter Licht, *Industrializing America: The Nineteenth Century* (Baltimore: Johns Hopkins University Press, 1995); George Rogers Taylor, *The Transportation Revolution, 1815–1860* (New York: Holt, Rinehart, and Winston, 1951); Charles Sellers, *The Market Revolution: Jacksonian America, 1815–1846* (New York: Oxford University Press, 1991); Stuart Bruchey, *Enterprise: The Dynamic Economy of a Free People* (Cambridge, MA: Harvard University Press, 1990).

21. Cf. Bruce, *Launching of Modern American Science.*

22. The best overview of these relations remains John M. Staudenmaier, *Technology's Storytellers: Reweaving the Human Fabric* (Cambridge, MA: MIT Press, 1985), esp. ch. 3.

23. On the substantial contributions of science, see Margaret C. Jacob, *Scientific Culture and the Making of the Industrial West* (New York: Oxford University Press, 1997); Larry Stewart, *The Rise of Public Science: Rhetoric, Technology, and Natural Philosophy in Newtonian Britain, 1660–1750* (Cambridge: Cambridge University Press, 1992); Barbara Whitney Keyser, "Between Science and Craft: The Case of Berthollet and Dyeing," *Annals of Science, 47* (1990):213–260; Carleton Perrin, "Of Theory Shifts and Industrial Innovations: The Relations of J. A. C. Chaptal and A. L. Lavoisier," *Annals of Science, 43* (1986):511–542; Neil McKendrick, "The Rôle of Science in the Industrial Revolution: A Study of Josiah Wedgwood as a Scientist and Industrial Chemist," in Mikulás Teich and Robert Young, eds., *Changing Perspectives in the History of Science: Essays in Honour of Joseph Needham* (London: Heinemann, 1973), 274–319; A. E. Musson and Eric Robinson, *Science and Technology in the Industrial Revolution* (Manchester: Manchester University Press, 1969). For minimalist contributions, see Charles Coulston Gillispie, "The Natural History of Industry," *Isis, 48* (1957):398–407; A. Rupert Hall, "What Did the Industrial Revolution in Britain Owe to Science?" in Neil McKendrick, ed., *Historical Perspectives: Studies in English Thought and Society in Honour of J. H. Plumb* (London: Europa, 1974), 129–151; John J. Beer, "Eighteenth-Century Theories on the Process of Dyeing," *Isis, 51* (1960):21–30.

24. George Meyer-Thurow, "The Industrialization of Invention: A Case Study from the German Chemical Industry," *Isis, 73* (1982):363–381; John J. Beer, *The Emergence of the German Dye Industry* (Urbana: University of Illinois Press, 1959); L. F. Haber, *The Chemical Industry during the Nineteenth Century: A Study of the Economic Aspects of Applied Chemistry in Europe and North America* (Oxford: Oxford University Press, 1958); George Wise, *Willis R. Whitney, General Electric, and the Origins of U.S. Industrial Research* (New York: Columbia University Press, 1985); Leonard S. Reich, *The Making of American Industrial Research: Science and Business at GE and Bell, 1876–1926* (Cambridge: Cambridge University Press, 1985); Harold C. Passer, *The Electrical Manufacturers, 1875–1900: A Study in Competition, Entrepreneurship, Technical Change, and Economic Growth* (Cambridge, MA: Harvard University Press, 1953); David S. Landes, *The Unbound Prometheus: Technological Change and Industrial Development in Western Europe from 1750 to the Present* (Cambridge: Cambridge University Press, 1969).

25. Lesley to John A. Lowell, 20 May 1864, Lesley Mss, APS.

26. On practical science, see Smith and Wise, *Energy and Empire;* and Margaret W. Rossiter, *The Emergence of Agricultural Science: Justus Liebig and the Americans, 1840–1860* (New Haven: Yale University Press, 1975). On the debilitating effects of practical science, see Bruce, *Launching of Modern American Science;* and George H. Daniels, *American Science in the Age of Jackson* (New York: Columbia University Press, 1968).

27. Alexis de Tocqueville, *Democracy in America,* ed. J. P. Mayer, trans. George Lawrence (New York: Anchor Books, Doubleday, 1967), vol., 2, book 1, chs. 9 and 10; Richard Harrison Shryock, "American Indifference to Basic Science during the Nineteenth Century," *Archives Internationales d'Histoire des Science, 28* (1948):3–18. Nathan Reingold, "American Indifference to Basic Research: A Reappraisal," in Daniels, *Nineteenth-Century American Science,* 38–62; Aldrich, *New York State Natural History Survey.*

28. Curiously, coal has not been studied much by historians of geology, who have

been much more interested in older rocks. James A. Secord, *Controversy in Victorian Geology: The Cambrian-Silurian Dispute* (Princeton: Princeton University Press, 1986); and Martin J. S. Rudwick, *The Great Devonian Controversy: The Shaping of Scientific Knowledge among Gentlemanly Specialists* (Chicago: University of Chicago Press, 1985).

29. The only study of the history of petroleum geology is Edgar Wesley Owen, *Trek of the Oil Finders: A History of Exploration for Petroleum* (Tulsa: American Association of Petroleum Geologists, 1975).

30. On systems, see Thomas Hughes, *Networks of Power: Electrification in Western Society, 1880–1930* (Baltimore: Johns Hopkins University Press, 1983).

31. Cf. Harold F. Williamson and Arnold R. Daum, *The American Petroleum Industry: The Age of Illumination, 1859–1899* (Evanston: Northwestern University Press, 1959); Paul H. Giddens, *Birth of the Oil Industry* (New York: Macmillan, 1938).

32. Michel Foucault, *The Order of Things: An Archeology of the Human Sciences* (New York: Vintage, 1973); Rachel Laudan, *From Mineralogy to Geology: The Foundations of a Science, 1650–1830* (Chicago: University of Chicago Press, 1987); Geoffrey C. Bowker and Susan Leigh Star, *Sorting Things Out: Classification and Its Consequences* (Cambridge, MA: MIT, 1999); John V. Pickstone, *Ways of Knowing: A New History of Science, Technology, and Medicine* (Chicago: University of Chicago Press, 2001).

Chapter 1. Geological Enterprise

1. The other major Nova Scotia coal fields were at Pictou on the Northumberland Strait and at Sydney on the northern shore of Cape Breton. In the early nineteenth century, Cape Breton was often treated separately. It was an island colony from 1783 (the partition of Nova Scotia at the end of the American Revolution) until 1820 (its political reattachment to Nova Scotia). The first detailed account of Nova Scotia's coal was Richard Brown, "Geology and Mineralogy," in Thomas C. Haliburton, ed., *An Historical and Statistical Account of Nova Scotia*, 2 vols. (Halifax: Joseph Howe, 1829), 2:414–453.

2. Suzanne Zeller, *Inventing Canada: Early Victorian Science and the Idea of a Transcontinental Nation* (Toronto: University of Toronto Press, 1987); Morris Zaslow, *Reading the Rocks: The Story of the Geological Survey of Canada, 1842–1972* (Toronto: Macmillan, 1975); Robert A. Stafford, *Scientist of Empire: Sir Roderick Murchison, Scientific Exploration, and Victorian Imperialism* (Cambridge: Cambridge University Press, 1989).

3. Abraham Gesner, *Fourth Report on the Geological Survey of the Province of New-Brunswick* (Saint John: Henry Chubb, 1842), 64.

4. During the American Revolution, Henry Gesner served with the King's Orange Rangers. The British government granted him a 400-acre tract in Cornwallis Valley, where he settled in 1786. Sarah Pineo was descended from French Huguenots. Anton Temple Gesner, *The Gesner Family of New York and Nova Scotia, Together with Some Notes Concerning the Families of Bogardus, Brower, Ferdon, and Pineo* (Middletown, CT: Pelton & King, 1912).

5. George W. Gesner, "Dr. Abraham Gesner, A Biographical Sketch," *Bulletin of the Natural History Society of New Brunswick, 14* (1896):3–11, 4; Joyce Barkhouse, *Abraham Gesner* (Don Mills, ON: Fitzhenry & Whiteside, 1980).

6. Abraham and Harriet Gesner's first son was Henry. They had five other sons:

George, Abraham Jr., Brower, Frederick, and Conrad. Two girls, Harriet and Elizabeth, along with another boy, Robert, died before reaching maturity.

7. Gesner's close friend and brother-in-law, William Bennet Webster (1798–1861), also practiced medicine (MD, University of Edinburgh) and took up geology around Cape Blomidon. Charles Lyell, "Notes on Some Recent Foot-Prints on Red Mud in Nova Scotia, Collected by W. B. Webster of Kentville," *Quarterly Journal of the Geological Society of London*, 5 (1849):344.

8. The reconnaissance was serialized in the *American Journal of Science*. Charles T. Jackson and Francis Alger, "A Description of the Mineralogy and Geology of a Part of Nova Scotia," *AJS*, *14* (1828):305–330; *15* (1829):132–160, 201–217. After visiting Gesner, their revised account appeared as "Remarks on the Mineralogy and Geology of Nova Scotia," *Memoirs of the American Academy of Arts and Sciences*, *1* (1832):217–330, and reprinted as *Remarks on the Mineralogy and Geology of the Peninsula of Nova Scotia, Accompanied by a Colored Map, Illustrative of the Structure of the Country, and by Several Views of its Scenery* (Cambridge, MA: E. W. Metcalf, 1832).

9. Jackson and Alger, "Mineralogy and Geology," *AJS*, *15*, 136.

10. Jackson and Alger, "Mineralogy and Geology," *AJS*, *15*, 136.

11. They even apologized for their geology; "we have perhaps exceeded the limits within which . . . we should have confined ourselves." Jackson and Alger, "Mineralogy and Geology," *Memoirs*, 325.

12. Rachel Laudan, *From Mineralogy to Geology: The Foundations of a Science, 1650– 1830* (Chicago: University of Chicago Press, 1987), 87–112; Martin Rudwick, "Minerals, Strata and Fossils," in N. Jardine, J. A. Secord, and E. C. Spary, eds., *Cultures of Natural History* (Cambridge: Cambridge University Press, 1996), 266–286. Werner used the granite example himself. Abraham Gottlob Werner, *Short Classification and Description of the Various Rocks,* trans. with an introduction and notes by Alexander M. Ospovat (New York: Hafner, 1971), 48.

13. Laudan, *From Mineralogy to Geology*, 113–137; Dennis R. Dean, *James Hutton and the History of Geology* (Ithaca: Cornell University Press, 1992).

14. Jackson and Alger, "Mineralogy and Geology," *AJS*, *15*, 137–138.

15. Werner, *Short Classification*, 19; Laudan, *From Mineralogy to Geology*, 138–179.

16. As in mineralogy and natural history more generally, Werner considered formations to be *species,* and a *species of formation* was identified by its predominant bed. Werner, *Short Classification*, 19–20; Laudan, *From Mineralogy to Geology*, 138–179.

17. Jackson and Alger, "Mineralogy and Geology," *AJS*, *15*, 155.

18. Jackson and Alger, "Mineralogy and Geology," *AJS*, *15*, 216.

19. Review of "Mineralogy and Geology of Nova Scotia," *AJS*, 22 (1832):167–169.

20. Ebenezer Emmons, "Notice of a Scientific Expedition," *AJS*, 30 (1836):330–354, 330, 350; Abraham Gesner, *Remarks on the Geology and Mineralogy of Nova Scotia* (Halifax: Gossip and Coade, 1836).

21. Abraham Gesner, *The Industrial Resources of Nova Scotia* (Halifax: A. & W. MacKinlay, 1849), 230.

22. Peter von Bitter, "Abraham Gesner (1797–1864), an Early Canadian Geologist—Charges of Plagiarism," *Geoscience Canada*, 4 (1977):97–100.

23. *Report of a Case, tried at Albert Circuit, 1852, before his Honor Judge Wilmot, and a Special Jury. Abraham Gesner vs. William Cairns. Copied from the Judge's Notes* (Saint John: William L. Avery, 1853), 116; Gesner, *Remarks*, ix. Gesner cited his sources for large

sections but not for sentences or phrases, a practice appropriate to contemporary conventions. Jackson and Alger, "Mineralogy and Geology," *AJS*, 15, 134; cf. Gesner, *Remarks*, 233.

24. He discovered only one new mineral, phrenite. Gesner, *Remarks*, 202, 216.

25. Gesner, *Remarks*, 225–226.

26. Jackson and Alger, "Mineralogy and Geology," *AJS*, 15, 141.

27. Jackson and Alger, "Mineralogy and Geology," *Memoirs*, 323.

28. Surveying by boat was also difficult. Jackson and Alger discussed the Bay of Fundy's dangerous tides, and Gesner described the pounding surf, treacherous undertow, and tricky currents. Gesner, *Remarks*, 188–190, 194–195, 205.

29. Gesner, *Remarks*, 1.

30. Another difficulty was "the length of the winter season." Gesner, *Remarks*, 111.

31. Gesner, *Remarks*, xii.

32. Gesner, *Remarks*, 38, 66.

33. Jackson and Alger, "Mineralogy and Geology," *Memoirs*, 297.

34. Gesner, *Remarks*, 66.

35. Gesner, *Remarks*, 80–84.

36. Gesner, *Remarks*, 153.

37. Gesner, *Remarks*, 159.

38. Gesner, *Remarks*, viii.

39. Gesner, *Remarks*, viii.

40. Gesner, *Remarks*, 122.

41. William Buckland (1784–1856) was professor of mineralogy and geology at Oxford University.

42. Gesner, *Remarks*, 106.

43. John William Dawson, *Acadian Geology* (Edinburgh: Oliver and Boyd, 1855), 5–6.

44. Gesner, *Remarks*, ix.

45. Charles T. Jackson and Francis Alger to the House of Assembly of Nova Scotia, 25 February 1840, reproduced in von Bitter, "Charges of Plagiarism," 97–98.

46. Jackson and Alger to Campbell, 10 June 1837, reproduced in Peter von Bitter, "Charles Jackson, M.D. (1805–1880) and Francis Alger (1807–1863)," *Geoscience Canada*, 5 (1978):79–82, 80–81.

47. Jackson and Alger to the House of Assembly, 25 February 1840, reproduced in von Bitter, "Charges of Plagiarism," 97–98.

48. Bluenoses held "an unreasoning hatred for Yankees." W. S. MacNutt, *New Brunswick: A History, 1784–1867* (Toronto: Macmillan, 1963), 225, 281.

49. Jackson to Silliman Sr., 24 March 1837, Silliman Mss, YUL.

50. From comments scribbled in the margins, it appears that the copy of Gesner's *Remarks* in the Harvard University Library is the one Jackson read. The pencil marks provide clues to points of possible plagiarism.

51. Benjamin Silliman Sr., "Address before the Association of American Geologists and Naturalists, assembled at Boston, April 24, 1842," *AJS*, 43 (1842):217–250, 234.

52. Charles T. Jackson, "Encouragement and Cultivation of the Sciences in the United States, Twenty-Fourth Anniversary Address, Before the American Institute, of the City of New-York, at the Tabernacle, on 16th of October, 1851," *Transactions of the American Institute for the year 1851* (Albany: Charles Van Bentuysen, 1852), 227–246, 237.

53. For an introduction to the major themes of the professionalization of American science, see, for example, Samuel Haber, *The Quest for Authority and Honor in the Amer-*

ican Professions, 1750–1900 (Chicago: University of Chicago Press, 1991); Nathan O. Hatch, ed., *The Professions in American History* (Notre Dame: University of Notre Dame Press, 1988); Joseph Ben-David, *The Scientist's Role in Society: A Comparative Study* (Chicago: University of Chicago Press, 1984); Gerald L. Geison, ed., *Professions and Professional Ideologies in America* (Chapel Hill: University of North Carolina Press, 1983); Burton J. Bledstein, *The Culture of Professionalism* (New York: W. W. Norton, 1976); Robert K. Merton, *The Sociology of Science: Theoretical and Empirical Investigations* (Chicago: University of Chicago Press, 1973); George H. Daniels, "The Process of Professionalization in American Science: The Emergent Period, 1820–1860," *Isis, 58* (1967):151–166; Nathan Reingold, "Definitions and Speculations: The Professionalization of Science in America in the Nineteenth Century," in Reingold, *Science, American Style* (New Brunswick: Rutgers, 1991), 24–53.

54. MacNutt, *New Brunswick*; T. W. Acheson, *Saint John: The Making of a Colonial Urban Community* (Toronto: University of Toronto Press, 1985); A. A. den Otter, *The Philosophy of Railways: The Transcontinental Railway Ideal in British North America* (Toronto: University of Toronto Press, 1997), 126–157.

55. Abraham Gesner, *First Report on the Geological Survey of New-Brunswick* (Saint John: Henry Chubb, 1839), 12, 63; Barkhouse, *Gesner*, 45.

56. Gesner expressed his "obligations" to Charles Simonds for assistance and to John R. Partelow (1796–1865), chair of the appropriations committee and member for Saint John County. Abraham Gesner, *Fourth Report on the Geological Survey of New-Brunswick* (Saint John: Henry Chubb, 1842), 4.

57. McNutt referred to this period as the "Age of Harmony." The years between 1837 and 1841 were also the time when Charles Simonds was at the height of his political influence. MacNutt, *New Brunswick, 259.*

58. Edward Bailey, *Geological Survey of Great Britain* (London: Thomas Murphy, 1952); John Smith Flett, *The First Hundred Years of the Geological Survey of Great Britain* (London: His Majesty's Stationery Office, 1937); James Secord, "The Geological Survey of Great Britain as a Research School," *History of Science, 24* (1986):223–275.

59. George P. Merrill, *The First One Hundred Years of American Geology* (New Haven: Yale University Press, 1924); Merrill, *Contributions to a History of American State Geological and Natural History Surveys*, U.S. National Museum, Bulletin No. 109 (Washington, DC: GPO, 1920); Mary C. Rabbitt, *Minerals, Lands, and Geology for the Common Defense and General Welfare*, vol. 1: *Before 1879* (Washington, DC: GPO, 1979); Michele L. Aldrich, *New York State Natural History Survey, 1836–1842: A Chapter in the History of American Science* (Ithaca, NY: Paleontological Research Institution, 2000), 267–277.

60. William Logan did not begin his survey of the provinces of Canada until 1841, and Joseph Beete Jukes did not begin surveying Newfoundland until 1839. After confederation in 1867, Logan's survey became the official Geological Survey of Canada. Zaslow, *Reading the Rocks;* Zeller, *Inventing Canada;* Stafford, *Scientist of Empire;* cf. David R. Oldroyd, *Thinking about the Earth: A History of Ideas in Geology* (London: Athlone, 1996), 108–130.

61. Gesner thought the appropriation was "truly embarrassing" when contrasted with the "great expense" made by "every State in the Union." Gesner, *First Report,* 6. James Hannay, *History of New Brunswick* (St. John: John A. Bowes, 1909), 64–65.

62. Gesner, *First Report,* 4.

63. Michele L. Aldrich, "American State Geological Surveys, 1820–1845," in Cecil J. Schneer, ed., *Two Hundred Years of Geology in America: Proceedings of the New Hamp-*

shire Bicentennial Conference on the History of Geology (Hanover, NH: University Press of New England, 1979); Paul Lucier, "A Plea for Applied Geology," *History of Science, 37* (1999):283–318.

64. Gesner, *First Report;* G. F. Matthew, "Abraham Gesner. A Review of His Scientific Work," *Bulletin of the Natural History Society of New Brunswick, 15* (1897):3–48.

65. Abraham Gesner, *Third Report on the Geological Survey of New-Brunswick* (Saint John: Henry Chubb, 1841), 31.

66. Edward Hitchcock, director of the Massachusetts survey, is often given credit for the distinctively American balance of "scientific" and "economical" geology. Hitchcock, *Report on the Geology, Mineralogy, Botany, and Zoology of Massachusetts* (Amherst: J. S. and C. Adams, 1833); Michele L. Aldrich, "Charles Thomas Jackson's Geological Surveys in New England, 1836–1844," *Northeastern Geology, 3* (1981):5–10; Aldrich, *New York State Natural History Survey;* Patsy Gerstner, *Henry Darwin Rogers, 1808–1866: American Geologist* (Tuscaloosa: University of Alabama Press, 1994).

67. MacNutt, *New Brunswick,* 258.

68. New Brunswick had twelve counties: Saint John, Charlotte, Westmorland, Kings, Queens, Sunbury, York, Carleton, Kent, Northumberland, Gloucester, and Restigouche.

69. Abraham Gesner, *Second Report on the Geological Survey of the Province of New-Brunswick* (Saint John: Henry Chubb, 1840), 25.

70. Gesner, *Second Report,* xi.

71. Gesner appended a letter about coal operations from Moses H. Perley (1804–1862), a young lawyer and the company's secretary, who was keen to encourage economic development and settlement of immigrants; he was also Gesner's friend. Gesner, *Third Report,* 81.

72. Gesner mentioned "this important [coal] formation" in his first report, but he was specifically instructed to report on the region "southward of the coal district." Gesner, *First Report,* 77.

73. Gesner, *Third Report,* 63.

74. Abraham Gesner, *New Brunswick; with Notes for Emigrants* (London: Simmond's & Ward, 1847), 57–71; MacNutt, *New Brunswick,* 265–276; John Francis Sprague, *The North Eastern Boundary Controversy and the Aroostook War* (Dover, ME: Observer Press, 1910); Henry S. Burrage, *Maine in the Northeastern Boundary Controversy* (Portland: Marks, 1919).

75. Gesner, *New Brunswick for Emigrants,* 60. Charles T. Jackson, *First Report on the Geology of the State of Maine* (Augusta: Smith & Robinson, 1837), *Second Report on the Geology of the State of Maine* (Augusta: Smith & Robinson, 1838), and *Third Report on the Geology of the State of Maine* (Augusta: Smith & Robinson, 1839); Ezekiel Holmes, *Report of an Exploration and Survey of the Territory on the Aroostook River* (Augusta: Smith & Robinson, 1839); Aldrich, "Jackson's Geological Surveys."

76. Abraham Gesner, *Report on the Geological Survey of New-Brunswick, with a Topographical Account of the Public Lands and the Districts Explored in 1842* (Saint John: Henry Chubb, 1843), 14.

77. Gesner used Jackson's work. Compare his description and woodcut of a lime kiln (*Topographical Account,* 87) with Jackson's (*Second Report . . . on Maine,* 119).

78. Gesner, *Fourth Report,* 64.

79. Gesner, *Topographical Account,* 4.

80. The plan was rejected by the institute members. Martin Hewitt, "Science as Spec-

tacle: Popular Scientific Culture in Saint John, New Brunswick, 1830–1850," *Acadiensis,* *18* (1988):91–119.

81. Gesner, *Topographical Account,* 3–4.

82. Gesner, *Topographical Account,* 11.

83. Gesner, *Topographical Account,* 5–8.

84. Gesner, *Third Report,* xiii.

85. Gesner, *Third Report,* xii.

86. Gesner, *Fourth Report,* 13–14.

87. On the investment "success" of Gesner's survey, see Hugh M. Grant, "Public Policy and Private Capital Formation in Petroleum Exploration," in Paul A. Bogaard, ed., *Profiles of Science and Society in the Maritimes prior to 1914* (Sackville, NB: Acadiensis Press, 1990):137–160, 141–142.

88. Gesner, *Third Report,* 25.

89. Gesner, *Third Report,* xiv.

90. Gesner, *Third Report,* 25–26.

91. Gesner, *Fourth Report,* 69.

92. MacNutt, *New Brunswick,* 213.

93. Gesner, *New Brunswick for Emigrants,* 340.

94. Gesner, *Third Report,* xii.

95. Gesner, *Fourth Report,* 4.

96. Gesner, *Topographical Account,* 54.

97. Gesner, *Second Report,* 3. On nomenclature, see Aldrich, *New York State Natural History;* Gerstner, *Henry Darwin Rogers;* and Patsy A. Gerstner, "Henry Darwin Rogers and William Barton Rogers on the Nomenclature of the American Paleozoic Rocks," in Schneer, *Two Hundred Years of Geology in America,* 175–186.

98. Gesner, *Topographical Account,* 66.

99. Zeller, *Inventing Canada.* Cf. Nancy Christie, "Sir William Logan's Geological Empire and the 'Humbug' of Economic Utility," *Canadian Historical Review, 75* (June 1994):161–204.

100. Gesner, *Fourth Report,* 54–62, 59; Gesner, *Topographical Account,* 56–59.

101. Gesner, *First Report,* 6.

102. Gesner, *Topographical Account,* 27.

103. Gesner to J. Forbes Rolph, 6 July 1838; Gesner to Rolph, 18 September 1838; SLB, LR4, (1838–1839), and SLB, LR5, (1839–1840), GSL.

104. Gesner to William Lonsdale, 15 January 1839, SLB, LR4, GSL.

105. Gesner to Lyell, 30 January 1840, SLB, LR5, GSL.

106. Charles Lyell, *Travels in North America, in the Years 1841–2; with Geological Observations on the United States, Canada, and Nova Scotia,* 2 vols. (New York: Wiley and Putnam, 1845), 2:149.

107. Lyell also identified the Gypsiferous formation at Horton Bluff. Charles Lyell, "On the upright Fossil-Trees found at different levels in the Coal Strata of Cumberland, Nova Scotia," *Proceedings of the GSL, 4* (1842–1845):313–315, reprinted in *AJS, 45* (1843):353–356.

108. Charles Lyell, "On the Coal-formation of Nova Scotia, and on the age and relative position of the Gypsum and accompanying marine limestones," *Proceedings of the GSL, 4* (1843):184–186; Abraham Gesner, "A Geological Map of Nova Scotia," *Proceedings of the GSL, 4* (1843):186–190. Cf. Lyell, *Travels in North America,* 2:148–149.

109. At Horton Bluff, Logan found "evidence of the age of the gypsiferous strata." William Edmond Logan, "On the Coal-fields of Pennsylvania and Nova Scotia," *Proceedings of the GSL, 3* (1842):707–712, 712. Roderick Impey Murchison, "Anniversary Address," *Proceedings of the GSL, 4* (1843):65–151, 125; H. S. Torrens, "William Edmond Logan's Geological Apprenticeship in Britain, 1831–1842," *Geoscience Canada, 26* (1999):97–110.

110. J. W. Dawson, "The Gypsum of Nova Scotia," *Proceedings of the Academy of Natural Sciences of Philadelphia, 3* (1847):272–274.

111. Abraham Gesner, "On the Gypsum of Nova Scotia," *Proceedings of the GSL, 5* (1849):129–130; Gesner, *Industrial Resources,* 237.

112. R. W. Ells, *A History of New Brunswick Geology* (Montreal: Gazette Printing, 1887).

113. See the section, "Bituminous Coal," in L. W. Bailey, *The Mineral Resources of the Province of New Brunswick* (Ottawa: S. E. Dawson, 1893), 59–68.

114. Matthew, "Abraham Gesner."

115. James C. Robb, "Short notice on the existence of Coal in New Brunswick," in J. F. W. Johnston, *Report on the Agricultural Capabilities of the Province of New Brunswick* (Fredericton: J. Simpson, 1849; 2nd ed., 1850), 16–18, 18.

116. Johnston, *Report,* 14. Johnston toured Nova Scotia, New Brunswick, Upper and Lower Canada, and several northeastern states. James F. W. Johnston, *Notes on North America: Agricultural, Economical, and Social,* 2 vols. (Boston: Charles C. Little and James Brown, 1851). Robb thought Johnston's book was "selfish and ungenerous"; Robb to Elizabeth Robb, 8 June 1851, in Alfred Goldsworthy Bailey, ed., *The Letters of James and Ellen Robb: Portrait of a Fredericton Family in Early Victorian Times* (Fredericton: Acadiensis Press, 1983), 108–109.

117. Gesner, *Topographical Account,* 4.

118. MacNutt, *New Brunswick,* 174.

119. Gesner, *New Brunswick for Emigrants,* 311.

120. Gesner had written to Harvey in 1837 requesting the position. Robb had been recommended by Thomas Thomson (1773–1852), Regius Professor of Chemistry in the University of Glasgow. Robb arrived in September 1837. L. W. Bailey, "Dr. James Robb, First Professor of Chemistry and Natural History in King's College, Fredericton, a Sketch of His Life and Labours," *Bulletin of the New Brunswick Natural History Society, 14* (1898–1904):1–15; Alfred G. Bailey, "Robb, James," *Dictionary of Canadian Biography,* vol. 9 (Toronto: University of Toronto Press, 1976), 665–667.

121. Charles E. Rosenberg introduced the term *scientist-entrepreneur* to describe members, especially directors, of agricultural experiment stations in the United States. My description of scientific entrepreneurs corresponds to Nathan Reingold's *practitioners,* but Reingold limited that category to scientists in government or educational institutions. Crosbie Smith and M. Norton Wise used scientific entrepreneur in similar ways to characterize the Scottish physicist William Thomson (1824–1907). Rosenberg, *No Other Gods: On Science and American Social Thought* (Baltimore: Johns Hopkins University Press, 1976), 135–152, 153–172; Reingold, "Definitions and Speculations: The Professionalization of Science in America in the Nineteenth Century," in Reingold, *Science, American Style* (New Brunswick: Rutgers University Press, 1991), 24–53; Smith and Wise, *Energy and Empire: A Biographical Study of Lord Kelvin* (Cambridge: Cambridge University Press, 1989).

122. Hewitt, "Science as Spectacle," 107–108; Hewitt, "Science, Popular Culture, and

the Producer Alliance in Saint John, N.B.," in Bogaard, *Profiles of Science and Society in the Maritimes prior to 1914*, 243–275, 270.

123. Abraham Gesner, *Report on the Geological Survey of Prince Edward Island* (Cornwallis, NS, 1846).

124. Abraham Gesner, *Report on the Londonderry Iron and Coal Deposits* (Halifax, 1846); Gesner, *Industrial Resources*, 293–295.

125. In his final report on New Hampshire, Jackson summarized nearly fifteen years' surveying experience in New England. Jackson, *Final Report on the Geology and Mineralogy of the State of New Hampshire* (Concord: Carroll and Baker, 1844); Aldrich, "Charles Thomas Jackson," 8; Merrill, *First One Hundred Years*, 204–206.

126. James A. Secord, introduction to Charles Lyell, *Principles of Geology* (London: Penguin, 1997); Secord, *Victorian Sensation: The Extraordinary Publication, Reception, and Secret Authorship of Vestiges of the Natural History of Creation* (Chicago: University of Chicago Press, 2000).

127. Gesner, *New Brunswick for Emigrants*, v.

128. Gesner, *Industrial Resources*, 18. Gesner's inclusion of a chapter on Native Americans resulted in his appointment by the Nova Scotia government as commissioner to the Indians.

129. Gesner, *New Brunswick for Emigrants*, 304.

130. Gesner, *New Brunswick for Emigrants*, 354. Robb reached the same conclusion, in Johnston, *Report*, 15; cf. Grant, "Public Policy and Private Capital Formation in Petroleum Exploration."

131. Gesner, *Industrial Resources*, 287.

132. Gesner, *Industrial Resources*, 10.

133. Gesner came very close to espousing "the philosophy of railways," as A. A. den Otter described it, although Gesner's sense of nationalism pertained strictly to the Maritimes; he had no desire to confederate with Upper and Lower Canada. Den Otter, *Philosophy of Railways*, 3–31. On American Whigs, see Hugh Richard Slotten, *Patronage, Practice, and the Culture of American Science: Alexander Dallas Bache and the U.S. Coast Survey* (Cambridge: Cambridge University Press, 1994); Daniel Walker Howe, *The Political Culture of the American Whigs* (Chicago: University of Chicago Press, 1979); cf. Nathan Reingold, "Between American History and History of Science," *Studies in History and Philosophy of Science, 27* (1996):115–129.

134. John Lauritz Larson, *Internal Improvement: National Public Works and the Promise of Popular Government in the Early United States* (Chapel Hill: University of North Carolina Press, 2001).

135. Gesner, *Industrial Resources*, 11.

136. Gesner, *Industrial Resources*, ii.

137. Gesner, *Industrial Resources*, 9.

138. Gesner, *Industrial Resources*, 14–15.

139. Gesner, *Industrial Resources*, 15, 11.

140. Gesner, *Industrial Resources*, 12.

141. Gesner, *Industrial Resources*, 11.

142. Gesner, *Industrial Resources*, 267.

143. Also called "The Nova Scotian and Cape Breton Mining Company." James M. Cameron, *The Pictonian Colliers* (Halifax: Nova Scotia Museum, 1974), 21–33.

144. Gesner, *Industrial Resources*, 230.

145. Gesner, *Industrial Resources*, 196.

146. C. Ochiltree Macdonald, *The Coal and Iron Industries of Nova Scotia* (Halifax: Chronicle, 1909), 137–139.

147. Richard Cowling Taylor, *Statistics of Coal* (Philadelphia: J. W. Moore, 1848), 195–196.

148. Gesner, *Industrial Resources,* 273.

149. On the transition of power from a Loyalist-mercantile elite to a "producers alliance" among artisanal and industrial groups, see Acheson, *Saint John;* Hewitt, "Science as Spectacle"; and Hewitt, "Science, Popular Culture and the Producer Alliance in Saint John, N.B." On industrialization, see L. D. McCann, "The Mercantile-Industrial Transition in the Metals Towns of Pictou County, 1857–1931," *Acadiensis, 10* (1981):29–64.

150. Gesner, *First Report,* 9.

Chapter 2. The Strange Case of the Albert Mineral

1. Charles Jackson, ["A new kind of fuel"], *Proceedings of the BSNH, 3* (April 1850):279–280. Reprinted as "On the Asphaltic Coal of New Brunswick," *AJS, 11* (1851):292–293; and "Asphaltum from New Brunswick," *Annual of Scientific Discovery for 1851,* 307. Different titles reflected Jackson's reclassifications. On shipment, see *Report of a Case, tried at Albert Circuit, 1852, before his Honor Judge Wilmot, and a Special Jury. Abraham Gesner vs. William Cairns. Copied from the Judge's Notes* (Saint John: William L. Avery, 1853), 111.

2. Jackson, ["New kind of fuel"], 280.

3. Expert witnessing went back at least a generation, but in Britain, there were legal precedents in the eighteenth century. Tal Golan, *Laws of Men and Laws of Nature: The History of Scientific Expert Testimony in England and America* (Cambridge, MA: Harvard University Press, 2004); Golan, ed., " Law and Science," special issue, *Science in Context, 12* (1999); Roger Smith and Brian Wynne, eds., *Expert Evidence: Interpreting Science in the Law* (London: Routledge, 1989); Christopher Hamlin, "Scientific Method and Expert Witnesses: Victorian Perspectives on a Modern Problem," *Social Studies of Science, 16* (1986):485–573; June Z. Fullmer, "Technology, Chemistry, and the Law in Early 19th-Century England," *T&C, 21* (1980):1–28; Kenneth Allen De Ville, "Fractured Confidence: Origins of American Medical Malpractice, 1790–1900" (PhD dissertation, Rice University, 1989); John M. Clarke, *James Hall of Albany* (Albany, 1921), 204–216.

4. Abraham Gesner, *Practical Treatise on Coal, Petroleum, and Other Distilled Oils* (New York: Baillière Brothers, 1860), 26–27, 74.

5. Lord Thomas Cochrane, 11th Earl of Dundonald, and H. R. Foxbourne, *The Life of Thomas, Lord Cochrane, 10th Earl of Dundonald, G.C.B.,* 2 vols. (London: R. Bentley, 1869); [Obituary, 10th Earl of Dundonald], *Scientific American, 3* (24 November 1860):346–347.

6. Gesner received the patent in February 1850. He applied in New Brunswick but was denied because of nonresidence. "A Patentee" [Abraham Gesner], *Gas Monopoly: Piracy of Patents and Farmers' Rights in which is contained a Reply to the Directors of the Halifax Gas Company and a Brief Account of the Asphaltum Mines of New Brunswick* (Halifax, 1851), 13, 20, 30.

7. "Illuminating Light Houses," *Scientific American, 7* (24 April 1852):251.

8. U.S. Patent 7052, "Manufacture of Illuminating Gas from Bitumen" (29 January 1850).

9. "Gas Light in Factories," *Scientific American, 7* (6 December 1851):93.

10. *AG-LI, 4* (15 June 1863):373.

11. "Burning Fluid," *Scientific American, 11* (2 August 1856):374; "Explosion of Burning Fluids," *Scientific American, 7* (19 June 1852):315.

12. "Give Us Cheap Gas," *Scientific American, 8* (11 December 1852):101.

13. "New York Gas," *Scientific American, 5* (29 September 1849):13.

14. "New Material for the Manufacture of Gas," and "Purification of Coal Gas," *Annual of Scientific Discovery for 1851*, 84–86.

15. Along the Atlantic seaboard, Nova Scotia coal was cheaper than Pennsylvania and Virginia coals because of heavy charges by railroad and canal companies. Also, after 1846, duties on imported coal were reduced. Richard Cowling Taylor, *Statistics of Coal* (Philadelphia: J. W. Moore, 1848), 201–203.

16. Charles Roome to Benjamin Silliman Jr., 21 April 1854, Silliman Mss, YUL.

17. Thomas Hughes, *Networks of Power: Electrification in Western Society, 1880–1930* (Baltimore: Johns Hopkins University Press, 1983); Wolfgang Schivelbusch, *Disenchanted Night: The Industrialization of Light in the Nineteenth Century* (Berkeley: University of California Press, 1988).

18. "Asphaltum from New Brunswick," *Annual of Scientific Discovery for 1851*, 307.

19. "Dr. Gesner's Patent Kerosene Gas Light," *Scientific American, 5* (16 February 1850):172; "New Kind of Gas," *Scientific American, 5* (9 February 1850):164.

20. *Scientific American, 5* (16 February 1850):172; "New Material for the Manufacture of Gas," *Annual of Scientific Discovery for 1851*, 84–85.

21. Louis I. Kuslan, "Benjamin Silliman, Jr.: The Second Silliman," in Leonard G. Wilson, ed., *Benjamin Silliman and His Circle: Studies on the Influence of Benjamin Silliman on Science in America* (New York: Science History Publications, 1979), 159–205; [James Dwight Dana], "Benjamin Silliman," *AJS, 29* (1885):85–92; Arthur W. Wright, "Benjamin Silliman, 1816–1885," *Biographical Memoirs. National Academy of Sciences, 7* (1913):115–141.

22. Kuslan, "Silliman," 183–185.

23. Benjamin Silliman Jr. and Charles H. Porter, "Notice of a Photometer and of Some Experiments Therewith upon the Comparative Power of Several Artificial Means of Illuminations," *AJS, 23* (1857):315–318.

24. Gesner made a second visit to New York City in December 1850, during which he might have filed another patent claim. "Kerosene Gas—Nova Scotia Going-A-Head of 'Old Mother,'" *Scientific American, 7* (7 December 1850):89. No record of the patent can be found. Most likely it was for a method to enrich illuminating gas by passing it through "hydro-carbon fluids," such as naphtha or perhaps the "fluids obtained from New Brunswick asphalt." "New Light—Kerosene Gas," *Scientific American, 9* (8 October 1853):29. Cf. Harold F. Williamson and Arnold R. Daum, *The American Petroleum Industry: The Age of Illumination, 1859–1899* (Evanston: Northwestern University Press, 1959), 777.

25. *Scientific American, 5* (16 February 1850):172; cf. Taylor, *Statistics of Coal*, 223, 243–244.

26. Gesner, *Practical Treatise*, 26.

27. Edward Allison was a partner of Cairns and an assignee of the license.

28. During the months of January and February 1851, Cairns raised and shipped more than 200 cauldrons of the Albert mineral. A cauldron holds between 2,000 and 3,000 pounds.

29. For the full text of Cairns's license and the crown reservation, see *Reports on the Geological Relations, Chemical Analyses, and Microscopic Examination of the Coal of the Albert Coal Mining Co., Situated in Hillsboro, Albert County, New Brunswick* (New York: George F. Nesbitt & Co., 1851), 29.

30. [Gesner], *Gas Monopoly,* 25.

31. [Gesner], *Gas Monopoly,* 23.

32. [Gesner], *Gas Monopoly,* 26.

33. "A Fellow of the Geological Society of London," [Abraham Gesner], *Review of "Reports on the Geological Relations, Chemical Analyses, and Microscopic Examination of the Coal of the Albert Coal Mining Company, Situated in Hillsboro, Albert County, New Brunswick," as Written and Compiled by Charles T. Jackson, M.D., of Boston* (New York: C. Vinten, 1852), 4; cf. *Report of a Case,* 125, 136.

34. [Gesner], *Gas Monopoly,* 7–8.

35. Gesner asked Jackson "to explore the district [and] to establish the character of the rocks in which the asphaltum is found." Gesner to Jackson, 25 March 1851, reprinted in *Reports for the Albert Coal Mining Co.,* 29. Jackson's asphaltum report appeared in the *British Colonist* (Halifax), 2 May 1850.

36. [Gesner], *Review of "Reports,"* 27.

37. *Report of a Case,* 116.

38. *Proceedings of the BSNH, 4* (16 April 1851):55–56.

39. [Gesner], *Review of "Reports,"* 5.

40. *Report of a Case,* 116.

41. *Reports for the Albert Coal Mining Co.,* 4.

42. Abraham Gesner, *Second Report on the Geological Survey of the Province of New-Brunswick* (Saint John: Henry Chubb, 1840), 67.

43. *Report of a Case,* 49.

44. Jackson's guide was Robert Foulis, a Saint John foundry owner and bitter enemy of Gesner. [Gesner], *Gas Monopoly,* 10–12.

45. *Reports for the Albert Coal Mining Co.,* 3.

46. *Proceedings of the BSNH, 4* (21 May 1851):64–65.

47. *Reports for the Albert Coal Mining Co.,* 11.

48. Emphasis added. *Reports for the Albert Coal Mining Co.,* 17.

49. *Reports for the Albert Coal Mining Co.,* 16.

50. *Reports for the Albert Coal Mining Co.,* 9.

51. *Proceedings of the BSNH, 4* (18 June 1851):73–74.

52. *Reports for the Albert Coal Mining Co.,* 8.

53. *Reports for the Albert Coal Mining Co.,* 15.

54. William Barton Rogers, former director of the Virginia Geological Survey, was the exception; *Proceedings of the BSNH, 4* (December 1851):169–170.

55. *Report of a Case,* 152; Julius H. Ward, *The Life and Letters of James Gates Percival* (Boston: Ticknor and Fields, 1866).

56. *Reports for the Albert Coal Mining Co.,* 35.

57. *Report of a Case,* 157.

58. *Reports for the Albert Coal Mining Co.,* 34.

59. *Reports for the Albert Coal Mining Co.,* 41–43.

60. [James Dwight Dana], "Reports on the Albert Coal Mine," *AJS, 13* (1852):276–277.

61. Gesner might have had the additional support of some prominent Pennsylvania

coal mining interests, including the Carey, Lea, and Wetherill families. Anthony F. C. Wallace, *St. Clair: A Nineteenth-Century Coal Town's Experience with a Disaster-Prone Industry* (Ithaca: Cornell University Press, 1988), 54–70.

62. Taylor, *Statistics of Coal,* 40.

63. Isaac Lea hired Taylor to survey 42,000 acres along the Susquehanna River. Richard C. Taylor, *Two Reports: on the Coal Lands, Mines and Improvements of the Dauphin and Susquehanna Coal Company, and the Geological Examinations, Present Condition and Prospects of the Stony Creek Coal Estate, etc.* (Philadelphia: E. G. Dorsey, 1840).

64. The second edition, published posthumously in 1853, included a biography of Taylor by Isaac Lea; cf. Isaac Lea, "Memoir of Richard Cowling Taylor," *Proceedings of the Academy of Natural Sciences of Philadelphia,* 5 (1850–1851):290–296.

65. Taylor, *Statistics of Coal,* 186.

66. "Joint Geological Report on the Asphalte Mine of Hillsborough, N.B. By Richd. C. Taylor and James Robb," in *Supreme Court, Halifax, N.S. Abraham Gesner vs. Halifax Gas-Light Company. Deposition of Richard C. Taylor, respecting the asphaltum mine at Hillsborough, in the County of Albert and Province of New Brunswick. Illustrated by a map and diagrams* (Philadelphia: King & Baird, 1851), 35–40. The joint report was also published as an appendix to [Gesner], *Gas Monopoly.*

67. *Deposition of Taylor,* 7.

68. *Deposition of Taylor,* 22.

69. Richard C. Taylor, "On a Vien [*sic*] of Asphaltum at Hillsborough, in Albert County, Province of New Brunswick," *Proceedings of the APS,* 5 (16 January 1852):241–243, 243.

70. *Report of a Case,* 150.

71. Robb thought a volcanic or igneous force arched the rocks. *Report of a Case,* 32.

72. *Deposition of Taylor,* 26.

73. *Deposition of Taylor,* 36; Taylor, "Vien of Asphaltum," 241–242.

74. *Deposition of Taylor,* 14.

75. Charles M. Wetherill, "On a New Variety of Asphalt (Melan-Asphalt)," *Transactions of the APS,* 10 (1853):353–358, 355.

76. Melan was from the Greek *melas* (gen. *melanos*), meaning black. Wetherill, "Melan-Asphalt," 358.

77. [Gesner], *Review of "Reports,"* 18.

78. [Gesner], *Review of "Reports,"* 17.

79. Wetherill, "Melan-Asphalt," 353.

80. [Gesner], *Gas Monopoly,* 5, 19.

81. [Gesner], *Gas Monopoly,* 15, 30–32.

82. The Halifax Gas Company comprised James B. Uniacke (President); J. H. Anderson; W. A. Black; James Donaldson; Andrew McKinlay; John Naylor; James Tremain; and John Tremain (Secretary). For more on Johnston's and Uniacke's governments, see J. Murray Beck, *Politics of Nova Scotia,* vol. 1: *1710–1896* (Tantallon, NS: Four East Publications, 1985).

83. In the Saint John trial, counsel for the defense made the same motion for dismissal, but Judge Wilmot decided there was evidence warranting the trial's continuation. *Report of a Case,* 60–61.

84. No transcript of the proceedings survives. [Gesner], *Gas Monopoly,* appendix B.

85. Gesner, *Practical Treatise,* 22.

86. James A. Secord, *Controversy in Victorian Geology: The Cambrian-Silurian Dispute* (Princeton: Princeton University Press, 1986); Martin J. S. Rudwick, *The Great Devonian Controversy: The Shaping of Scientific Knowledge among Gentlemanly Specialists* (Chicago: University of Chicago Press, 1985); Jan Golinski, *Making Natural Knowledge: Constructivism and the History of Science* (Cambridge: Cambridge University Press, 1998).

87. Isiah Deck testimony, *Report of a Case*, 128–129.

88. *Report of a Case*, 23.

89. "Deposition of Richard C. Taylor in *Abraham Gesner v. William Cairns*, 6 June 1851, given at Dorchester, New Brunswick," in *Report of a Case*, appendix A, 140–152. Taylor also gave a sworn deposition between 14 and 18 June 1851 for the Halifax case, which was published separately as *Deposition of Taylor*.

90. Gesner, *Second Report*, 66; Gesner, *Third Report on the Geological Survey of the Province of New-Brunswick* (Saint John: Henry Chubb, 1841), 27, 28; *Report of a Case*, 38, 62.

91. "Deposition of James Gates Percival in *Abraham Gesner v. William Cairns*, 22 August 1851, given in Dorchester, New Brunswick," in *Report of a Case*, appendix A, 152–167.

92. Antisell, Leidy, Hayes, and Deck also visited the mine during the trial. *Report of a Case*, 49.

93. *Report of a Case*, 32.

94. Jackson mentioned Louis Agassiz, but did not defer to him as the authority. *Report of a Case*, 115.

95. *Report of a Case*, 114.

96. *Report of a Case*, 113, 115.

97. *Report of a Case*, 100.

98. *Report of a Case*, 117.

99. *Report of a Case*, 76–77.

100. Jackson was upset by Bacon's absence: "I don't know why he is not here." *Report of a Case*, 117.

101. *Report of a Case*, 77.

102. *Report of a Case*, 101.

103. *Report of a Case*, 120.

104. [Gesner], *Review of "Reports,"* 27.

105. [Gesner], *Review of "Reports,"* 28–29.

106. *Report of a Case*, 129.

107. *Report of a Case*, 116. Silliman to Cook and Smith, 24 June 1852; and President of [Saint John] Mechanics' Institute to Silliman, 28 September 1852, Silliman Mss, YUL.

108. So too for Antisell. *Report of a Case*, 48.

109. *Report of a Case*, 108, 113, 142.

110. *Report of a Case*, 152.

111. *Report of a Case*, 123.

112. *Report of a Case*, 119.

113. *Report of a Case*, 108, 113, 142.

114. *Report of a Case*, 125.

115. *Report of a Case*, 117.

116. *Report of a Case*, 144.

117. The only other moment was Jackson's criticism of the New York geologist James

Hall: "I think Hall makes errors in his science." Jackson quickly qualified this: "I no doubt make a great many mistakes." *Report of a Case,* 116.

118. *Report of a Case,* 48–49.

119. *Report of a Case,* 87.

120. *Report of a Case,* 102.

121. Deck simply agreed. *Report of a Case,* 98, 128.

122. *Report of a Case,* 143.

123. *Report of a Case,* 119, 121.

124. *Report of a Case,* 102.

125. *Report of a Case,* 161.

126. *Report of a Case,* 147.

127. The term was Deck's. *Report of a Case,* 128.

128. *Report of a Case,* 89.

129. *Report of a Case,* 95.

130. *Report of a Case,* 40.

131. *Report of a Case,* 41.

132. *Report of a Case,* 136.

133. *Report of a Case,* 135. Lemuel Allan Wilmot had been the attorney general (1848–1851) before his appointment as a puisne judge in 1851. Between 1868 and 1873, Wilmot was the lieutenant governor of New Brunswick.

134. Richard Cowling Taylor, *Statistics of Coal,* revised by S. S. Haldeman (Philadelphia: J. W. Moore, 1855), xiii.

135. Gesner implied the money came from his own pocket, but given his modest finances, it most likely came from his wealthy Philadelphia backers. Gesner, *Practical Treatise,* 22.

136. *Gesner v. Cairns,* 7 N.B.R. (1853) 608.

137. *Gesner v. Gas Company,* 2 N.S.R. (1853) 75.

138. *Gesner v. Gas Company,* 2 N.S.R. (1853) 75.

139. *Gesner v. Cairns,* 7 N.B.R. (1853) 601.

140. *Gesner v. Gas Company,* 2 N.S.R. (1853) 75.

141. *Gesner v. Gas Company,* 2 N.S.R. (1853) 81.

142. *Gesner v. Cairns,* 7 N.B.R. (1853) 609.

143. *Gesner v. Gas Company,* 2 N.S.R. (1853) 85.

144. W. S. MacNutt, *New Brunswick: A History, 1784–1867* (Toronto: Macmillan, 1963), 383.

145. *Gesner v. Gas Company,* 2 N.S.R. (1853) 90.

146. [Gesner], *Review of "Reports,"* 20, 28, 29.

Chapter 3. The American Sciences of Coal

1. J. H. Calder, "The Carboniferous Evolution of Nova Scotia," and Andrew C. Scott, "The Legacy of Charles Lyell: Advances in our Knowledge of Coal and Coal-Bearing Strata," in Derek J. Blundell and Andrew C. Scott, eds., *Lyell: The Past Is the Key to the Present,* Special Publications, *143* (London: Geological Society, 1998), 261–302, 243–260; John J. Stevenson, "The Formation of Coal Beds: An Historical Summary of Opinion from 1700 to the Present Time," *Proceedings of the APS, 50* (1911):1–116; Karl Alfred von Zittel, *History of Geology and Paleontology to the End of the Nineteenth Century,* trans. Maria M. Ogilvie-Gordon (London: Walter Scott, 1901), 239–241.

2. Histories of mineralogy tend to focus on the seventeenth and eighteenth centuries, not the nineteenth. David R. Oldroyd, *Thinking about the Earth: A History of Ideas in Geology* (London: Athlone, 1996), chs. 1 and 9; Rachel Laudan, *From Mineralogy to Geology: The Foundations of a Science, 1630–1830* (Chicago: University of Chicago Press, 1987); John G. Burke, *Origins of the Science of Crystals* (Berkeley: University of California Press, 1966).

3. Daniel C. Gilman, *The Life of James Dwight Dana: Scientific Explorer, Mineralogist, Geologist, Zoologist, Professor in Yale University* (New York: Harper, 1899); Michael Pendergrast, "James Dwight Dana: The Life and Thought of an American Scientist" (PhD dissertation, UCLA, 1978); Margaret W. Rossiter, "A Portrait of James Dwight Dana," in Leonard G. Wilson, ed., *Benjamin Silliman and His Circle: Studies on the Influence of Benjamin Silliman on Science in America* (New York: Science History Publications, 1979), 105–127; Julie R. Newell, "James Dwight Dana and the Emergence of Professional Geology in the United States," *AJS, 297* (1997):273–282. Edward Lurie also wrote a biography, unpublished, which he kindly showed me.

4. Gloria Robinson, "Charles Upham Shepard," in Wilson, *Silliman and His Circle,* 85–103.

5. James Dwight Dana, *A System of Mineralogy, Comprising the Most Recent Discoveries* (New York: Wiley & Putnam, 1844), 128, iii.

6. Carl von Linné, *System of Nature,* trans. William Turton, 7 vols. (London: Lackington Allen, 1768), 7:9.

7. *Report of a Case, tried at Albert Circuit, 1852, before his Honor Judge Wilmot, and a Special Jury, Abraham Gesner vs. William Cairns. Copied from the Judge's Notes* (Saint John: William L. Avery, 1853), 35, 114.

8. James Dwight Dana, *A System of Mineralogy, Comprising the Most Recent Discoveries* (New York: George P. Putnam, 1850), 171.

9. James Dwight Dana, *A System of Mineralogy, Comprising the Most Recent Discoveries,* 2 vols. (New York: George P. Putnam, 1854), 1:6; cf. Amos Eaton, "Geological Equivalents," *AJS, 21* (1832):132–138, 134.

10. Emphasis added. Dana, *Mineralogy* (1850), 5; cf. James Dwight Dana, *A System of Mineralogy: Including an Extended Treatise on Crystallography* (New Haven: Durrie & Peck and Herrick & Noyes, 1837), iv.

11. On textbooks, see Thomas Kuhn, *The Structure of Scientific Revolutions* (Chicago: University of Chicago Press, 1970), ch. 2.

12. Dana, *Mineralogy* (1850), 5; cf. John G. Greene and John G. Burke, "The Science of Minerals in the Age of Jefferson," *Transactions of the APS, 68* (1978):107.

13. Dana, *Mineralogy* (1850), 170.

14. Dana, *Mineralogy* (1850), 18.

15. Dana, *Mineralogy* (1854), 1:244.

16. Dana, *Mineralogy,* (1850), 170.

17. Dana, *Mineralogy* (1844), 7.

18. Dana, *Mineralogy* (1850), 5.

19. Dana, *Mineralogy* (1850), 168.

20. Dana, *Mineralogy* (1850), 5.

21. "Whether a given substance is or is not coal, must remain a matter of opinion until philosophers have arrived at a true definition of coal." "The Evidence of Experts," *Chemical News, 5* (5 April 1862):183; cf. "The Torbanehill Mineral," *Chemical News, 3* (2 March 1861):217.

---

22. Wetherill had taken his PhD in chemistry at Göttingen with Friedrich Wöhler.

23. *Report of a Case,* 48.

24. Wetherill to Silliman, 21 June 1854, Silliman Mss, YUL.

25. T. Sterry Hunt, "On the Objects and Method of Mineralogy," *AJS,* 43 (1867):203–206; George J. Brush, "A Sketch of the Progress of American Mineralogy," *Proceedings of the AAAS, 31* (1883):1–28.

26. Dana, *System of Mineralogy* (1854), 5.

27. Abraham Gesner, *Practical Treatise on Coal, Petroleum, and Other Distilled Oils* (New York: Baillière Brothers, 1860), 22.

28. Robert Jameson, *System of Mineralogy,* 3 vols. (Edinburgh: William Blackwood, 1808), 3:70.

29. The conversion of plants into coal was a chemical process requiring no (Huttonian) heat. Abraham Gottlob Werner, *Short Classification and Description of the Various Rocks,* trans. with an introduction and notes by Alexander M. Ospovat (New York: Hafner, 1971); V. A. Eyles, "Abraham Gottlob Werner (1749–1817) and His Position in the History of the Mineralogical and Geological Sciences," *History of Science, 3* (1964):102–115.

30. Werner referred to anthracite as slaty- or glance-coal. Bituminous coal was black coal. Werner, *Short Classification,* 147; Jameson, *Mineralogy,* 3:178–179.

31. Jameson, *Mineralogy,* 3:178–179.

32. Jameson, *Mineralogy,* 3:157.

33. Phillips classified mineral coal as a Carbon Combustible. William Phillips, *An Elementary Introduction to the Knowledge of Mineralogy* (London: W. Phillips, 1823), 370–371.

34. Earlier, in 1808, the French geologist Jean-Baptiste d'Omalius d'Halloy had introduced a *terrain bituminifère* (Bituminous class of formations) that corresponded roughly to the Carboniferous. By 1831, d'Omalius had subdivided his Bituminous into an older *terrain anthraxifère* and a younger *terrain houiller* equating approximately with the Old Red Sandstone-Mountain Limestone and Coal Measures.

35. Amos Eaton, "Ought American Geologists to adopt the changes in the Science, proposed by Phillips and Conybeare?" *AJS,* 8 (1824):261–263. Cf. Martin J. S. Rudwick, *The Great Devonian Controversy: The Shaping of Scientific Knowledge among Gentlemanly Specialists* (Chicago: University of Chicago Press, 1985), 71.

36. William D. Conybeare and William Phillips, *Outlines of the Geology of England and Wales, with an Introductory Compendium of the General Principles of that Science, and Comparative Views of the Structure of Foreign Countries* (London: William Phillips, 1822), Preliminary Notice; cf. James A. Secord, *Controversy in Victorian Geology: The Cambrian-Silurian Dispute* (Princeton: Princeton University Press, 1986) 30.

37. England's coal fields fell "naturally" into three geographic districts: Great Northern, Central, and Western. Conybeare and Phillips, *Outlines,* 326.

38. Conybeare and Phillips, *Outlines,* 325–326.

39. Conybeare and Phillips, *Outlines,* 323.

40. Conybeare and Phillips, *Outlines,* 325.

41. Conybeare and Phillips, *Outlines,* 325.

42. Conybeare and Phillips, *Outlines,* 327.

43. Conybeare and Phillips, *Outlines,* 329.

44. Conybeare and Phillips, *Outlines,* 327.

45. Conybeare and Phillips, *Outlines,* 335.

46. Conybeare and Phillips, *Outlines,* 345.

47. The theory also rested on an analogy with "the actual order of things," the partial filling of extant estuaries. Conybeare and Phillips, *Outlines,* 344–346. Such theorizing ran counter to their self-avowed stricture against theory. Rudwick, *Devonian Controversy,* 24; Secord, *Controversy in Victorian Geology,* 30–32.

48. Robert Bakewell, *An Introduction to Geology,* second American from the fourth London edition, ed. Benjamin Silliman (New Haven: Hezekiah Howe, 1833), 101.

49. Leonard G. Wilson, "The Emergence of Geology as a Science in the United States," *Journal of World History, 10* (1967):416–437.

50. Hitchcock went to Yale in 1818 to study for the ministry, but in 1825, for reasons of health, he decided to take on the less arduous task of teaching. He later became college president. Gloria Robinson, "Edward Hitchcock," in Leonard G. Wilson, ed., *Benjamin Silliman and His Circle: Studies on the Influence of Benjamin Silliman on Science in America* (New York: Science History Publications, 1979), 49–83.

51. Edward Hitchcock, *Report on the Geology, Mineralogy, Botany, and Zoology of Massachusetts* (Amherst: J. S. and C. Adams, 1833), introduction; George P. Merrill, *Contributions to a History of American State Geological and Natural History Surveys,* U.S. National Museum, Bulletin No. 109 (Washington, DC: GPO, 1920), 149–158.

52. The other parts treated topographical geology, scientific geology, and animals and plants. Edward Hitchcock, *Report of a Geological Survey of Massachusetts: Part I. Economical Geology* (Amherst: J. S. and C. Adams, 1832), reprinted as "Geology of Massachusetts," *AJS, 22* (1832):1–70.

53. Hitchcock, "Geology of Massachusetts," 43.

54. James G. Percival, *Report on the Geology of the State of Connecticut* (New Haven: Osborn & Baldwin, 1842).

55. Hitchcock, *Report on . . . Massachusetts,* 277.

56. Hitchcock, *Report on . . . Massachusetts,* 282.

57. Jackson used "Transition" because he was certain the terms Cambrian and Silurian "will never be regarded in this country." Charles T. Jackson, *Report on the Geological and Agricultural Survey of the State of Rhode-Island, made under resolve of legislature in the year 1839* (Providence: B. Cranston, 1840), 38–40; "anthracite," 102–103.

58. Richard Cowling Taylor, *Statistics of Coal* (Philadelphia: J. W. Moore, 1848), 146.

59. Percival, *Report on . . . Connecticut,* 452.

60. Hitchcock, "Geology of Massachusetts," 227–228.

61. Amos Eaton, "Geological Podromus," *AJS, 17* (1830):63–68, 68; Samuel Reznick, *Education for a Technological Society: A Sesquicentennial History of Rensselaer Polytechnic Institute* (Troy: Rensselaer Polytechnic Institute, 1968).

62. Amos Eaton, *A Geological Nomenclature for North America; founded upon geological surveys, taken under the direction of the Hon. Stephen Van Rensselaer* (Albany: Packard and Van Benthusen, 1828), reprinted as "Geological Nomenclature, Classes of Rocks," *AJS, 14* (1828):144–159, 359–368.

63. Eaton grouped six other formations under two classes of Detritus. He preferred the old term Detritus to the new term Tertiary. Eaton, *Geological Nomenclature,* 25.

64. Eaton, *Geological Nomenclature,* 32.

65. Amos Eaton, "Observations on the Coal Formations in the State of New York; in connexion with the great Coal Beds of Pennsylvania," *AJS, 21* (1831):21–26.

66. Eaton, "Observations on the Coal," 23.

67. Eaton, "Observations on the Coal," 23.

68. Eaton, "Observations on the Coal," 26.

69. The first survey proposals emphasized coal exclusively. But the successful proposal listed coal with iron ore, gypsum, salt, and building stones. Casting a broader economic net secured the legislative votes. Michele L. Aldrich, *New York State Natural History Survey, 1836–1842: A Chapter in the History of American Science* (Ithaca, NY: Paleontological Research Institution, 2000).

70. Emphasis added. Timothy A. Conrad, "Second Annual Report," in *Third Annual Report of the New York State Natural History Survey*, 61; cf. Aldrich, *New York State Natural History Survey*, 142–146.

71. Charles Lyell, *Lectures on Geology* (New York: Greeley & McElrath, 1843), 36.

72. Taylor, *Statistics of Coal*, 19.

73. Patsy Gerstner's biography of Rogers is an excellent study of the politics and science of the Pennsylvania survey, and it deserved more praise than this fledgling *Isis* reviewer gave it. Gerstner, *Henry Darwin Rogers, 1806–1866: American Geologist* (Tuscaloosa: University of Alabama Press, 1994).

74. They also did research together. H. D. Rogers and A. D. Bache, "Analysis of some of the Coals of Pennsylvania," *Journal of the Academy of Natural Sciences of Philadelphia*, 7 (1834):158–177.

75. J. Peter Lesley, "Pennsylvania," in Merrill, *Contributions*, 428–456; cf. Anthony F. C. Wallace, *St. Clair: A Nineteenth-Century Coal Town's Experience with a Disaster-Prone Industry* (Ithaca: Cornell University Press, 1988), 202–204.

76. J. Peter Lesley continued to use them for decades, including during the Second Geological Survey of Pennsylvania, which he directed between 1874 and 1890.

77. In time, Rogers did discover unconformities, but he stuck by his decision to classify all formations as Secondary. J. Peter Lesley, *Historical Sketch of Geological Explorations in Pennsylvania and Other States* (Harrisburg: Board Commissioners for the Second Geological Survey, 1876), 43.

78. Leonard G. Wilson, *Lyell in America: Transatlantic Geology, 1841–1853* (Baltimore: Johns Hopkins University Press, 1998), 27–38.

79. Lyell's letter contained several references to Rogers about observations and locations, not theoretical points. On American worries about Lyell's possible plagiarism, see Gerstner, *Rogers*, 106–107; Robert H. Silliman, "The Hamlet Affair: Charles Lyell and the North Americans," *Isis*, 86 (1995):541–561. Cf. Wilson's defense of Lyell: Wilson, *Lyell in America*, 97–100.

80. The Rogers brothers presented two papers on coal. In the first, William showed that coal from Richmond was younger than the Carboniferous. William B. Rogers, "On the Age of the Coal Rocks of Eastern Virginia," and Henry D. Rogers, "An Inquiry into the Origin of the Appalachian Coal Strata, Bituminous and Anthracitic," in *Reports of the First, Second, and Third Meetings of the Association of American Geologists and Naturalists, at Philadelphia, in 1840 and 1841, and at Boston in 1842* (Boston: Gould, Kendall, & Lincoln, 1843), 298–301, 433–474.

81. Rogers, "Origin," 438.

82. Rogers, "Origin," 451.

83. Rogers, "Origin," 447.

84. Rogers singled out William Buckland, but then noted that Buckland had recently changed his mind. Rogers, "Origin," 448.

85. Charles Lyell, *Principles of Geology*, 3 vols. (reprint, Chicago: University of Chicago Press, 1990–1991), 1:187–189; cf. 2:239–245 (timber in the Mackenzie River).

86. Charles Lyell, *Elements of Geology*, 2 vols. (London: John Murray, 1841), 2:106–107.

87. Bakewell, *Introduction*, (1833), 113.

88. Bakewell, *Introduction*, (1833), 113.

89. Bakewell, *Introduction*, (1833), 115.

90. Rogers claimed to have been the first to unravel the coal field of Ashby-de-la-Zouch, but Bakewell did so in the first edition of his *An Introduction to Geology* (London: J. Harding, 1813).

91. Logan thought the presence of fireclay and *Stigmaria* made the drift theory "an unsatisfactory hypothesis." William Logan, "On the Characters of the Beds of Clay immediately below the Coal-Seams of South Wales, and on the occurrence of Boulders of Coal in the Pennant Grit of that district," *Transactions of the GSL*, 6 (1841):491–497, 494; summarized in *Proceedings of the GSL*, 3 (1840):275–277, 275.

92. Rogers, "Origin," 453.

93. Rogers, "Origin," 437.

94. Rogers, "Origin," 436.

95. Rogers, "Origin," 434.

96. Rogers, "Origin," 434.

97. Rogers, "Origin," 473.

98. John J. Stevenson (1841–1924), a member of the Second Geological Survey of Pennsylvania and an authority on coal geology, called it "classical." Stevenson, "The Formation of Coal Beds. I. An Historical Summary of Opinion from 1700 to the Present Time," *Proceedings of the APS* (1911):1–116; 21.

99. Lyell, *Lectures on Geology*, 34.

100. Lyell, *Lectures on Geology*, 37.

101. Taylor, *Statistics of Coal*, 74.

102. Taylor, *Statistics of Coal*, 75.

103. Taylor, *Statistics of Coal*, cvii–cviii.

104. Taylor, *Statistics of Coal*, civ.

105. Binney subsequently studied *Sigillaria* and *Stigmaria* in a Manchester coal field. Edward Binney, "Description of the Dunkinfield Sigillaria," *Quarterly Journal of the GSL*, 2 (1846):390–393.

106. Taylor, *Statistics of Coal*, xci. On the history of paleobotany and the centrality of the Carboniferous, see Henry N. Andrews, *The Fossil Hunters: In Search of Ancient Plants* (Ithaca: Cornell University Press, 1980).

107. Lyell, *Lectures on Geology*, 36.

108. Charles Lyell, "On the Probable Age and Origin of a Bed of Plumbago and Anthracite Occurring in Mica-Schist near Worcester, Massachusetts," *Quarterly Journal of the GSL*, 1 (1844):199–202, 201.

109. Charles Lyell, "On the Upright Fossil Trees Found at Different Levels in the Coal Strata of Cumberland, Nova Scotia," *AJS*, 45 (1843):353–356, 354. Cf. Abraham Gesner, *Remarks on the Geology and Mineralogy of Nova Scotia* (Halifax: Gossip and Coade, 1836), 153.

110. Lyell to his sister, 30 July 1842, in Katherine Murray Lyell, ed., *Life, Letters and Journals of Sir Charles Lyell, Bart*, 2 vols. (London: John Murray, 1881), 2:64–66.

111. Still, it was not until Lyell had visited the St. Etienne coal field near Lyon, France, the following year that he completely embraced the growth-on-the-spot theory. Lyell,

Travels in North America, in the years 1841–2; with Geological Observations on the United States, Canada, and Nova Scotia, 2 vols. (New York: Wiley and Putnam, 1845), 2:164.

112. Dawson and Lyell might have met in the summer of 1841, during Lyell's six-hour layover in Halifax en route to Boston. Susan Sheets-Pyenson, *John William Dawson: Faith, Hope, and Science* (Montreal: McGill-Queen's University Press, 1996), 24; Sheets-Pyenson, "Sir William Dawson: The Nova Scotia Roots of a Geologist's Worldview," in Paul A. Bogaard, ed., *Profiles of Science and Society in the Maritimes Prior to 1914* (Sackville, NB: Acadiensis Press, 1990), 83–99; cf. Wilson, *Lyell in America*, 124.

113. Cf. Michael Bartholomew, "The Singularity of Lyell," *History of Science, 17* (1979):276–293.

114. All Dawson's articles were listed and many summarized in John William Dawson, *Acadian Geology: An Account of the Geological Structure and Mineral Resources of Nova Scotia and Portions of the Neighboring Provinces of British America* (Edinburgh: Oliver and Boyd, 1855), 7–10.

115. This was Lyell's third American visit (August–December 1852). During his second visit (September 1845–June 1846), he toured the southern United States, the Mississippi valley, and Pennsylvania. He did not go to British North America. Wilson, *Lyell in America.*

116. Charles Lyell, William Dawson, et al., "On the Remains of a Reptile and a Land-shell in an Erect Tree in the Coal-measures of Nova Scotia," *Quarterly Journal of the GSL, 9* (1853):58–67.

117. William Dawson, "On the Albert Mine, Hillsborough, New Brunswick" *Quarterly Journal of the GSL, 9* (1853):107–114, 108.

118. Dawson, "Albert Mine," 108.

119. Dawson, *Acadian Geology,* 198.

120. Dawson, "Albert Mine," 108.

121. Dawson, "Albert Mine," 108.

122. Dawson, *Acadian Geology,* 203.

123. Dawson, *Acadian Geology,* 201.

124. Dawson, *Acadian Geology,* 203.

125. Dawson, *Acadian Geology,* 206.

126. Dawson, *Acadian Geology,* 210.

127. Dawson, *Acadian Geology,* 210.

128. Dawson, "Albert Mine," 110.

129. Dawson, *Acadian Geology,* 206–207.

130. Dawson, *Acadian Geology,* 198.

131. Six of fifteen chapters were devoted to the Carboniferous. Dawson, *Acadian Geology,* 148.

132. J. Peter Lesley, *Manual of Coal and Its Topography* (Philadelphia: J. B. Lippincott, 1856), 123–124. The structural method had much in common with the "geometrical" one of the English geologist Adam Sedgwick. Lesley, *Geological Explorations*, 19; cf. Secord, *Controversy in Victorian Geology*, 200, and R. H. Dott Jr., "The American Countercurrent—Eastward Flow of Geologists and Their Ideas in the Late Nineteenth Century," *Earth Sciences History, 9* (1990):158–162.

133. Lesley, *Geological Explorations*, 124.

134. Lesley, *Geological Explorations*, 79.

135. Lesley, *Geological Explorations*, 115.

136. Lesley's dramatic decision remains somewhat mysterious. He might not have been able to accept the church's doctrine on the age of the earth. Lyell thought this, but Lesley did not like Lyell. By the late 1840s, Lesley had adopted a theological stance embracing the divinity of humankind and God's role in creation. He accepted some progression of life forms as revealed in the fossil record, but Lesley never fully embraced Charles Darwin's theory of natural selection. Lesley, *Man's Origin and Destiny Sketched from the Platform of the Physical Sciences,* 2nd ed. (1868; Boston: Geo. H. Ellis, 1881), ch. 4.

137. Gerstner, *Rogers,* 202–205.

138. Lesley rejected Rogers's theory of mountain building for Élie de Beaumont's "theory of lateral pressure." Lesley, *Manual,* 181. Cf. Crosbie Smith, "William Hopkins and the Shaping of Dynamical Geology: 1830–1860," *BJHS,* 22 (1989):27–52.

139. Lesley, *Manual,* 207–208.

140. Lesley, *Man's Origin and Destiny,* 331–332.

141. Lesley, *Manual,* 189.

142. Lesley, *Manual,* 84.

143. Lesley, *Manual,* 66–69.

144. Lesley, *Manual,* 89.

145. Lesley, *Manual,* 89.

146. Lesley, *Manual,* 25.

147. Lesley, *Manual,* 25.

148. Lesley, *Geological Explorations,* 58–67.

149. Henry Darwin Rogers, *The Geology of Pennsylvania: A Government Survey,* 2 vols. (Edinburgh and London: W. Blackwood and Sons; Philadelphia: J. B. Lippincott, 1858), 2:946.

150. Dawson, *Acadian Geology,* 210.

151. Rogers, *Geology of Pennsylvania,* 2:946.

152. Rogers, *Geology of Pennsylvania,* 2:947.

153. Rogers, *Geology of Pennsylvania,* 2: 947.

154. Rogers, *Geology of Pennsylvania,* 2:947.

155. Rogers, *Geology of Pennsylvania,* 2:990.

156. Taylor, *Statistics of Coal,* vii.

157. Lesley, *Manual,* 22.

158. Lesley, *Manual,* 66–67.

159. Bakewell declared that in "more than a third of England" the search for coal was "useless." "[T]he knowledge of a negative fact becomes important," he explained, "when it saves us from loss of time, expense, and disappointment." Bakewell, *Introduction to Geology* (1833), 109.

160. Rogers, *Geology of Pennsylvania,* 2:1007–1009.

161. James Dwight Dana, *Manual of Geology: Treating of the Principles of the Science with special reference to American Geological History* (Philadelphia: Theodore Bliss, 1863), 325. Much of the book was written before June 1859, but the first "revised edition" was published in 1862 in New York and then republished in 1863 in Philadelphia. A new "revised edition" appeared in 1868; with the exception of an appendix, the text and pagination remained the same in the 1863 and 1868 editions.

162. Dana, *Geology* (1863), vii.

163. The first 1862 edition of the *Manual of Geology* might not have included the Carboniferous frontispiece; the 1863 and 1868 editions usually did. In the 1876 and subse-

quent editions (1880 and 1895), Dana replaced the Carboniferous with a morbid photograph of a human skeleton discovered in a cave in southern Europe. Dawson chose a picture of Devonian flora for the frontispiece of his *Acadian Geology* (1855). On the importance of frontispieces, see Martin J. S. Rudwick, "The Emergence of Visual Language for Geological Science, 1760–1840," *History of Science, 14* (1976):149–195.

164. Dana, *Geology* (1863), 369.

165. Such repeating series are called cyclothems, a term introduced by John Strong Newberry.

166. The best evidence of a "break" between the Subcarboniferous and Carboniferous was found by James Hall and other survey geologists in Iowa and Illinois, where coal lay unconformably upon limestone. Dana, *Geology* (1868), 320.

167. In the fourth edition of his *Manual of Geology* (1895), Dana attributed the introduction of the terms "Mississippian" (for the lower or limestone division) and "Pennsylvanian" (for the upper or coal and conglomerate division) to Henry S. Williams, *Correlation of the Devonian and Carboniferous,* USGS Bulletin 80, (Washington, DC: GPO, 1891). Dana, himself, introduced the term *Carbonic* for Carboniferous, but that term was never accepted. James Dwight Dana, *Manual of Geology* (New York: American Book Co., 1895), 631–632. William R. Brice, "Henry Shaler Williams (1847–1918) and the Pennsylvanian Period," paper presented at the "Conference on the History of Geologic Pioneers," Northeastern Science Foundation, Troy, NY, 3–5 August 2000.

168. Taylor, *Statistics of Coal,* 74. Cf. "Statistics of Coal," *AJS, 6* (1848):150–151.

169. Rogers, *Geology of Pennsylvania,* 2:1015.

170. Dana, *Geology* (1868), 352.

Chapter 4. Mining Science

1. Charles Jackson, "Remarks on Mining Operations, 1846," Jackson Mss in the Papers of the Association of American Geologists and Naturalists, Academy of Natural Sciences, Philadelphia.

2. This prosopographical method and its conclusions contrast with the biographical approaches of Colin A. Russell, *Edward Frankland: Chemistry, Controversy, and Conspiracy in Victorian England* (Cambridge: Cambridge University Press, 1996), and Jack Morrell, *John Phillips and the Business of Victorian Science* (Aldershot: Ashgate, 2005). Both Frankland and Phillips accepted far fewer commissions than their American counterparts. Frankland, especially, was on retainer for long periods to large chemical firms. Cf. Paul Lucier, "Commercial Interests and Scientific Disinterestedness: Consulting Geologists in Antebellum America," *Isis, 86* (1995):245–267.

3. Ethel M. McAllister, *Amos Eaton: Scientists and Educator, 1776–1842* (Philadelphia: University of Pennsylvania Press, 1941); Samuel Reznick, *Education for a Technological Society: A Sesquicentennial History of Rensselaer Polytechnic Institute* (Troy: Rensselaer Polytechnic Institute, 1968); and David Ian Spanagel, "Chronicles of a Land Etched by God, Water, Time, and Ice" (PhD dissertation, Harvard University, 1996).

4. Rensselaer was also the president of New York's canal commission between 1825 (the opening of the Erie Canal) and 1839 (his death). [Editorial], *AJS, 14* (1828):360.

5. Eaton's second duty was to the public. Eaton, "Geological Podromus," *AJS, 17* (1830):63–69, 68.

6. Amos Eaton, "General Geological Strata," *AJS*, *14* (1828):359–368, 360. Rensselaer defended Eaton against criticism for abusing such patronage. "Gen. Van Rensselaer's Note," *AJS*, *20* (1831):419–420, 419.

7. John F. Fulton and Elizabeth H. Thomson, *Benjamin Silliman, 1779–1864: Pathfinder in American Science* (New York: Henry Schuman, 1947); John C. Greene, "Protestantism, Science and American Enterprise: Benjamin Silliman's Moral Universe," in Leonard G. Wilson, ed., *Benjamin Silliman and His Circle: Studies on the Influence of Benjamin Silliman on Science in America* (New York: Science History Publications, 1979), 11–27; Chandos Michael Brown, *Benjamin Silliman: A Life in the Young Republic* (Princeton: Princeton University Press, 1989).

8. Benjamin Silliman, "Particulars Relative to the Lead-Mine near Northampton, (Massachusetts)," *American Mineralogical Journal*, *1* (1814):63–69.

9. Fulton and Thomson, *Silliman*, 197. Benjamin Silliman and Benjamin Silliman Jr., *Extracts from a report made to the Maryland and New York coal and iron company . . . in the county of Alleghany, state of Maryland* (London, 1839); and Benjamin Silliman, William B. Rogers, and Forrest Sheperd, *To the president and directors of the Walton Mining Company* (Fredericksburg, 1836).

10. Benjamin Silliman, *Report on the coal formation of the valleys of Wyoming and Lackawanna* (New Haven, 1830); cf. Silliman, "Notice of the Anthracite Region in the Valley of the Lackawanna and of the Wyoming on the Susquehanna," *AJS*, *18* (1830):308–328; and "Notes on a journey to Mauch Chunk and other Anthracite regions of Penn.," *AJS*, *19* (1831):1–21. On gold, see Benjamin Silliman, *Report on the gold mine in Culpepper county* (Fredericksburg, 1836); *Report of an examination of the gold districts of the Virginia and New England Mining Company* (Fredericksburg, 1836); and *To the president and directors of the Richmond Mining Company* (Richmond, 1836). Cf. Silliman, "Remarks on some of the Gold Mines, and on parts of the Gold Region of Virginia, founded on personal observations, made in the months of August and September, 1836," *AJS*, *32* (1837):98–130.

11. Greene, "Protestantism, Science, and American Enterprise," 19–22. On Silliman's "doctrine of service," see Brown, *Silliman*, 114.

12. A. Hunter Dupree, *Science in the Federal Government: A History of Politics and Activities* (Baltimore: Johns Hopkins University Press, 1986); Margaret W. Rossiter, *The Emergence of Agricultural Science: Justus Liebig and the Americans, 1840–1880* (New Haven: Yale University Press, 1975); for the Gilded Age, Charles E. Rosenberg, *No Other Gods: On Science and American Social Thought* (Baltimore: Johns Hopkins University Press, 1976), 135–195.

13. The literature on surveys is large. See, for example, George P. Merrill, *The First One Hundred Years of American Geology* (New Haven: Yale University Press, 1924); Merrill, *Contributions to a History of American State Geological and Natural History Surveys*, U.S. National Museum, Bulletin No. 109 (Washington, DC: GPO, 1920); Patsy Gerstner, *Henry Darwin Rogers, 1808–1866: American Geologist* (Tuscaloosa: University of Alabama, 1994); Michele L. Aldrich, *New York State Natural History Survey, 1836–1842* (Ithaca: Paleontological Research Institute, 2000); Anne Marie Millbrooke, "State Geological Surveys of the Nineteenth Century" (PhD dissertation, University of Pennsylvania, 1981); Julie Renee Newell, "American Geologists and Their Geology: The Formation of the American Geological Community, 1780–1865" (PhD dissertation, University of Wisconsin at Madison, 1993); Walter B. Hendrickson, "Nineteenth-Century State Geological Surveys: Early Government Support for Science," *Isis*, *52* (1961):357–371; Ger-

ald D. Nash, "The Conflict between Pure and Applied Science in Nineteenth-Century Public Policy: The California State Geological Survey, 1860–1874," *Isis, 54* (1963):217–228; Michele L. Aldrich, "American State Geological Surveys, 1820–1845," in Cecil J. Schneer, ed., *Two Hundred Years of Geology in America: Proceedings of the New Hampshire Bicentennial Conference on the History of Geology* (Hanover, NH: University Press of New England, 1979), 133–144; Stephen P. Turner, "The Survey in Nineteenth-Century American Geology: The Evolution of a Form of Patronage," *Minerva, 25* (1987):282–330.

14. J. Peter Lesley, *Historical Sketch of Geological Explorations in Pennsylvania and Other States* (Harrisburg: Board Commissioners for the Second Geological Survey, 1876), 101.

15. Lesley, *Geological Explorations*, 101.

16. John Lauritz Larson, *Internal Improvement: National Public Works and the Promise of Popular Government in the Early United States* (Chapel Hill: University of North Carolina Press, 2001); Douglass C. North, *The Economic Growth of the United States, 1790–1860* (New York: W. W. Norton, 1966).

17. Jackson, "Remarks on Mining."

18. John M. Clarke, *James Hall of Albany: Geologist and Paleontologist, 1811–1898* (Albany: privately published, 1921); J. J. Stevenson, "Memoir of James Hall," *Bulletin of the Geological Society of America, 61* (1895):425–451; and a "James Hall" special issue of *Earth Sciences History, 6* (1987):1–144.

19. Aldrich, *New York State Natural History Survey*, 105, 141.

20. Hall Mss, box 39: Mining Reports, NYSL.

21. Hall received $1,000 for his Minnesota survey. H. C. Lee to Hall, 19 April 1865, Hall Mss, NYSA. Hall, "On the Structure of the Mountains and Valleys in Tennessee, Northern Georgia, and Alabama," and "Remarks on the Geological Structure of Southern Minnesota," in *Proceedings of the AAAS, 15* (1866):105. For the Minnesota survey, see Merrill, *Contributions*, 239–242.

22. Biddle took over. Patsy Gerstner, *Henry Darwin Rogers, 1808–1866: American Geologist* (Tuscaloosa: University of Alabama Press, 1994).

23. Lesley, *Geological Explorations*, 122.

24. Sheafer rejoined the survey and was instrumental in getting funding to restart in 1851. Lesley, *Geological Explorations*, 126; Gerstner, *Rogers*, 174–179.

25. Lesley, *Geological Explorations*, 105, 186.

26. Mary Lesley Ames, *Life and Letters of Peter and Susan Lesley*, 2 vols. (New York: G. P. Putnam, 1909); B. S. Lyman, "Biographical Sketch of J. Peter Lesley," *Transactions of the American Institute of Mining Engineers, 34* (1903):726–739; J. J. Stevenson, "Memoir of J. Peter Lesley," *Bulletin of the Geological Society of America, 15* (1904):532–541; H. M. Chance, "A Biographical Notice of J. Peter Lesley," *Proceedings of the APS, 45* (1906):1–14; A. Geikie, "Notice of J. P. Lesley," *Quarterly Journal of the GSL, 60* (1904):xlix–lv.

27. Lesley to Joseph Lesley, 5 September 1857, Lesley Mss, APS.

28. J. Peter Lesley, *The Iron Manufacturer's Guide to the Furnaces, Forges, and Rolling Mills of the United States with Discussion of Iron as a Chemical Element, an American Ore, and a Manufactured Article, in Commerce and in History* (New York: John Wiley, 1859).

29. Charles Jackson, *Report on the Geological and Agricultural Survey of the State of Rhode-Island* (Providence: B. Cranston, 1840), 241–246.

30. Charles Jackson has been described as paranoid about priority and propriety.

Richard J. Wolfe, *Tarnished Idol: William Thomas Green Morton and the Introduction of Surgical Anesthesia* (San Anselmo, CA: Norman, 2000); Julie M. Fenster, *Ether Day: The Strange Tale of America's Greatest Medical Discovery and the Haunted Men Who Made It* (New York: HarperCollins, 2001).

31. The position was for a chemist to replace Benjamin Silliman Jr., who left Yale for the University of Louisville. Whitney was not offered the job because he was disliked by John P. Norton, the school director. Louis I. Kuslan, "The Rise of the Yale School of Applied Chemistry (1845–1856)," in Wilson, *Benjamin Silliman and His Circle*, 129–157, 139; Robert V. Bruce, *Launching of Modern American Science, 1846–1876* (Ithaca: Cornell University Press, 1988), 140.

32. Such exchanges have not received much attention from historians of science, although historians of medicine have used them and marketplaces as analytical tools for understanding the development of professional medicine and the relations between physicians and patients. Roy Porter, *Health for Sale: Quackery in England, 1660–1850* (Manchester: Manchester University Press, 1989); Dorothy Porter and Roy Porter, *Patient's Progress: Doctors and Doctoring in Eighteenth-Century England* (Stanford: Stanford University Press, 1989); W. F. Bynum and Roy Porter, eds., *Medical Fringe and Medical Orthodoxy, 1750–1850* (London: Croom Helm, 1987).

33. Letters might be considered the equivalent of contracts, but they were not always legally binding. See, for example, Stephen Wilson to Lesley, 3 and 4 March 1856, Lesley Mss, APS. I have found only one example of a legal contract between a consultant and a business—James Hall and the Empire State Iron and Coal Mining Company, located in northern Georgia. Hall did a survey in the autumn of 1865 for $2,000. Hall Mss, box 39: Mining Reports, NYSL.

34. Arthur Hill to Lesley, 5 and 16 April 1856, Lesley Mss, APS.

35. Henry A. DuBois to Hall, 11 January 1854, Hall Mss, NYSA.

36. Hall Mss, box 39: Mining Reports, NYSL.

37. Hall to Mr. Hunt, 20 April 1846, Hall Mss, NYSL.

38. Hotchkiss to Silliman, 15 May 1854, Lesley Mss, APS.

39. Silliman to Lesley, 16 May 1854, Lesley Mss, APS.

40. Hotchkiss to Lesley, 19 May 1854, Lesley Mss, APS.

41. Hotchkiss to Lesley, 26 May 1854; S. M. Newton to Lesley, [28 May 1854]; J. Hobbes to Joseph Lesley, 17 and 24 June 1854, Lesley Mss, APS.

42. Lesley did not go with Hall. Hall to Lesley, 28 May 1854, Lesley Mss, APS.

43. Hall to John E. Devlin, 23 October 1864; Hall to Lesley, 23 October 1864, Lesley Mss, APS.

44. Hall to William Hall, three undated letters, probably October 1845, Hall Mss, NYSL.

45. Hall's expenses for 28 July–21 October 1845 included: "one barrel of pork @ $13.00," "Fare to Detroit @ $3.00," "Oil, lamps, candles, nails, augur &c @ $4.50," "mittens for use in drilling @ $2.25," and "Fish purchased for several days @ $1.42." Hall Mss, box 39: Mining Reports, NYSL. Hall to Hunt, 20 April 1846, Hall Mss, NYSL. The pistol cost $16.00; no price was recorded for the bottles of brandy.

46. Emphasis added. Lesley, *Geological Explorations*, 111–112.

47. "Report of J. P. Lesley, Geologist, to Mr. Wm. E. S. Baker, Secy Duncannon Iron Co.," 25 October 1864, Lesley Mss, APS.

48. Lesley to Baker, 25 October 1864, Lesley Mss, APS.

49. Lesley to Charles Rann, 26 June 1865, Lesley Mss, APS.

50. Anthony F. C. Wallace, *St. Clair: A Nineteenth-Century Coal Town's Experience with a Disaster-Prone Industry* (Ithaca: Cornell University Press, 1988), 200–215.

51. Hodge to Hall, 5 December 1853, Hall Mss, NYSA. Cf. Josiah Dwight Whitney, *The Metallic Wealth of the United States, Described and Compared with that of other Countries* (Philadelphia: Lippincott, Grambo & Co., 1854), 129.

52. Lesley to L. H. Simpson, 27 June 1864. Lesley Mss, APS.

53. Lesley to L. A. Mackey, 11 July 1864. Lesley Mss, APS.

54. Richard Cowling Taylor, *Statistics of Coal* (Philadelphia: J. W. Moore, 1848), 90.

55. Taylor, *Statistics of Coal,* 94.

56. J. Peter Lesley, *A Collection of Occasional Surveys of Iron, Coal and Oil Districts in the United States, made during the last ten years* (Philadelphia: McCalla & Stavely, 1874).

57. Charles Jackson, *Reports on the Geological Relations, Chemical Analyses, and Microscopic Examination of the Coal of the Albert Coal Mining Co. situated in Hillsboro, Albert Co., New Brunswick* (New York: Nesbitt, 1851), 3.

58. Jackson, *Reports for the Albert Coal Mining Co.,* 8.

59. Jackson, *Reports for the Albert Coal Mining Co.,* 20.

60. Jackson, *Reports for the Albert Coal Mining Co.,* 23. Jackson described five other species: *Palaeoniscus Alberti* (for Albert County), *P. Brownii* (for Archibald Brown, chief miner), *P. Cook and Smithii* (for Cairns's American agents), *P. Allisonii* (for Edward Allison), and *P. Jacksonii.*

61. *Proceedings of the BSNH, 4* (18 June 1851):73–74.

62. [James Dwight Dana], "Reports on the Albert Coal Mine," *AJS, 13* (1852):276–277.

63. Jackson, *Reports for the Albert Coal Mining Co.,* 26.

64. There is no analogy to a paradigm shift in normal commercial science, although the case has been made. George Wise, "A New Role for Professional Scientists in Industry: Industrial Research at General Electric, 1900–1916," *T&C* (1980):408–429; Leonard S. Reich, *The Making of American Industrial Research: Science and Business at GE and Bell, 1876–1926* (Cambridge: Cambridge University Press, 1985). On normal science, see Thomas Kuhn, *Structure of Scientific Revolutions* (Chicago: University of Chicago Press, 1970).

65. Taylor, *Statistics of Coal,* 97.

66. Taylor, *Statistics of Coal,* 67, 69.

67. Taylor, *Statistics of Coal,* 97–98.

68. Henry Darwin Rogers, *The Geology of Pennsylvania: A Government Survey,* 2 vols. (Edinburgh and London: W. Blackwood and Sons; Philadelphia: J. B. Lippincott, 1858), 1:699–705; cf. Rogers, *Reports of Professor Henry D. Rogers on Wheatley, Brookdale, and Charleston Mines, Phoenixville, Chester County, Pennsylvania* (Philadelphia: T. K. & P. G. Collins, 1853); Gerstner, *Rogers,* 187–188.

69. "The Metallic Wealth," *AJS, 18* (1854):274–277, 277.

70. Whitney, *Metallic Wealth,* xxix.

71. Whitney, *Metallic Wealth,* xxxi, 320.

72. Whitney to Hall, 29 December 1853, Hall Mss, NYSA.

73. J. Peter Lesley, *Manual of Coal and Its Topography* (Philadelphia: J. B. Lippincott, 1856).

74. "Manual of Coal and its Topography," *Journal of the Franklin Institute, 33* (1857):70–72.

75. Review of *Manual of Coal, AJS, 22* (1856):302–303.

76. W. M. Davis, "Biographical Memoir of J. Peter Lesley, 1819–1903," *Biographical Memoirs. National Academy of Sciences, 8* (1915):152–240, 183. Merrill thought Lesley's book might have been "the most important" work on economic geology thus far to appear; Merrill, *American Geology,* 342. T. Sterry Hunt gave Lesley partial credit for developing the geosynclinal theory of mountain building; Hunt, "On some points in American Geology," *AJS, 31* (1861):392–410.

77. Lesley to Hall, 20 April 1856, Hall Mss, NYSA.

78. Hall Mss, box 39: Mining Reports, NYSL.

79. Bruce, *Launching of Modern American Science,* 136–137.

80. Lesley, "Report of Cape Breton Coal Beds for the International, Caledonia, Clyde and Blockhouse Coal Companies," 27 June 1865, Lesley Mss, APS.

81. Bruce, *Launching of Modern American Science,* 140.

82. Julius H. Ward, *The Life and Letters of James Gates Percival* (Boston: Ticknor and Fields, 1866), 481–488.

83. Whitney to Hall, 22 April 1854; Newberry to Hall, 25 January 1855, Hall Mss, NYSA.

84. Agassiz to Hall, [undated]; Agassiz to Hall, 14 March 1855, Hall Mss, NYSA.

85. Lesley was dean of the Towne Scientific School at the University of Pennsylvania from 1875 to 1883 but gave up the post to work full time on the geological survey.

86. Lesley to Benjamin S. Lyman, 19 December 1864, Lesley Mss, APS.

87. Newberry took the position, although Lesley recommended T. Sterry Hunt. F. A. P. Barnard to Lesley, 28 February 1866; Barnard to Lesley, 20 March 1866, Lesley Mss, APS.

88. It was a nonresident position with a salary of only $1,000; Hall advised Lesley to turn it down. Hall to Lesley, 6 August 1868; Lesquereux to Lesley, 27 December 1868, Lesley Mss, APS.

89. Hall's average income from consulting does not include two surveys for "Peuse" amounting to $1,500 and $2,000, which he added to the bottom of the list and which would have put his average income at more than $550 per year. Nor does it include the $13,400 Hall apparently made from land sales in Ohio. Hall Mss, box 39: Mining Reports, NYSL. Hall's salary as state paleontologist was $800 to $1,000 a year. Merrill, *Contributions,* 353, 361.

90. August S. Peabody to Lesley, 8 August 1854, Lesley Mss, APS.

91. Lesley to L. A. Mackey, 11 July 1864; August S. Peabody to Lesley, 8 August 1854, Lesley Mss, APS.

92. David Blair to Lesley, 14 and 21 July 1856, Lesley Mss, APS.

93. Lesley to William E. S. Baker, 25 October 1864, Lesley Mss, APS.

94. Agassiz to Lesley, 12 and 24 April 1865; Lesley to Agassiz, 14 April 1865, Lesley Mss, APS.

95. A. P. Wilson to Lesley, 31 May 1860, Lesley Mss, APS.

96. Lesley to Wilson, 2 June 1860, Lesley Mss, APS.

97. Lesley to Joseph Lesley, 2 June 1860, Lesley Mss, APS.

98. *Report of a Case, tried at Albert Circuit, 1852, before his Honor Judge Wilmot, and a Special Jury. Abraham Gesner vs. William Cairns. Copied from the Judge's Notes* (Saint John: William L. Avery, 1853), 116.

99. Hall to Lyell, 24 September 1854. On fees, see Hall to Lyell, 4 July 1854, Hall Mss, NYSA.

100. Hall to Lyell, 22 August 1854, Hall Mss, NYSA.

101. Lyell to Hall, 22 October 1854; Lyell to Hall, 4 November 1854, Hall Mss, NYSA.

102. Lesley to H. A. Hendry, 4 November 1864, Lesley Mss, APS.

103. August S. Peabody to Lesley, 8 and 14 August 1854, Lesley Mss, APS. Thomas S. Ridgeway, *Geological Report [on the Wood River Coal Mines]* (St. Louis, 1850).

104. Hall to L. M. Arnold, 23 July 1864, Hall Mss, NYSA.

105. Lesley to Prince & Co., 9 and [23] November 1864; Lesley to [Benjamin S.] Lyman, 19 December 1864, Lesley Mss, APS.

106. Scott and Wilson to Lesley, 19 January 1856 [two letters], Lesley Mss, APS.

107. R. Bruce Petrikin to Lesley, 28 January 1856, Lesley Mss, APS.

108. A. P. Wilson to Lesley, 19 February 1856, Lesley Mss, APS.

109. Silliman to Hall, 20 March 1855; Headley to Hall, 27 March and 18 April 1855; Headley to Silliman, 9 June 1855; Silliman to Hall, 28 June and 13 August 1855, Hall Mss, NYSA. Benjamin Silliman Jr. and James Hall, *Coal Lands in Kentucky. Report on the Bituminous Coal Lands upon the Waters of the Big Sandy River in Kentucky. Examined in 1854 for the Kentucky Mining and Manufacturing Company* (New York: Holman & Gray, 1855).

110. This case referred to Cobb & Company, but I found no further information on this engagement. Whitney to Hall, 7 February 1854, Hall Mss, NYSA.

111. See, for example, W. G. Snethen to Silliman Jr., 27 June 1854, Silliman Mss, YUL.

112. Jackson, "Remarks on Mining."

113. Emphasis added. *Proceedings of the BSNH*, 2 (1846):109–114. Jackson consulted for the Lake Superior Mining Company, the Phoenix Mining Company, the Pittsburgh and Isle Royal Mining Company, and the Boston Copper Company. All were reviewed favorably in Whitney, *Metallic Wealth*, 262, 266–267, 285.

114. Thomas Bouvé to Hall, 14 June 1849, Hall Mss, NYSL.

115. Wolfe, *Tarnished Idol*; Fenster, *Ether Day*.

116. Hall to Hunt, 20 April 1846, Hall Mss, NYSL.

117. Thomas M. Howe to Hall, 4 October 1845, Hall Papers, box 39: Mining Reports, NYSL.

118. Hall to William Hall, 1 October 1845, Hall Mss, NYSL. Two copies of Hall's application for a lease from the War Department are located under Hall to General John Stockton (superintendent of the Mineral Lands of Lake Superior), October 1845, Hall Mss, NYSL.

119. F. A Genth to Silliman Jr. 11 April 1854, Silliman Mss, YUL.

120. Cf. Gerald T. White, *Scientists in Conflict: The Beginnings of the Oil Industry in California* (San Marino: Huntington Library, 1968), 43; Louis I. Kuslan, "Benjamin Silliman, Jr., The Second Silliman," in Wilson, *Benjamin Silliman and His Circle*, 186; Bruce, *Launching of Modern American Science*, 141–142.

121. Silliman to Hall, 13 June 1854, Hall Mss, NYSA.

122. Silliman to Hall, 13 and 26 June 1854, Hall Mss, NYSA.

123. Silliman to Lesley, 24 June 1854, Lesley Mss, APS.

124. Silliman to Lesley, 5 August 1854, Lesley Mss, APS.

125. Lesley was also concerned about slavery, and Silliman assured him that in that part of Kentucky, slavery was a "nullity." Silliman to Lesley, 5 August 1854, Lesley Mss, APS.

126. Silliman to Lesley, 11 August 1854, Lesley Mss, APS.

127. Silliman to Lesley, 17 August 1854, Lesley Mss, APS.

128. Silliman to Lesley, 23 August 1854, Lesley Mss, APS.

129. Silliman to Hall, 24 August 1854; Silliman to William G. Forbes, 24 August 1854; William G. Forbes to Hall, 25 August and 4 and 5 September 1854, Hall Mss, NYSA.

130. Silliman to Hall, 26 February 1856, Hall Mss, NYSA. Silliman was referring to the Diamond Coal Company, for whom he did a survey and report along with John Addison Porter, professor of agricultural chemistry at Yale's School of Applied Chemistry. Porter replaced John Pitkin Norton, who had died in 1852. Silliman and Porter, *Report of an examination of the Ohio Diamond Coal Co. properties in Jefferson county. O.* (New Haven: T. J. Stafford, 1855). It is unknown whether Hall was interested in this company or not. In a letter to Lesley, Hall mentioned that he was going to Ohio "to look at a coal tract." He even invited Lesley to join him, but it seems Lesley was already engaged. Hall to Lesley, 28 May 1856, Lesley Mss, APS.

131. Silliman to Hall, 26 February 1856, Hall Mss, NYSA.

132. Briggs to Hall, 11 December 1837, Hall Mss, NYSL.

133. D. J. Gregory to Hall, 26 January 1838, Hall Mss, NYSL.

134. Merrill, *Contributions,* 394–397.

135. Hall to William Hall, 11 March 1845, Hall Mss, NYSL. The "New York" purchase was near Jackson Furnace; Portsmouth was the nearest town on the Ohio River. Campbell, Peters, and Co. to Hall, 8 January 1853, Hall Mss, NYSL.

136. "A Bill to Incorporate the Ohio Iron Manufacturing Company passed by the Ohio House of Representatives on March 6, 1845." The company had a capitalization of $300,000 in $100 shares. Hall Mss, box 39: Mining Reports, NYSL.

137. Merrill, *Contributions,* 275–277.

138. Merrill, *Contributions,* 106.

139. Campbell, Peters, and Co. to Hall, 8 January 1853, Hall Mss, NYSL. They paid $5 an acre for 2,000 acres.

140. Newberry to Hall, 19 April 1854, Hall Mss, NYSA.

141. Newberry to Hall, 18 December 1854, Hall Mss, NYSA.

142. Newberry to Hall, 25 January 1855, Hall Mss, NYSA.

143. Horsford to Hall, 31 May 1854, Hall Mss, NYSA.

144. "A memorandum of several sums of money received from various sources since the beginning of 1844," Hall to Unknown, December 1854, Hall Mss, NYSL.

145. Alexander W. Thayer to Hall, 7 December 1853, Hall Mss, NYSA.

146. Joseph to Peter Lesley, 18 July 1865, J. Peter and Joseph Lesley Correspondence, 1860–1887, APS.

147. The Yale chemist John Pitkin Norton (1822–1852) was one of the original directors of the company. Benjamin Silliman Jr. and J. D. Whitney, *Report of an Examination of the Bristol Copper Mine, in Bristol, Conn.* (New Haven, 1855); cf. Silliman and Whitney, "Notice of the Geological position and character of the Copper Mine at Bristol, Connecticut," *AJS,* 20 (1855):361–368; Kuslan, "Silliman," 195–205, 186–187.

148. An exception was Eben Horsford, who ran a very successful firm, the Rumford Chemical Co., maker of baking powder, but he had to give up his professorship at Harvard University in 1863 to do so. Margaret W. Rossiter, *The Emergence of Agricultural Science: Justus Liebig and the Americans, 1840–1880* (New Haven: Yale University Press, 1975).

149. Hodge to Hall, 14 December 1853, Hall Mss, NYSA.

150. Hodge to Hall, 5 December 1853, Hall Mss, NYSA.

151. Hodge to Hall, 5 December 1853, Hall Mss, NYSA. Jackson and Hodge even

worked together. *Prospectus of the Southern Gold Company: together with the reports of C. T. Jackson and James T. Hodge* (New York: George F. Nesbitt, 1854).

152. Lesley to L. H. Simpson, 27 June 1864, Lesley Mss, APS.

153. Whitney to Hall, 29 December 1854, Hall Mss, NYSA.

154. Mark Wahlgren Summers, *The Plundering Generation: Corruption and the Crisis of the Union, 1849–1861* (Oxford: Oxford University Press, 1987).

155. Jackson, "Remarks on Mining."

156. Silliman to Lesley, 24 June 1854, Lesley Mss, APS.

157. Hall to Charles Wilkes, 22 September 1846, Hall Mss, NYSL.

158. The report was on coal in Clinton County, Pennsylvania. A. G. Holmes to Hall, 9 August 1854, Hall Mss, NYSA.

159. Agassiz to Hall, [undated] and 14 March 1855, Hall Mss, NYSA.

160. Jackson, "Remarks on Mining."

161. Jackson, "Remarks on Mining."

162. For institutional science, see Robert K. Merton, "The Normative Structure of Science," in Merton, *The Sociology of Science: Theoretical and Empirical Investigations* (Chicago: University of Chicago Press, 1973), 267–278. The norms guiding nineteenth-century consultants were more characteristic of gentlemanly behavior of the seventeenth and eighteenth centuries. Steven Shapin, *A Social History of Truth: Civility and Science in Seventeenth-Century England* (Chicago: University of Chicago Press, 1994). The scientific and commercial communities were still small enough in midcentury America that recourse to impersonal or mechanically objective methods of ensuring trust were unnecessary. Theodore M. Porter, *Trust in Numbers: The Pursuit of Objectivity in Science and Public Life* (Princeton: Princeton University Press, 1995).

163. Thomas L. Haskell, "Professionalism versus Capitalism: R. H. Tawney, Emile Durkheim, and C. S. Peirce on the Disinterestedness of Professional Communities," in Haskell, ed., *The Authority of Experts: Studies in History and Theory* (Bloomington: Indiana University Press, 1984), 180–225.

164. Taylor, *Statistics of Coal,* 67.

Chapter 5. The Technological Science of Kerosene

1. "A Practical Treatise on Coal, Petroleum, and Other Distilled Oils," *AG-LJ,* 2 (15 December 1860):187.

2. The scholarly literature is interesting, important, and large. The best, albeit somewhat dated, review remains John M. Staudenmaier, *Technology's Storytellers: Reweaving the Human Fabric* (Cambridge, MA: MIT Press, 1985), 83–120.

3. Coal oil was more than a mere prelude to petroleum. Cf. Harold F. Williamson and Arnold R. Daum, *The American Petroleum Industry: The Age of Illumination, 1859–1899* (Evanston: Northwestern University Press, 1959); Paul H. Giddens, *The Birth of the Oil Industry* (New York: Macmillan, 1938). On British coal oil, see R. J. Forbes, *Studies in Early Petroleum History* (Leiden: E. J. Brill, 1958), 182–194; John Butt, "James Young, Scottish Industrialist and Philanthropist" (PhD dissertation, Glasgow University, 1964).

4. "Mineral Oils," *Scientific American,* 1 (23 July 1859):59.

5. "Coal Oil Manufacture," *Scientific American,* 2 (2 January 1860):3; "Coal-Oil and Its Manufacture," *AG-LJ,* 2 (1 December 1860):168; "Coal Oil," *AG-LJ,* 2 (2 July 1860):271. [Note: This issue had misnumbered pages; it should be 7].

6. Joyce Barkhouse, *Abraham Gesner* (Don Mills, ON: Fitzhenry & Whiteside, 1980), 60–61.

7. Gas prices were higher in New York than in Philadelphia; *Scientific American, 14* (25 December 1858):125.

8. "Gas Companies," *Scientific American, 4* (4 May 1861):276. The second largest was the New York Gas-Light Company. "Coal Gas," *Scientific American, 13* (26 September 1857):17.

9. "New Light—Kerosene Gas," *Scientific American, 9* (8 October 1853):29; A.C.F. [A. C. Ferris], "Petroleum Reminiscences and Early Oil Refining," *Oil, Paint & Drug Reporter,* 20 March 1889, 40; Albert Norton Leet, *Petroleum Distillation and Modes of Treating Hydrocarbons* (New York: Oil, Paint & Drug Publishing, 1884), 107; cf. Williamson and Daum, *American Petroleum,* 47.

10. George Weltden Gesner, "Kerosene—The Origin of the Name, the History of a Great Industry of Years Past, and the Possibility of its Revival," *Engineering and Mining Journal, 37* (9 February 1884):99–100; cf. Kendall Beaton, "Dr. Gesner's Kerosene: The Start of American Oil Refining," *Business History Review, 29* (1955):28–53; L. M. Cumming, "Abraham Gesner (1797–1864)—Author, Inventor, and Pioneer Canadian Geologist," *Proceedings of the Geological Society of Canada, 23* (1971):5–10.

11. The different spelling might have been a typographical error or a short-lived attempt to distinguish the oil from Kerosene gas. George Gesner, "Kerosene," 99.

12. Beaton, "Dr. Gesner's Kerosene," 38–39.

13. Beaton, "Dr. Gesner's Kerosene," 39.

14. Beaton, "Dr. Gesner's Kerosene," 41–43; Edwin G. Burrows and Mike Wallace, *Gotham: A History of New York City to 1898* (Oxford: Oxford University Press, 1999), 661.

15. U.S. Patents 11,203; 11,204; 11,205 for "Improvement in Kerosene Burning-Fluids" (27 June 1854).

16. In the public announcements of Gesner's patents the oil was misspelled "Kerocene," a typographical error. "No. 11,203," No. 11,204," "No. 11,205," *Report of the Commissioner of Patents for the Year 1854, Arts and Manufactures,* vol. 1: *Text* (Washington, DC: Beverly Tucker, 1855), 462–463. "Patent Claims," *Scientific American, 13* (17 October 1854):43.

17. *Report of a Committee appointed by the creditors of the North American Kerosene Gas Light Company* (New York, 1860); Beaton, "Dr. Gesner's Kerosene," 41–43, 51; Williamson and Daum, *American Petroleum,* 47–48; John Butt, "Legends of the Coal Oil Industry (1847–1864)," *Explorations in Entrepreneurial History, 2* (1964):16–30, 18.

18. "The Manufacture of Coal Oil—The First Patent," *Scientific American, 14* (12 February 1859):186.

19. Kendall Beaton, "Founders' Incentives: The Pre-Drake Refining Industry," in R. J. Forbes and Ralph W. Hidy, eds., *Oil's First Century* (Cambridge, MA: Harvard University Graduate School of Business Administration, 1960), 7–19; cf. Butt, "Legends," 18.

20. Abraham Gesner, *Practical Treatise on Coal, Petroleum, and Other Distilled Oils* (New York: Baillière Brothers, 1860), 38.

21. Thomas Antisell, *The Manufacture of Photogenic or Hydro-Carbon Oils, from Coal and Other Bituminous Substances, Capable of Supplying Burning Fluids* (New York: D. Appleton, 1859), 92.

22. Gesner, *Practical Treatise* (1860), 107.

23. "Artificial Illumination. Burning Fluids." *Scientific American, 13* (2 January 1858):133.

24. Beaton, "Dr. Gesner's Kerosene," 44–45.

25. U.S. Patent 12,612 for "Improvement in Processes for Making Kerosene" (27 March 1855); U.S. Patent 12,936 for "Improvement in Burning-Fluids" (22 May 1855); and U.S. Patent 12,987 for "Improvement in Burning-Fluid Compounds (29 May 1855).

26. Kerosene was first tried in common Argand lamps. Abraham Gesner, *A Practical Treatise on Coal, Petroleum, and Other Distilled Oils*, 2nd ed., revised and enlarged by George Weltden Gesner (New York: Baillière Brothers, 1865), 10–11; George Gesner, "Kerosene," 99; Beaton, "Dr. Gesner's Kerosene," 44–45; Williamson and Daum, *American Petroleum*, 30–32.

27. All three had to be Kentucky residents. The other two were George D. Prentice and Bryan R. Young. *Charter of the Breckenridge Cannel Coal Company, Granted by the General Assembly of Kentucky. Approved Feb. 9, 1854*, 9.

28. "Report of Commissioners," in *Charter*, 11–20, 14, 16. The analysis appeared in James Dwight Dana, *A System of Mineralogy*, 2 vols. (New York: Putnam, 1854), 2:27; and Benjamin Silliman, "Breckenridge Cannel Coal," *AG-LJ*, 2 (15 August 1860):58.

29. *Annual Report of the President and Directors of the Breckenridge Coal and Oil Company, for January, 1857*. (New York, 1857), 6–7.

30. "Oil from Coal," *Scientific American*, 11 (7 June 1856):312.

31. *Annual Report . . . of the Breckenridge Coal and Oil Company*, 8.

32. J. D. Van Slyck, *New England Manufacturers and Manufactories* (Boston: Van Slyck, 1879), 232–234; Merrill, "Reminiscences."

33. Atwood thought "Coup oil" was a stroke of chemistry just as the December 1852 coup d'état of Charles Louis-Napoléon Bonaparte (Napoléon III) was a stroke of state. U.S. Patent 9630 for "Improvement in Preparing Lubricating Oils" (29 March 1853). Stephen F. Peckham, *The Production, Technology, and Uses of Petroleum and Its Products*, 47th Cong., 2nd sess., H.R. Misc. Doc. 42, Part 10 (Washington, DC: GPO, 1884), 9–10; Merrill, "Reminiscences," 880.

34. Merrill estimated that Downer spent $102,000 to set the company straight. Merrill, "Reminiscences," 890.

35. Merrill, "Reminiscences," 890; Peckham, *Petroleum*, 10.

36. Peckham, *Petroleum*, 10.

37. "Artificial Illumination. Burning Fluids." *Scientific American*, 13 (2 January 1858):133; cf. Beaton, "Dr. Gesner's Kerosene," 46; Williamson and Daum, *American Petroleum*, 49–55.

38. Merrill, "Reminiscences," 884.

39. Peckham, *Petroleum*, 10; *Report of the Committee Appointed by the Creditors of the North American Kerosene Gas-Light Co.* (New York: Wm. C. Bryant, 1860).

40. "The Albert Coal," *Scientific American*, 1 (1 October 1859):221; Merrill, "Reminiscences," 884.

41. "Artificial Illuminants. Burning Fluids," *Scientific American*, 13 (2 January 1858): 133.

42. Merrill, "Reminiscences," 884.

43. "Are Coal Oils Explosive? Adulterations," *Scientific American*, 14 (4 December 1858):101.

44. "Mineral Oils," *Scientific American*, 1 (23 July 1859):59.

45. Merrill, "Reminiscences," 889.

46. Lance E. Davis, Robert E. Gallman, and Teresa D. Hutchins, "The Decline of U.S. Whaling: Was the Stock of Whales Running Out?" *Business History Review*, 62 (Winter 1988):569–595.

47. "Artificial Illumination. Burning Fluids," *Scientific American*, 13 (2 January 1858):133.

48. "Artificial Illumination. Burning Fluids," *Scientific American*, 13 (2 January 1858):133.

49. "Linseed Oils," *Scientific American*, 2 (24 March 1860):201–202.

50. "Lard Oil," *Scientific American*, 6 (23 August 1851):386.

51. "Lard Oil," *Scientific American*, 6 (23 August 1851):386.

52. "Burning Fluids," *Scientific American*, 11 (2 August 1856):374; "Artificial Illumination. Burning Fluids," *Scientific American*, 13 (2 January 1858):133; Williamson and Daum, *American Petroleum*, 33.

53. The *American Gas-Light Journal* ran gruesome biweekly accounts of burning fluid tragedies under the title "The Beauties of Burning Fluids." See, for example, the "Shocking Death [of Capt. Rufus Shepard] from Burning Fluid" *AG-LJ*, 2 (15 September 1860):86.

54. Rufus S. Merrill, "skilled mechanic and an ingenious inventor," lived in Williamsburg, Long Island, the same place as Gesner. During the Kerosene boom, he sold between 20,000 and 30,000 lamps. R.M. [Rufus Merrill], "Coal Oil. Secret Inventions," *Scientific American*, 1 (26 November 1859):350; Merrill, "Reminiscences," 883.

55. Charles M. Wetherill, "Of the Relative Cost of Illumination," *AG-LJ*, 1 (1 May and June 1860):230–231, 246–247.

56. "Our Artificial Light," *Scientific American*, 14 (25 December 1858):125; "Gas, Oil, and Burning Fluid," *Scientific American*, 14 (23 April 1859):273; "New York Markets," *Scientific American*, 1 (30 July, 6 August, 20 August 1859):75, 91, 123. Unfortunately, the column did not continue after 20 August to include coal oil prices.

57. Wetherill, "Cost of Illumination," 247.

58. "Our Artificial Light," *Scientific American*, 14 (25 December 1858):125.

59. "Are Coal Oils Explosive? Adulterations," *Scientific American*, 14 (4 December 1858):101.

60. "Quack Names for Burning Fluids," *Scientific American*, 13 (16 January 1858):150.

61. Antisell, *Hydro-Carbon Oils*, 133–135; "More About Coal-Oils and Coal," *Scientific American*, 1 (22 October 1859):270; "Coal Oil Manufacture," *Scientific American*, 2 (2 January 1860):3.

62. "Our Artificial Light," *Scientific American*, 14 (25 December 1858):125.

63. "More About Coal-Oils and Coal," *Scientific American*, 1 (22 October 1859):270.

64. Butt, "Legends," 25.

65. "More About Coal-Oils and Coal," *Scientific American*, 1 (22 October 1859):270.

66. For a critique of coal transportation costs, see "The Kanawha Coal Field," in *Cannel-Coal Company, of the Coal River, Virginia, chartered 19 March 1850* (New York, 1851).

67. "Artificial Illumination. Burning Fluids," *Scientific American*, 13 (2 January 1858):133.

68. "Coal Oil of Western Virginia," *Scientific American*, 13 (26 December 1857):123.

69. Antisell, *Hydro-Carbon Oils*, 133–135; "Coal Oil Commerce," *Scientific American*, 5 (9 November 1861):293.

70. Gesner, *Practical Treatise* (1860), 128–129.

71. "Kerosene Oils," *New York Commercial Advertiser*, 24 August 1859.

72. "Kerosene Oils," *New York Commercial Advertiser*, 24 August 1859.

73. "Coal-Oil," *AG-LJ*, 2 (2 July 1860):7, [misnumbered 271]; [Abraham Gesner],

"Coal-Oil and Its Manufacturers," *AG-LJ*, 2 (1 December 1860):168; cf. "Coal-Oil," *Annual of Scientific Discovery for 1861*, 87–88.

74. Gesner, *Practical Treatise* (1860), viii. And on the title page of his book, Gesner identified himself as "consulting chemist."

75. Dumas Grinand, *Report on Coal Oils of the Coal River and Kanawha Mining and Manufacturing Company* (New York, 1856), 7.

76. Neither Gesner nor his son George Weltden Gesner mentioned which companies by name. [Abraham Gesner], "Coal-Oil and Its Manufacturers," *AG-LJ*, 2 (1 December 1860):168; George Gesner, "Kerosene," 99.

77. Gesner, *Practical Treatise* (1860), 11.

78. Lesley to Hall, 20 April 1856, Lesley Mss, APS.

79. Lesley's other special study was to map the great Pittsburg coal bed. Lesley to Lesquereux, 15–16 January 1865, Lesley Mss, APS.

80. Newberry, "On the Mode of Formation of Cannel Coal," *AJS*, 23 (1857):212–215, reprinted as "On the Formation of Cannel Coal," *Annual of Scientific Discovery for 1857*, 329–331.

81. Newberry, "Formation of Cannel Coal," 212.

82. Newberry to Hall, 18 December 1854, Hall Mss, NYSA.

83. "Are Coal Oils Explosive? Adulterations," *Scientific American*, 14 (4 December 1858):101. For geological surveys, see "Cannel Coal and Its Oils," *Scientific American*, 1 (3 September 1859):151.

84. Robert Peter was a physician and professor in the Medical School of Lexington, Kentucky. Review of *Second Report of the Geological Survey of Kentucky*, *AJS*, 25 (1858):283–286. Robert Peter wrote the history of the Kentucky Survey in George P. Merrill, ed., *Contributions to a History of American State Geological and Natural History Surveys*, U.S. National Museum, Bulletin No. 109 (Washington, DC: GPO, 1920), 100–124.

85. "It has generally been believed that no other than the Breckenridge Cannel Coal could be profitably used, in this country, for [the manufacture of oil]." Robert Peter, "Second Chemical Report," in David Dale Owen, *Second Report of the Geological Survey of Kentucky, Made During the Years 1856 and 1857* (Frankfort: A. G. Hodges, 1857), 211–215, 213; "Table 2. Coals," 292–293.

86. Robert Peter, "Introductory Letter," in David Dale Owen, *Fourth Report of the Geological Survey in Kentucky, Made During the Years 1858 and 1859* (Frankfort: J. B. Major, 1861), 44–45.

87. Henry How, "On an Oil-Coal found near Pictou, Nova Scotia; and the Comparative Compositions of the Minerals often included in the term Coals," *AJS*, 30 (1860):74–79, 74.

88. George Gesner, "Kerosene," 99–100.

89. On charlatans, see Charles Coulston Gillispie, *Science and Polity in France at the End of the Old Regime*, 257–258.

90. "The Coal-Oil Man," *AG-LJ*, 2 (15 August 1860):58.

91. Original emphasis. "The Coal-Oil Man," *AG-LJ*, 2 (15 August 1860):58.

92. George Gesner, "Kerosene," 99.

93. [Abraham Gesner], "Coal-Oil and Its Manufacturers," *AG-LJ*, 2 (1 December 1860):168.

94. George Gesner, of course, recognized a difference between practical men and charlatans, who he called "chemists from Freiburg." "Kerosene," 99.

95. George Weltden Gesner's advertisement for his professional services appeared on the last page of Gesner, *Practical Treatise* (1865).

96. "Annual Statement of the Whale Fishery for 1860," *Scientific American, 4* (2 February 1861), 76; "Our Whaling Operations," *Scientific American, 2* (3 March 1860):154.

97. "The Whaling Business," *Scientific American, 4* (30 March 1861):196.

98. "Diminished Slaughter," *Scientific American, 4* (2 February 1861):68.

99. "Kerosene Oil-Works," *Scientific American, 2* (2 June 1860):363.

100. Peter Cooper Jr. bought the works. Another coal oil firm, Appleton and Cozzens, bought 10,000 tons of the Torbanehill mineral, which was on site. "Sale of the Kerosene Oil-Works," *AG-LJ, 1* (1 June 1860):250.

101. As part of its assets, the company listed 12,060 tons of "boghead coal" on site. Its liabilities included a $7,407.19 debt to James Young (presumably licensing fees) and $10,730 due to Philo T. Ruggles, president of the company. The creditors thought Gesner's three patents (purchased in 1854 for $12,477.42) were practically worthless. *Report of the Committee Appointed by the Creditors of the North American Kerosene Gas-Light Co.* (New York: Wm. C. Bryant, 1860), 2, 8–9. "Kerosene Oil-Works," *Scientific American, 2* (2 June 1860):363.

102. "Sale of the Kerosene Oil-Works," *AG-LJ, 1* (1 June 1860):250.

103. In 1863 Gesner's former company was renamed The New-York Kerosene Oil Works and was being run by Cozzens and Company. *Derrick's Hand-Book,* 1:14.

104. "New-York Kerosene Oil." Flier in the miscellaneous collections, Drake Well Museum Collections, Titusville, Pennsylvania.

Chapter 6. The Kerosene Cases

1. John Butt, "James Young, Scottish Industrialist and Philanthropist" (PhD dissertation, Glasgow University, 1964); R. J. Forbes, *Studies in Early Petroleum History* (Leiden: E. J. Brill, 1958), 182–194; Paul Lucier, "Court and Controversy: Patenting Science in the Nineteenth Century," *BJHS, 29* (1996): 139–154.

2. L. F. Haber, *The Chemical Industry during the Nineteenth Century: A Study of the Economic Aspects of Applied Chemistry in Europe and North America* (Oxford: Clarendon, 1958).

3. English Patent 12,359 (9 December 1848) for an improvement in manufacturing stannate of soda and potash.

4. James Oakes owned several coal mines, quarries, and iron foundries in Derbyshire and frequently consulted Playfair, his brother-in-law. Playfair did not think that Oakes had the know-how for chemical manufacturing. Butt, "James Young," 58.

5. "Evidence of James Young," *Binney v. Clydesdale* (1860), 159.

6. English Patent 13,292 (17 October 1850). The actual patent and its great wax seal are located in Young Mss, SUA.

7. The mineral was mined on the estate of William Gillespie and became the subject of a great deal of litigation and scientific controversy; see *Report of Trial before Lord Justice General in the Action . . . Mr. and Mrs. Gillespie . . . Against Messrs Russel & Son,* 29 July–4 August 1853, Court of Sessions (Edinburgh), 1854 and 1856–1857. Russel & Son were Young's coal agents, and they won the case. The mineral was declared "coal." *Report of the Trial* and other court documents are located in Young Mss, SUA.

8. *James Young and Others v. Steven White and Others,* Queen's Bench, 28 June 1854, Young Mss, SUA.

9. Colin A. Russell, *Edward Frankland: Chemistry, Controversy and Conspiracy in Victorian England* (Cambridge: Cambridge University Press, 1996).

10. "Lord Chief Justice Campbell to the Jury," *Young v. White*, 145, Young Mss, SUA.

11. U.S. Patent 8833 (23 March 1852).

12. Merrill, "Reminiscences," 882; John Butt, "Legends of the Coal Oil Industry (1847–1864)," *Explorations in Entrepreneurial History*, 2 (1964):24.

13. Merrill, "Reminiscences," 884.

14. Binney to Young, 6 January 1859, Young Mss, SUA.

15. Benedict and Boardman to Binney, 8 March 1859, Young Mss, SUA.

16. Benedict and Boardman to Binney, 20 November 1858, Young Mss, SUA.

17. "[W]e have no pecuniary interests, directly or indirectly, in the business of coal oil making. We know nothing of Mr. Young, or of the parties interested with him." "Young's Coal Oil Patent," *Scientific American*, 14 (5 March 1859):213. Young's patent was reprinted in "The Manufacture of Coal Oil. The First Patent," *Scientific American*, 14 (12 February 1859):186.

18. "Young's Coal Oil Patent," *Scientific American*, 14 (12 March 1859):221. Under the rubric "The Coal Oil Controversy," *Scientific American* published fifteen articles on Young's patent between March and December 1859. See "The Coal Oil Controversy," *Scientific American*, 14 (26 March 1859), 238.

19. "Young's Coal Oil Patent," *Scientific American*, 14 (5 March 1859):213.

20. R.M., "Coal Oil. Secret Inventions," *Scientific American*, 1 (26 November 1859): 350. R.M. was probably Rufus S. Merrill.

21. During the 1850s the business of patent soliciting expanded tremendously. There were nearly three dozen law firms in Washington, and perhaps twice that many elsewhere. See Robert C. Post, *Physics, Patents, and Politics: A Biography of Charles Grafton Page* (New York: Science History Publications, 1976), 160–161.

22. "Young's Coal Oil Patent," *Scientific American*, 14 (12 March 1859):221.

23. Benedict and Boardman to Binney, 22 April 1859; Binney to Young, 6 May 1859; and Benedict and Boardman Account to Young, 19 March 1860, Young Mss, SUA.

24. "The Manufacture of Coal Oil. The First Patent," *Scientific American*, 14 (12 February 1859):186.

25. "The Coal Oil Controversy," *Scientific American*, 14 (26 March 1859):238.

26. "The Coal Oil Controversy," *Scientific American*, 14 (26 March 1859):238.

27. On U.S. patent law, see Edward C. Waltersheid, *To Promote the Progress of Useful Arts: American Patent Law and Administration* (Littleton, CO: Fred B. Rothman, 1998); Steven Lubar, "The Transformation of Antebellum Patent Law," and K. J. Dood, "Pursuing the Essence of Inventions: Reissuing Patents in the 19th Century," in the special issue "Patents and Invention," *T&C*, 32 (1991):932–59, 999–1017. On British patent law, see H. I. Dutton, *The Patent System and Inventive Activity during the Industrial Revolution, 1750–1852* (Manchester: Manchester University Press, 1984); Christine MacLeod, *Inventing the Industrial Revolution: The English Patent System, 1660–1800* (Cambridge: Cambridge University Press, 1988).

28. Controversies over the precision of patent specifications are as old as patents themselves. The earliest examination of this issue involved James Watt and his patented steam engine; see Eric Robinson, "James Watt and the Law of Patents," *T&C*, 13 (1972), 115–139. On the social construction of patents and patent specifications, see Carolyn C. Cooper, *Shaping Invention: Thomas Blanchard's Machinery and Patent Management in Nineteenth-Century America* (New York: Columbia University Press, 1991).

29. This point was made explicit by the commissioner of patents, L. D. Gale, "On the Relations of the American Patent System to the Progress of Science," *Proceedings of the AAAS, 9* (1854):292–301.

30. Young to Binney, 14 February 1859, Young Mss, SUA.

31. Benedict and Boardman to Binney, 21 May 1859, Young Mss, SUA.

32. Agreement between Downer and Young, 1 March 1860, Young Mss, SUA.

33. Benedict and Boardman to Binney, 15 July 1859, Young Mss, SUA.

34. Benedict and Boardman to Binney, 8 March 1859, Young Mss, SUA.

35. Young to Binney, 23 March 1859, Young Mss, SUA.

36. Benedict and Boardman probably consulted Renwick's son, Edward Sabine Renwick (1823–1912) an engineer, inventor, and patent expert, who had for several years been a solicitor of patents in Washington, DC, before returning to New York City in the 1850s.

37. James Renwick, *Applications of the Science of Mechanics to Practical Purposes* (New York: Harper & Brothers, 1840); Robert V. Bruce, *The Launching of Modern American Science, 1846–1876* (Ithaca: Cornell University Press, 1988), 159.

38. This fee included payments to retain legal counsel. Benedict and Boardman to Binney 15 July 1859, Young Mss, SUA.

39. According to Benedict and Boardman, Gesner led "the van in the attack." Butt, "Legends of the Coal-Oil Industry," 26.

40. Benedict and Boardman to Binney, 13 October 1859, Young Mss, SUA.

41. Thomas Antisell, *The Manufacture of Photogenic or Hydro-Carbon Oils, from Coal and Other Bituminous Substances, Capable of Supplying Burning Fluid* (New York: D. Appleton, 1859), 15.

42. For very similar defenses of Young's patent, see Benjamin H. Paul, "On Destructive Distillation, considered in Reference to Modern Industrial Arts," *Chemical News, 7* (1863):282–283, 295–297; and *8* (1863):56–58, 78–79; Edward Frankland, "On Artificial Illumination," *Chemical News, 7* (1863):91–93, reprinted in *Journal of the Franklin Institute, 48* (1864):30–37.

43. Antisell, *Manufacture*, 15.

44. Reichenbach coined the term paraffine because the wax was unaffected by acids or alkalies, and it did not affect other substances, especially metals. Antisell, *Manufacture*, 11.

45. Selligue's French Patent 9159 (14 November 1838).

46. Antisell, *Manufacture*, 14–15.

47. The book misidentified Gesner as Solomon Gesner, an error that must be the printer's. Antisell, *Manufacture*, 16; cf. 84.

48. Abraham Gesner, *Practical Treatise on Coal, Petroleum, and Other Distilled Oils* (New York: Baillière Brothers, 1860), 8; cf. Abraham Gesner, *A Practical Treatise on Coal, Petroleum, and Other Distilled Oils,* 2nd ed., revised and enlarged by George Weltden Gesner (New York: Baillière Brothers, 1865), 9.

49. Charles A. Browne, "Storer, Francis Humphreys," *Dictionary of American Biography,* vol. 18 (New York: Charles Scribner's Sons, 1936), 94–95.

50. More than twenty pages in length, it amounted to an article unto itself. Frank H. Storer, "Review of Dr. Antisell's Work on Photogenic Oils, &c," *AJS, 30* (1860):112–121, 254–265, 112. Antisell thought it a "double handed review, the joint production of a young chemist and a more mature patent solicitor." See Thomas Antisell, "A Scientific Reviewer Reviewed," *AG-LJ, 2* (16 July 1860):25–26.

51. Storer, "Review," 112.

52. Storer, "Review," 260–262.

53. Storer, "Review," 117.

54. Storer, "Review," 117.

55. For a different perspective on "the paraffine wars," see Tal Golan, *Laws of Men and Laws of Nature: The History of Scientific Expert Testimony in England and America* (Cambridge, MA: Harvard University Press, 2004).

56. Stenhouse to Young, 14 September 1860; and John Stenhouse, "Report upon the Case E. W. Binney v. The Clydesdale Company," Young Mss, SUA.

57. Robert Warrington was the son of the chemist Robert Warrington (1807–1867), F. R. S. Binney also secured Frankland's and Warrington's services as witnesses for the case against the Columbian Oil Company. Binney to Frankland, 31 October and 3 November 1859; Frankland to Binney, 2 November 1859; and Binney to Young 3 November 1859, Young Mss, SUA. Frankland typically accepted retaining fees from large firms, a feature of his consulting practice that was not popular among American consultants; Russell, *Frankland*.

58. "Young's Coal Oil Patent Case," *Scientific American, 4* (5 January 1861):11.

59. The question mark is in the original. The *American Gas-Light Journal* took pleasure in lampooning the "very disinterested and chemically learned expert gentlemen." See "Coal Oil. Important Patent Decision," *AG-LJ, 2* (1 December 1860):172.

60. *Binney v. Clydesdale,* 1–7 November 1860, Court of Sessions, Edinburgh, Scotland. Cited in "The Paraffine Patent Case," *AG-LJ, 2* (1 January 1861):206.

61. "Coal Oil Patent. Important Case," *Scientific American, 3* (8 December 1860):378; "The Paraffine Patent Case," *AG-LJ, 2* (1 January 1861):206.

62. "Young's Coal Oil Patent Case," *Scientific American, 4* (5 January 1861):11.

63. "Coal-Oil. Important Patent Decision," *AG-LJ, 2* (1 December 1860):172.

64. E. W. Binney to E. Meldrum, 11 January 1859, cited in Butt, "Legends of the Coal-Oil Industry," 25.

65. See *Report of Trial James Young and Others v. Ebenezer Fernie and Others,* in Chancery, tried before Vice-Chancellor, Sir John Stuart, 29 February through 7 May 1864; and *Report of an Appeal, in the House of Lords, Fernie and Others v. Young and Others, 1864.* The trial received an extended review under "Young v. Fernie," *Chemical News, 9* (21 and 28 May and 4 June 1864):249–250, 262–264, and 273–276.

66. "Young v. Fernie," *Chemical News, 9* (28 May 1864):263.

67. "The Torbanehill Mineral," *Chemical News, 3* (2 March 1861): 217.

68. "Young v. Fernie," *Chemical News, 9* (21 May 1864): 250.

69. This was practically the same argument Charles Wetherill used at the Albert mineral trial.

70. "The Torbanehill Mineral," *Chemical News, 3* (2 March 1861):116.

71. Stuart was quoting directly; cf. Antisell, *Manufacture,* 14.

72. "Young v. Fernie," *Chemical News, 9* (4 June 1864):276.

73. On appeal, the case went to the House of Lords where it was heard on 19 April 1866. After four days of testimony, it was dismissed on 24 April 1866.

74. Baron Reichenbach, "Contribution towards the History of Paraffine," *Philosophical Magazine, 13* (1854):463–464, 464.

75. Binney to Young, 18 April 1860; and "Recognition of Baron Reichenbach," April 1860, Young Mss, SUA.

76. Edward Frankland, "On Artificial Illumination," *Chemical News, 7* (1863):91–93.

77. Steven Shapin, *A Social History of Truth: Civility and Science in Seventeenth-Century England* (Chicago: University of Chicago Press, 1994).

78. Charles T. Jackson, "Encouragement and Cultivation of the Sciences in the United States, Twenty-Fourth Anniversary Address, before the American Institute, of the City of New-York, at the Tabernacle, on 16th of October, 1851," *Transactions of the American Institute for the Year 1851*, 227–246, 237.

79. "Dr. Jackson's Address before the American Institute," *Scientific American, 7* (1 November 1851):51; and Junius Redivivus, "Dr. Jackson on Patents," *Scientific American, 7* (20 December 1851). Robert C. Post, "Science, Public Policy, and Popular Precepts: Alexander Dallas Bache and Alfred Beach as Symbolic Adversaries," in Nathan Reingold, ed., *The Sciences in the American Context: New Perspectives* (Washington, DC: Smithsonian Institution Press, 1979), 77–98; Post, *Physics, Patents, and Politics.*

80. Storer, "Review," 255.

81. Paraffine, "Coal-Oil," *AG-LJ, 2* (15 September 1860):86.

82. In Jackson's opinion, the answer was a national academy of sciences modeled on the French Academy. Scientific judges would examine discoveries, ascertain priority, determine value, and award a state pension. This would prevent abuses of the patent system and allow men of science to control, outright, the development of technology. Jackson, "Encouragement and Cultivation of the Sciences in the United States," 237.

83. This criticism of European-trained chemists might have sounded somewhat odd to Americans familiar with Antisell's own background; he was, after all, Irish.

84. Antisell, "Reviewer Reviewed," 25–26.

85. Gale added: "It is proper to say, that Dr Jackson did not avail himself of the benefits of the patent. He stated that his wish was to *secure the honor of discovery more than to obtain pecuniary reward.*" L. D. Gale, "On the Relations of the American Patent System to the Progress of Science," *Proceedings of the AAAS, 9* (1854):292–301, 294.

86. "Young's Coal Oil Patent Case," *Scientific American, 4* (5 January 1861):11.

87. "Coal-Oil," *AG-LJ, 2* (1 December 1860):172.

88. "The Torbanehill Mineral," *Chemical News, 3* (2 March 1861):217.

89. "The Evidence of Experts," *Chemical News, 5* (5 April 1862):183.

Chapter 7. The Rock Oil Report

1. Andrew Cone and Walter R. Johns, *Petrolia: A Brief History of the Pennsylvania Petroleum Region, Its Development, Growth, Resources, etc., from 1859 to 1869* (New York: D. Appleton, 1870), 50; S. J. M. Eaton, *Petroleum: A History of the Oil Region of Venango County, Pennsylvania* (Philadelphia: J. P. Skelly, 1866), 67–68.

2. Boverton Redwood, "Petroleum," *Encyclopædia Britannica*, 11th ed. (1910–1911), 316–322, 317. Most recently, Daniel Yergin began with "the matter of the missing $526.08," Silliman's consulting fee. Yergin, *The Prize: The Epic Quest for Oil, Money & Power* (New York: Touchstone, 1991), 19.

3. The largest collections of correspondence belong to Dr. Francis Beattie Brewer, George H. Bissell, and James M. Townsend, located at the Drake Well Museum in Titusville, Pennsylvania. The Brewer and Bissell letters have been published in Giddens, *Sources.* The Townsend letters along with personal histories by Brewer, Bissell, Townsend, and E. L. Drake were published in Giddens, *Documents.* Bissell's history first appeared in S. S. Hayes's "Report of the United States Revenue Commission on Petro-

leum as a Source of National Revenue," *House Executive Documents*, No. 51, 39th Cong., 1st sess., (February 1866):1–39.

4. There are different, older versions of this part of the story on the inception of the oil industry. Paul H. Giddens, *The Birth of the Oil Industry* (New York: Macmillan, 1938); J. Stanley Clark, *The Oil Century: From the Drake Well to the Conservation Era* (Norman: University of Oklahoma Press, 1958); Harold F. Williamson and Arnold R. Daum, *The American Petroleum Industry: The Age of Illumination, 1859–1899* (Evanston: Northwestern University Press, 1959); Ernest C. Miller, ed., *This Was Early Oil: The Contemporary Accounts of the Growing Petroleum Industry, 1848–1885* (Harrisburg: Pennsylvania Historical and Museum Commission, 1968). The most colorful is Hildegarde Dolson, *The Great Oildorado: The Gaudy and Turbulent Years of the First Oil Rush; Pennsylvania, 1859–1880* (New York: Random House, 1959).

5. Francis Beattie Brewer received his MD in 1843. J. T. Henry, *The Early and Later History of Petroleum with Authentic Facts in Regard to its Development in Western Pennsylvania* (Philadelphia: James B. Rodgers, 1873), 60, 383–396; cf. Giddens, *Birth*, 30.

6. Undated and unsigned letter in Giddens, *Sources*, 61.

7. Henry, *History*, 393.

8. According to Brewer, he first sent some of this "Creek oil" to Dixi Crosby and later visited him. Giddens, *Documents*, 46.

9. George Bissell graduated from Dartmouth College in 1845. Henry, *History*, 346–350.

10. Henry, *History*, 61.

11. Henry, *History*, 62. It is uncertain what was meant by "commercial quantities." Brewer stated that "the yield is abundant & is believed to be inexhaustible." Brewer to Eveleth and Bissell, undated, in Giddens, *Sources*, 12.

12. Brewer's Account, in Giddens, *Documents*, 46.

13. Brewer to Eveleth and Bissell, undated, in Giddens, *Sources*, 13–14. Cf. Brewer's Account, in Giddens, *Documents*, 47.

14. Crosby to Brewer, 11 September [1854], in Giddens, *Sources*, 14.

15. Giddens, *Birth*, 34.

16. Henry, *History*, 66–67.

17. Crosby to Brewer, 11 September [1854], in Giddens, *Sources*, 14.

18. Crosby to Brewer, 11 September [1854], in Giddens, *Sources*, 14.

19. Henry, *History*, 66–67.

20. Douglass C. North, *The Economic Growth of the United States, 1790–1860* (New York: W. W. Norton, 1966), 204–215.

21. Giddens, *Birth*, 35.

22. Henry, *History*, 69.

23. Bissell to Brewer, [November 1854], in Giddens, *Sources*, 22.

24. Henry, *History*, 69.

25. J. D. Whitney, *The Metallic Wealth of the United States, Described and Compared with that of Other Countries* (Philadelphia: Lippincott, Grambo & Co., 1854), xxvi–xxviii.

26. The Certificate of Incorporation is in the Drake Well Museum and was reprinted in Henry, *History*, 70–71.

27. Henry noted that on the board of trustees "all but Dr. Brewer . . . were mere lay-figures, occupying positions it was necessary for appearances' sake, that some one

should fill. Not more than one of them at most, represented stock held in his own right, stock for which he paid." Henry, *History*, 71.

28. Charles T. Jackson, "Remarks on Mining Operations, 1846," part of the Jackson Mss from the Papers of the Association of American Geologists and Naturalists, Academy of Natural Sciences, Philadelphia.

29. Crosby to Brewer, 11 September 1854, in Giddens, *Sources*, 14.

30. Article 2 of the Certificate of Incorporation; Henry, *History*, 70–71.

31. Williamson and Daum supported this assumption with a suggestion that Bissell saw Gesner's prospectus for the Kerosene Oil Company; see Williamson and Daum, *American Petroleum*, 65.

32. Hayes, "Report," 4; cf. Williamson and Daum, *American Petroleum*, 65.

33. Emphasis added. Henry, *History*, 61.

34. A. H. Crosby to Frank [Crosby], 29 September [1854], in Giddens, *Sources*, 12.

35. Eveleth to Brewer, 4 November 1854, in Giddens, *Sources*, 19.

36. Eveleth to Brewer, 4 November 1854, in Giddens, *Sources*, 20.

37. Henry, *History*, 72.

38. Thomas A. Gale, *The Wonder of the Nineteenth Century: Rock Oil, in Pennsylvania and Elsewhere* (Erie, PA: Sloan & Griffith, 1860).

39. Ebenezer Brewer to F. B. Brewer, 23 March 1855, in Giddens, *Sources*, 29.

40. Henry, *History*, 73.

41. Brewer's Account, in Giddens, *Documents*, 49.

42. Henry, *History*, 74.

43. Eveleth to Brewer, 4 November 1854, in Giddens, *Sources*, 19.

44. Bissell to Brewer, 6 November 1854, in Giddens, *Sources*, 20.

45. Silliman's title and position were somewhat confusing; he held the rank of professor in the Medical and Philosophical Departments at Yale, but he was not a member of the Academical Faculty; his salary, $1,000 a year, was $800 less than that of a regular professor. Louis I. Kuslan, "Benjamin Silliman, Jr.: The Second Silliman," in Leonard G. Wilson, ed., *Benjamin Silliman and His Circle: Studies on the Influence of Benjamin Silliman on Science in America* (New York: Science History Publications, 1979), 159–205, 174–177.

46. Benjamin Silliman Jr., *Report on the Rock Oil, or Petroleum, from Venango Co., Pennsylvania, with special reference to its use for illumination and other purposes* (New Haven: J. H. Benham, 1855), 3–4. A copy is on display at the Drake Well Museum. It was reprinted in Paul H. Giddens, *The First Scientific Analysis of Petroleum: A Chemical Classic that Touched Off an Industry* (Meadville, PA: Paul H. Giddens, 1949).

47. Silliman, *Report*, 4.

48. Silliman, *Report*, 5.

49. Silliman, *Report*, 5.

50. Silliman, *Report*, 6.

51. Bissell to Brewer, 6 November 1854, in Giddens, *Sources*, 21.

52. Bissell to Brewer, 6 November 1854, in Giddens, *Sources*, 21.

53. "Copy of an Extract from Professor Silliman's Letter, December 21, 1854," in Giddens, *Sources*, 23.

54. Sheldon to Silliman, 29 December 1854, Silliman Mss, YUL.

55. Eveleth to Brewer, 1 February 1855, in Giddens, *Sources*, 24.

56. Eveleth to Brewer, 8 February 1855, in Giddens, *Sources*, 25.

57. Eveleth to Brewer, 8 February 1855, in Giddens, *Sources*, 25.

58. Silliman to president and fellows, May 1855, cited in Kuslan, "Silliman," 176–177.

59. Silliman to Eveleth and Bissell, 1 March 1855, in Giddens, *Sources*, 62–63.

60. Judd's Patent Sixes Sperm Candles were the "*unit* adopted for comparison of intensities of illumination." Silliman, *Report*, 17; Silliman to Eveleth and Bissell, 1 March 1855, in Giddens, *Sources*, 62–63.

61. Eveleth to Brewer, 17 February 1855, in Giddens, *Sources*, 26.

62. Silliman, *Report*, 17.

63. The gas experiments were conducted in January 1856. B. Silliman Jr. and Chas. H. Porter, "Notice of a Photometer and of some experiments upon the comparative power of several artificial means of illumination," *AJS*, 23 (1857):315–317.

64. Silliman, *Report*, 17.

65. Silliman to Eveleth and Bissell, 1 March 1855, in Giddens, *Sources*, 62–63. Louis I. Kuslan, "The Yale School of Applied Chemistry," in Wilson, *Benjamin Silliman and His Circle*, 128–157. According to Margaret Rossiter, Benjamin Silliman Jr. and James Dana thought John Porter was too interested in commercial patrons. Rossiter, *The Emergence of Agricultural Science*, 137.

66. Kuslan, "Silliman," 188.

67. Sheldon to Brewer, 23 April 1855, in Giddens, *Sources*, 34–35.

68. Sheldon to Brewer, 23 April 1855, in Giddens, *Sources*, 34–35.

69. Sheldon to Brewer, 23 April 1855, in Giddens, *Sources*, 34–35.

70. Silliman, *Report*, 20.

71. Emphasis added. Eveleth to Brewer, 22 May 1855, in Giddens, *Sources*, 38.

72. Sheldon to Brewer, 11 May 1855, in Giddens, *Sources*, 36.

73. Eveleth to Brewer, 21 April 1855, in Giddens, *Sources*, 34.

74. Sheldon to Brewer, 4 September 1855, in Giddens, *Sources*, 52.

75. Ebenezer Brewer to F. B. Brewer, 4 June 1855, in Giddens, *Sources*, 42.

76. Sheldon to Brewer, 11 May 1855, in Giddens, *Sources*, 36.

77. Henry, *History*, 75.

78. Sheldon to Brewer, 11 May 1855, in Giddens, *Sources*, 37.

79. Sheldon to Brewer, 11 May 1855, in Giddens, *Sources*, 37.

80. Sheldon to Bissell, 30 July 1855, in Giddens, *Sources*, 66.

81. Sheldon to Brewer, 3 July 1855, in Giddens, *Sources*, 45–46.

82. Eveleth to Brewer, 28 May 1855, in Giddens, *Sources*, 40.

83. Eveleth to Brewer, 17 June 1855, in Giddens, *Sources*, 43.

84. Sheldon to Brewer, 3 July 1855, in Giddens, *Sources*, 46.

85. Eveleth to Brewer, 25 June 1855, in Giddens, *Sources*, 44.

86. Cf. Williamson and Daum, *American Petroleum*, 67.

87. Eveleth to Brewer, 25 June 1855, in Giddens, *Sources*, 44–45.

88. Sheldon to Bissell, 30 July 1855, in Giddens, *Sources*, 65–66.

89. Eveleth to Brewer, 23 July 1855, in Giddens, *Sources*, 47.

90. Sheldon to Brewer, 30 July 1855, in Giddens, *Sources*, 66.

91. Sheldon to Eveleth, 25 July 1855, in Giddens, *Sources*, 64.

92. Eveleth to Brewer, 23 July 1855, in Giddens, *Sources*, 47.

93. Eveleth to Brewer, 20 August 1855, in Giddens, *Sources*, 51.

94. "Articles of Association of the Pennsylvania Rock Oil Company of Connecticut," in Giddens, *Documents*, 129–132.

95. On Pennsylvania's laws of incorporation, see Eaton, *Petroleum*, 73–84.

96. Henry, *History*, 83.

97. Eveleth to Brewer, 20 August 1855, in Giddens, *Sources,* 51.

98. Henry, *History,* 83.

99. Townsend to Brewer, 8 January 1858, in Giddens, *Documents,* 135–137.

100. "Articles of Association of the Seneca Oil Company," in Giddens, *Documents,* 139–140.

101. Townsend Account, Townsend Papers, Drake Well Museum Collections.

102. Drake to W. A. Ives, 16 August 1858, in Giddens, *Documents,* 144–146.

103. Townsend's Account, Townsend Papers, Drake Well Museum Collections.

104. Henry, *History,* 86.

105. Henry, *History,* 83.

106. Silliman, *Report,* 3.

107. At least six different dates have been given for the "discovery." Uncle Billy finished his work on Saturday, 27 August, but his results were probably not known until the following day. Drake did not visit the well until Monday, 29 August. For more on Drake's success see Ernest C. Miller, *Pennsylvania's Oil Industry* (Gettysburg: Pennsylvania Historical Association, 1974), 19.

108. Henry, *History,* 29.

109. Henry, *History,* 30.

110. The well-to-do used gas lighting. Henry, *History,* 31.

111. Henry, *History,* 31.

112. Silliman, *Report,* 21.

113. Henry, *History,* 36.

Chapter 8. The Elusive Nature of Oil and Its Markets

1. Samuel W. Tait, *The Wildcatters: An Informal History of Oil-Hunting in America* (Princeton: Princeton University Press, 1946), 74–77; Harold F. Williamson and Arnold R. Daum, *The American Petroleum Industry: The Age of Illumination, 1859–1899* (Evanston: Northwestern University Press, 1959), 90; and Keith L. Miller, "Petroleum in America," and "Petroleum Geology to 1920," in Gregory A. Good, ed., *Science of the Earth: An Encyclopedia of Events, People, and Phenomena,* 2 vols. (New York: Garland, 1998), 2:670–673, 673–675.

2. Marius R. Campbell, "Historical Review of Theories Advanced by American Geologists to Account for the Origin and Accumulation of Oil," *Economic Geology,* 6 (1911):363–395, 364.

3. Paul H. Giddens, *The Birth of the Oil Industry* (New York: Macmillan, 1938), 64; Brian Black, *Petrolia: The Landscape of America's First Oil Boom* (Baltimore: Johns Hopkins University Press, 2000), 46.

4. Edmund Morris, *Derrick and Drill, or an Insight into the Discovery, Development, and Present Condition and Future Prospects of Petroleum, in New York, Pennsylvania, Ohio, West Virginia, &c.* (New York: James Miller, 1865), 159.

5. Oil barrels were nonstandardized and held between forty and forty-two gallons.

6. "Articles of Agreement," 14 November 1859, Edwin L. Drake Papers, Drake Well Museum Collections, Titusville, Pennsylvania.

7. "Mr. E. L. Drake in Account with S. M. Kier," Drake Papers, Drake Well Museum Collections.

8. Copy of the Seneca Oil Company trade card, Drake Well Museum Collections.

9. L. Eastman to Drake, 27 February 1860; and Drake to Leonard, 12 February 1860,

Drake Papers, Drake Well Museum Collections. Cf. "A New Source of Wealth," *Scientific American*, 1 (24 December 1859):413.

10. "Memorandum of an Agreement," 12 March 1860, Drake Papers, Drake Well Museum Collections.

11. Drake to Townsend, 11 March 1860, in Giddens, *Documents*, 162–163. Drake or F. C. Ford, the master mechanic, reported this information to the local newspaper, the *Erie Observer;* it was reprinted in Thomas A. Gale, *The Wonder of the Nineteenth Century: Rock Oil, in Pennsylvania and Elsewhere* (Erie, PA: Sloan & Griffith, 1860), 48.

12. "Coal-Oil," *AG-LJ*, 2 (2 July 1860):271.

13. Cited in "Coal-Oil," *AG-LJ*, 2 (15 January 1861):215.

14. Gale, *Wonder*, 13.

15. Gale, *Wonder*, 58.

16. Gale, *Wonder*, 58.

17. Gale, *Wonder*, 70–71.

18. Gale, *Wonder*, 5.

19. J. S. Newberry, "Rock Oils of Ohio," in *The Ohio Agricultural Report for 1859*, 5, 13.

20. J. Peter Lesley, "Coal Oil," in *Report of the Commissioner of Agriculture for the Year 1862*, 37th Cong., 3rd sess., House of Representatives Executive Document, No. 78 (Washington, DC: GPO, 1863), 429–447, 441.

21. Gale, *Wonder*, 22.

22. Andrew Cone and Walter R. Johns, *Petrolia: A Brief History of the Pennsylvania Petroleum Region, Its Development, Growth, Resources, etc., from 1859 to 1869* (New York: Appleton, 1870), 116.

23. Gale, *Wonder*, 19. According to the *Springfield (Mass.) Republican*, "The word *boring*, which is universally used, conveys a very erroneous idea of the process of making an oil-well, which is not boring in any sense, but drilling." Quoted in Morris, *Derrick and Drill*, 85.

24. Literally, it cost $2.12½ per foot; one bit equals one-eighth of a dollar. Gale, *Wonder*, 25. Newberry estimated $250 for a 100-foot well. See Newberry, "Rock Oils," 11.

25. Gale, *Wonder*, 31.

26. Gale, *Wonder*, 32–33; Newberry, "Rock Oils," 12.

27. Gale, *Wonder*, 35.

28. Newberry, "Rock Oils," 12.

29. J. S. Newberry, "The First Oil Well. The Birth of a Great Industry," *Harper's Magazine*, 81 (October 1890):723–729, 728.

30. S. J. M. Eaton, *Petroleum: A History of the Oil Region of Venango County, Pennsylvania* (Philadelphia: J. P. Skelly, 1866), 84.

31. Newberry, "Rock Oils," 5.

32. The article was signed "Medicus"; most historians assume it was the physician Francis Brewer. Reprinted in Ernest C. Miller, *This Was Early Oil: Contemporary Accounts of the Growing Petroleum Industry, 1848–1885* (Harrisburg: Pennsylvania Historical and Museum Commission, 1968), 29.

33. The analogy between fluids in the earth and in the human body is very old. Carolyn Merchant, *The Death of Nature: Women, Ecology, and the Scientific Revolution* (San Francisco: Harper & Row, 1980).

34. Eaton, *Petroleum*, 255.

35. Newspaper cutting in the Townsend Papers, Drake Well Museum Collections.

36. Gale, *Wonder*, 33.

37. Gale, *Wonder*, 76.

38. Gale, *Wonder*, 76.

39. Eaton, *Petroleum*, 90.

40. Eaton, *Petroleum*, 86.

41. Eaton made no explicit references to any books (except the Bible). However, many of his explanations drew directly on Rogers. Eaton, *Petroleum*, iii–iv, 254.

42. Gale, *Wonder*, 18, 21.

43. Henry Rogers's brother William in Cambridge, Massachusetts, sent him regular updates.

44. Henry Rogers, "On the Distribution and Probable Origin of the Petroleum, or Rock Oil, of Western Pennsylvania, New York, and Ohio," *Proceedings of the Philosophical Society of Glasgow, 4* (1860):355–359, 356–357. Cf. Edgar Wesley Owen, *Trek of the Oil Finders: A History of Exploration for Petroleum* (Tulsa: American Association of Petroleum Geologists, 1975), 60.

45. Rogers, "Distribution," 358.

46. [Editorial], *AG-LJ, 2* (15 March 1861):284.

47. "Petroleum. Its Sources. Various Theories," *Scientific American, 7* (19 July 1862):37.

48. J. T. Henry, *The Early and Later History of Petroleum with Authentic Facts in Regard to its Development in Western Pennsylvania* (Philadelphia: James B. Rogers, 1873), 82.

49. Newberry, "Rock Oils," 6; cf. J. S. Newberry, "The Oil Wells of Mecca," *Canadian Naturalist and Geologist, 5* (1860):325–326.

50. Newberry, "Rock Oils," 4–5.

51. Abraham Gesner, *Practical Treatise on Coal, Petroleum, and Other Distilled Oils* (New York: Baillière Brothers, 1860), 32.

52. Alexander Winchell, *First Biennial Report of the Progress of the Geological Survey of Michigan* (Lansing: Hosmer & Kerr, 1861), 73.

53. E. B. Andrews, "Rock Oil, its Geological Relations and Distribution," *AJS, 32* (1861):85–93.

54. Andrews, "Rock Oil," 92.

55. Andrews, "Rock Oil," 92–93.

56. Andrews, "Rock Oil," 92.

57. "Historical and Scientific Facts About Petroleum," *Scientific American, 6* (1 February 1862):71.

58. Abraham Gesner, "On the Petroleum-Springs in North America," *Quarterly Journal of the GSL, 18* (1862):3–4, 4.

59. T. Sterry Hunt, "Notes on the History of Petroleum, or Rock Oil," *Canadian Naturalist and Geologist, 6* (1861):241–255, 248, reprinted in *Smithsonian Institution Annual Report for 1861*, 319–329; and *Chemical News, 4* (1862):5–6, 18–19, 35–36. Abstracted as "On the Formation of Petroleum and Allied Substances from Woody Fibre or Animal Tissue," *AG-LJ, 3* (15 March 1862):278.

60. *Derrick's Hand-Book of Petroleum: A Complete Chronological and Statistical Review of Petroleum Developments from 1859 to 1889*, 2 vols. (Oil City, PA: Derrick Publishing Company, 1898), 1:18.

61. Hunt, "Notes on Petroleum," 250, 249.

62. Hunt, "Notes on Petroleum," 242.

63. Hunt, "Notes on Petroleum," 249–250.

64. Eaton, *Petroleum*, 85.

65. Cone and Johns, *Petrolia*, 116.

66. Gale, *Wonder*, 34.

67. Cone and Johns, *Petrolia*, 119–120.

68. Gale, *Wonder*, 76–77.

69. Gale, *Wonder*, 23.

70. Rogers, "Distribution," 357.

71. Eaton, *Petroleum*, 72.

72. Cone and Johns, *Petrolia*, 13.

73. Hayes, "Report," 5.

74. Andrews, "Rock Oil," 88.

75. Morris, *Derrick and Drill*, 150.

76. Hayes, "Report," 5.

77. Eaton, *Petroleum*, 150–151.

78. Eaton, *Petroleum*, 150–151; cf. Henry, *History*, 335–343; Cone and Johns, *Petrolia*, 17, 158–160.

79. Eaton, *Petroleum*, 142.

80. Morris, *Derrick and Drill*, 39; Eaton, *Petroleum*, 147.

81. Eaton, *Petroleum*, 147.

82. Cone and Johns, *Petrolia*, 153.

83. Lesley, "Coal Oil," 433.

84. Cone and Johns, *Petrolia*, 153; Giddens, *Birth of the Oil Industry*, 121.

85. Eaton, *Petroleum*, 149.

86. Cone and Johns, *Petrolia*, 73–74. Cf. Williamson and Daum, *American Petroleum*, 114.

87. Morris, *Derrick and Drill*, 40.

88. Eaton, *Petroleum*, 79. Oil sold as low as ten cents a barrel; Cone and Johns, *Petrolia*, 73.

89. Eaton, *Petroleum*, 147; Cone and Johns, *Petrolia*, 160; Black, *Petrolia*.

90. Lesley, "Coal Oil," 433; Hayes, "Report," 6.

91. Cited in Henry, *History*, 232.

92. Lesley, "Coal Oil," 439.

93. Eaton, *Petroleum*, 255.

94. E. W. Evans, "On the Action of Oil-Wells," *AJS*, *38* (1864):159–166; Evans, "On the Oil-Producing Uplift of West Virginia," *AJS*, *42* (1866):335–343. Evans consulted for the West Virginia and Ohio Petroleum Company.

95. Eaton, *Petroleum*, 153–154.

96. T. Sterry Hunt, "Contributions to the Chemical and Geological History of Bitumens, and Pyroschists or Bituminous Shales," *AJS*, *35* (1863):157–171, 166.

97. Hunt excluded marine plants. Hunt, "Contributions," 168.

98. Hunt, "Contributions," 167.

99. [Henry D. Rogers], "Coal and Petroleum," *Harper's Magazine*, *27* (1863):259–264, 261.

100. Rogers, "Coal and Petroleum," 263.

101. Henry D. Rogers, "On Petroleum," *Philosophical Society of Glasgow*, *6* (1865):48–60, 51. Written in Boston, the article was based on newspaper accounts and on William Wright, *The Oil Regions of Pennsylvania: Showing Where Petroleum is Found; How it is Obtained, and at What Cost, with Hints for Whom it may Concern* (New York: Harper & Brothers, 1865).

102. Rogers, "Coal and Petroleum," 261, 263.

103. Lesley, "Coal Oil," 445.

104. Lesley, "Coal Oil," 443.

105. Lesley, "Coal Oil," 445.

106. Lesley, "Coal Oil," 442.

107. Lesley, "Coal Oil," 444.

108. Lesley, "Coal Oil," 433.

109. Merrill, "Reminiscences," 885.

110. Gale, *Wonder*, 79–80.

111. Gale, *Wonder*, 40.

112. Hayes, "Report," 10.

113. Gesner, *Practical Treatise*, 75.

114. Gesner, *Practical Treatise*, 121.

115. Gale, *Wonder*, 40–41. Cf. "Petroleum is not Coal Oil," *Scientific American*, 7 (20 September 1862):185.

116. Merrill, "Reminiscences," 884.

117. Henry, *History*, 321.

118. Young left the United States in June 1860; he did not accompany Downer.

119. Cited in Williamson and Daum, *American Petroleum*, 108.

120. Gesner was very optimistic about future supplies. Like Rogers, he envisioned petroleum as being continually produced. "[T]he oil springs [will] continue their supplies until all the chemical agencies operating upon thick strata of coal beneath have subsided, and the once woody fibre of coal has ceased to expel carburetted hydrogen." Gesner, *Practical Treatise*, 32.

121. Gesner, *Practical Treatise*, 107.

122. Williamson and Daum, *American Petroleum*, 111.

123. Allen Norton Leet, *Petroleum Distillation and Modes of Testing Hydro-Carbons* (New York, 1884).

124. Merrill, "Reminiscences," 884.

125. Wright, *Oil Regions*, 201–202.

126. George Weltden Gesner (1829–1904) continued in the oil business after his father returned to Nova Scotia in 1863. The following year Abraham Gesner died.

127. Cone and Johns, *Petrolia*, 575.

128. Henry, *History*, 37.

129. "Gas from Petroleum," *Scientific American*, 8 (23 May 1863):324; "Petroleum Gas," *Scientific American*, 8 (13 June 1863):378; "Improvements in Gas made from Petroleum or other Hydro-carbons," *Scientific American*, 9 (25 July 1863):56.

130. Cited in Williamson and Daum, *American Petroleum*, 220.

131. Wright, *Oil Regions*, 205; Williamson and Daum, *American Petroleum*, 104–111; Paul Lucier, "Petroleum. What Is It Good For?" *American Heritage of Invention and Technology*, 7 (1991):56–63.

132. Williamson and Daum, *American Petroleum*, 120–121.

133. *Derrick's Hand-Book*, 1:707–711; Williamson et al., *American Petroleum*, 120–121.

134. Henry, *History*, 221; *Derrick's Hand-Book*, 1:40.

135. Eaton, *Petroleum*, 92.

136. Eaton, *Petroleum*, 191.

137. Morris, *Derrick and Drill*, 139.

138. William Culp Darrah, *Pithole, the Vanished City: A Story of the Early Days of the Petroleum Industry* (Gettysburg, PA: privately published, 1972), 124–137.

139. Crocus [C. C. Leonard], *The History of Pithole* (Pithole City, PA: privately published, 1867), 8.

140. Quoted in Morris, *Derrick and Drill*, 73. Darrah cited the same quotation but from the *Oil City Register*, 2 March 1865; Darrah, *Pithole*, 10–13.

141. Morris, *Derrick and Drill*, 75–76.

142. Eaton, *Petroleum*, 202.

143. Wright, *Oil Regions*, 78.

144. Morris, *Derrick and Drill*, 84; cf. Eaton, *Petroleum*, 81.

145. Wright, *Oil Regions*, 244.

146. Wright, *Oil Regions*, 77.

147. John William Dawson, *Acadian Geology. The Geological Structure, Organic Remains, and Mineral Resources of Nova Scotia, New Brunswick, and Prince Edward Island* (London: Macmillan, 1868), 247.

148. The peak years were 1865 and 1866 during which 20,500 tons were exported to the United States. The price at the Hillsborough wharf varied from $15 to $20 per ton. After 1870, production fell to an average of not more than 6,000 tons per year. L. W. Bailey and R. W. Ells, "Report on the Lower Carboniferous Belt of Albert and Westmorland Counties, New Brunswick, including the Albert Shales," in Alfred R. C. Selwyn, *Report of Progress for 1876–77* (Ottawa: Geological Survey of Canada, 1878), 351–395, 392; and L. W. Bailey, *The Mineral Resources of the Province of New Brunswick* (Ottawa: S. E. Dawson, 1893), 69.

149. W. S. MacNutt, *New Brunswick: A History, 1784–1867* (Toronto: Macmillan, 1963), 383.

150. French to Townsend, 14 October 1862, Townsend Papers, Drake Well Museum Collections.

151. Agassiz to Lesley, 12 and 24 April 1865; Lesley to Agassiz 14 April 1865, Lesley Mss, APS.

152. L. W. Bailey, *On the Mines and Minerals of New Brunswick, with an Account of the Present Condition of Mining Operations in the Province* (Fredericton: G. E. Fenety, 1864), 52.

153. *Report of a Case, tried at Albert Circuit Court, 1852, before his Honor Judge Wilmot, and a Special Jury. Abraham Gesner vs. William Cairns. Copied from the Judge's Notes* (Saint John: William L. Avery, 1853), 35, 41.

154. *Report of a Case*, (Silliman) 91, (Hayes) 96.

155. The Westmorland Petroleum Company, founded in 1863, was the first firm to bore for oil in New Brunswick. The company owned land on the eastern shore of the Petitcodiac River, across from Albert Mines, and sunk four wells; but in 1864, with the glut from Oil Creek, the company went bankrupt. Hugh M. Grant, "Public Policy and Private Capital Formation in Petroleum Exploration," in Paul A. Bogaard, ed., *Profiles of Science and Society in the Maritimes Prior to 1914* (Sackville, NB: Acadiensis Press, 1990):137–160, 146.

156. Hitchcock had visited the region in 1861, presumably as part of his official duties with the Maine survey. When he returned as a consultant, he was not allowed to examine the Albert mine. Charles H. Hitchcock, "The Albert Coal, or Albertite, of New Brunswick," *AJS*, 39 (1865):267–273, 267, 271.

157. Hitchcock, "Albertite," 271.

158. Hitchcock, "Albertite," 270.

159. Hitchcock, "Albertite," 272.

160. Hitchcock, "Albertite," 272.

161. Gesner, *Practical Treatise,* 28.

162. J. Peter Lesley, "Notice of a Remarkable Coal Mine or Asphalt Vein, Cutting the Horizontal Coal-Measures of Wood County, Western Virginia," *Proceedings of the APS, 9* (1862–1864):183–197, 183.

163. Lesley, "Asphalt Vein," 183.

164. Lesley, "Asphalt Vein," 186.

165. Lesley's *entire* APS presentation was reprinted in *Prospectus of the Grahame Crystallized Rock Oil Company, Ritchie County, West Virginia* (Baltimore: John Murphy, 1865), 18–36.

166. Lesley, "Coal Oil," 447.

167. Lesley to Agassiz, 14 April 1865, Lesley Mss, APS.

168. Henry Wurtz, *Report upon a Mineral Formation in West Virginia, by Professor Henry Wurtz to the Ritchie Mineral, Resin, and Oil Co., Baltimore, MD.* (New York: Francis & Loutrel, 1865), 9.

169. Wurtz, *Report,* 9.

170. Wurtz, *Report,* 9.

171. Wurtz, *Report,* 7.

172. Wurtz, *Report,* 16.

173. Extracts of Lesley's APS presentation also appeared in Henry, *History,* 116–118.

174. Abraham Gesner, *A Practical Treatise on Coal, Petroleum, and Other Distilled Oils,* 2nd ed., revised and enlarged by George Weltden Gesner (New York: Baillière Brothers, 1865), 27; cf. Wurtz, *Report upon a Mineral Formation in West Virginia,* 29; (Lesley's extract), 30–31.

175. Morris Zaslow, *Reading the Rocks: The Story of the Geological Survey of Canada, 1842–1972* (Ottawa: Macmillan, 1975), 95–96.

176. W. L. Morton, *Henry Youle Hind, 1823–1908* (Toronto: University of Toronto Press, 1980).

177. Southern New Brunswick comprised the counties of Charlotte, Saint John, Kings, and Albert. Joseph Whitman Bailey, *Loring Woart Bailey: The Story of a Man of Science* (Saint John: J&A McMillan, 1925).

178. Review of "A Preliminary Report on the Geology of New Brunswick," *AJS, 40* (1865):142.

179. Bailey, *Mines and Minerals of New Brunswick,* 51.

180. Bailey's change of mind might have been strongly influenced by the observations and opinions of his two colleagues George F. Matthew and C. F. Hartt. L. W. Bailey, *Observations on the Geology of Southern New Brunswick* (Fredericton: G. E. Fenety, 1865), 105, 108.

181. Dawson, *Acadian Geology,* 240, 247, 248.

182. Lesley, "Coal Oil," 442.

Chapter 9. The Search for Oil and Oil-Finding Experts

1. Hall to anonymous, 8 March 1865, Hall Mss, BO 579, NYSA.

2. Hall to Lesley, 24 February 1865 and 24 December 1864, Lesley Mss, APS.

3. Hall to Lesley, 2 March 1865, Lesley Mss, APS.

4. Hall to Lesley, 2 March 1865, Lesley Mss, APS. Hall's friend was Spencer Gooding of Canandaigua, for whom he prepared small maps showing petroleum outcrops in northeastern Ohio and in Chautauqua and Cattaraugus counties, New York. Gooding to Hall, 27 February 1865; Hall to Gooding, 15 March 1865, Hall Mss, NYSA.

5. Hall to Lesley, 24 February 1865, Lesley Mss, APS.

6. Hall to Lesley, 2 March 1865, Lesley Mss, APS.

7. Hall to Lesley, 2 March 1865, Lesley Mss, APS.

8. Hall to Lesley, 24 February 1865, Lesley Mss, APS.

9. Edmund Morris, *Derrick and Drill, or an Insight into the Discovery, Development, and Present Condition and Future Prospects of Petroleum in New York, Pennsylvania, Ohio, West Virginia, &c.* (New York: James Miller, 1865), 70.

10. "Pithole," *The Nation, 1* (21 September 1865):370–372; William Culp Darrah, *Pithole, the Vanished City: A Story of the Early Days of the Petroleum Industry* (Gettysburg, PA: privately published, 1972).

11. Morris, *Derrick and Drill,* 192–193. "Mud everywhere, illimitable, unfathomable." J. H. A. Bone, *Petroleum and Petroleum Wells* (Philadelphia: J. B. Lippincott, 1865), 66.

12. Morris, *Derrick and Drill,* 69.

13. Morris, *Derrick and Drill,* 15.

14. Morris, *Derrick and Drill,* 71. "Oil Stock Excitement," *Scientific American, 12* (2 January 1865):8; "The Petroleum Oil Interests," *Scientific American, 12* (9 January 1865):23.

15. Darrah, *Pithole,* 38. In 1865, prices *in greenbacks* were 250% higher than prices *in gold.* Andrew Cone and Walter R. Johns, *Petrolia: A Brief History of the Pennsylvania Petroleum Region, Its Development, Growth, Resources, etc., from 1859 to 1869* (New York: D. Appleton, 1870), 79–82.

16. Morris, *Derrick and Drill,* 116.

17. S. J. M. Eaton, *Petroleum: A History of the Oil Regions of Venango County, Pennsylvania* (Philadelphia: J. P. Skelly, 1866), 239; Cone and Johns, *Petrolia,* 80; cf. 500 in Morris, *Derrick and Drill,* 257.

18. Eaton, *Petroleum,* 236–244.

19. Morris, *Derrick and Drill,* 257.

20. William Wright, *The Oil Regions of Pennsylvania: Showing Where Petroleum is Found; How it is Obtained, and at What Cost, with Hints for Whom it may Concern* (New York: Harper & Brothers, 1865), 100.

21. Morris, *Derrick and Drill,* 263.

22. Morris, *Derrick and Drill,* 264.

23. Morris, *Derrick and Drill,* 97.

24. Wright, *Oil Regions,* 253.

25. Morris, *Derrick and Drill,* 259.

26. Morris, *Derrick and Drill,* 97.

27. Darrah, *Pithole,* 7–13.

28. Morris, *Derrick and Drill,* 257.

29. Eaton, *Petroleum,* 245.

30. This song appeared about 1865 and was published in New York City. The original sheet music is in the Drake Well Museum Collections, Titusville, Pennsylvania, and it was reprinted in slightly modified form in Giddens, *Documents,* 298.

31. Wright, *Oil Regions,* 60, 215.

32. Wright, *Oil Regions*, 206–227.

33. "Petroleum Stock Swindle," *Scientific American*, 12 (21 January 1865):48. For bibliographies see Stephen F. Peckham, *The Production, Technology, and Uses of Petroleum and Its Products*, 47th Cong., 2nd sess., H.R. Misc. Doc. 42, Part 10 (Washington, DC: GPO, 1884); Giddens, *Sources*; Ernest C. Miller, *A Guide to the Early History of Petroleum* (Pittsburgh: Historical Society of Western Pennsylvania, 1969); Edward Benjamin Swanson, *A Century of Oil and Gas in Books: A Descriptive Bibliography* (New York: Appleton-Century-Crofts, 1960); Miller, *Pennsylvania's Oil Industry* (Gettysburg: Pennsylvania Historical Association, 1974).

34. Morris, *Derrick and Drill*, 259; cf. Eaton, *Petroleum*, 245–246.

35. Wright, *Oil Regions*, 218; Bone, *Petroleum and Petroleum Wells*, 147.

36. Morris, *Derrick and Drill*, 263.

37. Wright, *Oil Regions*, 206, 213.

38. Wright, *Oil Regions*, 79.

39. Morris, *Derrick and Drill*, 115.

40. Morris, *Derrick and Drill*, 15; Bone, *Petroleum and Petroleum Wells*, 56.

41. Morris, *Derrick and Drill*, 253–254.

42. Cf. Samuel W. Tait, *The Wildcatters: An Informal History of Oil-Hunting in America* (Princeton: Princeton University Press, 1946).

43. Wright, *Oil Regions*, 61; "Oil Smellers," *Scientific American*, 14 (17 March 1866):181; Bone, *Petroleum and Petroleum Wells*, 35.

44. Eaton, *Petroleum*, 89.

45. Cone and Johns, *Petrolia*, 121.

46. Wright, *Oil Regions*, 62–63.

47. Cone and Johns, *Petrolia*, 120.

48. Wright, *Oil Regions*, 63.

49. Morris, *Derrick and Drill*, 125.

50. Morris, *Derrick and Drill*, 125; Eaton, *Petroleum*, 89–90; Cone and Johns, *Petrolia*, 120.

51. Cone and Johns, *Petrolia*, 120.

52. Morris, *Derrick and Drill*, 125.

53. *Titusville Morning Herald*, 3 February 1868 and 26 May 1869; Cone and Johns, *Petrolia*, 121.

54. Cone and Johns, *Petrolia*, 120.

55. Morris, *Derrick and Drill*, 125.

56. Eaton, *Petroleum*, 90.

57. Morris, *Derrick and Drill*, 19.

58. Morris, *Derrick and Drill*, 262.

59. *By-Laws of the Hard Pan Oil Company, of Oil City, Pa, together with the Prospectus and By-Laws for the Regulation of the Board of Directors* (New York, 1866).

60. *The Shreve Farm Oil Company* (New York, 1865), 6.

61. *The Rockbottom Oil Company of Pennsylvania* (New York, 1865).

62. *Boston Petroleum Company* (New York, 1865), 4.

63. Review of "Mineral Oil," *AJS*, 41 (1866):284–285, 284.

64. *Prospectus of the First National Petroleum Company* (New York, 1865), 6.

65. *Prospectus of the Boston Petroleum Oil Company, on Duck Creek and Vicinity* (Boston, [1865]), 1.

66. *The Harmon Petroleum Company of New-York* (New York, 1865), 6.

67. There were more than twenty pages of advertisements for oil companies, land agents, engineers, lawyers, etc., at the back of the atlas. F. W. Beers, *Atlas of the Oil Region of Pennsylvania* (New York: F. W. Beers, A. D. Ellis, & G. G. Soule, 1865).

68. *Organization and Reports: Little Kanawha and Elk River Petroleum and Mining Company, of West Virginia* (New York, 1865); *Statement of the Kentucky and West Virginia Oil and Coal Company* (1865); *Braxton County Petroleum and Mining Company of West Virginia* (New York, 1865); and *West Virginia Oil and Coal Company* (New York, 1865).

69. See, for example, *Prospectus of the Buena Vista Oil Company* (New York, 1865); *Prospectus of the Lennox Oil Company* (New York, 1865); *Prospectus of the Palmer Oil Company* (New York, 1865); and *Prospectus of the Wilcox Oil Company* (New York, 1865).

70. Wright, *Oil Regions*, 223.

71. Wright, *Oil Regions*, 224.

72. Morris, *Derrick and Drill*, 96.

73. Lesley to Whitney, 4 November 1864, Lesley Mss, APS.

74. Lesley to Whitney, 4 November 1864, Lesley Mss, APS.

75. George P. Merrill, *Contributions to a History of American State Geological and Natural History Surveys*, U.S. National Museum, Bulletin No. 109 (Washington, DC: GPO, 1920), 466–471.

76. *Prospectus and Geological Report upon the Portion of the Lands of the Tennessee Petroleum and Mining Co.* (Chicago, 1866), 5, 11.

77. *Tennessee Petroleum*, 13.

78. *Prospectus of the New York and Pennsylvania Petroleum, Mining and Manufacturing Company* (New York, 1865); *The Parent Oil, Coal and Land Association of the Guyandotte* (New York, 1865); *The Homowack Oil and Mineral Company* ([New York], 1865). Cf. Merrill, *Contributions*, 135–137, 302–307, 502–505.

79. *The Kentucky National Petroleum & Mining Company* (Cincinnati, 1865), 9.

80. *Kentucky Petroleum*, 17.

81. *Prospectus of the Neff Petroleum Company, with the Geological Reports of Prof. A. Winchell, Prof. J. S. Newberry, Prof. H. L. Smith* (Gambier, OH: Western Episcopalian, 1866).

82. *Neff Petroleum*, 24–30, 26.

83. Merrill, *Contributions*, 203–231.

84. Edgar Wesley Owen, *Trek of the Oil Finders: A History of Exploration for Petroleum* (Tulsa: Association of American Petroleum Geologists, 1975), 71.

85. *Neff Petroleum*, 15.

86. *Oil Lands of Crocus Creek, Cumberland County, Ky. Report of Dr. J. S. Newberry* (Cleveland, 1865); cf. "Mineral Oil. Prospectus of the Indian Creek and Jack's Knob (Cumberland and Clinton counties, Kentucky) Coal, Salt, Oil, etc. Company, with a Geological Report on the Lands," *AJS, 41* (1866):284–285.

87. *Neff Petroleum*, 22.

88. See, for example, *Report of E. B. Andrews, Professor of Natural Science and Practical Geology, Marietta College, Ohio, to the West Virginia Oil & Oil Land Co.* (Detroit: Advertiser and Tribune Company, 1866); and *Prospectus of the Ohio Petroleum Company* (New York, 1864). Cf. Owen, *Trek of the Oil Finders*, 68.

89. *West Virginia Oil*, 7.

90. Owen, *Trek of the Oil Finders,* 69.

91. Owen, *Trek of the Oil Finders,* 69; cf. E. B. Andrews, ["Ohio Oil Lands"], *AG-LJ* (1 June 1861).

92. *Neff Petroleum,* 35–43.

93. Lesley, an ardent Union and Lincoln man, was enraged by the *Mercury*'s unpatriotic and "slanderous" attacks on the Sanitary Commission. Lesley to *Philadelphia Mercury,* 20 June 1864, Lesley Mss, APS.

94. Lesley to P. W. Sheafer, 6 December 1864, Lesley Mss, APS.

95. J. P. Lesley, "Coal Oil," *Report of the Commissioner of Agriculture for the Year 1862,* 429–447. Lesley considered this article "of a very different order from opinions respecting particular tracts." Lesley to P. W. Sheafer, undated [1864], Lesley Mss, APS.

96. Norton to Lesley, 13 February 1865, Lesley Mss, APS.

97. Lesquereux to Lesley, 9 January 1865, Lesley Mss, APS.

98. Lesley to Lesquereux, 31 December 1864, Lesley Mss, APS.

99. A list of Lesley's and Silliman's engagements, along with those of other petroleum geologists, is found in Paul Lucier, "Scientists and Swindlers: Coal, Oil, and Scientific Consulting in the American Industrial Revolution, 1830–1870" (PhD dissertation, Princeton University, 1994), appendices.

100. Lesley to H. A. Hendry, 4 November 1864; Lesley to [Benjamin] Lyman, 16 January 1865, Lesley Mss, APS.

101. Lesley to Lyman, 19 December 1864, Lesley Mss, APS.

102. Lesley to Lesquereux, 15–16 January 1865, Lesley Mss, APS.

103. Lesley to John P. Prince, 8 November 1864, Lesley Mss, APS.

104. "[Report to Chresworth C. Spanger. Esq. on the Warner Tract] situated on Beaver Creek. Beaver County, Pennsylvania," 20 December 1864, Lesley Mss, APS.

105. Lesquereux to Lesley, 5 January 1865, Lesley Mss, APS.

106. Lesley to Lesquereux, 15–16 January 1865, Lesley Mss, APS.

107. Lesquereux to Lesley, 9 January 1865; Lesley to Lesquereux, 15–16 January 1865, Lesley Mss, APS.

108. Lesley to Alexander Agassiz, 14 April 1865, Lesley Mss, APS.

109. Lesley to Joseph Lee, 8 February 1865, Lesley Mss, APS.

110. Newberry, "Mineral Oil," *AJS, 41* (1866):284.

111. "Report on Paint Lick Fork of Sandy River in Eastern Kentucky," Lesley Mss, APS. Cf. *Report on Lands on Paint Lick Fork of Sandy River in Eastern Kentucky* ([Philadelphia], 1865); *Proceedings of the APS, 10* (April 1865):33–68.

112. "Continuation of the Osceola—Moshannon Report," 27 December 1864, Lesley Mss, APS.

113. Lesley to Charles Rann, 26 June 1865, Lesley Mss, APS.

114. Wright, *Oil Regions,* 242.

115. "Continuation of the Osceola—Moshannon Report," 27 December 1864, Lesley Mss, APS.

116. Lesley to Lesquereux, 15–16 January 1865, Lesley Mss, APS.

117. Lesley to Hall, 25 January 1865, Hall Mss, NYSA.

118. Lesley to W. B. Ogden, 2 March 1865, Lesley Mss, APS. Cf. *Geological Report on the Brady's Bend Land and Coal Company Land in Armstrong County, Pennsylvania* (Philadelphia, 1865); *Proceedings of the APS, 10* (April 1865):227–242.

119. Lesley to W. B. Ogden, 2 March 1865, Lesley Mss, APS.

120. Lesley to Samuel Bost, 17 September 1865; Lesley to J. W. Harden, 17 September 1865, Lesley Mss, APS.

121. Lesley to David S. Brown, 2 March 1865, Lesley Mss, APS.

122. Lesquereux to Lesley, 2 May 1865, Lesley Mss, APS.

123. Lesquereux to Lesley, 28 April 1865, Lesley Mss, APS.

124. Lesquereux to Lesley, 4 May 1865, Lesley Mss, APS.

125. Lesquereux to Lesley, 25, 26, 28 May and 2 June 1865, Lesley Mss, APS.

126. Lesley to David S. Brown, 16 June 1865, Lesley Mss, APS.

127. Lesquereux to Lesley, 26 June 1865, Lesley Mss, APS.

128. "Appendix to Report on Slippery Rock Creek showing certain details of the geology in the neighborhood necessary to the perfect understanding of the underground structure," 10 July 1865, Lesley Mss, APS.

129. Lesquereux to Lesley, 16 July 1865, Lesley Mss, APS.

130. *Proceedings of the APS, 10* (May 1865):110.

131. Lesquereux to Lesley, 26 June 1865, Lesley Mss, APS.

132. *Report on a Geological Survey of the Lands of the Columbus National Petroleum Company* (Columbus, OH, 1865), cited in Owen, *Trek of the Oil Finders,* 77–78.

133. Lesquereux to Lesley, 18 August 1865, Lesley Mss, APS.

134. Alexander Winchell, "On the Oil Formation in Michigan and Elsewhere," *AJS, 39* (1865):350–353.

135. Leo Lesquereux, "On the Fucoids in the Coal Formation," *Transactions of the APS, 13* (1866):313; cf. "On the Fucoids in the Coal Formation," *AJS, 42* (1866):264.

136. J. P. Lesley, [Petroleum in the Eastern Coal-Field of Kentucky], *Proceedings of the APS, 10* (April 1865):33–69; cf. "Report on Lands on Paint Lick Fork of Sandy River in Eastern Kentucky;" and [Preliminary Report on Brady's Bend], Lesley to W. B. Ogden, 2 March 1865, Lesley Mss, APS.

137. Winchell, "On the Oil Formation in Michigan and Elsewhere"; J. M. Safford, "Note on the Geological Position of Petroleum Reservoirs in Southern Kentucky and in Tennessee," *AJS, 42* (1866):104–107; J. S. Newberry, "Mineral Oil. Prospectus of the Indian Creek and Jack's Knob (Cumberland and Clinton Counties, Kentucky) Coal, Salt, Oil, etc. Company, with a Geological Report on the Lands," *AJS, 41* (1866):284–285.

138. E. B. Andrews, "Petroleum in its Geological Relations," *AJS, 42* (1866):33–43. Cf. Andrews, "Distribution of Bitumen in Palæozoic Rocks," and "Section of Strata through South-eastern Ohio, and the Western Part of Virginia," *Proceedings of the AAAS, 15* (1866):29–30.

139. C. H. Hitchcock, "Petroleum in North America," *Geological Magazine, 4* (1867):34–37.

140. Cf. Owen, *Trek of the Oil Finders,* 95–96; Edward Orton, *Report of the Occurrence of Petroleum, Natural Gas and Asphalt Rock in Western Kentucky, based on examinations made in 1888 and 1889. Geological Survey of Kentucky* (Frankfort, 1891), 27–102; James Dwight Dana, aided by George Jarvis Brush, *A System of Mineralogy,* 5th ed. (New York: John Wiley & Son, 1868), 723–727.

141. Andrews, "Petroleum in its Geological Relations," 39.

142. Hitchcock, "Petroleum in North America," 36.

143. *Neff Petroleum,* 18.

144. *Neff Petroleum,* 18.

145. *Neff Petroleum,* 19.

146. Wright also suggested a "petroleus" formation. Wright, *Oil Regions,* 242.

147. *Neff Petroleum,* 19.

148. Andrews, "Petroleum in its Geological Relations," 39.

149. H. D. Rogers, "Two Lectures on Coal and Petroleum," *Geological Magazine, 3* (1866):258–259.

150. Hitchcock, "Petroleum in North America," 95.

151. *Proceedings of the APS, 10* (December 1865):190.

152. *Proceedings of the APS, 10* (April 1865):60.

153. *Neff Petroleum,* 19.

154. *Neff Petroleum,* 15.

155. *Proceedings of the APS, 10* (April 1865):62.

156. Lesquereux to Lesley, 5 January 1865, Lesley Mss, APS.

157. Winchell and Newberry also advocated a distillation theory. *Neff Petroleum,* 12–13, 18–20.

158. Andrews, "Petroleum in its Geological Relations," 40–43.

159. T. Sterry Hunt, *Geological Survey of Canada. Report of Progress from 1863 to 1866* (1866), 233–262.

160. *Proceedings of the APS, 10* (April 1865):43, 53; Hitchcock, "Petroleum in North America," 37.

161. Hitchcock, "Petroleum in North America," 37.

162. Harold F. Williamson and Arnold R. Daum, *The American Petroleum Industry: The Age of Illumination, 1859–1899* (Evanston: Northwestern University Press, 1959), 90; Tait, *The Wildcatters,* 79. On wildcatting as the antithesis to science, see Roger M. Olien and Diana Davids Olien, *Wildcatters: Texas Independent Oilmen* (Austin: Texas Monthly, 1984); Sam T. Mallison, *The Great Wildcatter* (Charleston, WV: Education Foundation of West Virginia, 1953); Ruth Sheldon Knowles, *The Greatest Gamblers: The Epic of American Oil Exploration* (New York: McGraw-Hill, 1959); and James A. Clark and Michel T. Halbouty, *The Last Boom* (New York: Random House, 1972).

163. Wright, *Oil Regions,* 212.

164. Wright, *Oil Regions,* 263.

165. *Proceedings of the APS, 10* (April 1865):227–230.

166. H. P. Cushing, "Peter Neff," *Bulletin of the Geological Society of America, 15* (1904):541–544.

167. Orton, *Report on the Occurrence of Petroleum,* 76–77.

Chapter 10. California Crude

1. George P. Merrill, ed., *Contributions to a History of American State Geological and Natural History Surveys,* U.S. National Museum, Bulletin No. 109 (Washington, DC: GPO, 1920), 30.

2. George P. Merrill, *The First One Hundred Years of American Geology* (New Haven: Yale, 1924), 407–411; Merrill, *Contributions,* 27–40; Gerald Nash, "The Conflict between Pure and Applied Science in Nineteenth-Century Public Policy: The California State Geological Survey, 1860–1874," *Isis, 54* (1963):174–185; William H. Goetzmann, *Exploration and Empire: The Explorer and the Scientist in the Winning of the American West* (New York: Knopf, 1966), 355–389; Gerald T. White, *Scientists in Conflict: The Beginnings of the Oil Industry in California* (San Marino: Huntington Library, 1968), 20. Cf. Robert Harry Block, "The Whitney Survey of California, 1860–74: A Study of Environmental

Science and Exploration" (PhD dissertation, University of California, Los Angeles, 1982).

3. An argument might be made for New York; cf. Merrill, *Contributions*, 537–538.

4. Edwin Tenney Brewster, *Life and Letters of Josiah Dwight Whitney* (Boston: Houghton Mifflin, 1909), 184.

5. Merrill, *Contributions*, 31.

6. Merrill, *Contributions*, 36.

7. *AJS, 30* (1860):424, 157.

8. Whitney, *An Address Delivered before the Legislature of California, . . . Mar. 12, 1861* (San Francisco: Towne & Bacon, 1861), 44–47.

9. Stephen M. Testa, "Josiah D. Whitney and William P. Blake: Conflicts in Relation to California Geology and the Fate of the First Geological Survey," *Earth Sciences History, 21* (2002):46–76.

10. Whitney, *An Address Delivered . . . Mar. 12, 1861*, 34.

11. "Letter of the State Geologist relative to the progress of the State Geological Survey," cited in *AJS, 34* (1862):157.

12. George J. Brush, "Review and reprint of *Annual Report of the State Geologist of California for the Year 1862*," *AJS, 36* (1863):118–122, 119.

13. Brush, "Review of *Annual Report . . . for the Year 1862*," *AJS, 36* (1863):118–122, 122. George Gervis Brush (1831–1912), professor of metallurgy in the Sheffield School at Yale, had been engaged by Whitney to examine California ores.

14. *Brief Report of J. D. Whitney, State Geologist of California, on the progress of the Geological Survey [for 1863]*, cited in *AJS, 37* (1864):427–431, 427.

15. *Brief Report of J. D. Whitney, State Geologist of California, on the progress of the Geological Survey [for 1863]*, cited in *AJS, 37* (1864):427–431, 429.

16. Nash, "Conflict between Pure and Applied," 181; White, *Scientists in Conflict*, 47; Goetzmann, *Exploration and Empire*, 379–383.

17. "An Agreement between Thomas A. Scott and Benj. Smith Lyman in regard to an expedition to Arizona," 24 December 1863, B. S. Lyman Mss, APS. The draft dated 21 December 1863 listed all the expedition members.

18. Joseph Lesley was secretary of the Pennsylvania Railroad from May 1869 through March 1881; Scott was president from June 1874 through June 1880. George H. Burgess and Miles C. Kennedy, *Centennial History of the Pennsylvania Railroad Company* (Philadelphia: Pennsylvania Railroad Co., 1949).

19. "Agreement," 24 December 1863, Lyman Mss, APS.

20. Hodge to Hall, 20 August 1864, Hall Mss, NYSA.

21. Brush to Brewer, 7 March 1864, Brewer Mss, YUL.

22. Eliot Lord, *Comstock Mining and Miners*, United States Geological Survey, (Washington, DC: GPO, 1883), 131–180, 177.

23. Lord, *Comstock*, 165–171.

24. Ralston founded the Bank of California in June 1864. Otis E. Young Jr., *Western Mining* (Norman: University of Oklahoma Press, 1970); George D. Lyman, *Ralston's Ring: California Plunders the Comstock Lode* (New York: C. Scribner, 1937).

25. Ralston to Silliman, 10 February and 11 January 1875, Silliman Mss, Smithsonian Institution, National Museum of American History, Washington, DC.

26. Benjamin Silliman Jr., "Notes on the New Almaden Quicksilver Mines," *AJS, 38* (1864):190–194. Young, *Western Mining*, 118–121. On the assumed Scott-Silliman engagement, see White, *Conflict*, 47.

27. *Prospectus of the Empire Gold & Silver Mining Co. of New York* (New York, 1864), 36–38.

28. Hodge to Hall, 20 August 1864, Hall Mss, NYSA.

29. Whitney, however, did take the stand as an expert witness, without pay, in some litigation. Whitney to Brewer, 18 April 1864, cited in Clark C. Spence, *Mining Engineers & the American West: The Lace-Boot Brigade* (New Haven: Yale University Press, 1970), 86, 203.

30. Merrill, *Contributions*, 32.

31. William Blake, "Preliminary Geological Report of the U.S. Pacific Railroad Survey, under the command of Lieut. R. S. Williamson," *AJS, 19* (1855):433–434. Blake, *Geological Report*, in *Exploration and Surveys for a Railroad Route from the Mississippi River to the Pacific Ocean*, 33rd Cong., 2nd sess., Sen. Ex. Doc. No. 78, Pt. II (1857), Vol. V. Cf. John R. Coash, "Perceptions of the Early Railroad Surveys in California," *Earth Sciences History, 11* (1992):40–44.

32. Thomas Antisell, *Geological Report*, in *Exploration and Surveys for a Railroad Route from the Mississippi River to the Pacific Ocean*, 33rd Cong., 2nd sess., Sen. Ex. Doc. No. 78, Pt. II (1857), Vol. VII, 107–114.

33. Josiah Dwight Whitney, *Geology*, vol. 1: *Report of Progress and Synopsis of the Field-Work, From 1860 to 1864* (Philadelphia: Caxton Press, 1865), 115–116.

34. White, *Conflict*, 56.

35. Silliman's letter with original italics is in *A Description of the Recently Discovered Petroleum Region in California, With a Report on the Same by Professor Silliman* (New York: Francis & Loutrel, 1865), 2.

36. "Report of Professor Silliman upon the Conway & Company Petroleum Property, Santa Barbara County, California," in *San Buenaventura Pacific Petroleum Co., of New York. Brief Account of One Hundred Thousand Acres of Oil Lands in California* (New York: Barnes & Martin, 1865):13–24, 13, 24.

37. "Report of Professor Silliman," 15, 18.

38. Silliman noted that petroleum in Italy and Rangoon might be as young as in California. "Report of Professor Silliman," 14.

39. Benjamin Silliman, "On some of the Mining Districts of Arizona near the Rio Colorado, with remarks on the Climate, &c," *AJS, 61* (1866):289–308.

40. White, *Conflict*, 61.

41. *Description of the . . . Petroleum Region in California*, 18.

42. *Description of the . . . Petroleum Region in California*, 19.

43. *Description of the . . . Petroleum Region in California*, 20.

44. White, *Conflict*, 71–73; Nash, "Pure and Applied Science," 181; Robert V. Bruce, *The Launching of Modern American Science, 1846–1876* (Ithaca: Cornell University Press, 1988), 315.

45. White, *Conflict*, 72–73.

46. According to Louis I. Kuslan, Silliman sold shares in the Wheatley Silver Lead Mines at Phoenixville, Pennsylvania, in February 1864 for $130,000, which seems way out of line with his stated income. Louis I. Kuslan, "Benjamin Silliman, Jr.: The Second Silliman," in Leonard G. Wilson, ed., *Benjamin Silliman and His Circle: Studies on the Influence of Benjamin Silliman on Science in America* (New York: Science History Publications, 1979), 184.

47. These lands are now part of Ventura County. *Professor Silliman's Report upon the Oil Property of the Philadelphia and California Company, of Philadelphia, Situated in*

Santa Barbara and Los Angeles Counties, California (Philadelphia: E. C. Markey & Son, 1865), 3. For Cresson, see Bruce Sinclair, *Philadelphia's Philosopher Mechanics: A History of the Franklin Institute, 1824–1865* (Baltimore: Johns Hopkins University Press, 1974).

48. *Silliman's Report [to] the Philadelphia and California Company,* 5.

49. *Silliman's Report [to] the Philadelphia and California Company,* 9.

50. *Silliman's Report [to] the Philadelphia and California Company,* 13.

51. *Silliman's Report [to] the Philadelphia and California Company,* 15.

52. *Professor Silliman's Report upon the Oil Property of the Pacific Coast Petroleum Company, of New-York, Situated in San Luis Obispo County, California* (New York: William A. Wheeler, 1865), 7, 11.

53. Silliman was nowhere near "rediscovering the anticline theory." Cf. White, *Conflict,* 78.

54. *Silliman's Report [to] the Pacific Coast Petroleum Company,* 14.

55. *Silliman's Report [to] the Pacific Coast Petroleum Company,* 8–9. Exciting excerpts, especially Silliman's claim of "fabulous wealth in the best of oil," were published under "Extraordinary Discoveries of Petroleum in California,: *Scientific American,* 12 (14 January 1865):33.

56. Cited in White, *Conflict,* 79.

57. Whitney to William Whitney, 1 March 1865, Whitney Mss, YUL.

58. Brewer to *Springfield Republican,* 21 March 1865, Brewer Mss, YUL. D. L. Harris might have been spurred to write to Brewer by "Petroleum and Asphaltum," *Scientific American,* 12 (21 January 1865):55.

59. "The California Oil is Not Asphaltum," Silliman to Harris, 8 April 1865, Brewer Mss, YUL. Reprinted as "Professor Silliman on Petroleum in California," *Scientific American,* 12 (29 April 1865):274.

60. Benjamin Silliman, "Examination of Petroleum from California," *AJS, 39* (1865):341–343; C. M. Warren, "On a Process of Fractional Condensation: applicable to the Separation of Bodies having small differences between their Boiling-points," *AJS, 39* (1865):327–340. Maisch's results appeared in *Silliman's Report [to] the Philadelphia and California Petroleum Company,* 30–31.

61. Silliman lost an $8,000 fee. White, *Conflict,* 250.

62. *Report of the Directors of the California Petroleum Company to the Stockholders, at the Annual Meeting Held at Philadelphia, March 14, 1866* (New York: Francis & Loutrel, 1866), 6.

63. *Mining and Scientific Press,* 22 April and 17 June 1865.

64. Cited in Andrew D. Rodgers III, *John Torrey: A Story of North American Botany* (Princeton: Princeton University Press, 1942), 287.

65. *Report of the Directors of the California Petroleum Co.,* 4. Torrey also did an assay of ore from Arizona; Silliman, "On Some of the Mining Districts of Arizona," 303.

66. *Report of the Directors of the California Petroleum Co.,* 5.

67. *Report of the Directors of the California Petroleum Co.,* 5–6.

68. Whitney, *Geology,* 115.

69. Whitney, *Geology,* 117.

70. Whitney, *Geology,* 118.

71. Whitney, *Geology,* 114–115. Whitney added that "the originators of these schemes knew that they were deceiving" (116).

72. *Silliman's Report [to] the Philadelphia and California Petroleum Company,* 3.

73. Andrew Cone and Walter R. Johns, *Petrolia: A Brief History of the Pennsylvania*

Petroleum Region, Its Development, Growth, Resources, etc., from 1859 to 1869 (New York: D. Appleton, 1870), 83–84. "Hundreds of companies . . . go down beneath the crash, and sink in merited oblivion." J. H. A. Bone, *Petroleum and Petroleum Wells* (Philadelphia: J. B. Lippincott, 1865), 148.

74. Stephen F. Peckham Manuscripts are in the Thomas Bard Papers, Huntington Library.

75. Peckham Mss, Huntington Library.

76. Emphasis in original. Thomas Bard's 1866 Diary, entry dated 29 January, in Thomas Bard Papers, Huntington Library.

77. Peckham Mss, Huntington Library.

78. William H. Hutchinson, *Oil, Land, and Politics: The California Career of Thomas Robert Bard,* 2 vols. (Norman: University of Oklahoma Press, 1965).

79. S. F. Peckham, "On the Supposed Falsification of Samples of California Petroleum," *AJS, 43* (1867):345–351, 346; cf. S. F. Peckham, "American Asphaltum," *American Chemist* 2 (July 1873):6–9, 6.

80. Peckham, "On the Supposed Falsification," 348.

81. This was an average: .861 (Silliman), .863 (Warren), .864 (Maisch). Peckham, "On the Supposed Falsification," 347.

82. Peckham, "On the Supposed Falsification," 350.

83. Peckham, "On the Supposed Falsification," 349, 350.

84. Peckham, "On the Supposed Falsification," 351.

85. B. Silliman, "On Naphtha and Illuminating Oil from Heavy California Tar (Maltha)," *AJS, 43* (1867):242–246, 245–246.

86. Silliman, "Heavy California Tar (Maltha)," 244–245. Cf. S. F. Peckham, "On the Distillation of Dense Hydrocarbons at High Temperatures, technically termed 'Cracking,'" *AJS, 47* (1869):9–16.

87. Silliman, "Heavy California Tar (Maltha)," 242, 245; Wolfgang König, "Science-Based Industry or Industry-Based Science? Electrical Engineering in Germany before World War I," *T&C, 37* (1996):70–101.

88. *Lewis R. Ashurst, et al. v. Thomas A. Scott, et al.* in the Supreme Court in and for the Eastern District of Pennsylvania. February 1867. A copy is located in the Brewer Mss, YUL.

89. *Eugene T. Lynch v. John B. Church, et al.,* New York, 1867.

90. S. J. M. Eaton, *Petroleum: A History of the Oil Region of Venango County, Pennsylvania* (Philadelphia: J. P. Skelly, 1866), 246.

91. *Ashurst, et al. v. Scott et al.,* 8.

92. Maisch's analysis also "exercised some effect in procuring subscriptions to said stock." *Ashurst, et al. v. Scott et al.,* 9–10.

93. White, *Conflict,* 254.

94. *Ashurst, et al. v. Scott et al.,* 8.

95. Silliman to President Woolsey, 23 June 1870, cited in White, *Conflict,* 153.

96. *Reports on the Gold and Silver Deposits at Quail Hill, Calaveras Co., Cal.* (San Francisco, 1867). Crossman to Silliman, 14 June 1869, Silliman Mss, Smithsonian Institution.

97. *The Empire Mining Company, Ophir Hill, Grass Valley* (San Francisco, 1867). Crossman to Silliman, 30 March 1870, Silliman Mss, Smithsonian Institution.

98. Crossman to Silliman, 14 June 1869, Silliman Mss, Smithsonian Institution.

99. "California Petroleum," *San Francisco Bulletin,* 3 April 1867; cf. Marcelin Ber-

thelot, "Sur l'origine des carbures et des combustibles minéraux," *Annales de Chimie et Physique, 9* (1866):481–483.

100. "Notes on the origin of Bitumens, together with experiments upon the formation of Asphaltum; by S. F. Peckham. Communicated by Prof. J. D. Whitney, Chief of the Geological Survey of California," *Proceedings of the APS, 10* (1868):445–462; cf. "Notes on the Origin of Bitumens," *AJS, 48* (1869):131–133. Dana was probably the reviewer.

101. Peckham, "Origin of Bitumens," 462, 445.

102. Peckham, "Origin of Bitumens," 451. On American responses to Berthelot, see Stephen F. Peckham, *Production, Technology, and Uses of Petroleum and Its Products* (Washington, DC: GPO, 1884), 61, 68; and Edward Orton, *Report of the Occurrence of Petroleum, Natural Gas and Asphalt Rock in Western Kentucky, based on examinations made in 1888 and 1889. Geological Survey of Kentucky* (Frankfort, 1891), 32.

103. Peckham, "Origin of Bitumens," 452.

104. S. F. Peckham, "American Asphaltum," *American Chemist, 2* (1873):6–9, 7.

105. The oxygen came either from direct contact with the atmosphere or indirectly through solution in rain water.

106. Peckham, "American Asphaltum," 7.

107. James Dwight Dana aided by George Jarvis Brush, *A System of Mineralogy,* 5th ed. (New York: John Wiley & Son, 1868), preface.

108. Dana, *System of Mineralogy* (1868), 722. For comparative tables, see 756–758.

109. Dana, *System of Mineralogy* (1868), 724.

110. Dana, *System of Mineralogy* (1868), 723, 728–729. On nomenclature, see xxix–xxxiv.

111. Dana, *System of Mineralogy* (1868), 722.

112. Dana also singled out Bathvillite, an altered lump of resin occurring in the Torbanite on lands adjoining Torbanehill, in Bathville, Scotland. Torbanite and Bathvillite belonged to the Succinite [Amber] Group of the Oxygenated Hydrocarbons. Dana, *System of Mineralogy* (1868), 742, 755.

113. Peckham, "American Asphaltum," 8.

114. S. F. Peckham, "On the probable origin of Albertite and allied minerals," *AJS, 48* (1869):362–370, 364–366.

115. James Dwight Dana, *Manual of Coal* (New York: Ivison, Blakeman, Taylor, and Co., 1876), 316.

116. Whitney to William Whitney, 13 April 1868, Whitney Mss, YUL.

117. Established in 1865, Whitney had accepted the position during his first East Coast sojourn. In 1875, it closed for lack of students. Nash, "The Conflict between Pure and Applied Science"; Peggy Champlin, *Raphael Pumpelly: Gentleman Geologist of the Gilded Age* (Tuscaloosa: University of Alabama Press, 1994), 80–81, 89.

118. Merrill, *Contributions,* 33–34.

119. Agassiz to Bache, 23 May 1863, in Nathan Reingold, ed., *Science in Nineteenth-Century America: A Documentary History* (Chicago: University of Chicago Press, 1964), 208–209.

120. Henry Darwin Rogers was not chosen. Lesley to Wilson, 7 March 1863, Lesley Mss, APS.

121. Henry to Stephen Alexander, 9 March 1863, in Reingold, *Science,* 204.

122. Leidy to Hayden, 28 April 1863, in Reingold, *Science,* 209.

123. Whitney to George Brush, 8 April 1863, Whitney Mss, YUL.

124. In February and March 1872, Silliman made a follow-up investigation of the Emma mine and other properties in Utah and California for $10,000. Silliman, "Geological and Mineralogical Notes on some of the Mining Districts of Utah Territory, and especially those of Wahsatch and Oquirrh Ranges of Mountains," *AJS, 3* (1872):195–201; and "Mineralogical Notes on Utah, California, and Nevada, with a Description of Priceite, a New Borate of Lime," *AJS, 6* (1873):18–22. *Report of Professor B. Silliman on the Emma Silver Mine* (London: W. Brown, 1872).

125. "Emma Mine Investigation," *House Report* 579, 44th Cong., 1st sess. (1875–1876), 634.

126. *Mining and Scientific Press,* 28 March 1874.

127. Minutes of the meeting of the Council of the Academy, 28 October 1873. "Excerpts from the Meetings of the Council of the Academy" are part of the "Silliman-Whitney Controversy" file at the National Academy of Sciences, Washington, DC.

128. Silliman to Hilgard, 10 November 1873, S-W NAS.

129. Whitney to the Council of the National Academy of Sciences, 18 October 1873, William Dwight Whitney Mss, YUL. Subsequent quotations are taken from this document.

130. Silliman to Hilgard, 10 February 1874, S-W NAS.

131. Silliman to Hilgard, 5 March 1874, S-W NAS.

132. Silliman to Hilgard, 5 March 1874, S-W NAS.

133. Silliman submitted at least thirty more documents. Silliman to Hilgard, 14 and 25 February, 11 March 1874; Hilgard Memoranda, 18 February and 3 March, 1874, S-W NAS.

134. William Whitney to Hilgard, 20 February 1874, S-W NAS. William continued to complain about the council in contrast to Silliman who praised it for its "good faith fairmindedness and justness." William Whitney to Hilgard, 23 and 25 February 1874; Silliman to Hilgard, 24 February 1874, S-W NAS.

135. Silliman to Hilgard, 10 March 1874, S-W NAS.

136. Minutes of Council, 21 April 1874, S-W NAS.

137. Silliman to Hilgard, 8 June and 16 July 1874, S-W NAS.

138. Henry to Silliman, 7 August 1874, S-W NAS.

139. Silliman to Hilgard, 10 March 1874, S-W NAS.

140. Silliman to Hilgard, 24 February 1874, S-W NAS.

141. Silliman to Hilgard, 16 March 1874, S-W NAS.

142. Silliman to Hilgard, 16 March 1874, S-W NAS.

143. Godkin to William Whitney, [2 letters dated] 16 February 1874, William Dwight Whitney Mss, YUL.

144. Silliman to Hilgard, 28 March 1874, S-W NAS.

145. Gibbs to Hilgard, undated, S-W NAS.

146. William Whitney to Hilgard, 31 March 1874, S-W NAS.

147. Silliman to Hilgard, 16 July 1874, S-W NAS.

148. William Whitney to Hilgard, 3 December 1874, S-W NAS.

149. Silliman to Hilgard, 19 December 1874, S-W NAS.

150. The "petroleum draft" is part of the William Brewer Papers, YUL. All subsequent quotations refer to this document.

151. Gibbs to Hilgard, 26 December 1874, S-W NAS.

152. Barnard to Hilgard, 29 December 1874, S-W NAS.

153. LeConte to Hilgard, 26 December 1874, S-W NAS.

154. Minutes of the Meeting of the Council of the Academy, 30 December 1874, S-W NAS; cf. Rexmond C. Cochrane, *The National Academy of Sciences: The First Hundred Years, 1863–1963* (Washington, DC: National Academy of Sciences, 1978), 120–124.

155. Silliman wanted the charges dismissed entirely, not a "*Not Proven*" verdict. Silliman to Hilgard, 24 February 1874, S-W NAS.

156. Henry to Hilgard, 31 December 1874, S-W NAS.

157. Silliman to Hilgard, 31 December 1874, S-W NAS.

158. Larry Owens, "Pure and Sound Government: Laboratories, Gymnasia, and Playing Fields in Nineteenth-Century America," *Isis, 76* (1985):182–194; David A. Hollinger, "Inquiry and Uplift: Late Nineteenth-Century American Academics and the Moral Efficacy of Scientific Practice," in Thomas L. Haskell, ed., *The Authority of Experts* (Bloomington: Indiana University Press, 1984), 142–156.

159. Henry to Hilgard, 31 December 1874, S-W NAS.

160. White adopted this "downgraded" interpretation. White, *Conflict.*

161. Peckham, *Petroleum,* 12.

162. Peckham to Silliman, 21 September 1881, Silliman Mss, YUL.

163. Peckham, *Petroleum,* 27.

164. George E. Webb, *Science in the American Southwest: A Topical History* (Tucson: University of Arizona Press, 2002), ch. 2.

Epilogue

1. "American Influence in Civilization," *Popular Science Monthly, 13* (1878):495–497, 495. For other comments on the deleterious effects of the Americanization of science, see Emil Du Bois-Reymond, "Civilization and Science," *Popular Science Monthly, 13* (1878):391–396; and Jeffrey Allan Johnson, *The Kaiser's Chemists: Science and Modernization in Imperial Germany* (Chapel Hill: University of North Carolina Press, 1990), 159–160.

2. Mark Wahlgren Summers, *The Era of Good Stealings* (New York: Oxford University Press, 1993); Eric Foner, *Reconstruction: America's Unfinished Revolution, 1863–1877* (New York: Harper & Row, 1988).

3. Andrew D. White, "Science and Public Affairs," *Popular Science Monthly, 2* (1873):736–739, 738.

4. Robert V. Bruce, *The Launching of Modern American Science, 1846–1876* (Ithaca: Cornell University Press, 1988), 342.

5. Simon Newcomb, "Abstract Science in America," *North American Review, 122* (1876):88–123; "Professor Newcomb on American Science," *Popular Science Monthly, 6* (1876):238–244, 240.

6. Joseph Henry, "On the Importance of the Cultivation of Science," *Popular Science Monthly, 2* (1873):641–650; for other positive evaluations of American geology and chemistry, see John W. Draper, "Science in America," *Popular Science Monthly, 10* (1877):313–326; Benjamin Silliman, "American Contributions to Chemistry," *American Chemist, 5* (1875):70–114.

7. Henry, "Cultivation of Science," 645.

8. Cited in Bruce, *Launching of Modern American Science,* 315.

9. National Academy of Sciences, *Proceedings,* April 1878, 132. On the postbellum

years and the crises within the academy, see Rexmond C. Cochrane, *The National Academy of Sciences: The First Hundred Years, 1863–1963* (Washington, DC: National Academy of Sciences, 1978), 100–133.

10. H. A. Rowland, "A Plea for Pure Science," *Science, 29* (1883):242–250, 242. This interesting footnote was omitted in the reprint of Rowland's collected papers; cf. Rowland, *The Physical Papers of Henry Augustus Rowland* (Baltimore: Johns Hopkins University Press, 1902), 596.

11. Rowland, "Plea," 242, 243.

12. George H. Daniels, "The Pure-Science Ideal and Democratic Culture," *Science, 156* (1967):1699–1705; Daniel J. Kevles, *The Physicists: The History of a Scientific Community in Modern America* (New York: Alfred A. Knopf, 1971), ch. 4; Daniel J. Kevles, Jeffrey L. Sturchio, and P. Thomas Carroll, "The Sciences in America, circa 1880," *Science, 209* (1980):27–32; Spencer R. Weart, "The Rise of 'Prostituted' Physics," *Nature, 262* (1976):13–17; Ronald Kline, "Construing 'Technology' as 'Applied Science': Public Rhetoric of Scientists and Engineers in the United States, 1880–1945," *Isis, 86* (1995):194–221.

13. Sydney Ross, "Scientist: The Story of a Word," *Annals of Science, 18* (1962):65–85.

14. In the cynical interpretation, antebellum men of science preached the gospel of utility in order to hoodwink Americans into granting them more authority and autonomy. George H. Daniels, *American Science in the Age of Jackson* (New York: Columbia University Press, 1968); Burton J. Bledstein, *The Culture of Professionalism: The Middle Class and the Development of Higher Education in America* (New York: W. W. Norton, 1976).

15. John W. Servos, "Mathematics and the Physical Sciences in America, 1880–1930," *Isis, 77* (1986):611–629.

16. Laurence R. Veysey, *The Emergence of the American University* (Chicago: University of Chicago Press, 1965); Joseph Ben-David, *The Scientist's Role in Society: A Comparative Study* (Chicago: University of Chicago Press, 1984), ch. 8; Larry Owens, "Pure and Sound Government: Laboratories, Playing Fields, and Gymnasia in the Late-Nineteenth-Century Search for Order," *Isis, 76* (1985):182–194; Robert E. Kohler, "The Ph.D. Machine: Building on the Collegiate Base," *Isis, 81* (1990):638–662.

17. David A. Hollinger, "Inquiry and Uplift: Late Nineteenth-Century American Academics and the Moral Efficacy of Scientific Practice," in Thomas L. Haskell, ed., *The Authority of Experts: Studies in History and Theory* (Bloomington: Indiana University Press, 1984), 142–156; Daniel J. Kevles, "American Science," in Nathan O. Hatch, ed., *The Professions in American History* (Notre Dame: University of Notre Dame Press, 1988), 107–125. On professionalization in the early nineteenth century, see George H. Daniels, "The Professionalization of American Science: The Emergent Period, 1820–1860," *Isis, 58* (1967):151–166; Nathan Reingold, "Definitions and Speculations: The Professionalization of Science in America in the Nineteenth Century," in Reingold, *Science, American Style* (New Brunswick: Rutgers University Press, 1991), 24–53; and Jan Golinski, *Making Natural Knowledge: Constuctivism and the History of Science* (Cambridge: Cambridge University Press, 1998), esp. ch. 2.

18. Robert Harrison Shryock, "American Indifference to Basic Science during the Nineteenth Century," *Archives Internationales d'Histoire des Sciences, 28* (1948–1949):3–18; cf. Nathan Reingold, "American Indifference to Basic Research: A Reappraisal," in George H. Daniels, ed., *Nineteenth-Century American Science: A Reappraisal* (Evanston:

Northwestern University Press, 1972), 38–62; Margaret W. Rossiter, *The Emergence of Agricultural Science: Justus Liebig and the Americans, 1840–1880* (New Haven: Yale University Press, 1975).

19. Such commercialization would also apply to such a successful inventor-entrepreneur as Thomas Edison. David A. Hounshell, "Edison and the Pure Science Ideal in America," *Science, 207* (1980):612–617.

20. Rowland, "Plea," 244.

21. Hatch, *The Professions in American History*. On the bankruptcy of professional models, see Gerald L. Geison, introduction to Geison, ed., *Professions and Professional Ideologies in America* (Chapel Hill: University of North Carolina Press, 1983).

22. Rowland, "Plea," 243, 244, 246.

23. Rowland, "Plea," 246, 247. Cf. Henry, "Cultivation of Science," 646.

24. Rowland, "Plea," 246.

25. Chemistry was the largest discipline in America with roughly ten times as many practitioners as in physics. Benjamin Silliman, "American Contributions to Chemistry," *American Chemist, 5* (1875):70–114; Edward H. Beardsley, *The Rise of the American Chemistry Profession, 1850–1900* (Gainesville: University of Florida Press, 1964); Arnold Thackray, Jeffrey L. Sturchio, P. Thomas Carroll, and Robert Bud, *Chemistry in America, 1876–1976: Historical Indicators* (Dordrecht: D. Reidel, 1985).

26. Ira Remsen, "The Relation between Chemical Science and Chemical Industry," *An Address Delivered Before the Chemical and Natural History Society of the Lehigh University* (Bethlehem, PA: 1888), 9.

27. Remsen, "Chemical Science and Chemical Industry," 8–9.

28. Remsen, "Chemical Science and Chemical Industry," 13, 11.

29. Owen Hannaway, "The German Model of Chemical Education in America: Ira Remsen at Johns Hopkins (1876–1913)," *Ambix, 23* (1976):145–164.

30. David Cahan, "Helmholtz and the Shaping of the American Physics Elite in the Gilded Age," *Historical Studies in the Physical and Biological Sciences, 35* (2004):1–34.

31. Over the past generation or two, historians of technology have thoroughly disassembled the linear model connecting science to technology. The latest in this deconstruction project is Karl Grandin, Nina Worms, and Sven Widmalm, eds., *The Science-Industry Nexus: History, Policy, Implications* (Sagamore Beach, ME: Science History Publications, 2005). For an example of a cultural analysis of this nexus, see Emily Thompson, *The Soundscape of Modernity: Architectural Acoustics and the Culture of Listening in America, 1900–1933* (Cambridge, MA: MIT Press, 2002).

32. John W. Draper, "Science in America," *Popular Science Monthly, 10* (1877):313–326, 322–323; Henry, "Cultivation of Science," 648.

33. Edward Orton, "Ohio," in George P. Merrill, ed., *Contributions to a History of American State Geological and Natural History Surveys*, U.S. National Museum, Bulletin No. 109 (Washington, DC: GPO, 1920), 387–427, 401. Orton succeeded Newberry as state geologist in 1882. The second Ohio survey (1869–1874) received $333,892; it was the second largest state survey of the late nineteenth century.

34. J. Peter Lesley "Pennsylvania," in Merrill, *Contributions*, 428–456, 435, 436.

35. Lesley, "Pennsylvania," in Merrill, *Contributions*, 436.

36. Lesley, "Pennsylvania," in Merrill, *Contributions*, 436.

37. Mary C. Rabbitt, *Minerals, Lands, and Geology for the Common Defense and General Welfare*, vol. 1: *Before 1879* (Washington, DC: GPO, 1979), 283–284.

38. The bitter irony was that King was financially unsuccessful. Thurman Wilkins and Caroline Lawson Hinkley, *Clarence King: A Biography,* revised and enlarged (Albuquerque: University of New Mexico Press, 1988).

39. Lesley, "Pennsylvania," in Merrill, *Contributions,* 403.

40. Lesley, "Pennsylvania," in Merrill, *Contributions,* 443.

41. Brewer to Whitney, 28 February 1866, cited in Clark C. Spence, *Mining Engineers & the American West: The Lace-Boot Brigade, 1849–1933* (New Haven: Yale University Press, 1970), 38.

42. Peggy Champlin, *Raphael Pumpelly: Gentleman Geologist of the Gilded Age* (Tuscaloosa: University of Alabama Press, 1994), 89.

43. Spence, *Mining Engineers & the American West;* Kathleen H. Ochs, "The Rise of American Mining Engineers: A Case Study of the Colorado School of Mines," *T&C, 33* (1992):278–301; Logan Hovis and Jeremy Mouat, "Miners, Engineers, and the Transformation of Work in the Western Mining Industry," *T&C, 37* (1996):429–456.

44. Daniel H. Calhoun, *The American Civil Engineer: Origins and Conflict* (Cambridge, MA: MIT Press, 1960).

45. A. B. Parsons, ed., *Seventy-Five Years of Progress in the Mineral Industry, 1871–1946* (New York: American Institute of Mining and Metallurgical Engineers, 1947), 407, 432–433.

46. Edwin T. Layton Jr., "Mirror-Image Twins: The Communities of Science and Technology in Nineteenth-Century America," *T&C, 12* (1971):562–580; Edwin T. Layton Jr., *The Revolt of the Engineers: Social Responsibility and the American Engineering Profession* (Cleveland: Case Western Reserve University Press, 1971); Bruce Sinclair, "Episodes in the History of the American Engineering Profession," in Hatch, *Professions in American History,* 127–144; Samuel Haber, *The Quest for Authority and Honor in the American Professions, 1750–1900* (Chicago: University of Chicago Press, 1991), 303–304; Terry S. Reynolds, ed., *The Engineer in America: A Historical Anthology from Technology and Culture* (Chicago: University of Chicago Press, 1991); Ronald Kline, "Construing 'Technology' as 'Applied Science': Public Rhetoric of Scientists and Engineers in the United States, 1880–1945," *Isis, 86* (1995):194–221.

47. The classic analysis of the difficulties engineers face in balancing professional and commercial demands is Layton, *Revolt of the Engineers.*

48. T. C. Mendenhall, "The Relations of Men of Science to the General Public," *Proceedings of the AAAS, 40* (1891):1–15, 8.

49. Merrill, *Contributions,* 435.

50. Robert H. Thurston, "The Mission of Science," *Proceedings of the AAAS, 34* (1885):227–253, 251, 239.

51. "Prostitution" does not appear in the pronouncements of late nineteenth-century scientists or engineers; it does in the historical literature. Hannaway, "The German Model of Chemical Education in America," 157; Weart, "The Rise of 'Prostituted' Physics"; P. Thomas Carroll, "American Science Transformed," *American Scientist, 74* (1986):466–485, 474; Michael Aaron Dennis, "Accounting for Research: New Histories of Corporate Laboratories and the Social History of American Science," *Social Studies of Science, 17* (1987):479–518. On engineers as corporate tools, see David F. Noble, *America by Design: Science, Technology, and the Rise of Corporate Capitalism* (New York: Alfred A. Knopf, 1977).

52. Brewer, "Petroleum Draft," Brewer Mss, YUL.

53. Benjamin Silliman, preface to *Combustion of Wet Fuel: Testimony Given in the Case of the Moses Thompson Wet Fuel Furnace* (New York: Evening Post, 1872).

54. Sinclair, "Engineering," 132; cf. Hollinger, "Inquiry and Uplift"; Kevles, "American Science," on the origin and persistence of the pure-science-to-applied technology model.

55. Contrast this vision of Americanization with commercialization-as-prostitution. Weart, "The Rise of 'Prostituted' Physics"; Dennis, "Accounting for Research."

Essay on Sources

Manuscript Sources

The most important sources for any study of consulting are the papers of the men of science, and to a large extent, this book is based on manuscript sources. The most useful are found in the J. Peter Lesley Papers at the American Philosophical Society, Philadelphia, which contain Lesley's correspondence with capitalists and mining companies, as well as many drafts of his consulting reports, along with maps and bills for his professional services. The American Philosophical Society also holds the papers of Lesley's nephew and sometimes assistant, Benjamin Smith Lyman, and of his brother Joseph. The next best records of consulting are found in the two collections of James Hall's papers in the New York State Library and New York State Archives, respectively. There is a third, and until this study unknown, Hall collection in the New York State Archives containing material from the 1860s, during which Hall worked in Canada and on petroleum. Other material on Hall's (and Eben Horsford's) commercial interests can be found in the Eben Horsford Papers at Rensselaer Polytechnic Institute, Troy, New York. Yale University houses several key manuscript collections, including the Silliman Family Papers, the Dana Family Papers, the William Henry Brewer Papers (containing the "Petroleum Draft"), and the William D. Whitney Family Papers (containing Josiah Whitney's infamous indictment of Benjamin Silliman). The National Academy of Sciences in Washington, DC, has a separate collection aptly entitled the Silliman-Whitney Controversy containing correspondence of the council and minutes of its meetings. A small collection of Benjamin Silliman Jr.'s correspondence is located at the Dibner Library of the History of Science and Technology at the Smithsonian Institution. The Huntington Library, San Marino, California, holds significant material on American science and the California oil boom in, respectively, the W. J. Rhees collection and the Thomas Bard Papers, which includes Stephen F. Peckham's letters and journals.

The records of other consulting chemists and geologists are of lesser importance. Some correspondence of Henry Darwin Rogers is located at the Massachusetts Institute of Technology in the William Barton Rogers Papers and the Rogers Family Papers. Richard Cowling Taylor's papers are in the Academy of Natural Sciences in Philadelphia, but they lack correspondence relating to the Albert mineral trials. The Academy of Natural Sciences does have the American Association of Geologists and Naturalists Papers, which includes Charles Jackson's insightful "Remarks on Mining Operations, 1846." Other papers of Jackson are located in the Massachusetts Historical Society, Boston, but these deal largely with the ether controversy. There are very few letters of Abraham Gesner; a handful relating to the New Brunswick Geological Survey are preserved in the Secretary's Letterbooks of the Geological Society of London. The James Young Papers in the University of Strathclyde, Scotland, are the best resource for following the legal and commercial controversies surrounding Kerosene and Paraffine. The Drake Well Museum in Titusville, Pennsylvania, has the best sources on the Pennsylvania oil boom, including the letters and reminiscences of Francis Beattie Brewer, George H. Bissell, James M. Townsend, and Edwin L. Drake, along with the unique photographs of the John A. Mather collection. The Brewer and Bissell letters were published in Paul H. Giddens, ed., *The Beginnings of the Petroleum Industry: Sources and Bibliography* (Harrisburg: Pennsylvania Historical Commission, 1941). The Townsend and Drake letters and the personal histories of Brewer, Bissell, Townsend, and Drake were published in Paul H. Giddens, ed., *Pennsylvania Petroleum, 1750–1872: A Documentary History* (Titusville: Pennsylvania Historical and Museum Commission, 1947).

Primary Sources

The next most important sources for this study are published consulting reports. Many of these can be located by searching library collections under the names of particular men of science or the companies for whom they consulted. Other reports are not so easy to find; they are interleaved in untitled business prospectuses or with company bylaws and are found in the archives among the papers of men of science. The Drake Well Museum contains the richest collection of oil company prospectuses and consulting reports along with pamphlets, maps, atlases, and newspapers relating to the oil boom. The Hagley Museum and Library in Wilmington, Delaware, has a smaller collection of coal and oil prospectuses. Yale University has a number of Silliman's reports, but the DeGolyer Library at Southern Methodist University owns a complete set of Silliman's California oil reports.

The other published sources for nineteenth-century American science and technology fall into three categories: journals, books, and government reports. Much of the geology and chemistry of coal and petroleum appeared first as articles in the *American Journal of Science* or in the proceedings and/or transactions of scientific societies, such as the American Association for the Advancement of Science, the American Philosophical Society, the Boston Society of Natural History, and the Geological Society of London. Charles T. Jackson and Francis Alger's survey of Nova Scotia, for example, was first serialized in the *American Journal of Science* and then revised for the *Memoirs of the American Academy of Arts and Sciences* before being reprinted as *Remarks on the Mineralogy and Geology of Nova Scotia, Accompanied by a Colored Map, Illustrative of the Structure of the Country, and by Several Views of its Scenery* (Cambridge, MA: E. W. Metcalf, 1832). More popular media, such as *Scientific American*, the *American Gas-Light Journal, Harper's New Monthly Magazine*, and *Popular Science Monthly*, carried the latest developments in science and in the coal, kerosene, and oil businesses. One of the best sources on the Kerosene and Paraffine legal cases was the *London Chemical News*. The records of the various court cases can be located in any good law library. I found the essential *Report of a Case, tried at Albert Circuit, 1852, before his Honor Judge Wilmot, and a Special Jury. Abraham Gesner vs. William Cairns. Copied from the Judge's Notes* (Saint John: William L. Avery, 1853) and the reports from the supreme courts at the University of New Brunswick in Fredericton. The only copy I know of Abraham Gesner's *Gas Monopoly: Piracy of Patents and Farmers' Rights in which is contained a Reply to the Directors of the Halifax Gas Company and a Brief Account of the Asphaltum Mines of New Brunswick* (Halifax, 1851) is bound with other miscellaneous pamphlets in the Firestone Library at Princeton University.

The place to begin any history of coal geology is Abraham Gottlob Werner, *Short Classification and Description of the Various Rocks*, trans. with introduction and notes

by Alexander M. Ospovat (New York: Hafner, 1971), and William D. Conybeare and William Phillips, *Outlines of the Geology of England and Wales, with an Introductory Compendium of the General Principles of that Science, and Comparative Views of the Structure of Foreign Countries* (London: William Phillips, 1822). Robert Bakewell's *An Introduction to Geology* (London: J. Harding, 1813; 2nd ed., 1815), provided a sound practical critique of Werner and his Independent Coal formation. The third edition of Bakewell's *Introduction* (1828), however, included a separate Coal formation and, more significantly, was edited by Benjamin Silliman (New Haven: Hezekiah House, 1829) and used as a textbook for many American geologists.

The most comprehensive book on coal in the first half of the nineteenth century was Richard Cowling Taylor's *Statistics of Coal* (Philadelphia: J. W. Moore, 1848). A second, posthumous edition was issued by S. S. Haldeman in 1855. J. Peter Lesley's *Manual of Coal and Its Topography* (Philadelphia: J. B. Lippincott, 1856) dealt almost exclusively with Pennsylvania coal, and John William Dawson's *Acadian Geology* (Edinburgh: Oliver and Boyd, 1855; 2nd ed., 1868; 3rd ed., 1878), treated Nova Scotia and New Brunswick. The best source on antebellum American mining is Josiah Dwight Whitney, *The Metallic Wealth of the United States, Described and Compared with that of Other Countries* (Philadelphia: Lippincott, Grambo & Co., 1854). James Dwight Dana's *Manual of Geology: Treating of the Principles of the Science with special reference to American Geological History* (Philadelphia: Theodore Bliss, 1863), and subsequent editions, summarized the latest developments and discoveries in the earth sciences with particular attention to American contributions. Likewise, James Dwight Dana's *System of Mineralogy* (New Haven: Durrie & Peck, 1837), and subsequent editions, provided an overview of that science and American work. Charles Lyell's changing views of American geology can be traced through his numerous publications: *Travels in North America, in the Years 1841–2; with Geological Observations on the United States, Canada, and Nova Scotia*, 2 vols. (New York: Wiley and Putnam, 1845); *Principles of Geology*, 7th ed. (London: John Murray, 1847) and subsequent editions; *A Second Visit to the United States*, 2 vols. (London: John Murray, 1849); and *A Manual of Elementary Geology* (London: John Murray, 1851) and subsequent editions.

The other, obvious sources for any history of American geology are the reports (annual and final) of the numerous geological surveys (provincial, state, and federal). This study began in British North America with Abraham Gesner's unofficial *Remarks on the Geology and Mineralogy of Nova Scotia* (Halifax: Gossip and Coade, 1836) and his official *[First–Fourth] Report on the Geological Survey of the Province of New Brunswick* (Saint John: Henry Chubb, 1839–1842) and *Report on the Geological Survey of New Brunswick, with a Topographical Account of the Public Lands and the Districts Explored in 1842* (Saint John: Henry Chubb, 1843). Gesner's mature views on British North American geology and industry appeared in *New Brunswick; with Notes for Emigrants* (London: Simmonds & Ward, 1847) and *The Industrial Resources of Nova Scotia* (Halifax: A. & W. MacKinlay, 1849).

State survey reports contained not only scientific findings but often a wealth of information regarding mining, agriculture, and other commercial prospects along with key companies, capitalists, and public officials. In this regard, the surveys that were most helpful in tracing the development of early American coal developments were Edward Hitchcock on Massachusetts (1830), James Gates Percival on Connecticut (1835), Charles T. Jackson on Rhode Island (1839), and David Dale Owen on Kentucky (1856). The essential source on Pennsylvania coal is Henry Darwin Rogers, *[First–Sixth] An-*

nual *Report of the Geological Survey of Pennsylvania* (Harrisburg: various publishers, 1836–1842), and Rogers, *The Geology of Pennsylvania: A Government Survey,* 2 vols. (Edinburgh and London: W. Blackwood and Sons; Philadelphia: J. B. Lippincott, 1858). There were numerous publications of the Second Geological Survey of Pennsylvania, 1874–1888, under J. Peter Lesley's direction, which are often cataloged under the name of the assistant geologist assigned to a particular region or under special subject headings such as bituminous coal, anthracite, and petroleum. A very useful overview of survey geology can be found in Lesley's *Historical Sketch of Geological Explorations in Pennsylvania and Other States* (Harrisburg: Board of Commissioners for Second Geological Survey, 1876). For California, see Josiah Dwight Whitney, *Geology,* vol. 1: *Report of Progress and Synopsis of the Field-Work, From 1860 to 1864* (Philadelphia: Caxton Press, 1865).

Besides journals, the best contemporary sources for the coal oil industry are Thomas Antisell, *The Manufacture of Photogenic or Hydro-Carbon Oils, from Coal and Other Bituminous Substances, Capable of Supplying Burning Fluids* (New York: D. Appleton, 1859), and Abraham Gesner, *Practical Treatise on Coal, Petroleum, and Other Distilled Oils* (New York: Baillière Brothers, 1860). The second edition of *A Practical Treatise* (1865) was revised by Gesner's son, George Weltden, and revealed the dramatic changes in the oil industry wrought by petroleum.

There was a staggering number of books, pamphlets, and articles published during or just after the oil boom. The first was a slim but insightful record of the first year: Thomas A. Gale, *The Wonder of the Nineteenth Century: Rock Oil, in Pennsylvania and Elsewhere* (Erie, PA: Sloan & Griffith, 1860). Other eyewitness accounts of the first years along Oil Creek include S. J. M. Eaton, *Petroleum: A History of the Oil Region of Venango County, Pennsylvania* (Philadelphia: J. P. Skelly, 1866), and J. H. A. Bone, *Petroleum and Petroleum Wells* (Philadelphia: J. B. Lippincott, 1865). Edmund Morris, *Derrick and Drill, or an Insight into the Discovery, Development, and Present Condition and Future Prospects of Petroleum in New York, Pennsylvania, Ohio, West Virginia, &c* (New York: James Miller, 1865), is a very helpful compilation of contemporary newspaper articles. William Wright, *The Oil Regions of Pennsylvania: Showing Where Petroleum is Found; How it is Obtained, and at What Cost, with Hints for Whom it may Concern* (New York: Harper & Brothers, 1865), is the sharpest account of the oil business, especially swindling. Andrew Cone and Walter R. Johns, *Petrolia: A Brief History of the Pennsylvania Petroleum Region, Its Development, Growth, Resources, etc., from 1859 to 1869* (New York: Appleton, 1870), and J. T. Henry, *The Early and Later History of Petroleum with Authentic Facts in Regard to its Development in Western Pennsylvania* (Philadelphia: James B. Rodgers, 1873), are both well-known and well-respected histories. Another valuable source is *Derrick's Hand-Book of Petroleum: A Complete Chronological and Statistical Review of Petroleum Developments from 1859 to 1900,* 2 vols. (Oil City, PA: Derrick Publishing Company, 1898).

There are also a few key government reports on the oil industry: J. P. Lesley, "Coal Oil," *Report of the Commissioner of Agriculture for the Year 1862* (Washington, DC: GPO, 1863), 429–447; S. S. Hayes, "Report on the United States Revenue Commission on Petroleum as a Source of National Revenue," *House Executive Documents,* No. 51, 39th Cong., 1st sess. (February 1866):1–39; and Stephen F. Peckham, *The Production, Technology, and Uses of Petroleum and Its Products,* 47th Cong., 2nd sess., H.R. Misc. Doc. 42, Part 10 (Washington, DC: GPO, 1884). Peckham interviewed many of the key players, especially those involved in the Silliman-Whitney controversy. Peckham's report also contained an unequaled, year-by-year bibliography on coal, kerosene, and petroleum.

Secondary Sources

Information about the principal men of science can be found in such standard sources as the *Dictionary of American Biography, Dictionary of American Scientific Biography, Dictionary of Canadian Biography, Dictionary of National Biography, Dictionary of Scientific Biography,* and the *Biographical Dictionary of American Science.* Most of the leading consultants were also members of the National Academy of Sciences and thus have entries in the academy's *Biographical Memoirs.* The exception is Josiah Whitney, who resigned from the academy. His best biography is Edwin T. Brewster's *Life and Letters of Josiah Dwight Whitney* (Boston: Houghton Mifflin, 1909). Others in this genre are Mary Lesley Ames, *Life and Letters of Peter and Susan Lesley,* 2 vols. (New York: G. P. Putnam, 1909), and Daniel C. Gilman, *The Life of James Dwight Dana: Scientific Explorer, Mineralogist, Geologist, Zoologist, Professor in Yale University* (New York: Harper, 1899). John M. Clarke's *James Hall of Albany: Geologist and Paleontologist, 1811–1898* (Albany: privately published, 1923) is of limited value, although Clarke was quite level-headed about Hall's consulting.

In general, there are surprisingly few recent biographical studies of the leading men of science of mid-nineteenth-century America, with the notable exception of Joseph Henry. Chandos Michael Brown's *Benjamin Silliman: A Life in the Young Republic* (Princeton: Princeton University Press, 1989) is richly detailed and thoughtful but covers only Silliman's youth and early career up to 1818. Leonard G. Wilson's edited volume *Benjamin Silliman and His Circle: Studies on the Influence of Benjamin Silliman on Science in America* (New York: Science History Publications, 1979) contains informative, but short, biographies of Edward Hitchcock, Charles Upham Shepard, James Dwight Dana, and Benjamin Silliman Jr. along with an insightful history of the Yale School of Applied Chemistry by Louis I. Kuslan. For Dana, see Michael Prendergast, "James Dwight Dana: The Life and Thought of an American Scientist" (PhD dissertation, UCLA, 1978), and Julie R. Newell, "James Dwight Dana and the Emergence of Professional Geology in the United States," *American Journal of Science, 297* (1997):273–282. Michele L. Aldrich, "Charles Thomas Jackson's Geological Surveys in New England, 1836–1844," *Northeastern Geology, 3* (1981):5–10, is practically the only study of Jackson's geology. Patsy Gerstner's *Henry Darwin Rogers, 1808–1866: American Geologist* (Tuscaloosa: University of Alabama Press, 1994) is an excellent history of the Pennsylvania Geological Survey and Rogers's geology. Joyce Barkhouse's *Abraham Gesner* (Don Mills, ON: Fitzhenry & Whiteside, 1980) is a serviceable summary of Gesner's life, but it should be supplemented with George W. Gesner, "Dr. Abraham Gesner, a Biographical Sketch," *Bulletin of the Natural History Society of New Brunswick, 14* (1896):3–11, and G. F. Matthew, "Abraham Gesner: A Review of His Scientific Work," *Bulletin of the Natural History Society of New Brunswick, 15* (1897):3–48.

As a scientific practice, consulting has not gone unnoticed by American historians. Discussions appear in Carroll Pursell, "Science and Industry," in George H. Daniels, ed., *Nineteenth-Century American Science: A Reappraisal* (Evanston: Northwestern University Press, 1972), 231–248, Clark A. Elliott, "Models of the American Scientist: A Look at

---

Collective Biography," *Isis, 73* (1982):77–93, and Julie Renee Newell, "American Geologists and Their Geology: The Formation of the American Geological Community, 1780–1865" (PhD dissertation, University of Wisconsin at Madison, 1993), but most historians treat consulting as secondary employment (providing supplemental income) to teaching and government positions. The traditional political economy of science is explained in Howard S. Miller, *Dollars for Research: Science and Its Patrons in Nineteenth-Century America* (Seattle: University of Washington Press, 1970).

Recently, historians of British science have begun to investigate in more detail a broad range of activities under the rubrics "commercial science" or "career-making." James A. Secord's *Victorian Sensation: The Extraordinary Publication, Reception, and Secret Authorship of Vestiges of the Natural History of Creation* (Chicago: University of Chicago Press, 2000) is the best recent study. In Victorian science, there is a key social distinction between gentlemen and men of lower social class; see Aileen Fyfe, "Conscientious Workmen or Booksellers' Hacks? The Professional Identities of Science Writers in the Mid-Nineteenth Century," *Isis, 96* (2005):192–223; Ruth Barton, "'Men of Science': Language, Identity, and Professionalization in the Mid-Victorian Scientific Community," *History of Science, 41* (2003):73–119; and Adrian Desmond, *Huxley: The Devil's Disciple* (London: Michael Joseph, 1994). Colin A. Russell, *Edward Frankland: Chemistry, Controversy and Conspiracy in Victorian England* (Cambridge: Cambridge University Press, 1996), and Jack Morrell, *John Phillips and the Business of Victorian Science* (Aldershot: Ashgate, 2005), discuss the consulting practices of two prominent Victorian men of science. Russell's typology of Playfair's commissions focuses on long-term retainers with large firms, a very different model from the American practice. Morrell's story revises, but does not entirely replace, the more common interpretation of a lack of overlap between the "art" of mining and the "science" of geology as argued in Roy Porter, "The Industrial Revolution and the Emergence of the Science of Geology," in Mikuláš Teich and Robert Young, ed., *Changing Perspectives in the History of Science,* (London: Heinemann, 1973), 320–343, and in Porter, "Gentlemen and Geology: The Emergence of a Scientific Career, 1660–1920," *Historical Journal, 21* (1978):809–836. For an example of such overlap—although in continental Europe, not Britain—see Theodore M. Porter, "The Promotion of Mining and the Advancement of Science: the Chemical Revolution of Mineralogy," *Annals of Science, 38* (1981):543–570.

In Britain, geology might be the exception for there are several examples of how Victorian science expressed in its language, its content, and its social organization the British industrial society of which it was an integral part. See, for example, Arnold W. Thackray, "Natural Knowledge in Cultural Context: The Manchester Model," *American Historical Review, 69* (1974):672–709; Robert H. Kargon, *Science in Victorian Manchester: Enterprise and Expertise* (Baltimore: Johns Hopkins University Press, 1977); Morris Berman, *Social Change and Scientific Organization: The Royal Institution, 1799–1844* (Ithaca: Cornell University Press, 1978); Robert Bud and Gerrylynn K. Roberts, *Science versus Practice: Chemistry in Victorian Britain* (Manchester: Manchester University Press, 1984); and Timothy L. Alborn, "The Business of Induction: Industry and Genius in the Language of British Scientific Reform, 1820–1840," *History of Science, 34* (1996):91–121. The best study of the balance between theory and application, science and industry, is in Crosbie Smith and M. Norton Wise, *Energy and Empire: A Biographical Study of Lord Kelvin* (Cambridge: Cambridge University Press, 1989).

By contrast, historians of American science have had a very conflicted relationship with the practical and commercial side of their science. The debate goes all the way back to the 1830s and Alexis de Tocqueville's *Democracy in America,* although the classic statement was Richard Harrison Shryock's "American Indifference to Basic Science during the Nineteenth Century," *Archives Internationales d'Histoire des Sciences, 28* (1948):3–18, which was subsequently developed into a sort of naive Baconianism by George H. Daniels, *American Science in the Age of Jackson* (New York: Columbia University Press, 1968). Nathan Reingold made a forceful response to Shryock, which can be followed through his articles republished in Reingold, *Science: American Style* (New Brunswick: Rutgers University Press, 1991). A different take on national style can be found in Hugh Richard Slotten, *Patronage, Practice, and the Culture of American Science: Alexander Dallas Bache and the U.S. Coast Survey* (Cambridge: Cambridge University Press, 1994). But the tendency to devalue practical and commercial science in favor of theory and "pure" science remains and can be seen in Robert V. Bruce, *The Launching of Modern American Science, 1846–1876* (Ithaca: Cornell University Press, 1988). This flawed dichotomy also underpins Gerald D. Nash, "The Conflict between Pure and Applied Science in Nineteenth-Century Public Policy: The California State Geological Survey, 1860–1874," *Isis, 54* (1963):217–228.

Consulting was part of the commercial-capitalist culture of nineteenth-century America, a culture that has been extensively discussed in, for example, Alfred D. Chandler, *The Visible Hand: The Managerial Revolution in American Business* (Cambridge, MA: Belknap, 1977); Stuart Bruchey, *Enterprise: The Dynamic Economy of a Free People* (Cambridge, MA: Harvard University Press, 1990); and Charles Sellers, *The Market Revolution: Jacksonian America, 1815–1846* (New York: Oxford University Press, 1991). For economic cycles and developments, there are three classic treatments: George Rogers Taylor, *The Transportation Revolution, 1815–1860* (New York: Holt, Rinehart, and Winston, 1951); Douglass C. North, *The Economic Growth of the United States, 1790–1860* (New York: W. W. Norton, 1966); and Peter Temin, *The Jacksonian Economy* (New York: W. W. Norton, 1969). An excellent survey of the literature can be found in Walter Licht, *Industrializing America: The Nineteenth Century* (Baltimore: Johns Hopkins University Press, 1995). A different perspective on coal, petroleum, and industrialization is in Robert B. Gordon and Patrick M. Malone, *The Texture of Industry: An Archeological View of the Industrialization of North America* (New York: Oxford University Press, 1994). The best study of early national political economy, especially the spoils system, is John Lauritz Larson, *Internal Improvement: National Public Works and the Promise of Popular Government in the Early United States* (Chapel Hill: University of North Carolina, 2001). Significantly, Larson does not mention science at all. Other helpful, though largely political interpretations, of corruption are Mark Wahlgren Summers, *The Plundering Generation: Corruption and the Crisis of the Union, 1849–1861* (New York: Oxford University Press, 1987), and Summers, *The Era of Good Stealings* (New York: Oxford University Press, 1993).

For the Canadian side of this story, there are a couple of useful political histories: W. S. MacNutt, *New Brunswick: A History, 1784–1867* (Toronto: Macmillan, 1963), and J. Murray Beck, *Politics of Nova Scotia,* vol. 1: *1710–1896* (Tantallon, NS: Four East Publications, 1985). For a more recent study of politics, economy, and culture, see T. W. Acheson, *Saint John: The Making of a Colonial Urban Community* (Toronto: University of Toronto Press, 1985), and Paul A. Bogaard, ed., *Profiles of Science and Society in the Maritimes Prior to 1914* (Sackville, NB: Acadiensis Press, 1990). An excellent analysis of the belief in technological progress that held sway in the Maritimes and Canada in the nineteenth century is in A. A. den Otter, *The Philosophy of Railways: The Transcontinental Railway Ideal in British North America* (Toronto: University of Toronto Press, 1997).

Because most consultants either were geologists or involved with mining, the practice contributed to the development of geology, a science about which there is a very large and sophisticated historical literature. The best place to begin is David R. Oldroyd, *Thinking about the Earth: A History of Ideas in Geology* (London: Athlone, 1996), and Rachel Laudan, *From Mineralogy to Geology: The Foundations of a Science, 1650–1830* (Chicago: University of Chicago, 1987). Much of the history of geology, however, has been told from the British perspective as an intellectual pursuit and social activity among gentlemen. See, for example, the classic Charles Coulston Gillispie, *Genesis and Geology: A Study of the Relations of Scientific Thought, Natural Theology, and Social Opinion in Great Britain, 1790–1850* (Cambridge, MA: Harvard University Press, 1951); Roy Porter, *The Making of Geology: Earth Science in Britain, 1660–1815* (Cambridge: Cambridge University Press, 1977); Nicolas A. Rupke, *The Great Chain of History: William Buckland and the English School of Geology* (Oxford: Oxford University Press, 1983); Martin J. S. Rudwick, *The Great Devonian Controversy: The Shaping of Scientific Knowledge among Gentlemanly Specialists* (Chicago: University of Chicago, 1985); James A. Secord, *Controversy in Victorian Geology: The Cambrian-Silurian Dispute* (Princeton: Princeton University Press, 1986); and Simon J. Knell, *The Culture of English Geology, 1815–1851: A Science Revealed through Its Collecting* (Aldershot: Ashgate, 2000).

Studies of American geology are not nearly so numerous or comprehensive. A good review of the historiographical terrain, though somewhat dated, is in Mott T. Greene, "History of Geology," *Osiris*, 1 (1985):97–116. George P. Merrill, *The First One Hundred Years of American Geology* (New Haven: Yale University Press, 1924), remains the best survey of the field, although it lacks vital source notes. Merrill's edited volume *Contributions to a History of American State Geological and Natural History Surveys*, U.S. National Museum, Bulletin No. 109 (Washington, DC: GPO, 1920), includes the legislation enabling each survey along with brief histories, some of which were written by survey geologists. Mary C. Rabbitt's *Minerals, Lands, and Geology for the Common Defense and General Welfare*, vol. 1: *Before 1879* (Washington, DC: GPO, 1979) is also a useful survey. Walter B. Hendrickson's "Nineteenth-Century State Geological Surveys: Early Government Support of Science," *Isis*, 52 (1961):357–371, provides a good introduction to the political economy of survey geology, as does Stephen P. Turner, "The Survey in Nineteenth-Century American Geology: The Evolution of a Form of Patronage," *Minerva*, 25 (1987):282–330. For a different perspective on nature-as-resource, see Benjamin R. Cohen, "Surveying Nature: Environmental Dimensions of Virginia's First Scientific Survey, 1835–1842," *Environmental History*, 11 (2006):37–69. Another good overview is in Anne Marie Millbrooke, "State Geological Surveys of the Nineteenth Century" (PhD dissertation, University of Pennsylvania, 1981), but the best study of survey geology is Michele L. Aldrich, *New York State Natural History Survey, 1836–1842: A Chapter in the History of American Science* (Ithaca, NY: Paleontological Research Institution, 2000).

For a more general understanding of American geological theory and practice (in addition to surveys), see Leonard G. Wilson, "The Emergence of Geology as a Science in the United States," *Journal of World History*, 10 (1967):416–437; Cecil J. Schneer, "Ebenezer Emmons and the Foundations of American Geology," *Isis*, 60 (1969):439–450; David J. Krause, "Testing a Tradition: Douglass Houghton and the Native Copper of Lake Superior," *Isis*, 80 (1989):622–639; and R. H. Dott Jr., "The American Countercurrent—Eastward Flow of Geologists and Their Ideas in the Late-Nineteenth Century," *Earth Sciences History*, 9 (1990):158–162. There are also some insightful articles in Cecil J. Schneer, ed., *Toward a History of Geology* (Cambridge, MA: MIT Press, 1969), and Schneer, ed., *Two Hundred Years of Geology in America: Proceedings of the New Hampshire Bicentennial Conference on the History of Geology* (Hanover, NH: University Press of New England, 1979). Histories of coal and petroleum geology are practically nonexistent. The rare exception is Edgar Wesley Owen's *Trek of the Oil Finders: A History of Exploration for Petroleum* (Tulsa: American Association of Petroleum Geologists, 1975), but it is thoroughly internalist.

The best sources on Canadian geology are Morris Zaslow, *Reading the Rocks: The Story of the Geological Survey of Canada, 1842–1972* (Toronto: Macmillan, 1975); Suzanne Zeller, *Inventing Canada: Early Victorian Science and the Idea of a Transcontinental Nation* (Toronto: University of Toronto Press, 1987); and William E. Eagan, "The Canadian Geological Survey: Hinterland between Two Metropolises," *Earth Sciences History*, 12 (1993):99–106. Susan Sheets-Pyenson's *John William Dawson: Faith, Hope, and Science* (Montreal: McGill-Queen's University Press, 1996) does not go into much detail about Dawson's Acadian geology.

Charles Lyell's role in the development of American geology has generated sort of a cottage industry of scholarship. See, for example, Robert H. Silliman, "Agassiz vs. Lyell: Authority in Assessment of the Diluvium-Drift Problem by North American Geologists with Particular Reference to Edward Hitchcock," *Earth Sciences History*, 13 (1994):180–186; Silliman, "The Hamlet Affair: Charles Lyell and the North Americans," *Isis*, 86 (1995):541–561; Robert H. Dott Jr., "Lyell in America—His Lectures, Field Work, and Mutual Influences, 1841–1853," *Earth Sciences History*, 15 (1996):101–140; and Leonard G. Wilson, *Lyell in America: Transatlantic Geology, 1841–1853* (Baltimore: Johns Hopkins University Press, 1998). There are also good articles by R. H. Dott Jr., A. C. Scott, and J. H. Calder in Derek J. Blundell and Andrew C. Scott, ed., *Lyell: The Past Is the Key to the Present* (London: Geological Society, 1998).

In contrast to geology, the literature on nineteenth-century American chemistry is embarrassingly meager, a predicament thoroughly explored in John W. Servos, "History of Chemistry," *Osiris*, 1 (1985):132–146. A brief account of the science can be found in Edward H. Beardsley, *The Rise of the American Chemistry Profession, 1850–1900* (Gainesville: University of Florida Press, 1964), and a statistical analysis of the discipline is presented in Arnold Thackray, Jeffrey L. Sturchio, P. Thomas Carroll, and Robert Bud, *Chemistry in America, 1876–1976: Historical Indicators* (Dordrecht: D. Reidel, 1985). A more integrated history combining theoretical developments with larger cultural changes is in Margaret W. Rossiter, *The Emergence of Agricultural Science: Justus Liebig and the Americans, 1840–1880* (New Haven: Yale University Press, 1975). Agricultural chemistry serves as a very useful case study of the relations between science, technology, and the economy in Charles E. Rosenberg, *No Other Gods: On Science and American Social Thought* (Baltimore: Johns Hopkins University Press, 1976).

Consultants were active in the courtroom as well as in the field and laboratory. Expert witnessing is a key theme in the burgeoning literature on law and science as shown in Tal Golan, *Laws of Men and Laws of Nature: The History of Scientific Expert Testimony in England and America* (Cambridge, MA: Harvard University Press, 2004); Golan, ed., "Law and Science," a special issue of *Science in Context*, 12 (1999); Sheila Jasanoff, *Science at the Bar: Law, Science, and Technology in America* (Cambridge, MA: Harvard University Press, 1997); Carol A. G. Jones, *Expert Witness: Science, Medicine, and the Prac-*

---

tice of Law (Oxford: Oxford University Press, 1994); Roger Smith and Brian Wynne, eds., *Expert Evidence: Interpreting Science in the Law* (London: Routledge, 1989); and June Z. Fullmer, "Technology, Chemistry, and the Law in Early 19th-Century England," *Technology and Culture, 21* (1980):1–28. Two other works on the role of experts are especially clear and analytical: Christopher Hamlin, *A Science of Impurity: Water Analysis in Nineteenth Century Britain* (Bristol: Adam Hilger, 1990), and Hamlin, *What Becomes of Pollution? Adversary Science and the Controversy on Self-Purification of Rivers in Britain, 1850–1900* (New York: Garland, 1987).

Unlike expert witnessing, histories of intellectual property have tended to focus on developments outside the courtroom, in particular on the role of patents in innovation. For the British patent system, the standard works are H. I. Dutton, *The Patent System and Inventive Activity during the Industrial Revolution, 1750–1852* (Manchester: Manchester University Press, 1984), and Christine MacLeod, *Inventing the Industrial Revolution: The English Patent System, 1660–1800* (Cambridge: Cambridge University Press, 1988). For the United States, see Carolyn C. Cooper, *Shaping Invention: Thomas Blanchard's Machinery and Patent Management in Nineteenth-Century America* (New York: Columbia University Press, 1991), the special issue "Patents and Invention," *Technology and Culture, 32* (1991), and Robert C. Post's *Physics, Patents, and Politics: A Biography of Charles Grafton Page* (New York: Science History Publications, 1976).

Consultants were also active in fashioning a professional identity. But the practice of taking fees for expertise does not fit the standard models of scientific professionalization as displayed by university professors. The classic description of this sort of profession is in Joseph Ben-David, *The Scientist's Role in Society: A Comparative Study* (Chicago: University of Chicago, 1984). For early American science, the traditional formulation is in George H. Daniels, "The Process of Professionalization in American Science: The Emergent Period, 1820–1860," *Isis, 58* (1967):151–166. A slightly different organizational model of professional science is in Sally Gregory Kohlstedt, *The Formation of the American Scientific Community: The American Association for the Advancement of Science, 1848–1860* (Urbana: University of Illinois, 1976). Consulting is more in line with the kind of professionalization models applied to lawyers and doctors. A good introduction to these professions is in Paul Starr, *The Social Transformation of American Medicine* (New York: Basic Books, 1982); Nathan O. Hatch, ed., *The Professions in American History* (Notre Dame: University of Notre Dame Press, 1988); Samuel Haber, *The Quest for Authority and Honor in the American Professions, 1750–1900* (Chicago: University of Chicago, 1991); Burton J. Bledstein, *The Culture of Professionalism: The Middle Class and the Development of Higher Education in America* (New York: W. W. Norton, 1976); Gerald L. Geison, *Professions and Professional Ideologies in America* (Chapel Hill: University of North Carolina Press, 1983); and Thomas L. Haskell, *The Authority of Experts: Studies in History and Theory* (Bloomington: Indiana University Press, 1984).

Of course, there were big potential problems with taking money for science. Historians of medicine have thus far done the most careful studies of the behavior of professionals within a commercial culture. See, for example, Roy Porter, *Health for Sale: Quackery in England, 1660–1850* (Manchester: Manchester University Press, 1989); W. F. Bynum and Roy Porter, eds., *Medical Fringe and Medical Orthodoxy, 1750–1850* (London: Croom Helm, 1987); and Anne Digby, *Making a Medical Living: Doctors and Patients in the English Market for Medicine, 1720–1911* (Cambridge: Cambridge University Press, 1994).

A large area of science and technology scholarship deals with scientific authority, objectivity, and ethics. The starting point for any discussion of scientific norms is Robert K. Merton, *The Sociology of Science: Theoretical and Empirical Investigations* (Chicago: University of Chicago, 1973). Recent work that has informed this study includes Steven Shapin, *A Social History of Truth: Civility and Science in Seventeenth-Century England* (Chicago: University of Chicago Press, 1994); Theodore M. Porter, *Trust in Numbers: The Pursuit of Objectivity in Science and Public Life* (Princeton: Princeton University Press, 1995); and Jan Golinski, *Making Natural Knowledge: Constructivism and the History of Science* (Cambridge: Cambridge University Press, 1998).

The relations of science, technology, and industry have been one of the most abiding and central topics for scholars in many fields. The best overview remains John M. Staudenmaier, *Technology's Storytellers: Reweaving the Human Fabric* (Cambridge, MA: MIT Press, 1985), esp. ch. 3. As Staudenmaier points out, the relations are very interesting, but they also are very problematic, a point also made in Rachel Laudan, "Natural Alliance or Forced Marriage? Changing Relations between Histories of Science and Technology," *Technology and Culture, 36* (1995):S17–S28. Scholars tend to focus on either of two historical periods: the late eighteenth century (the "first industrial revolution") or the late nineteenth century (the "second industrial revolution"). Whether and how science contributed to the first industrial revolution is a matter of some debate, a summary of which can be found in Joel Mokyr, "Editor's Introduction: The New Economic History and the Industrial Revolution," in *The British Industrial Revolution: An Economic Perspective* (Boulder, CO: Westview Press, 1998), and in Mokyr, *The Gifts of Athena: Historical Origins of the Knowledge Economy* (Princeton: Princeton University Press, 2002). The argument for the absence of science in industrialization is staked out in A. Rupert Hall, "What Did the Industrial Revolution Owe to Science?" in Neil McKendrick, ed., *Historical Perspectives: Studies in English Thought and Society in Honour of J. H. Plumb* (London: Europa, 1974):129–151. A persuasive rebuttable is put forward in Neil McKendrick, "The Rôle of Science in the Industrial Revolution: A Study of Josiah Wedgwood as a Scientist and Industrial Chemist," in Mikuláš Teich and Robert Young, ed., *Changing Perspectives in the History of Science: Essays in Honour of Joseph Needham* (London: Heinemann, 1973):274–319; and the classic argument for interaction is in A. E. Musson and Eric Robinson, *Science and Technology in the Industrial Revolution* (Manchester: Manchester University Press, 1969). More recent analyses of the commercial interests and activities of eighteenth-century natural philosophers include Larry Stewart, *The Rise of Public Science: Rhetoric, Technology, and Natural Philosophy in Newtonian Britain, 1660–1750* (Cambridge: Cambridge University Press, 1992), and Margaret C. Jacob, *Scientific Culture and the Making of the Industrial West* (New York: Oxford University Press, 1997). For an earlier period, see Pamela H. Smith, *The Business of Alchemy: Science and Culture in the Holy Roman Empire* (Princeton: Princeton University Press, 1994).

There is much less debate about whether late nineteenth-century industry, especially emerging technologies in chemicals and electricity, were "based" on science. The classic statement of the Second Industrial Revolution thesis is found in David S. Landes, *The Unbound Prometheus: Technological Change and Industrial Development in Western Europe from 1750 to the Present* (Cambridge: Cambridge University Press, 1969). On the rise of science-based chemical industry, see, for example, Ernst Homburg, "The Emergence of Research Laboratories in the Dyestuffs Industry, 1870–1900," *British Jour-*

nal for the History of Science, 25 (1992):91–111; and Georg Meyer-Thurow, "The Industrialization of Invention: A Case Study of the German Chemical Industry," Isis, 73 (1982):363–381. For an insightful critique of the entire idea of science-based industry, see Wolfgang König, "Science-Based Industry or Industry-Based Science? Electrical Engineering in Germany before World War I," Technology and Culture, 37 (1996):70–101.

Much of the work on the relations of science and industry in the United States has concentrated on the late nineteenth and twentieth centuries. The most recent studies are found in Karl Grandin, Nina Worms, and Sven Widmalm, eds., The Science-Industry Nexus: History, Policy, Implications (Sagamore Beach, ME: Science History Publications, 2005). For an older review of the literature, see Michael Aaron Dennis, "Accounting for Research: New Histories of Corporate Laboratories and the Social History of American Science," Social Studies of Science, 17 (1987):479–518, and John Kenly Smith Jr., "The Scientific Tradition in American Industrial Research," Technology and Culture, 31 (1990):121–131. On the rise of the electrical industries, see Leonard S. Reich, The Making of Industrial Research: Science and Business at GE and Bell, 1876–1926 (Cambridge: Cambridge University Press, 1985), and George Wise, Willis R. Whitney, General Electric, and the Origins of U.S. Industrial Research (New York: Columbia University Press, 1985).

To the extent that consulting represented a workable combination of theory and application, this study builds on the numerous studies of engineering. A good general introduction remains Edwin T. Layton Jr., The Revolt of the Engineers: Social Responsibility and the American Engineering Profession (Cleveland: Case Western Reserve University Press, 1971). Layton also set out the framework for explaining the converging professional trajectories of nineteenth-century scientists and engineers in Edwin T. Layton Jr., "Mirror-Image Twins: The Communities of Science and Technology in 19th-Century America," Technology and Culture, 12 (1971):562–580, and in Layton, "American Ideologies of Science and Engineering," Technology and Culture, 17 (1976):688–700. This study puts forth a different hypothesis. The practice of consulting also challenges some of the standard depictions of applied science; see, for example, David F. Noble, America by Design: Science, Technology, and the Rise of Corporate Capitalism (New York: Alfred A. Knopf, 1977). Ronald Kline's "Construing 'Technology' as 'Applied Science': "Public Rhetoric of Scientists and Engineers in the United States, 1880–1945," Isis, 86 (1995):194–221, is also a useful corrective to the overuse of these unexamined categories. Two other sophisticated analyses of the relations among science, engineering, and society are Emily Thompson, The Soundscape of Modernity: Architectural Acoustics and the Culture of Listening in America, 1900–1933 (Cambridge, MA: MIT Press, 2002), and Bruce Sinclair, Philadelphia's Philosopher Mechanics: A History of the Franklin Institute, 1824–1865 (Baltimore: Johns Hopkins University Press, 1974).

Consultants served in the role of short-term scientific advisers to industry, particularly coal, kerosene, and petroleum companies. The importance of these minerals to industrialization is obvious and has been studied extensively by historians and economists alike. For coal, a traditional place to begin is Howard N. Eavenson, The First Century and a Quarter of the American Coal Industry (Pittsburgh: privately published, 1942); another usual starting point is Alfred Chandler Jr., "Anthracite Coal and the Beginnings of the 'Industrial Revolution' in the United States," Business History Review, 46 (1972):141–181. More recent studies include Sean Patrick Adams, "Old Dominions and Industrial Commonwealths: The Political Economy of Coal in Virginia and Pennsylvania, 1810–1875" (PhD dissertation, University of Wisconsin at Madison, 1999); Donald

L. Miller and Richard E. Sharpless, *The Kingdom of Coal: Work, Enterprise, and Ethnic Communities in the Mine Fields* (Philadelphia: University of Pennsylvania Press, 1985); and Anthony F. C. Wallace, *St. Clair: A Nineteenth-Century Coal Town's Experience with a Disaster-Prone Industry* (Ithaca: Cornell University Press, 1988). Wallace is especially notable for his incorporation of geology in his analysis.

Scholarship on the kerosene industry is sparse largely because of its short life. See Kendall Beaton, "Dr. Gesner's Kerosene: The Start of American Oil Refining," *Business History Review*, 29 (1955):28–53; John Butt, "James Young, Scottish Industrialist and Philanthropist" (PhD dissertation, Glasgow University, 1964); Butt, "Legends of the Coal Oil Industry (1847–1864)," *Explorations in Entrepreneurial History*, 2 (1964):16–30; and Wolfgang Schivelbusch, *Disenchanted Night: The Industrialization of Light in the Nineteenth Century* (Berkeley: University of California Press, 1988).

In contrast, histories of the petroleum industry are too numerous to count. However, most of these histories focus on the twentieth century; the first years along Oil Creek, although certainly mentioned, are not treated in any depth. See, for example, Ron Chernow, *Titan: The Life of John D. Rockefeller* (New York: Vintage, 1998), or Daniel Yergin, *The Prize: The Epic Quest for Oil, Money, & Power* (New York: Touchstone, 1991). The standard overview of the early industry remains Harold F. Williamson and Arnold R. Daum, *The American Petroleum Industry: The Age of Illumination, 1859–1899.* (Evanston: Northwestern University Press, 1959). For Oil Creek, the most detailed history is still Paul H. Giddens, *The Birth of the Oil Industry* (New York: Macmillan, 1938), and Giddens, *Early Days of Oil: A Pictorial History of the Beginnings of the Industry in Pennsylvania* (Princeton: Princeton University Press, 1948). These should be supplemented with William Culp Darrah, *Pithole, the Vanished City: A Story of the Early Days of the Petroleum Industry* (Gettysburg, PA: privately published, 1972). An indispensable historiographical tool is Ernest C. Miller, *A Guide to the Early History of Pennsylvania Petroleum* (Pittsburgh: Historical Society of Western Pennsylvania, 1969). For the early California petroleum industry, the best account is in William H. Hutchinson, *Oil, Land, and Politics: The California Career of Thomas Robert Bard,* 2 vols. (Norman: University of Oklahoma Press, 1965). Gerald T. White's *Scientists in Conflict: The Beginnings of the Oil Industry in California* (San Marino: Huntington Library, 1968) is an excellent study of a clash of personalities, although White did not explore the scientific questions, meaning the geology and chemistry of petroleum, or the larger contexts of consulting, scientific ethics, or the professionalization of American science. More recently, the analytical tools of environmental history have been brought to bear on the early years of the oil industry in Brian Black, *Petrolia: The Landscape of America's First Oil Boom* (Baltimore: Johns Hopkins University Press, 2000).

Index

Academy of Natural Sciences of Philadelphia, 44, 51, 56, 145
Agassiz, Alexander, 128, 239, 241
Agassiz, Louis, 28, 54, 127, 138, 239, 261,
275, 300, 342 n. 94
Albert asphaltum, 41–42, 45–48, 51–65,
173, 239, 298
Albert coal, 49–51, 52, 54–65, 74, 97, 102,
123, 149, 169, 173, 199, 229, 242, 289,
295; production and price, 238–239,
377 n. 148
Albert Coal Co., 151, 199, 229, 230
Albert mine (New Brunswick), 48–49,
56, 59, 65, 68, 95, 123, 128, 157, 169, 239,
298
Albert mineral, 6, 7, 74, 94, 100–103, 123,
124, 145, 146, 151, 153, 155, 158, 169, 173,
177, 196, 229, 280; anticlinal axes of, 51–
53, 55, 59, 240–242; controversy surrounding, 41–68, 69, 95, 298, 299; as
false coal, 104; fusibility of, 49, 50, 59–
60, 62, 64, 71, 75; and petroleum, 41, 48,
71, 95, 239–243, 298; vegetable structure, 49, 51, 59–60, 97
Albert Mining Co., 66
Albertine, 241
Albertite, 97, 240, 241, 242, 243, 257, 268,
289, 294, 297, 298, 299, 311
Albion Mines (Pictou, Nova Scotia), 37
Alger, Francis, 12, 30; on Gesner's plagiarism, 20; *Mineralogy and Geology of
Nova Scotia*, 13, 19
Alladin Co. (coal oil), 229
Allegheny River and basin (Pa.), 155, 191,
213, 217, 228, 234, 238, 245, 262, 263
Alluvial, class of formations, 77
American Association for the Advance-
ment of Science, 56, 157, 183, 311, 315,
322

American Gas-Light Journal, 143, 156,
158–159, 161, 177, 182, 183, 217
American Institute of Electrical Engi-
neers, 321
American Institute of Mining Engineers,
311, 321
American Institute of New York City,
149, 181
Americanization of science, 313, 318,
395 n. 55
American Journal of Science, 20, 45, 58, 72,
110, 115, 123–126, 174, 183, 196, 199, 204,
205, 259, 265, 275, 277, 288, 291, 292
American Philosophical Society, 56, 240,
242, 264, 285
American Society of Civil Engineers, 321
American Society of Mechanical Engineers, 321, 322
Anasphaltic coal, 84
Anderson, Thomas, 176
Anderson's Institute (Glasgow, Scotland), 163, 164
Andrews, E. B., 224, 227, 240, 242, 258–
261, 265, 318; anticlinal theory of petroleum occurrence, 218–221, 259, 261,
267, 271; distillation theory of origin of
petroleum, 218, 226, 269; fissure theory
of petroleum occurrence, 218–219, 262
Antisell, Thomas, 3, 56–63, 174, 175, 177,
182, 280; *The Manufacture of Photogenic or Hydro-Carbon Oils,* 146, 156,
171–173; at U.S. Patent Office, 171, 179,
183; and Young's Paraffine patent, 173
Appalachian coal fields, 87, 92, 99, 103,
106, 126, 155, 217, 227, 240, 261
Argillite formation, of Nova Scotia, 17,
18, 84
Aroostook War, 25
Ashburner, William, 276

Ashurst, Lewis, 293
Asphalte Rock, 145, 149
Asphaltic coal, 48, 74, 97
Asphaltic limestone, 172
Asphalt Mining and Kerosene Co., 145–146
Asphaltum, 44, 54, 61, 70–75, 97, 146, 150, 241, 295–297; in California, 280–281, 283–290, 296. *See also* Bitumen; Tar; Trinidad pitch
Association of American Geologists and Naturalists, 58, 87
Atlas of the Oil Region of Pennsylvania, 254, 381 n. 67
Atwood, Luther, 150, 151, 156, 195, 261 n. 33
Atwood, William, 150, 229
Austen, John H. and George W., 147
Austin, William R., 150
Authorship (in popular writing), 32, 33, 111

Bache, Alexander Dallas, 2, 86, 300, 301
Bacon, John, Jr., 51, 60
Bailey, Loring Woart, 31, 41, 242, 243, 378 n. 180
Baker, William E. S., 120
Bakewell, Robert, 89, 93, 350 n. 159
Bank of California, 279
Bank of New Brunswick, 22
Bankruptcy: in coal gas, 147; in coal mining, 27; in coal oil (kerosene), 161; of New Brunswick oil companies, 377 n. 155; of New York City joint-stock companies, 201, 203, 204; of Pennsylvania, 35; of professional geologists, 135
Bard, Thomas R., 291
Barnard, F. A. P., 302, 308
Barnsdall well (Oil Creek, Pa.), 211, 213
Bathvillite, 389 n. 112
Beach, Alfred Ely, 181
Beers, F. W., 254
Benedict and Boardman (New York City law firm), 166–168, 170–171
Berthelot, Marcelin, 295
Berzelius, J. J., 168
Biddle, Nicholas, 113, 117
Binney, Edward William, 93, 163, 164, 166, 169, 177, 180, 348 n. 105

Birdseye, Summers, and Johnson (New York City law firm), 170
Bischoff, Carl Gustav, 220
Bissell, George H., 190, 191, 192, 193, 194, 195, 196, 197, 198, 199, 200, 201, 202, 203, 204–205, 369 n. 9, 370 n. 31
Bitumen: as class of minerals, 208, 295–296; as constituent of bituminous substances, 41, 42, 44, 45, 70, 71, 74–76, 91, 103–104, 146, 149; as mineral, 172, 229, 280
Bitumenization: and coal genesis, 81, 345 n. 29; of fish, 157, 242; of plants, 78–79, 90, 95, 97, 101, 102, 119
Bituminous shales, 95, 150, 168, 172, 174, 178, 218, 226, 227; and petroleum, 281, 289
Bituminous substances, as family of minerals, 69, 70–72, 164, 172–178, 182, 196, 217
Black Slate formation (Ohio), 217, 218, 220, 269
Blake, William P., 276, 279, 280, 294, 301, 304, 321
Bliss, Henry B., 55
Bliss, Jonathan, 55
Bodie, gold mining district (Calif.), 279, 285, 304
Boom/bust. *See* Oil boom/fever; Oil bust
Booth, James, 51
Booth, Newton, 300, 301
Boring. *See* Drilling for petroleum
Boston, Mass., coal imports in, 11
Boston Gas-Light Co., 41, 47, 48, 50, 151, 173, 230
Boston Society of Natural History, 41, 48, 51, 58, 123
Bowles, Samuel, 304–306
Brady's Bend Iron Co., 262, 263, 265, 269–270
Brande, William Thomas, 176
Breckenridge coal, 75, 76, 103, 149, 158, 177
Breckenridge Coal and Oil Co. (Ky.), 149, 194
Brewer, Ebenezer, 190, 194, 201
Brewer, Francis Beattie, 190, 191, 192, 193, 194, 195, 197, 199, 200, 201, 202, 204, 214, 369 nn. 5, 11

Brewer, Watson and Co., 190, 204
Brewer, William H., 276, 280, 287, 288,
    293, 294, 299, 300, 309, 310, 320, 323;
    and California oil swindle, 307–308
Briggs, Caleb, 132–133, 139
Bristol (copper) Mining Co., 135, 151
Brongniart, Adolphe, 93
Brongniart, Alexandre, 89, 93
Buckland, William, 19, 332 n. 41, 347 n. 84
Bunsen, Robert, 173
Burdette mine (Nevada City, Calif.), 302,
    304
Burning fluid, 43, 145, 147, 153, 362 n. 53
Burning Spring Creek (W.V.), 218, 219
Burning well (Oil Creek, Pa.), 222

Cairns, William, 46–48, 50–68, 123, 151,
    339 n. 28
Cairo asphalt, 240, 242. *See also* Ritchie
    mineral
Caldwell well (Oil Creek, Pa.), 222
Caledonian Coal Oil Co. (New Bruns-
    wick), 239
California: coal in, 277; gold in, 211, 239,
    245, 274, 276, 277, 285; petroleum in,
    291–297, 308; survey of, 273, 274–278,
    280, 290
California Petroleum Co., 282–283, 286,
    290, 291, 293, 307
Cambrian system, 346 n. 57; in New
    Brunswick, 24, 28; in Nova Scotia, 37
Campbell, Colin, 19, 20
Campbell, Dugald, 176
Campbell, James, 168
Campbell, Marius R., 208–209
Camphene, 43, 147, 153, 154, 198
Canada: Confederation, 38, 243, 333 n. 60;
    Upper and Lower, 35, 39
Canals, 4, 35, 38, 111, 155
Capitalists, 4, 6, 23, 39, 111, 112, 116, 120,
    126–130, 134, 136, 138, 140, 189, 194, 244,
    245, 253, 262, 274, 278, 311, 320, 321, 323;
    Albany, 112, 117, 139; Boston, 33, 114–
    115, 130, 239, 276, 279; British, 26, 27, 37,
    38; Gesner's critique of, 34–35; Halifax,
    27, 33; New Haven, 110, 132, 191, 195,
    199–203; New York City, 191, 200, 250,
    279; Philadelphia, 85, 133, 240; Saint

John, 27; San Francisco, 279, 294; and
    stock companies, 192, 247, 286–287
Carbon: as chemical constituent of coal,
    52, 74–78, 100; as class of minerals, 74,
    75, 296, 345 n. 33; and oil, 210
Carbonaceous, as class of minerals, 65
Carboniferous, class of formations, 78–
    79, 84–85
Carboniferous Limestone formation, 79
Carboniferous period, 79, 90, 105
Carboniferous system, 77, 85–86, 99, 104,
    105–106, 345 n. 34, 351 n. 166; Eaton's
    rejection, 84; of New Brunswick, 25; of
    Nova Scotia, 37, 94, 95, 97; of Pennsyl-
    vania, 263; and petroleum, 226, 257,
    265
Chandler, Charles F., 315, 320
Chemical analyses, 3, 111, 191; of Albert
    mineral, 41–42, 52, 54, 62; of cannel
    coal, 157–158; of petroleum, 189, 195–
    200, 206, 228
Chemical industry, 5, 45, 143, 146–147,
    156, 163, 179, 206, 317
Chemical mineralogy, 58, 70, 73, 74, 100,
    296–297
Chemical research, 162–165, 178–181, 189,
    292–293
Chemists: practical, 58, 150, 163, 178, 180,
    195, 228, 230, 288; scientific, 180, 228,
    292
Chemung Group (N.Y.), 216, 217, 218, 227
Chonestes mesoloba, 264
Church, John, 281, 293
Civil War, 33, 112, 127, 222, 230, 233, 246,
    290, 313, 319
Classification, scientific, and controver-
    sies surrounding, 6, 7, 42–43; of coal,
    69–78, 103, 169, 174, 176, 178; of knowl-
    edge, 179–185; of minerals, 13, 70–73,
    73–76, 296; of petroleum/hydrocar-
    bons, 294–299
Classification, of rocks, 13–14, 296; fossil
    method, 28, 51, 98, 100; lithological
    method, 28, 30; from oldest to
    youngest, 77, 83, 86; structural
    method, 87, 98, 100
Clay Slate, district of Nova Scotia, 14–15,
    17–18, 24

Clydesdale Chemical Co., 176, 178, 180, 184

Coal, 5, 6, 7, 69–70, 124, 172–174, 209, 214, 224, 227, 241, 268; and American geology, 79, 85; basins of, 49, 52, 54, 59, 77, 78, 87; cannel, 41, 44, 75, 76, 97, 149, 155–159, 165, 169, 174, 177, 180, 297; false, 100, 104, 298; hydrogenous, 103; and progress, 11, 19, 34, 38–40; relation of, to commerce and science, 103–104; "true," 77, 80, 100, 102, 104, 288. *See also* Bitumenization; Classification, scientific; Subcarboniferous systems; Torbanehill mineral; Torbanite; *and individual states and provinces*

Coal, anthracite, 72, 76, 84, 99, 102, 103, 110, 345 n. 30; vs. bituminous, 87, 91–93; in Massachusetts, 80–81, 94; in Pennsylvania, 80, 87, 91–93, 113, 119, 122, 138; in Rhode Island, 80–81; in transition formations, 77–78, 80–81. *See also* De-bitumenization theory of anthracite

Coal, bituminous, 43, 70–72, 75–79, 84, 90, 93, 99, 102, 103, 146, 150, 157, 165, 169, 175, 180, 345 n. 30; and anthracite, 87, 91–93; estuary or drift theory of origin of, 79, 89, 346 n. 47, 348 n. 91. *See also* Appalachian coal fields; Bitumenization; Peat-bog theory of origin of coal

Coal gas, 155, 165, 166, 194, 239. *See also* gas lighting

Coal measures, 76–81, 84, 86, 94, 104, 106, 196, 345 n. 34; American and British correlations, 81, 84; of New Brunswick, 24, 28, 30, 52, 54, 59; of Nova Scotia, 18

Coal oil, 6, 75, 103, 154, 193, 210, 291; and boom/fever, 144–145, 151, 156, 158, 160, 166, 170, 177, 211, 233; controversy surrounding, 146, 147, 163, 167, 168, 171, 174, 175, 179, 183, 184, 185; conversion from coal to petroleum, 227–233, 238, 239; crude, 147, 149, 155, 157, 160, 164, 165, 211, 217; and Gesner's patent gas, 43–46, 54, 144–146, 173; and Gesner's patent oil(s), 7, 143–149, 150–151, 153, 156, 160–162, 166, 173, 177, 194, 197, 206,

209, 229, 233; "history of the art," 171–173; industry, 143, 144, 146–147, 149, 155, 156, 164, 166, 173, 209, 211, 227, 229, 232; lamps, 147, 152, 153, 287; patents, 146, 163–165, 171; plant design, 148; refining of, 147, 166. *See also* Gesner, Abraham; Paraffine; Young, James

Coal tar, 150, 172

Cochrane, Archibald, 9th Earl of Dundonald, 172, 178

Cochrane, Thomas, 10th Earl of Dundonald, 38, 43, 45, 46, 145, 172, 173

Colebrooke, William, 25, 26, 27, 31

Collier, Peter, 288

Columbia Oil Co. (Brooklyn), 157, 167, 169, 170

Columbia University, 58, 127, 170; School of Mines, 258, 320

Colza oil, 152, 198

Combustibles, as class of minerals, 65, 69, 345 n. 33

Comstock Lode (Nev.), 278–279, 280

Cone, Andrew, 220, 221, 232, 246, 251, 252

Connecticut, survey of, 80

Consultant, as analytical and descriptive category, 327 n. 10

Consultants and fraud, 245, 293, 310, 313; and consulting reports, 302–303; and mineral samples, 48, 292–294, 308; and mining, 137–138, 140; and oil wells, 250–251; and patents, 175; and stock, 201, 248, 293

Consultants and consulting, 108, 160, 170, 252, 288, 301, 317; and American identity, 3, 5; boom period of, 111; chemists, 7, 59, 143, 151, 156, 159, 163, 165, 173, 181, 189, 316, 327 n. 10; on coal oil, 144, 156–158; as commercial patronage, 2, 3, 4; engagements, 5, 7, 8, 23, 33, 58, 61, 109, 111, 116–121, 126, 139, 173, 189, 199, 265, 273; ethics, 7, 8, 66–68, 110–113, 130–140, 183, 184, 259, 270, 273, 276, 280, 293, 303, 308, 310, 314, 315, 320; fees of, 3, 7, 61, 116, 121, 126–131, 176, 260, 285, 302–304, 314, 320; fieldwork, 118–121, 265; general vs. specialist, 63; geologists, 3, 7, 109, 115, 125, 156, 256, 258, 276, 320, 327 n. 10; local, 216, 276; oil-finding, 244, 249, 254, 270; origins of, 109–110; on petroleum, 245, 256–265, 269–272, 280, 318; practical, 178, 253, 254; as profession, 57, 111–116; and prostitution of science, 183, 323, 394 n. 51, 395 n. 55; and public lecturing, 1, 32, 33, 61, 111, 138, 287, 293; relations among science, technology, and industry, 5, 241; reports, 51, 109, 115, 117, 120–124, 138, 189–190, 196, 241, 254, 293, 303, 304, 307–308; and science, 112–113, 124–126, 273; scientific, 1, 2, 12, 21, 32, 67, 111, 112, 116, 118, 130, 133, 229; and scientific controversies, 55–56, 66, 68, 95, 171, 184, 196, 241; as witnesses, 4, 6, 42, 61, 66–68, 95, 165, 169, 176, 178, 179, 183, 184, 279, 316, 338 n. 3. *See also* Science, scientists, and scientific authority

Consultants and corruption, 136–138, 274, 304, 306, 310, 313, 316, 318, 323; of coal oils, 154; political, 31, 133, 314; of samples, 48, 292–294, 308

Conway, Edward, 280, 281, 284

Conybeare, William Daniel, 78, 79, 87; *Outlines of the Geology of England and Wales,* 18, 77

Cooke, Jay, 314

Cooke, Josiah P., 173

Cooper, James G., 276

Cornell University, 127, 314, 322

Corniferous Limestone formation (N.Y.), 220

Corning, A. J., 293

Coup oil, 150, 195, 197, 361 n. 33

Cresson, John C., 285, 286, 293

Cretaceous system, 284, 289

Cronise, W. H. V., 294

Crosby, Albert H., 190–193

Crosby, Dixi, 190, 193, 369 n. 8

Crossley well (Oil Creek, Pa.), 212

Crown reservations, in British land grants, 46, 47, 65

Cumming, James, 59

Dana, James Dwight, 48, 106, 157, 174, 275, 298, 300, 389 n. 112; on American science, 73, 105; chemical system of mineralogy, 73–74; *Manual of Geology,* 105, 297, 350 nn. 161, 163; natural-historical system of mineralogy, 71–73; *System of Mineralogy,* 71–76, 296

Davis, William Morris, 126

Dawson, John William, 19, 37, 94–95; *Acadian Geology,* 97–98, 102, 243; and Albertite, 243, 298; and Carboniferous system of Nova Scotia, 97; and Gypsiferous formation of Nova Scotia, 30; as Lyell's protégé, 94

De-bitumenization theory of anthracite, 91–94, 217

Deck, Isiah, 57–63

De la Beche, Henry Thomas, 22, 51

Dendrerpeton acadianum, 95, 96

Den Otter, A. A., 337 n. 133

De Verneuil, Édouard, 30

Devonian system, 264, 268; in California, 284; in Nova Scotia, 37; and petroleum, 226, 257, 261, 266, 281, 289, 295

Diamonds, 74, 80, 296

Distillation: of bituminous shale to produce Albertite, 95; of coal to produce (crude) oil, 143, 146–147, 168, 172, 175, 176, 178, 180, 182, 229; of coal to produce petroleum, 216–217, 225–227; fractional, 195, 197, 200, 228, 291–292, 297; practical vs. scientific, 288; of Silliman's sample of California petroleum, 288, 292

Divining rods, 209, 235, 251–252, 254

D'Omalius d'Halloy, Jean-Baptiste, 345 n. 34

Downer, Samuel Jr., 150, 151, 154, 155, 167, 170, 224, 229, 230, 293, 376 n. 118

Downer Co., 161, 166; oil works in Portland, Maine, 151, 229; oil works in Waltham, Maine, 149–151, 154, 195, 228, 229, 293; petroleum refinery in Corry, Pa., 230, 232

Drake, Edwin L., 203–208, 210, 211, 213, 215, 228

Drake well (Oil Creek, Pa.), 211–216, 220, 233, 236, 239

Drillers (oil well borers), 205, 209, 221; as practical oil men, 216, 217

Drilling for petroleum, 203–207, 213–215,

Drilling for petroleum (*continued*) 218–220, 223, 234–238; costs, 213, 214, 373 n. 24; on hillsides, 234, 235; spring pole method of, 213, 214; success rates, 221, 234; tool strings, 206, 207, 213, 218, 224, 238; and "wildcatting," 238, 250–251. *See also* Oil wells

DuBois, Henry A., 117

Duffy, John and Peter, 41, 48, 49

Eagle, Horatio, 145

East Albert Co. (New Brunswick), 239, 240

Eaton, Amos, 81–84, 104, 109–110, 112; *A Geological Nomenclature for North America*, 81–84

Eaton, S. J. M., 215, 216, 220, 221, 222, 224, 225, 234, 246, 248, 251, 252; theory of oil cavities, 225–226

Egbert farm (Oil Creek, Pa.), 222, 253

Élie de Beaumont, Léonce, 98, 115, 350 n. 138

Ellet, William H., 57–63

Emma (silver) mine (Utah), 301, 302, 304, 390 n. 124

Emmons, Ebenezer, 15, 112

Empire (gold) mine (Bodie, Calif.), 294, 295, 302

Empire well (Oil Creek, Pa.), 222

Engineers, 5; chemical, 144; civil, 254–256, 321; consulting on petroleum, 254–256; electrical, 321; mechanical, 321, 322; mining, 3, 108, 111–112, 120, 254–256, 276, 305, 311, 320, 321. *See also* Science, scientists, and scientific authority

Enniskillen, Canada West (Ontario), petroleum in, 213, 214, 219–221, 225, 228, 257–261, 281

Entogaea, class of minerals, 71

Epigaea, class of minerals, 71

Erie Canal (N.Y.), 81, 119

Evans, E. W., and theory of oil cavities, 225, 226, 227

Eveleth, Jonathan G., 191–204, 214

Fairfield, John, 25

"Famous Oil Firms" (song), 248

Fernie, Ebenezer W., 177; *James Young v. Ebenezer Fernie*, 177–79, 180, 184

Field, Stephen J., 275

Fireclay, 89, 90, 93

Fleury, A. L., 167

Formation, definition of, 13, 14, 17, 76, 331 n. 16

Formation VIII (shales of Pa.), 218, 227

Formation IX (sandstones of Pa.), 218, 227

Formation XI (limestones of Pa.), 86, 100, 106, 121. *See also* Proto-Carboniferous formation; Subcarboniferous system

Formation XII (Great Conglomerate of Pa.), 86, 90, 100, 106, 263

Formation XIII (Great Coal of Pa.), 86, 91, 100, 106, 263

Fossils: in classifying formations, 14, 28, 51, 98, 100; coal plants, 19, 30, 49, 52, 59, 77, 79, 80, 84, 93, 100, 101, 263, 264; fish, 49, 59, 68, 97, 123, 157, 298, 355 n. 60; Hall's collections of, 112–113; marine shells, 79; reptiles, 95, 96; upright trees, 18–19, 30, 89, 91, 93, 94, 164

Foster, John Wells, 136, 275

Foulis, Robert, 340 n. 44

Frankland, Edward, 165, 176, 180, 181, 351 n. 2, 367 n. 57

Franklin Institute (Philadelphia, Pa.), 52, 57, 86, 285

Frazier well (Pithole Creek, Pa.), 235, 254

Freiberg, Königliche Sächsische Bergakademie, 13, 71, 173

Fucoides Cauda-galli, 264

Funk well (Oil Creek, Pa.), 222

Fyfe, Andrew, 165

Gabb, William M., 276

Gale, Leonard D., 183, 368 n. 85

Gale, Thomas A., 194, 211–216, 221, 228, 229

Gardner, James T., 276

Gas lighting (via coal gas), 6, 38, 43–46, 58, 75, 93, 103, 124, 144, 147, 155, 165, 166, 194, 199, 293

Geikie, Archibald, 178

General Mining Association (Nova Scotia), 37, 39, 47, 113

Genth, Frederick, 131

Geological Society of London, 30, 86, 87, 90, 94, 219; Gesner's fellowship at, 29

Geological surveys, 2, 7, 32–35, 58, 126; of Canada, 31, 219, 243, 333 n. 60; as education, 1, 320; federal, 33, 115, 319, 320; of Great Britain, 22, 51, 178; and mining interests, 12, 27, 108–109, 277; for states, 10, 23, 31, 33, 35, 111, 115, 273–275, 318–320. *See also* Consultants and consulting: engagements; *individual states and provinces*

Geology, 314; American, 79, 85, 109, 111–112, 300–301, 318; causes in, 17, 19, 28; of coal, 6, 85, 99–100, 104, 105, 113, 120, 126, 143, 164, 170, 179, 276; economical, 80, 305, 318; and progress, 38–40; and religion, 19, 28, 29; scale of American, 81, 89, 225, 269, 277; scientific, 318, 334 n. 66; structural, 6, 78, 87, 89, 94, 98, 101–102, 240, 263, 349 n. 132. *See also* Coal; Petroleum

Geology and petroleum, 6, 7, 179, 196, 208, 209, 215, 216, 245, 264–269, 270, 289, 299; capstones, 217, 268; migration, 220, 265, 268; natural gas, 225–227, 232; scale of American fields, 225, 269; surface indications, 202, 217, 220, 250, 268, 286, 290

Geology and petroleum, theories of occurrence: anticlines, 218–220, 225, 257, 261, 267; big cavities, 225–227, 238, 262; fissures (vertical), 215–221, 225, 234, 238, 262, 268; sandstones (sandrocks), 216, 217, 227, 235–236, 238, 260–265, 268; shales, 281; stratigraphical diversity, 196, 216–217, 265; veins (horizontal), 215–216, 234

Geology and petroleum, theories of origin, 7, 196, 209, 217, 269; distillation, 196, 217–218, 226–227, 269, 295, 296; inorganic, 295; in situ theory of animal decomposition, 220, 225, 227, 295–296; in situ theory of plant decomposition, 196, 220, 264, 288, 295–296; relation to coal or Carboniferous, 196, 217, 227, 239, 261

George Miller and Co., 150

Gerstner, Patsy, 347 n. 73

Gesner, Abraham, 3, 11, 42, 66, 67, 76, 94, 95, 104, 114, 143, 147, 151, 154, 158, 162, 163, 170, 173, 177, 178, 181, 183, 193, 194, 197, 240, 242, 243, 277, 287, 330–331 n. 6, 336 n. 120, 337 n. 133, 339 n. 24, 376 n. 126; *Abraham Gesner v. Halifax Gas-Light Company*, 46, 54–55; *Abraham Gesner v. William Cairns*, 46, 55–63; as fellow of the Geological Society of London, 29; and General Mining Association, 37–38; and "Great New Brunswick Coal Field," 21, 23–27, 31, 37, 41, 66; and kerosene gas, 38, 43–45, 144–146; and kerosene oil, 146–149, 151, 156–157, 166–168; and Native Americans, 26, 29; and New Brunswick survey, 12, 21–32, 56, 111; and petroleum, 218, 219, 228–230, 376 n. 120; and plagiarism, 15, 20, 332 n. 50, 334 n. 77; *Practical Treatise on Coal, Petroleum and Other Distilled Oils*, 143, 144, 147, 156, 173, 229, 230, 240, 242; *Remarks on Geology and Mineralogy of Nova Scotia*, 15, 19, 332 n. 50

Gesner, George Weltden, 158, 159, 230, 232, 331 n. 6, 364 n. 95, 376 n. 126

Gesner, Harriet (Webster), 330 n. 6

Gesner, Henry, 12, 330 n. 4

Gibbs, Wolcott, 300, 302, 305, 306, 308, 315

Gillespie, William, 177, 364 n. 7

Glendon Oil Co. (Boston), 167, 169, 170

Gneiss formation (Nova Scotia), 18

Godkin, E. L., 305–306

Gold, 65, 110, 136, 274, 284, 285, 287, 294, 295, 304, 311

Gooding, Spencer, 379 n. 4

Gould and Curry v. North Potosi, 279

Graham, Thomas, 163, 165, 169

Grahame Crystallized Rock Oil Co. (N.Y.), 241

Grahamite, 241–242. *See also* Ritchie mineral

Graphite, 72, 74, 296

Granite: as formation, 14–15, 17–18, 24, 26; as rock, 13, 296; as system of formations, 37

Grass Valley, gold mining district (Calif.), 294, 295

Gray, John H., 64

Graywacke formation: of New Brunswick, 24, 28; of Nova Scotia, 17–18, 84

Great Britain, 26, 29, 33–34, 40; free trade debate in, 21; and geological fieldwork, 17; men of science in, 1, 2, 3; mining expertise of, 3, 108

Great Western Co. (Kanawha, Va.), 155

Greenough, George, 78

Grinand, Dumas, 156

Gypsiferous formation (Nova Scotia), 30, 94, 95, 335 n. 107, 336 n. 4

Gypsum, 14, 18, 26, 30

Hager, Albert D., 257

Halifax Gas-Light Co., 38, 42, 45, 47, 48, 50, 54, 55, 61, 95, 123, 128, 136, 341 n. 82

Hall, James, 2, 3, 87, 112–113, 115–117, 121, 125–139, 157, 200, 216, 244, 245, 268, 275, 276, 279, 302, 322, 342–343 n. 117, 351 n. 166, 379 n. 4; Lake Superior engagement, 118–119, 139, 354 n. 45

Hall, William, 119, 131

Hamilton Group (N.Y.), 216, 218, 220

Hare, Robert, 181

Harmonial well (Pleasantville, Pa.), 252

Harris, D. L., 287

Harvard University, 54, 58, 134, 173, 276, 300; School of Mining and Practical Geology, 299, 321, 390 n. 117

Harvey, John, 21, 22, 25, 27, 34, 45, 336 n. 120

Haüy, René-Just, 72

Havens, Rensselaer, 203, 204

Hayden, Ferdinand V., 319

Hayes, Augustus A., 3, 50, 54, 57–63, 70, 74, 75, 151, 239

Headley, Stephen F., 129, 132, 200

Henwood, W. J., 27

Henry, J. T., 191, 193, 203, 204, 205, 206, 217, 232

Henry, Joseph, 138, 199, 200, 300, 301, 305, 306, 310, 314, 320

Hibbard farm (Oil Creek, Pa.), 192, 203

Hilgard, Julius E., 305, 306

Hind, Henry Youle, 242, 243, 304

Hitchcock, Charles H., 243, 265, 269, 297, 377 n. 156; and Albertite, 240, 257; and anticlinal theory of petroleum occurrence, 267

Hitchcock, Edward, 79, 80, 81, 84, 104, 240, 334 n. 66, 346 n. 50

Hodge, James T., 113, 121, 135, 136, 139, 278, 279, 287

Hoffmann, Charles F., 276, 299

Horsford, Eben, 134, 358 n. 148

Hotchkiss, William B., 117–118

Howe, Joseph, 95

Hubbard, Oliver P., 190, 191

Hunt, T. Sterry, 208, 219–227, 233, 242, 260, 281, 304, 321; and anticlinal theory of petroleum occurrence, 221, 225, 267; and in situ theory of origin of petroleum, 220, 225, 269, 295

Hutton, James, 13

Hydro-Carbon Gas Co. (Manchester and Salford, England), 165, 169, 176, 178–180

Hydrocarbons, 157, 218, 226, 243, 292; chemistry of, 296–299, 308, 310; as class of minerals, 75, 296, 299

Hydrocarbonaceous shales, 296–297

Hypogaea, class of minerals, 71, 72, 74

Illinois, deposits in: coal, 94, 100, 106; lead, 127; petroleum, 257

Independent Coal formation, 76–78, 84

Invention and inventors, 162, 163, 172, 179–183, 185

Iowa: coal in, 100, 106, 134; survey of, 115, 275

Iron ore, 40, 132; in New Brunswick, 22, 23, 25, 26; in Nova Scotia, 33

Ives, William A., 203

Jackson, Charles Thomas, 3, 11, 12, 30, 38, 54, 58, 71, 74, 100, 104, 108–109, 124, 138, 139, 151, 173, 181–183, 192, 196, 242, 304, 308, 340 n. 35, 342–343 n. 117, 368 n. 82; and Albert mineral, 41–43, 47–51, 57–63; and California petroleum, 288–289; and fossil fish, 97, 123, 157, 355 n. 60; on Gesner's plagiarism, 15, 20, 332 n. 50, 334 n. 77; on Gesner's salted sample,

48; and Lake Superior survey, 33, 112–115, 125, 130–133, 136, 275, 357 n. 113; and Maine survey, 25, 31, 58, 111, 344 n. 77; *Mineralogy and Geology of Nova Scotia*, 13, 19; mining engineers, 111–112, 130; New England geology, 337 n. 125; New Hampshire survey, 33, 58, 114, 115; patents of, 368 nn. 82, 85; Rhode Island survey, 33, 58, 80, 114; and Transition formations, 346 n. 57

James Lee and Co. (N.Y.), 167–168

Jameson, Robert, 76, 77

Jennings, Isaiah, 153

Jersey well (Oil Creek, Pa.), 253

Johns, Walter R., 220, 221, 232, 246, 251, 252

Johns Hopkins University, 315–318

Johnston, James F. W., 31, 336 n. 116

Johnston, James W., 55

Joint-stock companies, 4, 144, 145, 149, 191, 192, 201, 202, 246–247, 278. *See also* Stocks

Judd's Patent Sixes Sperm Candles, 198, 371 n. 60

Jukes, Joseph Beete, 333 n. 60

Kanawha River (W.V.), 156, 204, 205, 219, 222, 229, 230

Kane, Robert, 176

Kentucky: coal in, 87, 133, 149, 155; petroleum in, 225, 228, 246, 254, 256, 258, 261, 265, 267

Kentucky Mining and Manufacturing Co., 129, 131, 132, 133, 134, 200

Kerosene. *See* Coal oil

Kerosene cases. *See* Coal oil: controversy surrounding

Kerosene Oil Co., 143, 144, 146, 147, 149, 154–157, 160–163, 166, 167, 170, 193, 364 n. 101, 370 n. 31

Kier, Samuel, 204, 205, 210, 211

King, Clarence, 276, 301, 319, 320, 394 n. 38

King's College (Fredericton, New Brunswick), 31, 32, 57

King's College (Windsor, Nova Scotia), 158

Klippart, John H., 318

Kopp, Emile, 173

La Brea, Calif., 280, 281

Lafayette Gas Co. (Ind.), 154

Lard oil, 150, 152–153, 154, 155, 160

Laurent, Auguste, 172

Lawyers, 2, 60, 116, 128, 129, 139, 146, 164, 183, 316

Lea, Isaac, 341 nn. 61, 63

LeConte, John L., 293, 302, 309

Leet, Allen Norton, 230

Leidy, Joseph, 3, 56–63, 97, 100, 300

Lepidodendron, 18

Lesley, J. Peter, 3, 111, 113–114, 116–138, 157, 222, 224, 233, 238–242, 244, 256, 259, 268–273, 278, 297, 299, 302, 319, 321, 322, 350 n. 136, 356 n. 85, 382 n. 93; and Albertine, 241; on anticlinal theory of petroleum occurrence, 227, 267; consulting on petroleum, 260–264; *Manual of Coal and its Topography*, 99, 103, 125–126; on origin of petroleum, 227, 261, 269; and Ritchie mineral, 240–241; and second Pennsylvania survey, 127, 268; and topographical geology, 98; on true and false coal, 104

Lesley, Joseph, 114, 134, 278, 385 n. 18

Lesley, Joseph, Jr., 119

Lesley, Susan (Lyman), 98

Lesquereux, Leo, 119, 126, 135, 157, 260–264, 268, 269, 295

Letheby, Henry, 176

Liebig, Justus von, 60, 115, 168

Lincoln, Abraham, 233

Lincoln, Levi, 80

Linneaus, Carl, 70, 72

Linseed oil, 152, 154

Logan, William, 30, 90, 333 n. 60, 336 n. 109, 348 n. 91

Londonderry Iron Works, 37

Lowell Institute, lectures at, 85, 87, 91, 94

Lower Carboniferous system: of Nova Scotia, 95, 97; of Pennsylvania, 100; and petroleum, 261. *See also* Mississippian system; Subcarboniferous system

Lower Devonian system, as source of petroleum, 216, 261

Lower Silurian system, as source of petroleum, 261, 265

Lucesco Co. (Kiskiminitas, Pa.), 155, 211, 229

Lyell, Charles, 11, 17, 33, 37, 87, 94, 95, 128, 129, 335 n. 107, 349 n. 115; on America, 29, 30, 85; estuary theory of coal, 89; and Gesner, 29, 30; on growth-on-the-spot theory of coal, 91–93, 348 n. 111

Lyman, Benjamin Smith, 119, 260, 278, 280, 321

Lyman, David H., 203, 204

Lynch, Eugene, 293

Lyon, Shorb and Co. (Pittsburgh), 122

Maine: logging in, 25; survey of, 25, 31, 58, 111, 240

Maisch, John, 288, 290, 291

Mammoth Gas Bubble Co., 244, 245

Manhattan Gas-Light Co., 44, 47, 59, 144

Maple Shade well (Oil Creek, Pa.), 222, 253

Maps, contour, 6, 99, 101, 106

Marcy, William, 117

Maryland, coal in, 87, 99, 110, 125

Maryland Mining Co., 125

Massachusetts: coal in, 80, 94, 106; survey of, 58, 80, 111, 114, 334 n. 66

Massachusetts Institute of Technology, 173, 320

Mather, William W., 2, 3, 132–133, 135, 277, 322

Matthew, George F., 31, 378 n. 180

McIntyre, Archibald, 112, 133

McNeill, Duncan, 176–177

Meldrum, Edward, 164–166

Merrill, Joshua, 150, 151, 153, 154, 228, 229, 293

Merrill, Rufus, 153, 167, 168, 362 n. 54

Metals, 125, 284; as class of minerals, 65, 74; occurrence in coal veins, 61, 125

Mica Slate formation, of Nova Scotia, 18

Michigan: coal in, 94, 102, 106; copper in, 114, 115, 133, 239; petroleum in, 257, 258, 265, 271; survey of, 58, 114, 115, 133

Microscope/microscopy, 60, 63, 97, 100

Middle Carboniferous system, of Nova Scotia, 94, 97

Millstone Grit formation, 78, 79, 82–84, 86, 106

Mineralogy, 13, 58, 70, 75, 98, 344 n. 2; and American science, 73, 75

Mines and mining, 4, 35, 213, 204, 274, 284–285; capital needs in, 26–27; examination of, 2, 27, 58, 108; and risk, 135, 139, 195, 262; and science, 26–27, 108–109, 135, 321

Mining and Scientific Press, 280, 288, 301

Mississippian system, 106, 351 n. 167

Missouri: coal in, 94, 102, 106; lead in, 211; survey of, 113

Mohs, Friedrick, 71

Mother Lode (gold region of Calif.), 274, 294

Mountain Limestone formation, 78, 79, 84, 86, 106, 345 n. 34; of New Brunswick, 24. *See also* Mississippian system

Mowbray, George, 210

Murchison, Roderick Impey, 28, 30

Muspratt, James, 163

Naphtha, 71, 72, 75, 145, 286, 292, 295

National Academy of Sciences, 7, 264, 274, 295, 300, 301, 302, 303, 304, 305, 309, 310, 311, 314, 315, 320, 323

Native Americans, 17, 29, 277, 337 n. 128

Natural history, 60, 79, 85, 158, 287, 300; as system of mineral classification, 70–73

Neff, Peter, and Neff Petroleum Co., 257, 258, 264, 268, 271

New Almaden (quicksilver) mines (Calif.), 277, 279

Newark Coal Oil Co. (Ohio), 154, 155

Newberry, John Strong, 3, 134, 157, 213–215, 221, 257–261, 265, 271, 320; and anticlinal theory of petroleum occurrence, 267; and distillation theory of petroleum, 217, 218, 226, 269; and physical theory of oil wells, 265–269; and second survey of Ohio, 318

New Brunswick: economy of, 21, 27, 34; government of, 21–23, 47, 242, 333 n. 57; "Great Coal Field" in, 12, 21–27, 31, 34, 37, 39–41, 51, 66, 80, 102, 106, 155, 289; land grants and crown reservations in, 46–47, 65–66; survey of, 12, 21–32, 56, 111, 159, 242

Newcomb, Simon, 314

New Hampshire, survey of, 33, 58, 114, 115

New Haven Gas-Light Co., 45, 47, 59, 135, 199

New Idria (quicksilver) mines (Calif.), 277

New Red Sandstone formation: along Connecticut River, 80, 81; of New Brunswick, 24, 28, 30; of Nova Scotia, 17, 18, 37

New York: coal in, 81, 84–87, 104; petroleum in, 244, 254; survey of, 84–85, 111, 112, 132, 216, 254

New York and New Haven Railroad Co., 201, 203

New York and Nova Scotia Gold Mining Co., 302

New York Oil Co. (New Galilee, Pa.), 155, 157

New York system, 85, 216

Nitrogen, as chemical component: of coal, 52, 75; of petroleum, 295, 296

Noble well (Oil Creek, Pa.), 222, 254

North America, British: climate of, 34; coal in, 37, 106; immigration to, 29, 34, 39; need for capital in, 27

North American Co. (Kiskiminitas, Pa.), 155, 211, 229

North American Kerosene and Gas-Light Co. See Kerosene Oil Co.

Norton, Charles Eliot, 260

Norton, John Pitkin, 354 n. 31, 358 n. 130, 358 n. 147

Nott, Eliphalet, 135, 181

Nova Scotia: coal in, 26, 40, 44, 45, 80, 100, 102, 106, 339 n. 15; and Gesner's geology, 17–18; gold in, 280; government of, 20, 37, 55; and Jackson and Alger's geology, 14–15, 17, 58. See also General Mining Association

Nova Scotian and Cape Breton Mining Co., 337 n. 143

Oakes, James, 364 n. 4

Odling, William, 176

Ohio: coal in, 87, 99, 103, 129, 157, 211; first survey of, 132–134, 139, 277; petroleum

in, 213, 214, 217, 221, 226, 229, 240, 246, 254, 258, 261, 264–267, 271, 284; second survey of, 318, 319; third survey of, 271

Ohio Diamond Coal Co., 132, 199

Oil. See Petroleum

Oil boom/fever: and boomtowns, 245, 246; from 1859–1862, 213, 221; from 1864–1866, 233–234, 244–249, 256, 258, 259, 264, 265, 268, 273, 288, 290, 293, 307, 317; and gold rush, 245. See also Petrolia; Pithole

Oil busts: in 1862, 224; in 1866, 290, 295, 388 n. 73

Oil City, Pa., 191, 245, 246

Oil Creek, Pa., 190, 191, 202, 205, 211–225, 245, 250, 259–263, 270, 284, 296; anticlinal structures at, 227, 262, 267; "bottom lands" at, 221, 224, 234, 238; environmental degradation at, 224; high lands at, 225, 234, 238; production at, 219, 224, 235; refineries at, 229, 230; sandrocks at, 260, 262, 263, 265; and transportation, 228, 229

Oil wells, 223, 247, 250; clustering of, 213, 218, 221, 224, 253; costs of, 214, 256; depth of, 214, 224; flowing, 221–227, 289, 290; operators of, 215, 222, 233, 251; pumping, 210, 214, 221; records of, 6, 262, 263, 268; and steam engines, 214, 224; as "unit" of trade in industry, 247–248

Olcott, Thomas, 117, 139

Old Red Sandstone formation, 78, 79, 345 n. 34; of New Brunswick, 24, 52, 54; of Nova Scotia, 17, 18

Orton, Edward, 271, 318

Pacific Coast Petroleum Co. (New York City), 285, 286, 287, 288

Paint Lick Fork (Ky.), 261, 262, 264, 267

Paleoniscus, 123, 126, 157

Paleontology, 94, 98, 100, 112, 113, 157, 264, 276, 278, 299

Panic of 1857, 203, 314

Paraffin(e), as chemical compound, 145, 172, 175, 178, 179, 180, 291, 296, 297, 366 n. 44

Paraffine, Young's patent oil, 146, 158, 162, 171–179, 206; English patent for, 164; U.S. patent for, 166, 167

Partelow, John R., 333 n. 56

Patents, 6, 8, 146, 162, 180, 181, 185, 229, 232; for gas lighting, 44–45; nullifying, 167–168, 174–177, 182–184; and science, 4, 45, 182; specifications, 168, 171, 174, 175, 178, 365 n. 28; writing of, 163–165, 167

Patents, infringement of, 146, 165; and Young's Paraffine, 166–171, 174–177, 232

Patronage, of science: in New Brunswick, 27, 31; in Nova Scotia, 15, 19–20; private (commercial/capitalist), 2, 3, 67, 198–199, 301; public (government), 1, 3, 12, 35, 300. *See also* Consults and consulting

Peat-bog theory of origin of coal, 89–90, 93–94

Peckham, Stephen F., 290, 297, 298, 308, 310; and salted sample of California oil, 291–293; and theory of bitumens, 295–296

Pennsylvania: bankruptcy in, 35; coal in, 11, 85–94, 98–106, 289; first survey of, 22, 85–94, 98–106; petroleum in, 220, 226, 229, 232, 233, 256, 257, 267, 270, 284, 319; second survey of, 127, 268, 318, 319

Pennsylvania, classification of formations: Rogers's first nomenclature (Formation I–XIII), 86, 88, 347 n. 76; second nomenclature, 88

Pennsylvania system, 106, 351 n. 167

Pennsylvania Railroad, 114, 278

Pennsylvania Rock Oil Co., 135, 190; of Connecticut, 201, 202; as "The Fancy Stock Company," 194; of New York, 192–194, 369–370 n. 27

Penny, Frederick, 51, 176

Percival, James G., 50–51, 57, 58, 60, 61, 80, 81, 100, 104, 127

Perley, Moses H., 334 n. 71

Peter, Robert, 157–158, 363 n. 84

Peter O'leum (swindler), 249–251

Petroleum (oil), 5, 71, 95, 146, 150, 161, 172, 177, 191, 204, 205, 208, 243, 288; and civilization, 206; exploration, 209, 210, 221, 234, 238, 245, 253; for gas manufacturing, 199; industry, 189, 224, 318, 319; as lamp oil, 193, 196, 197, 211, 228, 232; as lubricant, 193, 195, 196, 197, 210, 228; as medicine, 190, 193, 197; as "natural" analog to "artificial" coal oil, 163, 179, 193, 217, 218, 226, 233, 269; prices, 224, 233; production, 211, 219, 224, 233, 311; profitability of, 213–215; refining of, 198, 207, 211, 228, 229, 230, 230–231, 291; relation to asphaltum, 196, 242; relation to coal, 196, 239, 242; storage of, 210, 214; stratigraphical referents for, 217, 220, 261, 262, 265; as substitute for crude coal oil, 227; transportation, 228, 229; varieties of, 228, 291–297, 308

Petrolia (Pa.), 246, 247, 249, 250, 252, 254, 256, 262, 270

Philadelphia and California Petroleum Co. (Philadelphia), 285, 286–287, 288, 290, 293

Philadelphia Gas Co., 44, 51, 52

Philbrick, Samuel R., 150, 151

Phillips, John, 351 n. 2

Phillips, William, 78, 29, 87, 345 n. 33; *Outlines of the Geology of England and Wales*, 18, 77

Phillips well, Nos. 1 and 2 (Oil Creek, Pa.), 222

Photometric experiments, 154, 199, 371 n. 63

Phrenite, 332 n. 24

Physicians, 2, 58, 316; and science, 56, 57

Pierce, Benjamin, 302

Pierpont, Asahel, 202, 203

Pineo, Sarah, 12, 330 n. 4

Pitch, 145, 146; Pittinea, 71

Pitch coal, 97

Pithole (Pa.), 234, 235, 236, 245, 252, 254

Pittsburg coal bed, 89, 99, 100

Plagiarism, 12, 48, 242; Gesner of Jackson and Alger, 15, 20, 332 n. 50, 334 n. 77

Plant oils, 144, 150, 152, 155, 197, 233

Playfair, Lyon, 163–164, 165, 169, 176, 364 n. 4

Poole, Henry, 113

Popular Science Monthly, 212, 214

Portage Group (NY), 216, 217, 218, 228

Porter, John Addison, 199, 358 n. 130

Porter, Roy, 325 n. 1

Powell, John Wesley, 319

Powell, Lazarus, 149

Primary (Primitive), class of formations, 78, 82–85; district of Nova Scotia, 17–18; of New Brunswick, 28; of Nova Scotia, 14–15

Prince of Wales mine (New Brunswick), 239

Princess Alexandra mine (New Brunswick), 128, 239

Proto-Carboniferous formation (Pa.), 100

Pseudo-Coal Measures (Nova Scotia), 95

Quail Hill (gold) mine (Calif.), 294

Quartz Rock formation (Nova Scotia), 14–15, 17–18

Railroads, 4, 35, 38, 111, 150, 155, 177, 211, 299, 314, 322, 337 n. 133

Ralston, William C., 79, 385 n. 24

Ramsay, Andrew, 178

Rapeseed oil, 152

Raymond, Rossiter W., 301

Red Sandstone district (Nova Scotia), 14–15

Reed, William, 234

Reed well (Cherry Run, Pa.), 234

Reichenbach, Karl von, 172, 175, 176, 178, 180, 181, 182, 366 n. 44

Reingold, Nathan, 336 n. 121

Rémond, Auguste, 276

Remsen, Ira, 317–318

Rensselaer Polytechnic Institute, 109, 112, 132, 134

Renwick, Edward Sabine, 366 n. 36

Renwick, James, 170

Resins, 296; as class of minerals, 74, 75; pine tar, 153

Richardson, Charles, 255–256

Richter, Theodor, 173

Ridgeway, T. S., 129

Ritchie mineral, 240–241

Robb, James C., 32, 55, 56–63, 70, 71, 95, 239, 240, 242, 297, 336 n. 116; on Great New Brunswick Coal Field, 31, 57

Rockefeller, John D., 207

Rogers, Henry Darwin, 3, 85, 106, 157, 216–218, 221, 264, 269, 277, 320, 347 n. 77, 374, n. 43; and coal classification, 103; and de-bitumenization theory of anthracite, 91, 94, 217, 226; and distillation theory of petroleum, 216–217, 226–227; *The Geology of Pennsylvania*, 98, 99, 101, 104, 125, 216; and "growth on the spot" theory of coal, 89–90, 93–94; and law of gradation, 91, 217, 226; on origin of Appalachians, 87, 101, 217; and Pennsylvania survey, 86, 113; on scale of American coal, 87–89; on scale of American petroleum, 225; "searching for coal," 104–105

Rogers, William Barton, 86, 87, 300, 320, 347 n. 80, 374 n. 43

Roome, Charles, 44

Rosenberg, Charles E., 336 n. 121

Rosin oil, 153, 154

Rouse, Henry, 222

Rowland, Henry A., 315, 316, 317, 318, 320

Safford, James M., 256–257, 265

Saint Clair Railroad and Coal Co., 117–118, 138

Saint John Mechanics' Institute, 26, 33, 61, 334 n. 80

Saliferous Rock formation (N.Y.), 84

Saussure, Nicholas Théodore de, 172

Schieffelin Brothers of New York City, 210, 211, 228

Science, scientists, and scientific authority, 162, 171, 173, 307; applied, 274, 322, 361; and commerce/industry, 4, 5, 6, 114, 122, 126, 160, 181, 234, 272, 292, 310, 311, 316, 317, 319, 321, 322; and disinterestedness, 8, 61–62, 66–68, 130, 134, 136, 138–140, 149, 160, 163, 181, 184, 250, 270, 276, 304, 309, 311; and entrepreneurship, 1, 2, 3, 4, 6, 12, 32, 43, 45, 61, 110, 111, 143, 203, 311, 336 n. 121; and

Science, scientists, and scientific authority (*continued*)
ethics, 7, 8, 60–62, 66–68, 110–114, 130–140, 183, 184, 259, 270–273, 276, 280, 292, 293, 303, 308, 310, 314, 315, 320; and improvement/progress, 35, 38, 39, 40, 313; and law, 66–68; and money, 1, 2, 8, 67, 310, 314, 316, 317; and moral character, 308–312, 315; and motivation, 66–67, 114–115, 163, 180–185, 309, 310, 312, 314; and patents, 181, 183; and petroleum, 206, 270, 272; practical vs. theoretical, 5–6, 114, 313–315; and profitability, 270–271, 304, 311–312; public vs. private, 273, 320; "pure," 125, 131, 183, 184, 274, 303, 309, 315–318; and technology, 8, 143–144, 170, 179, 181, 315, 317; and trustworthiness, 67, 136, 140, 271, 276
Scientific American, 43, 44, 144, 151, 154, 155, 156, 157, 160, 167, 168, 171, 177, 181, 183, 217, 219
Scott, Thomas A., 278–280, 284, 285, 288, 293, 385 n. 18
Secondary, class of formations, 14, 15, 76–84, 87, 93; Lyell on, 17; in New Brunswick, 28
Sedgwick, Adam, 28, 58, 349 n. 132
Selligue, Alexandre François, 172, 174
Seneca Oil Co. (Conn.), 204, 210, 211
Seward, William, 112
Sheafer, Peter W., 113, 114, 119, 126, 259, 293, 353 n. 24
Sheldon, Anson, 191, 192, 195, 199, 200, 201, 202, 203
Shepard, Charles U., 70
Shumard, George G., 257
Sigillaria, 90, 93, 96
Silliman, Benjamin, Jr., 3, 7, 43, 44, 51, 57–63, 70, 75, 110, 115–118, 125, 129–135, 138, 151, 152, 174, 190, 191, 194, 202, 206, 207, 239, 241, 246, 260, 275, 290, 294, 299–315, 321, 322; and California oil, 273, 281, 291, 303, 307, 387 n. 54; consulting in the West, 278–285, 386 n. 46; New Haven Gas-Light Co., 45; *Report on the Rock Oil, or Petroleum, of Venango Company, Pennsylvania*, 89, 196–200, 203–205, 210, 280, 311; and salted sample of California oil, 286–287, 292; at University of Louisville, 149, 354 n. 31; at Yale, 149, 195, 370 n. 45
Silliman, Benjamin, Sr., 3, 20, 51, 70, 79, 81, 89, 109–111, 115, 125, 135, 200, 275, 285, 300, 303
Silurian system, 265, 346 n. 57; in California, 281, 284; in New Brunswick, 24, 28; in Nova Scotia, 37; and petroleum, 226, 295
Silver, 65, 278, 279, 287, 295, 301, 311
Simonds, Charles, 21, 333 nn. 56, 57
Smith, Crosbie, 336 n. 121
Smith, Hamilton L., 257
Smith, William, 51
Smith, William A. ("Uncle Billy"), 205
Smith and Rollins (Boston law firm), 170
Smithsonian Institution, 138, 199, 300, 306
Speculation and speculators, 39, 47, 48, 114, 115, 182, 192, 249, 273, 319; in coal oil, 151–152; in mineral lands, 133, 213, 221, 234, 246, 248; in mining, 27, 66, 85, 127, 130, 139, 177, 278; in petroleum, 211, 213, 234, 246–248, 256, 288, 289, 290, 307; and science, 121, 307
Sperm whale oil, 43, 152, 154, 196, 198
Springfield Republican, 287, 288, 306
Spring pole, method of boring for oil, 213–214
Standard Oil Co., 207
Stenhouse, John, 165, 175, 176
Stevenson, John J., 348 n. 98
Steves, John, 46
Stigmaria, 90, 93, 97
Stocks: "fancy," 137, 138, 194, 201, 248, 251; oil, 46–48, 192, 193, 201–202, 234, 287; promotion/selling of, 111, 191–193, 246, 247, 251, 286. *See also* Joint-stock companies
Storer, Francis (Frank) H., 171, 173–175, 176, 181, 183, 320–321; attack on Antisell and Young's patent, 174–175; and California asphaltum, 280, 290, 292
Stuart, John, 177–179
Subcarboniferous system, 104, 106, 264,

297, 351 n. 166; in Pennsylvania, 100, 263. *See also* Mississippian system

Swindlers and swindling, 7, 115, 130, 137–139, 144, 158–160, 192, 194, 245, 248–254, 256, 260, 300, 314, 363 n. 94; in California petroleum, 273, 306–309; in stock, 201, 290, 293; and Whitney's model of, 301–303. *See* Peter O'leum

Sydney Mines (Cape Breton), 37, 330 n. 1

Sylvic oil, 153

Tar, 280, 286, 292, 295. *See also* Pitch; Trinidad pitch

Taylor, Alfred Swaine, 176

Taylor, Richard Cowling, 56, 57, 59, 69, 81, 93, 95, 104, 106, 122, 124, 138, 140, 240; on the Albert mine, 51–52; *Statistics of Coal,* 51, 69, 124

Technology, 6, 8, 38, 103–104, 182; and coal/gas, 229, 232; and industrial innovation, 4, 209, 210, 232; and risk, 38–39; and science, 7, 143–144, 233, 393 n. 31; and swindling, 250

Telegraph, 35, 38, 119, 247, 250, 315, 322

Tennessee: coal in, 217, 256; petroleum in, 257, 258, 265

Tertiary system, 28, 78, 84, 289

Teschemacher, James E., 57–63

Thomson, Thomas, 336 n. 120

Thomson, William (Lord Kelvin), 336 n. 121

Thurston, Robert H., 322

Tionesta Sandstone (Pa.), 263, 264

Titusville, Pa., oil rush in, 190, 191, 194, 202–205, 210, 215, 252

Topographical geology and topography, 91, 99, 113, 276–277; as American invention, 98; and surveying, 35, 38. *See also* Geology: structural; Maps, contour

Torbanehill mineral, 75, 100, 155, 156, 158, 165, 169, 174, 175, 177, 364 n. 7; as coal, 178; as separate mineral species, 178; 297, 389 n. 112

Torbanite, 297, 389 n. 112

Torrey, John, 3, 57–63, 288, 300, 304, 308

Toshach, William, 256

Townsend, James M., 191, 195, 202, 203, 204, 211, 239

Transition, as class of formations, 14, 15, 77, 78, 80–87, 93; in New Brunswick, 28; in Nova Scotia, 17

Trap, formation: of New Brunswick, 24; of Nova Scotia, 13, 37

Treadwell, Daniel, 169

Trinidad pitch, 43, 46, 157

Uniacke, James B., 55, 95, 341 n. 82

Uniformitarianism, 89, 94

United States Army Laboratory (Philadelphia, Pa.), 288, 290

United States Chemical Co., 150

United States Geological Survey, 319–320

United States Patent Office, 146

United States well (Pithole Creek, Pa.), 235, 254

University of Edinburgh, 77, 331 n. 7

University of Glasgow, 98, 216, 336 n. 120

University of Göttingen, 315, 345 n. 22

University of New Brunswick (Fredericton), 31, 242

University of Pennsylvania (Philadelphia), 86, 98, 127, 285, 321

Upper Carboniferous system: of Nova Scotia, 94, 97; of Pennsylvania, 100; and petroleum, 217. *See also* Pennsylvania system

Upper Devonian system, and petroleum, 217, 261

Ure, Andrew, 51

Van Rensselaer, Stephen, III, 81, 109–110, 351 n. 4

Vegetable oils. *See* Plant oils

Virginia: coal in, 155, 157; copper in, 200; petroleum in, 220; survey of, 86, 111

Warren, Cyrus M., 288, 291, 293, 321

Warrington, Robert, 176, 367 n. 57

Washoe, silver mining region (Nev.), 276, 278, 279, 294

Waverly Group (Ohio), 217, 218, 264, 265

Webster, Isaac, 12

Webster, William Bennet, 331 n. 7
Webster-Ashburton Treaty, 25
Weed, Thurlow, 117, 139
Werner, Abraham Gottlob, 13, 78–79, 87; classification of rock formations, 76–77, 331 n. 16; Neptunist theory, 13–14
Westmoreland coal field, 25. *See also* New Brunswick: "Great Coal Field" in
Westmorland Petroleum Co., 377 n. 155
West Virginia: coal in, 217; petroleum in, 213, 218, 221, 229, 230, 240, 246, 254, 261, 265
Wetherill, Charles, 52, 57–63, 75, 100, 153–154, 297, 298, 345 n. 22; melan-asphalt, 54
Whale oil, 144, 150, 152, 154, 155, 160, 296. *See also* Sperm whale oil
Wheatley, Charles M., 125, 132
Wheeler, George M., 319
Whigs (U.S. political party), 35, 39, 117
White, Andrew D., 314
Whitney, Josiah Dwight, 3, 7, 115, 129, 135, 137, 138, 192, 256, 273, 279–280, 285, 291, 305–311, 314–316, 320, 321, 323, 354 n. 31, 386 n. 29, 389 n. 117; California survey, 274–278, 284, 287, 289, 290, 299–301; charges Silliman of swindling, 301–304; *The Metallic Wealth of the United States,* 125, 136, 275

Whitney, William Dwight, 293, 299, 300, 302, 305–307, 309, 310
Whittlesey, Charles W., 133
Wilmot, Lemuel Allan, 55, 64, 341 n. 83, 343 n. 133
Wilson, A. P., 128
Wilson, Henry, 300
Winchell, Alexander, 218, 257, 259, 265, 268, 271
Wöhler, Friedrich, 345 n. 22
Wright, William, 238, 244, 245, 248–251, 256, 262, 270, 309, 310
Wurtz, Henry, 241–242, 297
Wyeth, John, 278, 284

Yale University, 43–45, 57, 58, 70, 79, 81, 110, 138, 149, 195–199, 206, 285, 293, 299–304, 346 n. 50, 370 n. 45; School of Applied Chemistry, 241, 358 n. 130; Sheffield Scientific School, 115, 276, 287, 288, 294

Young, James, 146, 156, 162–180, 206, 229, 232, 292, 376 n. 118; *James Young v. Ebenezer Fernie,* 177–179, 180, 184, 367 n. 73; Paraffine patent (U.S.), 146–147, 164, 232

Zu Salm, Hugo, 81